T0309652

HIGH p_T PHYSICS IN THE HEAVY ION ERA

Aimed at graduate students and researchers in the field of high energy nuclear physics, this book provides an overview of the basic concepts of large transverse momentum particle physics, with a focus on pQCD phenomena. It examines high p_T probes of relativistic heavy ion collisions, and will serve as a handbook for those working on RHIC and LHC data analyses.

Starting with an introduction and review of the field, the authors look at basic observables and experimental techniques, concentrating on relativistic particle kinematics, before moving on to a discussion about the origins of high p_T physics. The main features of high p_T physics are placed within a historical context and the authors adopt an experimental outlook, highlighting the most important discoveries leading up to the foundation of modern QCD theory. Advanced methods are described in detail, making this book especially useful for newcomers to the field.

JAN RAK is a Senior Research Scientist in the Department of Physics at Jyväskylä University, Finland. He is also project leader of ALICE/CERN for Finland, and was involved in the CERES/SPS experiment at CERN, Switzerland and the PHENIX experiment at RHIC, USA. His achievements include the discovery of jet quenching in heavy ion collisions at RHIC and the development of the new methods of high p_T particle correlations that are now widely adopted by the heavy ion community.

MICHAEL J. TANNENBAUM is a Senior Scientist within the Physics Department at Brookhaven National Laboratory where he researches elementary particle and relativistic heavy ion physics. He has more than 300 scientific publications and his achievements include measurement of the statistics of the muon, and discoveries in hard scattering of quarks and gluons at the CERN-ISR which he applied to the discovery of jet quenching in collisions of nuclei at RHIC.

CAMBRIDGE MONOGRAPHS ON PARTICLE PHYSICS, NUCLEAR PHYSICS AND COSMOLOGY

1. K. Winter (ed.): *Neutrino Physics*
2. J. F. Donoghue, E. Golowich and B. R. Holstein: *Dynamics of the Standard Model*
3. E. Leader and E. Predazzi: *An Introduction to Gauge Theories and Modern Particle Physics, Volume 1: Electroweak Interactions, the 'New Particles' and the Parton Model*
4. E. Leader and E. Predazzi: *An Introduction to Gauge Theories and Modern Particle Physics, Volume 2: CP-Violation, QCD and Hard Processes*
5. C. Grupen: *Particle Detectors*
6. H. Grosse and A. Martin: *Particle Physics and the Schrödinger Equation*
7. B. Anderson: *The Lund Model*
8. R. K. Ellis, W. J. Stirling and B. R. Webber: *QCD and Collider Physics*
9. I. I. Bigi and A. I. Sanda: *CP Violation*
10. A. V. Manohar and M. B. Wise: *Heavy Quark Physics*
11. R. K. Bock, H. Grote, R. Frühwirth and M. Regler: *Data Analysis Techniques for High-Energy Physics, Second edition*
12. D. Green: *The Physics of Particle Detectors*
13. V. N. Gribov and J. Nyiri: *Quantum Electrodynamics*
14. K. Winter (ed.): *Neutrino Physics, Second edition*
15. E. Leader: *Spin in Particle Physics*
16. J. D. Walecka: *Electron Scattering for Nuclear and Nucleon Scattering*
17. S. Narison: *QCD as a Theory of Hadrons*
18. J. F. Letessier and J. Rafelski: *Hadrons and Quark-Gluon Plasma*
19. A. Donnachie, H. G. Dosch, P. V. Landshoff and O. Nachtmann: *Pomeron Physics and QCD*
20. A. Hoffmann: *The Physics of Synchroton Radiation*
21. J. B. Kogut and M. A. Stephanov: *The Phases of Quantum Chromodynamics*
22. D. Green: *High P_T Physics at Hadron Colliders*
23. K. Yagi, T. Hatsuda and Y. Miake: *Quark-Gluon Plasma*
24. D. M. Brink and R. A. Broglia: *Nuclear Superfluidity*
25. F. E. Close, A. Donnachie and G. Shaw: *Electromagnetic Interactions and Hadronic Structure*
26. C. Grupen and B. A. Shwartz: *Particle Detectors, Second edition*
27. V. Gribov: *Strong Interactions of Hadrons at High Energies*
28. I. I. Bigi and A. I. Sanda: *CP Violation, Second edition*
29. P. Jaranowski and A. Królak: *Analysis of Gravitational Wave Data*
30. B. L. Ioffe, V. S. Fadin and L. N. Lipatov: *Quantum Chromodynamics: Perturbative and Nonperturbative Aspects*
31. J. M. Cornwall, J. Papavassiliou and D. Binosi: *The Pinch Technique and its Applications to Non-Abelian Gauge Theories*
32. J. Collins: *Foundations of Perturbative QCD*
33. Y. V. Kovchegov and E. Levin: *Quantum Chromodynamics at High Energy*
34. J. Rak and M. J. Tannenbaum: *High p_T Physics in the Heavy Ion Era*

HIGH p_T PHYSICS IN THE HEAVY ION ERA

JAN RAK
University of Jyväskylä, Finland

MICHAEL J. TANNENBAUM
Brookhaven National Laboratory, New York

CAMBRIDGE
UNIVERSITY PRESS

University Printing House, Cambridge CB2 8BS, United Kingdom

One Liberty Plaza, 20th Floor, New York, NY 10006, USA

477 Williamstown Road, Port Melbourne, VIC 3207, Australia

314-321, 3rd Floor, Plot 3, Splendor Forum, Jasola District Centre, New Delhi - 110025, India

79 Anson Road, #06-04/06, Singapore 079906

Cambridge University Press is part of the University of Cambridge.

It furthers the University's mission by disseminating knowledge in the pursuit of education, learning and research at the highest international levels of excellence.

www.cambridge.org
Information on this title: www.cambridge.org/9780521190299

First published 2013

A catalogue record for this publication is available from the British Library

Library of Congress Cataloging in Publication data
Rak, Jan. 1962–
High p_T physics in the heavy ion era / Jan Rak, University of Jyväskylä, Finland,
Michael J. Tannenbaum, Brookhaven National Laboratory, New York.
pages cm. – (Cambridge monographs on particle physics, nuclear physics, and cosmology ; 34)
Includes bibliographical references and index.
ISBN 978-0-521-19029-9
1. Heavy ion collisions. 2. Quantum chromodynamics. 3. Particles (Nuclear physics)
I. Tannenbaum, Michael J. II. Title.
QC794.8.H4R35 2013
539.7′3–dc23
2012042710

ISBN 978-0-521-19029-9 Hardback

Contents

Preface

Commissioning of the Large Hadron Collider (LHC) in late summer 2009 opened a new, long awaited, era of high energy particle physics. The scientific quest of the LHC is the completion of the Standard Model (SM) of particles and forces and the search for novel phenomena beyond the SM. Exciting physics questions such as the existence of the Higgs boson, the missing link of the SM, supersymmetric particles, extra dimensions and many others are expected to be answered at the LHC. Another experimental effort follows the direction of exploration of hot and dense nuclear matter created in ultra-relativistic nuclear collisions. It is believed that such excited nuclear medium forms a soup of deconfined quarks and gluons known as a Quark Gluon Plasma (QGP) [1] and provides an ideal laboratory to study the many-body aspects of Quantum ChromoDynamics (QCD).

High p_T particle production played a key role in the foundation of QCD as a theory of the strong interaction. Shortly after the discovery of point-like constituents inside the proton in Deeply Inelastic Scattering (DIS) experiments at the Stanford Linear Accelerator Center (SLAC) [2] and the observation of particle production at large transverse momenta in $p + p$ collisions at the Intersecting Storage Rings (ISR) at CERN [3], QCD emerged as a mathematically consistent theory [4]. After Feynman rules had been established for the Yang–Mills theory [5] it became possible to expand the amplitudes in the dimensionless coupling constant of the theory, α_s, and the era of perturbative Quantum ChromoDynamics (pQCD) began.

The pQCD regime is characterized by a vanishing coupling constant $\alpha_s \to 0$ which provides amazingly accurate description of hard scattering phenomena characterized by particle production at high transverse momenta. Cross sections agree over many orders of magnitude and at the time of writing this book there are no experimental data in disagreement with pQCD calculations. This has led to the situation where the pQCD phenomena are not the primary focus of experimental effort at the LHC, and the main effort is invested in tuning of Monte Carlo generators like PYTHIA, HERWIG and many others.

The focus of this book is to guide a reader through the main experimental milestones in the QCD discovery era and to share the excitement. Knowledge of methods to analyze the data should not be forgotten in favor of Monte Carlo generators. This book is meant for postgraduate students and any students whose research is focused on high energy experiments at the Relativistic Heavy Ion Collider (RHIC) in the USA and the Large Hadron Collider (LHC) at CERN, Switzerland. It concentrates on the experimental viewpoint with the expectation of a reasonable familiarity with the theory by the reader. Appendices on probability and statistics, methods of Monte Carlo calculations, and fits including systematic errors are included so that the full toolbox of analysis will be available to the reader. The emphasis of the book is not only on the generally accepted theory and understanding of high p_T physics but also on the historical ideas and results that led up to the present understanding – the why and the how rather than simply the facts. Clearly, not every experimental or theoretical work on the subject can be quoted or discussed in the confines of a book, so the authors have concentrated on results which are generally accepted to have had a major influence on the field or which have had a major influence on the authors, with apologies to those not quoted.

1

Introduction and overview

1.1 Elementary particle physics

Elementary particle physics is the study of the fundamental constituents of matter and the forces between them. It is also called High Energy Physics (HEP) because in order to study fundamental particles with smaller and smaller sizes, shorter and shorter wavelength probes are required which correspond to higher and higher energy.

The field of high energy physics has proceeded for the past ~60 years in a typical sequence: a new accelerator opened up a new range of available energy (or type of accelerated particle, e.g. colliding beams of positrons and electrons), and coupled with new detector technology – which enabled improved or previously impossible measurements to be made – rapidly yielded discoveries soon after it started up. The hadron accelerators which have had the most influential impact on the modern high energy particle and heavy ion physics discussed in this book are shown in Figure 1.1. The upper branch shows four major generations of p–$p(\overline{p})$ colliders starting from the CERN Intersecting Storage Rings (ISR), the first hadron collider, while the lower branch depicts the major heavy ion facilities in the USA and Europe. The AGS at Brookhaven National Laboratory (BNL) ran for p–p physics from 1960–2002 and is now the injector to RHIC. The fixed target p–p programs of Fermilab (1972–) and the CERN-SPS (1976–) are also not shown. The CERN-SPS fixed target light ion program started in the same year (1986) as that at the AGS, but is plotted starting in 1994 when Pb beams were first provided.

1.2 The fundamental constituents of matter and their interactions

In brief, the properties of elementary particles must be conserved when they interact with each other. This leads to the study and classification of interactions according to conservation laws. The properties must be intrinsic properties of the

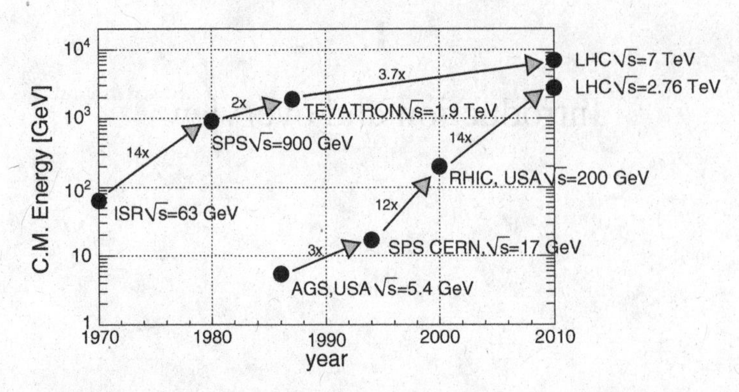

Figure 1.1 Most influential accelerators for high p_T p–$p(\overline{p})$ and heavy ion physics with their starting dates (dots) for the relevant activity.

particle itself and not just of a particular frame of reference. This leads to the study of invariance and symmetry principles. For instance, all the interactions (or forces) of nature conserve energy ($E = Mc^2$), a scalar; momentum ($\mathbf{p} = M\mathbf{v}$), a vector; and angular momentum, a vector ($\mathbf{L} = \mathsf{I} \cdot \boldsymbol{\omega}$), which may be in a different direction than the angular velocity vector $\boldsymbol{\omega}$ due to the inertia tensor, I; here M is the relativistic mass of a particle, $\mathbf{v}$ its velocity and c is the speed of light in vacuum. The relativistic formulas are used since high energies imply relativistic velocities. It is important to note that the relativistic mass of a particle is not an intrinsic property of a particle since it is frame dependent. This leads to the concept of invariant mass or rest mass (m) of a particle, where $mc^2 = \sqrt{E^2 - (pc)^2}$ is the invariant mass of the particle, which does not depend on the particle velocity $\mathbf{v}$ and is the same as the mass of the particle measured in its rest frame.

The properties by which elementary particles are classified are the following:

(i) rest mass, m, where mc^2 is measured in units of electronvolts ($1\,\text{eV} = 1.6 \times 10^{-19}$ J), or million electronvolts (MeV), etc.,

(ii) spin, which is the intrinsic angular momentum of a particle as seen in its rest frame in units of Planck's constant ($\hbar = 6.6 \times 10^{-16}$ eV s),

(iii) interaction strength for the fundamental interactions, for example electric charge in units of the proton charge ($e = 1.6 \times 10^{-19}$ C),

(iv) symmetry properties – these are observed symmetries which are not necessarily understood and sometimes lead to grouping of elementary particles in multiplets.

From now on, we shall use the standard units of HEP, with $\hbar = c = 1$, $\alpha \equiv e^2/(\hbar c) \simeq 1/137$.

At our present state of knowledge there are four fundamental forces of nature: gravity, electromagnetism, the weak interaction responsible for radioactive decay and the strong interaction or nuclear force which binds nuclei. In quantum field theory, forces correspond to the exchange of fundamental objects or quanta, with the "range" or effective distance over which the force acts being inversely proportional to the rest mass of its quantum. The quanta of all known forces have integer spin and are called bosons; the sources of the forces are fundamental particles with "charge," which have half-integer spin and are called fermions. The designations fermion and boson relate to the Fermi–Dirac or Bose–Einstein "statistics" of the particle, a property of the quantum mechanical wave function. Bosons have symmetric wave functions, while fermions have anti-symmetric wave functions which leads to the Pauli exclusion principle [6–8].

The quantum of electromagnetism is the photon, which itself has zero electric charge, so that photons do not interact directly with each other. The photon has zero rest mass which means that the range of electromagnetism is effectively infinite. The force of electromagnetism operates by the exchange of virtual photons from particles with electric charge, for example electrons. The word "virtual" means that the particle is off its mass shell, i.e. the relation between the energy and momentum of the exchanged or propagating virtual photon does not give the rest mass: $E^2 - p^2 \neq m^2$. The quanta of the force of gravitation are called gravitons for which the coupling constant or "charge" is the (gravitational) mass of the particle. Although gravity is the main force we experience in our daily lives, the quantum theory of gravity remains to be worked out at present. Since gravity is irrelevant to elementary particle physics, apart from the fact that the (inertial) mass is a fundamental property of all particles, it will be largely ignored here.[1]

The changing concept of the strong and weak interactions, and what constitutes an "elementary particle," are intertwined with the principal issues of this book and will be discussed in due course.

1.3 A new paradigm for the structure of matter

Before 1968, matter was thought to be composed of atoms with a positively charged nucleus of small diameter ($\lesssim 10 \times 10^{-15}$ m) at the center surrounded by a cloud of negatively charged electrons at a much larger radius ($\sim 100 \times 10^{-12}$ m). This is called the Rutherford–Bohr model of matter because Ernest Rutherford [9]

[1] The equivalence of inertial and gravitational masses is known as the weak equivalence principle and was first demonstrated by Eötvös in an experiment using a torsion balance in 1889. This equivalence is responsible for the universal free fall velocity in vacuum independent of mass, first demonstrated by Galileo. So far as we know, this has not been demonstrated for quarks.

discovered the nucleus in 1911, and Neils Bohr, in 1913, discovered a quantum based model of the atomic electrons which explained the empirical Rydberg formula for the spectral lines of hydrogen [10–12]. In 1932, Chadwick discovered the neutron [13], and it became generally accepted [14, 15] that the nucleus was composed of a collection of protons and neutrons held together by a nuclear or "strong" force which counteracted the electrical repulsion of the positively charged protons. Also, in the period 1930–1933, Pauli proposed the existence of a massless spin 1/2 particle [16, 17] to solve the problem of apparent energy non-conservation in the radioactive β decay of a nucleus, $A \rightarrow B + e^-$, which showed a continuous energy spectrum of e^-, in contrast to α decay of a nucleus, $C \rightarrow D + \alpha$, where the alpha particle has a unique energy, $E_\alpha = E_C - E_D$. In 1934, Fermi [18, 19] then completed the picture of the weak interaction as the force responsible for the β radioactive decay of nuclei via the point-like interaction $n \rightarrow p^+ + e^- + \nu$, with a universal coupling constant G_F. He also named Pauli's particle the neutrino.

In the 1950s, systematic measurements of the radii and internal charge distribution, or "form factor," of nuclei by Robert Hofstadter led to his discovery that the proton itself had a finite radius [20]. Then, in the early 1970s, it became clear that the nucleon was not an elementary particle but was composed of a substructure of three valence quarks confined into a bound state by a strong interaction, Quantum ChromoDynamics (QCD), which is mediated by the exchange of color-charged vector gluons [21]. In sharp distinction to the behavior of the uncharged quanta of Quantum Electrodynamics (QED), the color-charged gluons of QCD interact with each other. This leads to the property of asymptotic freedom [22, 23], the reduction of the effective coupling constant at short distances, and is believed to provide the confinement property at long distances where the quarks and gluons behave as if attached to each other by a color string. It is worth reviewing some developments leading up to and immediately following the discovery of QCD.

1.4 The particle zoo

Before World War II, when the few existing accelerators had too low an energy for the production of new particles in nucleon–nucleon collisions, high energy physics was largely an occupation of cosmic ray physicists, who made some important and fundamental discoveries including:

- the positron, the anti-particle of the electron;
- the muon, a particle with the same quantum numbers and weak and electromagnetic coupling constants as the electron, with the same property of no strong interaction but with a rest mass 207 times larger;
- the π-meson, originally thought to be the quantum of the strong interaction [24];

the "strange" particles which are produced in pairs at a rate corresponding to the strong interaction, but decay "slowly" at a rate corresponding to the weak interaction and appear as Vs in photographs [25].

In the early 1960s, with the construction of proton accelerators with energies well above the threshold for antiproton production, a veritable "zoo" of new particles and resonances was discovered [26]. Gell-Mann [27] and Ne'eman [28] noticed that particles sharing the same quantum numbers (spin, parity) follow the symmetry of the mathematical group SU(3) which is based on three elementary generators, up, down, strange, or u, d, s, with spin 1/2 and fractional electrical charge [29,30], which Gell-Mann called quarks. Mesons are described as states made of a quark–antiquark ($q\bar{q}$) pair and baryons as states of three quarks (qqq). This led to the prediction of a new baryon, the Ω^- (sss) with strangeness −3, which was observed shortly thereafter [31]. However, the Ω^- had a problem: three identical s quarks in the same state, apparently violating the Pauli exclusion principle. To avoid this problem, it was proposed [32] that quarks come in three "colors," i.e. distinguishing characteristics which would allow three otherwise identical quarks to occupy the same state (formally, para-Fermi statistics of rank 3). A major breakthrough was the realization that the fundamental SU(3) symmetry of nature was not the original three quarks uds (now called "flavor"), but the three colors, and that color-charged gluons are the quanta of the "asymptotically-free" strong interaction which binds hadrons [21]. The fourth "charm" or c quark, proposed to explain the absence of certain channels in weak decays of strange particles [33], thus had no problem fitting into this scheme – the quark symmetry became groups of doublets, ud, cs.

Elegant as these theories were, there are many other beautiful theories from this period that were not confirmed by experiment. It is the experimental results which sometimes lead and sometimes follow the theory that give the true picture of nature when in agreement. We would not go as far as Charles Peyrou in describing the cosmic ray discoveries who said: "Two of the discoveries (in cosmic ray physics) were predicted by theory: the positron and the π meson but in no case was the hand of the experimenter guided by the the theorist" [25]. From the point of view of the experimentalist, those were the good old days: "when the experiments were made of wood and the physicists were made of steel" [34]. Times are different now, as illustrated by the motivation for the next stage of discovery which came from a different direction and with a strong theory–experiment interplay with lots of discoveries on both sides.

1.5 The first high p_T physics, the search for the W boson

In the year 1960, Lee and Yang proposed the intermediate bosons $W^{\pm}$ as the quanta that transmit the weak interaction [35,36], and experiments were proposed to detect them with high energy neutrinos [37, 38]. Neutrino beams at the new BNL-AGS and CERN-PS accelerators provided the first opportunity to study weak interactions at high energy, whereas previously weak interactions had only been studied via radioactive decay. The first round of high energy neutrino experiments led to the discovery of a second neutrino [39] that coupled only to muons, in addition to the original (now electron) neutrino from β decay which coupled to electrons. This led to the concept of families of leptons with conserved lepton number (and later by analogy to generations or "flavors" of quarks). The $W^{\pm}$ was not observed in these experiments and only modest limits on the mass of the W (>2 GeV) were obtained due to the relatively low energies of the neutrino beams. However, it was soon realized that the intermediate bosons that mediate the weak interaction might be produced more favorably in nucleon–nucleon collisions [40–42] than in neutrino interactions. The signature of the heavy $W^{\pm}$ would be given by the two-body semi-leptonic decay:

$$W^+ \to e^+ + \nu_e \quad \text{or} \quad W^+ \to \mu^+ + \nu_\mu, \tag{1.1}$$

which would create a flux of seemingly direct leptons at large transverse momenta. However, the transverse momentum spectrum of single leptons from hadron collisions would be composed of the unavoidable, but smoothly falling, background from the decays of short lived hadrons, which decreased exponentially with increasing p_T according to the Cocconi formula, e^{-6p_T} [43, 44], upon which would be superimposed a peak at lepton transverse momentum

$$p_T = \frac{1}{2} M_W \tag{1.2}$$

for the assumed isotropic decay, where M_W is the mass of the intermediate boson. This beautiful idea, elegant in its simplicity and qualitativeness, soon became the stimulus of a large body of work both experimental and theoretical.

However, an objection was raised to the simple idea, when it was realized [45] that the interpretation of such experiments, particularly those with null results [46–48], would be impossible unless the form factor for the production of the intermediate boson were known. This form factor could be deduced [49] from the measurement of lepton pair production in hadron collisions; but the existence of such lepton pairs would create additional background in the single lepton spectrum, thus making the detection of a peak more difficult, if not impossible.

This new idea established the fundamental relationship between single lepton and lepton pair experiments and indicated the importance of doing both types of measurements. Nevertheless, the cross sections for producing the $W^{\pm}$ in hadron collisions remained a priori unknown until the complementary results of the measurement of di-lepton production in proton–nucleus collisions at BNL [50]

$$p + A \to \mu^+ + \mu^- + \text{anything} \tag{1.3}$$

and deeply inelastic electron–proton scattering (DIS) at SLAC [51–54]

$$e + p \to e + \text{anything} \tag{1.4}$$

paved the way for a revolution in the concept of the structure of the proton as a composite formed of quarks and gluons and the strong interactions as mediated by color-charged gluons exchanged by color-charged quarks and gluons.

There were four key experimental observations that made the composite theory of hadrons believable:

(1) the discovery of point-like constituents ("partons") inside the proton, in deeply inelastic (large energy loss, ν, large four-momentum transfer, Q) electron–proton (ep) scattering (DIS) at the Stanford Linear Accelerator Center (SLAC) [51–54];
(2) the complementary discovery of di-lepton production in $p + A$ collisions [50] which showed that the partons of DIS could annihilate into a $\mu^+\mu^-$ pair with electromagnetic strentgh [54];
(3) the observation of enhanced particle production at large transverse momenta (p_T) in p–p collisions at the CERN-Intersecting Storage Rings (ISR) [3, 55–57] and Fermilab [58, 59] in experiments searching for single $e^\pm$ from the $W^\pm$, which proved that the partons of DIS interacted much more strongly with each other than the electromagnetic scattering observed at SLAC;
(4) the observation of the J/Ψ, a narrow bound state of $c\bar{c}$, in both $p +$ Be collisions at the BNL-AGS [60], and in e^+e^- annihilations at SLAC [61], shortly followed by observation of the Ψ', a similar state with higher mass [62] which corresponded to $c\bar{c}$ bound states in a simple Couloumb-like potential with a string-like linear confining potential [63–66].

These discoveries turned Gell-Mann and Zweig's quarks from mere mathematical concepts to the fundamental constituents of matter, the components of the nucleon [67–71] and led to an entirely new vision of the strong interaction.

1.6 From Bjorken scaling to QCD to the QGP

The fundamental idea to emerge from DIS, which was the basis of much of the subsequent theoretical developments leading to QCD, was the concept of Bjorken scaling [72] which indicated that protons consist of point-like objects (partons). The structure function $F_2(Q^2, \nu)$ which describes the inelastic ep scattering cross section was predicted to "scale" [72], i.e. to be a function only of the ratio of the variables, Q^2/ν, which was observed [2, 51, 53], where Q^2 is the four-momentum transfer squared and ν is the energy transfer from the electron to the target. The deeply inelastic scattering of an electron from a proton is simply incoherent quasi-elastic scattering of the electron, which exhibits the standard elastic scattering recoil energy loss, $\nu = Q^2/2(Mx)$, from point-like partons of effective mass Mx, where M is the rest mass of the proton. Thus the Bjorken "x" is the fraction of the nucleon momentum (or mass) carried by the parton. Similar ideas for the scaling of longitudinal momentum distributions in p–p collisions were also given [73, 74]. However, these ideas related to the "soft" (low p_T) particle production rather than the large p_T or "hard scattering" processes described by Bjorken [75, 76].

Bjorken scaling was the basis of QCD [21], the MIT Bag model [77] and also led to the conclusion [78] that "superdense matter (found in neutron-star cores, exploding black holes, and the early big-bang universe) consists of quarks rather than of hadrons," because the hadrons overlap and their individuality is confused. Collins and Perry [78] called this state "quark soup" but used the equation of state of a gas of free massless quarks from which the interacting gluons acquire an effective mass, which provides long-range screening. They anticipated superfluidity and superconductivity in nuclear matter at high densities and low temperatures. They also pointed out that for the theory of strong interactions (QCD), "high density matter is the second situation where one expects to be able to make reliable calculations – the first is Bjorken scaling." In the Bjorken scaling region, the theory is asymptotically free at large momentum transfers while in high-density nuclear matter long-range interactions are screened by many-body effects, so they can be ignored and short distance behavior can be calculated with the asymptotically free QCD and relativistic many-body theory. Shuryak [79] codified and elaborated on these ideas and provided the name "QCD plasma," or "Quark–Gluon Plasma" (QGP) for "this phase of matter," a plasma being an ionized gas.

1.7 Relativistic heavy ion collisions and the QGP

It was soon realized that the collisions of relativistic heavy ions could provide the means of obtaining superdense nuclear matter in the laboratory [80–83]. The kinetic energy of the incident projectiles would be dissipated in the large volume of nuclear matter involved in the reaction. The system is expected to come to equilibrium, thus heating and compressing the nuclear matter so that it undergoes a phase transition from a state of nucleons containing bound quarks and gluons to a state of deconfined quarks and gluons, the quark–gluon plasma, in chemical and thermal equilibrium, covering the entire volume of the colliding nuclei or a volume that corresponds to many units of the characteristic length scale. In the terminology of high energy physics, this is called a "soft" process, related to the QCD confinement scale [84]

$$\Lambda_{QCD}^{-1} \simeq (0.2\,\mathrm{GeV})^{-1} \simeq 1\,\mathrm{fm}. \tag{1.5}$$

One of the nice features of the search for the QGP is that it requires the integrated use of many disciplines in physics: high energy particle physics, nuclear physics, relativistic mechanics, quantum statistical mechanics, and recently, string theory [85, 86]. From the point of view of an experimentalist there are two major questions in this field. The first is how to relate the thermodynamic properties (temperature, energy density, entropy, viscosity ...) of the QGP or hot nuclear matter to properties that can be measured in the laboratory. The second question is how the QGP can be detected.

One of the major challenges in this field is to find signatures that are unique to the QGP so that manifestations of this new state of matter can be distinguished from the "ordinary physics" of colliding nuclei (without the production of a QGP). Another more general challenge is to find effects which are specific to $A+A$ collisions, such as collective or coherent phenomena, in distinction to cases for which $A+A$ collisions can be considered as merely an incoherent superposition of nucleon–nucleon collisions [87–89]. Hence it is important to understand the underlying high energy physics of nucleon–nucleon collisions in order to interpret clearly the results of $A+A$ collisions. This makes the field of RHI physics one of the only places where the older results of high energy physics in p–p collisions can be applied, while the elementary particle physics moves on to higher energy frontiers to gather new knowledge, for example the answers to the following questions.

(i) Are there undiscovered principles of nature: new symmetries, new physical laws?
(ii) Does the Higgs boson exist as a real particle? Does the Higgs mechanism give mass to the quarks and leptons as well as to the fundamental bosons?
(iii) Do quarks and leptons have a finite size? Is there a fundamental length?
(iv) How can we solve the mystery of dark energy?
(v) Are there extra dimensions of space?
(vi) Do all the forces become one?
(vii) Why are there so many kinds of particles?
(viii) What is dark matter? How can we make it in the laboratory?
(ix) What are neutrinos telling us?
(x) How did the universe come to be?
(xi) What happened to the antimatter?

1.8 High energy physics and techniques in the RHI physicist's toolkit

One must not forget that in addition to QCD [90], which is the underlying theory of the QGP, one of the important legacies of high energy physics to relativistic heavy ion physics is the catalog of elementary particles and quanta. It is important to realize, as discussed above, that the definition of the "elementary" particles and fundamental interactions is time dependent according to our state of knowledge.

Before 1930, the table of elementary particle was very simple:

$$(p^+) \qquad (e^-)$$

$$\text{proton} \quad \text{electron.}$$

After 1934, a table of the elementary particles could be represented as:

$$\begin{pmatrix} p^+ \\ n^0 \end{pmatrix} \qquad \begin{pmatrix} e^- \\ \nu^0 \end{pmatrix}$$

$$\text{nucleon} \quad \text{lepton.}$$

There are two sets of two particles: nucleons (proton and neutron) and leptons (electron, neutrino), with electric charges indicated. One set has strong interactions, the other does not. In each set one particle has electric charge, while the other is neutral. All particles participate in the weak interaction via the decay $n \rightarrow p+e+\nu$. The grouping in doublets is significant because the strong interaction was found to be "charge independent," the same between n–p, p–p, n–n, so that the proton and neutron were taken to be two different states of the same particle, the nucleon [91], in analogy to the up and down states of spin 1/2 particles. This was called "isotopic spin" by Wigner [92]. The corresponding effect for the leptons is called "weak isospin" since it only involves the weak interaction.

The most up to date table of elementary particles, circa 2009, is both quantitatively and qualitatively different, but retains the doublet structure (Figure 1.2 [93, 94]). There are now six "flavors" of leptons and six "flavors" of quarks grouped in doublets forming three "generations." There is another set of anti-particles for each quark and lepton, not shown. Each of the quarks and antiquarks comes in three colors (red, green, blue) which are the "charges" of the strong interaction, mediated by the gluon (g) which comes in eight color charges (e.g. blue–anti-red, ...). The leptons do not carry color and hence do not participate in the strong interaction. The four other force carriers are γ, Z^0, W^+, W^- of the now unified weak and electromagnetic (electroweak) interaction [95–97] in which the photon (γ) still couples only to electric charge and retains its zero rest mass while the $W^\pm$ and Z^0 masses of $\sim$100 GeV correspond to the short range ($\sim 2 \times 10^{-18}$ m) of the weak interaction. All leptons and quarks participate in the weak interaction with

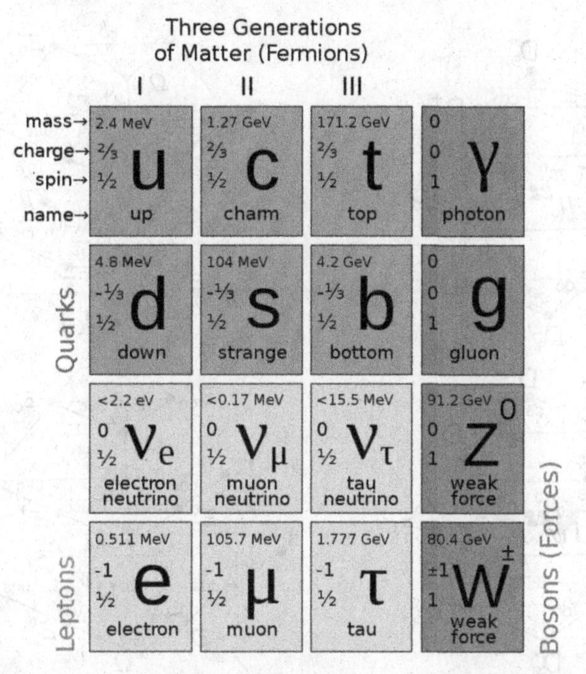

Figure 1.2 Circa 2009 schematic representation [93, 94] of the "elementary particles and forces" of nature with the invariant mass, electric charge, spin and name (flavor) indicated.

the same universal semi-weak coupling constants, but with some small mixing for the quarks [98] and, recently, for the neutrinos.

Roughly half of the entries in Figure 1.2 (the "color-charged" quarks and gluon) are not observable as real particles on their mass shell but are "confined" due to the nature of the strong force so that only color-neutral states can be physically observed. For the confined particles, the mass in Figure 1.2 is called the bare or "current-quark" mass which is used in the calculation or measurement of scattering phenomena. However, when bound in the observable "colorless" mesons ($q\bar{q}$) and baryons (qqq), the quarks take on a "constituent-quark" mass of 1/3 the nucleon (or 1/2 the ρ-meson) mass, which is called "chiral-symmetry breaking" [99,100]. Even with the quantum mechanical conservation laws for the wave functions of observable states [84, 101, 102], there are many possible ways to combine $q\bar{q}$ and qqq, so the table of actually observed particles is, to say the least, very rich (see Figure 1.3).

It may be interesting to note that of all the "elementary particles" shown in Figure 1.2 only the quarks and gluons are considered to be the components of the quark–gluon plasma. However, they can be observed when they "freeze out" to

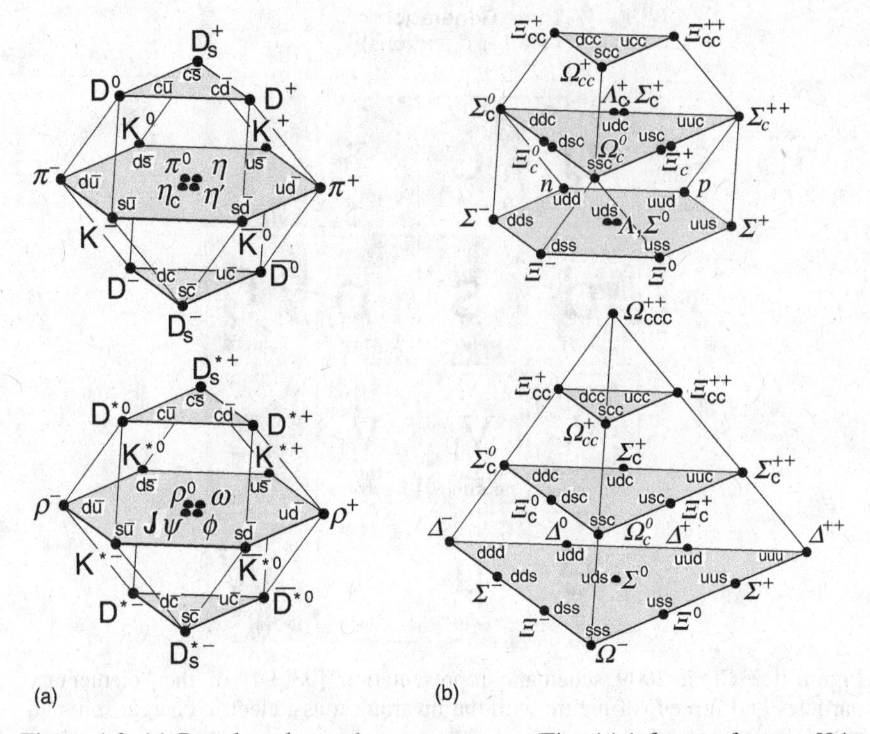

Figure 1.3 (a) Pseudoscalar and vector mesons (Fig. 14.1 from reference [84, 102]) made up of u, d, s and c quarks. Note that the nonets of light mesons occupy the central hexagons with the most relevant particles for later discussions at the center. (b) Baryons (Fig. 14.4 from reference [84, 102]) made up of u, d, s and c quarks. The lowest lying baryon octet and decuplet are shown as the lowest shaded hexagon and triangle, respectively.

the observable particles in Figure 1.3. Fortunately, for the purposes of the subject matter of this book, only very few of the particles in Figure 1.3 [84, 102] are relevant.

Many searches for the quark–gluon plasma have been performed in $A + A$ collisions over the past three decades, which cover the experimental RHI programs at the Berkeley Bevalac, the Brookhaven AGS, the CERN-SPS, RHIC at Brookhaven and recently the Large Hadron Collider (LHC) at CERN. The purpose of this book is to illustrate the use of the techniques that were developed in the 1970s in the elementary particle physics of p–p collisions, with emphasis on high p_T physics. These techniques have turned out to be vital to the discovery and measurement of the properties of the QGP in the relativistic heavy ion physics of $A + A$ collisions. However, the development of and history of these techniques is not well known to the present generation of high energy or relativistic heavy ion physicists. As emphasized by Van Hove [103], much could be gained in this regard by studying the history of strong interactions for the past 40 years. An understanding of the properties of high energy p–p and $p + A$ interactions is vital to the ability to distinguish the "ordinary physics" of relativistic nuclear interactions from the signatures of production of a new phenomenon like the QGP. It is also possible that the "ordinary physics" may in itself be quite interesting. Furthermore, by studying the hard won knowledge of the past, one might hope to avoid some pitfalls in the future.

2

Basic observables

2.1 Observables

The challenge of RHI collisions can be understood from Figure 2.1, which shows events from p–p and Au+Au collisions at nucleon–nucleon center-of-mass energy $\sqrt{s_{NN}}$=200 GeV at RHIC as observed in the STAR and PHENIX detectors. It would appear to be a daunting task to reconstruct all the particles in such events. Consequently, it is more common to use single particle or multiparticle inclusive variables to analyze these reactions.

This is in stark contrast to the situation that prevailed in high energy physics in the 1960s. It had already been observed in cosmic rays that when high energy nucleons collide inelastically, the predominant mode of dissipating the energy is by multiple particle production.[1] One of the important regularities observed was that the particles produced have limited transverse momentum with respect to the collision axis, exponentially decreasing as e^{-6p_T} [43, 44]. In this era, experiments studied mainly exclusive reactions in which all particles on individual events were fully reconstructed. The name derives from the fact that a specific reaction was selected, thus **excluding** a whole class of other reactions. Most of our well known particles were discovered in this way. A typical example from this period is cascade (Ξ^-) production in a bubble chamber, Figure 2.2. In this photograph [107], a 2.3 GeV/c K^- meson interacts with a proton in the liquid hydrogen to produce a Ξ^-, a K^0 and a π^+. The Ξ^- decays into a Λ^0 and a π^-. The K^0 and Λ^0 decay in flight to form the two beautiful V^0s in the picture. All the charged tracks were measured and the event reconstructed fully kinematically, so that the masses of the K^0, Λ^0 and Ξ^- could be determined.

[1] In the early 1950s it was not at all clear that a nucleon–nucleon collision could lead to the production of more than one meson ("multiple production") or whether the multiple meson production observed in nucleon–nucleus interactions was the result of several successive nucleon–nucleon collisions, with each collision producing only a single meson ("plural production") [104, 105]. The issue was decided when multiple meson production was first observed in 1954 in collisions between neutrons with energies up to 2.2 GeV, produced at the Brookhaven Cosmotron, and protons in a hydrogen filled cloud chamber [106].

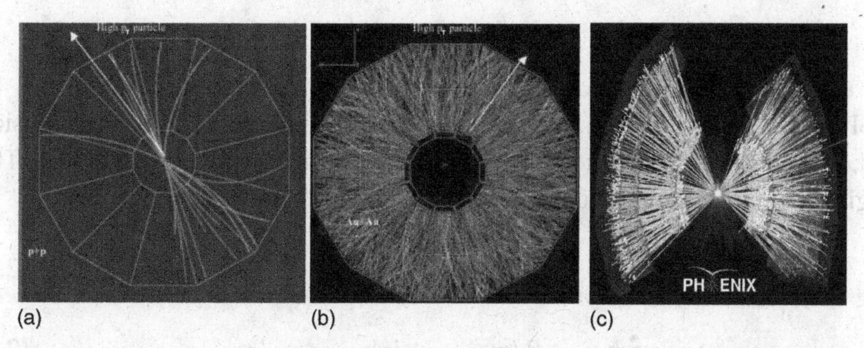

Figure 2.1 (a) A p–p collision in the STAR detector viewed along the collision axis; (b) Au+Au central collision at $\sqrt{s_{NN}} = 200$ GeV in the STAR detector; (c) Au+Au central collision at $\sqrt{s_{NN}} = 200$ GeV in the PHENIX detector.

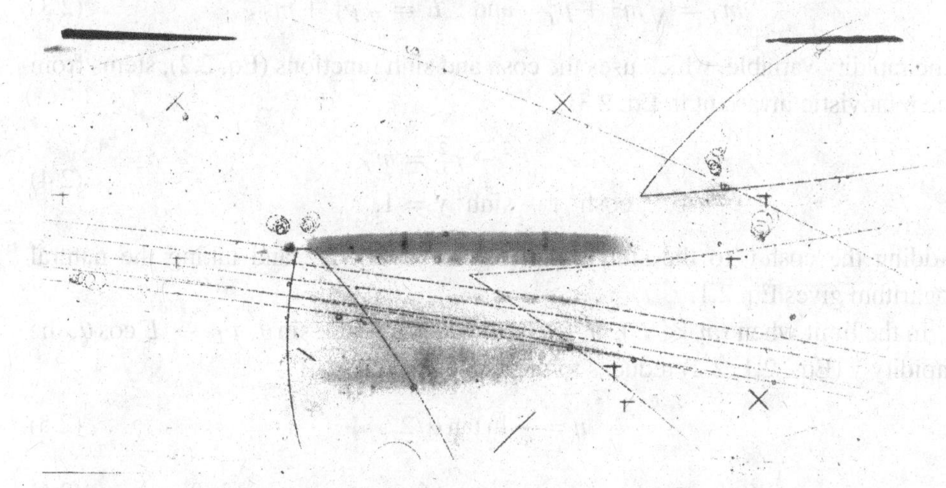

Figure 2.2 Courtesy Brookhaven National Laboratory [107]. Five K^- beam particles pass from left to right. The uppermost beam track interacts to produce a Ξ^- which is the short stub between the two downward-going tracks. The smudges are reflections from the bubble chamber window.

Such measurements became nearly impossible when the center-of-mass energies rose above 20 GeV in the early 1970s. Then, single particle inclusive measurements were introduced [73, 74, 108–110]. A single particle "inclusive" reaction involves the measurement of just one particle coming out of a reaction,

$$a + b \to c + \text{anything}.$$

The terminology comes from the fact that all final states with the particle c are summed over, or **included**. A "semi-inclusive" reaction [111] refers to the measurement of all events of a given topology or class, for example

$$a + b \to n_1 \text{ particles of type } 1 + \text{anything}.$$

2.1.1 Kinematics

For any observed particle of momentum p, energy E, the momentum can be resolved into components transverse (p_T) and longitudinal (p_L) to the collision axis; and in many cases the mass (m) of the particle can be determined. The longitudinal momentum is conveniently expressed in terms of the rapidity (y)

$$y = \ln\left(\frac{E + p_L}{m_T}\right) \tag{2.1}$$

$$\cosh y = E/m_T \qquad \sinh y = p_L/m_T \tag{2.2}$$

where

$$m_T = \sqrt{m^2 + p_T^2} \quad \text{and} \quad E = \sqrt{p_L^2 + m_T^2}. \tag{2.3}$$

The rapidity variable, which uses the cosh and sinh functions (Eq. 2.2), stems from the relativistic invariant in Eq. 2.3:

$$\begin{aligned} E^2 - p_L^2 &= m_T^2 \\ \cosh^2 y - \sinh^2 y &= 1. \end{aligned} \tag{2.4}$$

Adding the $\cosh y$ to the $\sinh y$ expressions in Eq. 2.2 and taking the natural logarithm gives Eq. 2.1.

In the limit when ($m \ll E$), $p \to E$, $m_T \to p_T \to E\sin\theta$, $p_L \to E\cos\theta$, the rapidity y (Eqs. 2.1, 2.2) reduces to the pseudorapidity (η)

$$\eta = -\ln\tan\theta/2 \tag{2.5}$$

$$\cosh\eta = \csc\theta \qquad \sinh\eta = \cot\theta \qquad \tanh\eta = \cos\theta, \tag{2.6}$$

where θ is the polar angle of emission. The transverse momentum, p_T, is conserved in a Lorentz transformation along the collision axis. The rapidity variable has the useful property that it transforms linearly under a Lorentz transformation so that the invariant differential single particle inclusive cross section becomes:

$$\frac{Ed^3\sigma}{dp^3} = \frac{Ed^3\sigma}{p_T dp_T dp_L d\phi} = \frac{d^3\sigma}{p_T dp_T dy d\phi} \tag{2.7}$$

where

$$dy = \frac{dp_L}{E}. \tag{2.8}$$

For any collision, the center-of-mass (c.m.) system – in which the momenta of the incident projectile and target are equal and opposite – is at rapidity Y^{cm}. The total energy in the c.m. system is denoted $\sqrt{s}$, which, evidently, is also the "invariant mass" of the c.m. system. For a collision of an incident projectile of energy E_1, mass m_1, in the "laboratory system," where the target, of mass m_2, is at rest, as in a fixed-target experiment:

$$s = m_1^2 + m_2^2 + 2E_1 m_2. \tag{2.9}$$

The c.m. rest frame moves in the laboratory system (along the collision axis) with a velocity corresponding to:

$$\gamma^{cm} = \frac{E_1 + m_2}{\sqrt{s}} \quad \text{and} \quad Y^{cm} = \cosh^{-1} \gamma^{cm}. \tag{2.10}$$

Another useful quantity is Y^{beam}, the rapidity of the incident particle in the laboratory system

$$Y^{beam} = \cosh^{-1} \frac{E_1}{m_1}, \tag{2.11}$$

and note that for equal mass projectile and target:

$$Y^{cm} = Y^{beam}/2. \tag{2.12}$$

In the region near the projectile or target rapidity, the Feynman x fragmentation variable [73] is also used:

$$x_F = 2p_L^*/\sqrt{s}, \tag{2.13}$$

where p_L^* is the longitudinal momentum of a particle in the c.m. frame.

2.2 Inclusive single particle reactions

2.2.1 Typical quantities measured

In an inclusive measurement, the transverse momentum distributions are determined for the different particle types, and the average transverse momentum, $\langle p_T \rangle$, or the mean transverse kinetic energy, $\langle m_T \rangle - m$, is taken as a measure of the temperature, T, of the reaction. The charged particle multiplicity, either over all space, or in restricted intervals of rapidity, is taken as a measure of entropy.

Single particle inclusive measurements are presented in terms of the (Lorentz) invariant single particle inclusive differential cross section:

$$\frac{Ed^3\sigma}{dp^3} = \frac{d^3\sigma}{p_T dp_T dy d\phi} = \frac{1}{2\pi} f(p_T, y). \tag{2.14}$$

A uniform azimuthal distribution is usually assumed, so the integral over azimuth is simply 2π

$$\frac{d^2\sigma}{p_T dp_T dy} = f(p_T, y). \tag{2.15}$$

The distributions in p_T are measured as a function of rapidity and integrated over p_T to find the average single particle multiplicity density in rapidity:

$$\bar{\rho}_1(y) = \frac{d\langle n(y)\rangle}{dy} = \frac{1}{\sigma_I}\int dp_T\, p_T\, d\phi\, Ed^3\sigma/dp^3 = \frac{1}{\sigma_I}\int dp_T\, p_T\, f(p_T, y). \tag{2.16}$$

Note that the common notation is simply $\rho(y)$. This may be further integrated over all the kinematically possible rapidity to find the overall mean multiplicity per interaction, $\langle n\rangle$:

$$\langle n\rangle = \frac{1}{\sigma_I}\int dy\, dp_T\;\; p_T\;\; f(p_T, y) = \int dy\, \rho(y). \tag{2.17}$$

An important point to remember about inclusive single particle cross sections is illustrated in Eq. 2.17. Integrals of the single particle inclusive cross section (Eq. 2.14) are not equal to σ_I the interaction cross section, but rather equal to the mean multiplicity times the interaction cross section: $\langle n\rangle \times \sigma_I$.

2.2.2 The rapidity density and the rapidity plateau

The shape and evolution with $\sqrt{s}$ of the charged particle density in rapidity, dn/dy, provide a graphic description of high energy collisions. Data from a classical measurement in a streamer chamber from p–p collisions at the CERN-ISR [112] are shown in Figure 2.3. Regions of nuclear fragmentation take up the first 1–2 units around the projectile and target rapidity and are particularly apparent at low multiplicities. If the center-of-mass energy is sufficiently high, a central plateau is exhibited. It is clear that the shape of the dn/dy distribution in rapidity changes as a function both of the observed charged particle multiplicity (in the interval $|\eta| < 4$) and of the c.m. energy of the collision, $\sqrt{s}$.

2.2.3 Limited transverse momentum

The inclusive charged particle p_T spectra, measured near the rapidity of the nucleon–nucleon c.m. system, y_{cm}^{NN} (or 90° in the c.m. system in HEP jargon), are shown in Figure 2.4. This figure [113] includes data over nearly the full available range in c.m. energy $\sqrt{s} > 20$ GeV. The "high" p_T or "hard scattering" region, above 1 to 2 GeV/c, shows an enormous variation with c.m. energy, while the

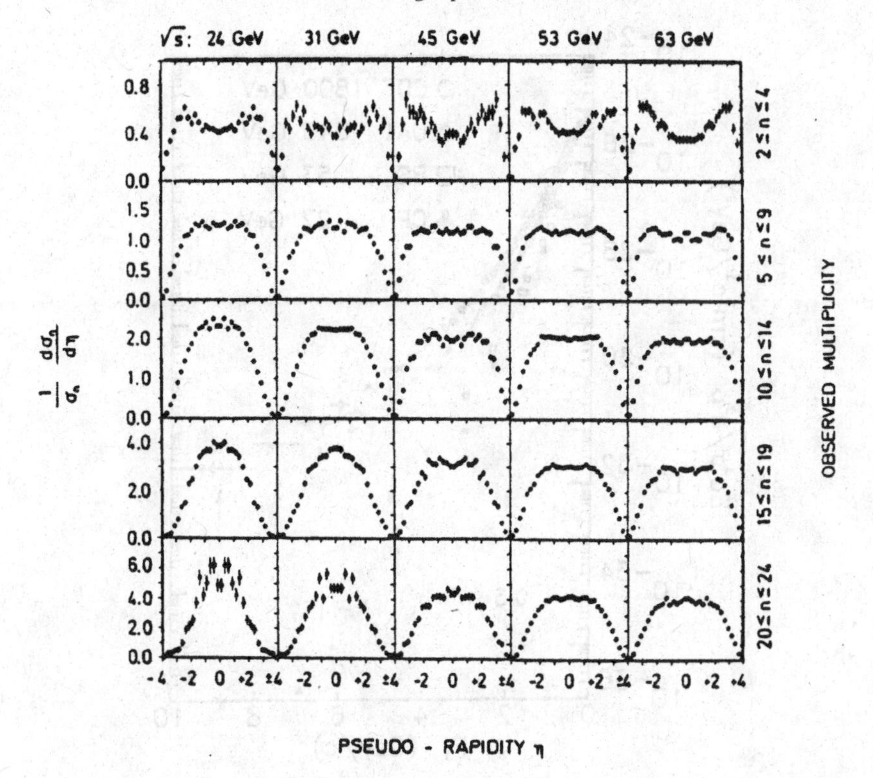

Figure 2.3 Measurements in a streamer chamber at the CERN-ISR of the normalized charged particle densities (corrected for acceptance up to $|\eta| \simeq 4$) in various intervals of the total observed multiplicity, as a function of the c.m. energy, $\sqrt{s}$, of the p–p collisions [112].

region below 1 GeV/c, the "soft" physics region which dominates the spectrum, remains essentially unchanged and is reasonably characterized over the full c.m. energy range by:

$$f(p_T, y) = A \exp(-bp_T), \tag{2.18}$$

with $b = 6\,(\text{GeV}/c)^{-1}$ and $\langle p_T \rangle = 2/b$ in agreement with the Cocconi formula [43, 44] from cosmic rays. This is the reason high energy physicists describe particle production as "longitudinal phase space." As the c.m. energy increases, the $\langle p_T \rangle$ remains relatively constant, while the central plateau (Figure 2.3) tends to expand to fill the available phase space in rapidity. Even at a fixed c.m. energy, the p_T and rapidity distributions are nearly independent. The $\langle p_T \rangle$ is 0.333 GeV/c at y_{cm}^{NN}, and decreases slightly – by less than 10% on the central plateau – as y moves away from y_{cm}^{NN}, then drops to ~220 MeV/c near the projectile rapidity [114–116]. The mean transverse momentum as a function of rapidity is defined as

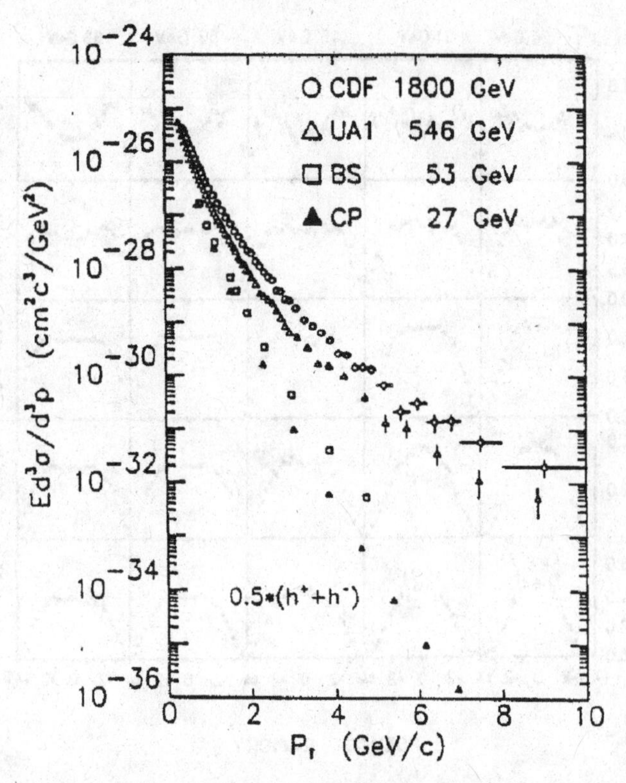

Figure 2.4 The transverse momentum dependence of invariant cross sections of charged-averaged hadrons $(h^+ + h^-)/2$, near y_{cm}^{NN}, in nucleon–(anti)nucleon collisions over the available range of c.m. energies. Typical data shown are from a fixed target ($\sqrt{s} = 27$ GeV), ISR (53 GeV), CERN-SPS collider (546 GeV), and FERMILAB collider (1800 GeV) [113].

$$\langle p_T \rangle|_y = \frac{\int dp_T \; p_T \; p_T \; f(p_T, y)}{\int dp_T \; p_T \; f(p_T, y)}. \tag{2.19}$$

The same equation applies to any rapidity region by integrating over rapidity first in both integrals. Note that in Lorentz invariant terms, this is the natural definition of $\langle p_T \rangle$ so that the $|_y$ is not usually explicitly indicated.

2.2.4 A warning – beware the seagull effect

It took particle physicists many years to become comfortable with the differential $dy = dp_L/E$ in the invariant cross section. Much effort was spent by particle physicists on detailed studies before it became clear that rapidity and p_T were the convenient independent variables for describing p–p collisions. There is much work in the literature studying the dependence of the mean transverse momentum on p_L^*, or Feynman x (Eq. 2.13). This creates an artificial variation of $\langle p_T \rangle$ called the **seagull effect** [117], which in simple terms is a consequence of the fact that the transverse momentum cannot exceed the momentum (in the appropriate c.m. system).

The mean transverse momentum at fixed p_L^* is defined by using the **non-invariant** cross section as the probability distribution:

$$\langle p_T \rangle|_{p_L^*} = \frac{\int p_T \; p_T \, dp_T \; d^3\sigma/dp^3}{\int \; p_T \, dp_T \; d^3\sigma/dp^3} = \frac{\int p_T \; p_T \, dp_T \; f(p_T, y)/E}{\int \; p_T \, dp_T \; f(p_T, y)/E}. \tag{2.20}$$

This is equivalent to computing the $\langle p_T \rangle$ of the distribution f/E where f is the invariant cross section. For large values of p_L^*, the energy E becomes equal to p_L^*, which for the purposes of Eq. 2.20 is a constant, so the true $\langle p_T \rangle$ is obtained. For $x_F = 0$, the weighting distribution is f/m_T which is obviously steeper in p_T than the unweighted f, resulting in a lower $\langle p_T \rangle$. As x_F increases away from $x_F = 0$, the $\langle p_T \rangle$ increases to its true value, producing a beautiful but irrelevant and confusing seagull drawing. (It may prove comforting to some to know that there was at least one thing on which Feynman's intuition was wrong!) This paragraph is placed here as a warning that the definition and plots of $\langle p_T \rangle$ from the original particle physics work in the 1960s and 1970s may not mean the same thing as we understand them today.

2.2.5 Do particle physics spectra prefer p_T or m_T?

From the point of view of relativistic kinematics, m_T rather than p_T would seem to be the preferred variable, since under a Lorentz transformation

$$E^2 - p_L^2 = m_T^2 = p_T^2 + m^2 \tag{2.21}$$

is conserved. In fact, the signature of a thermalized system would be a common m_T distribution – for all species of particles – whose inverse slope parameter would represent the temparature. Although the p_T spectra of pions are generally (and conveniently) described as $\exp(-6p_T)$ (as above), in actual fact the data are better represented as a function of m_T. All spectra in p_T fall below an exponential for values of $p_T \sim m$, so that the spectra are better represented [118] by exponentials in m_T (see Figure 2.5). There is absolutely no change in the differential cross section for this change of variables, since $p_T \, dp_T = m_T \, dm_T$. Shuryak and Zhirov [119] noted the irony that the "pure thermodynamical description of these spectra 'worked too well', providing a good fit to data with no sign of collective effects."

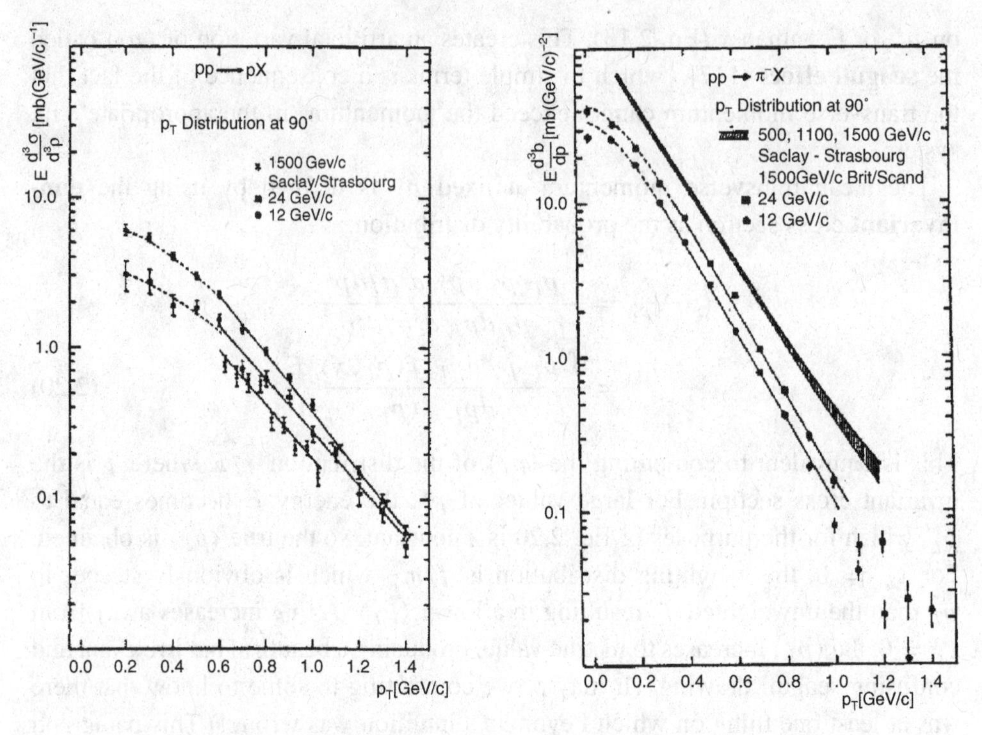

Figure 2.5 Invariant cross sections at 90° in the c.m. system for proton and π^- production from p–p collisions [118]. The solid lines are exponential fits in p_T, while the dashed lines are exponential in m_T.

2.2.6 Dependence of $\langle n \rangle$ on c.m. energy – the dn/dy plateau

The multiplicity density in rapidity, dn/dy (Eq. 2.16), is one of the principal descriptive variables in both high energy and relativistic heavy ion physics. The mean multiplicity in p–p collisions increases roughly logarithmically with $\sqrt{s}$. In the early 1970s, it was thought [120] that this could be explained if dn/dy on the rapidity plateau reached a constant or limiting value, so that $\langle n \rangle$, the integral of the distribution (Eq. 2.17), would just increase as $\ln s$, the available rapidity range in the c.m. system of the collision:

$$y^*(m)|_{\max} \approx \ln \frac{\sqrt{s}}{m}. \tag{2.22}$$

The first real evidence for the rapidity plateau in p–p collisions came from the CERN-ISR [120] and is well illustrated by the streamer chamber result shown above (see Figure 2.3), where the charged particle density in pseudorapidity, $dn/d\eta$, was presented as a function of the observed charged particle multiplicity n in the range $|\eta| < 4$. With increasing multiplicity at fixed $\sqrt{s}$, the widths of

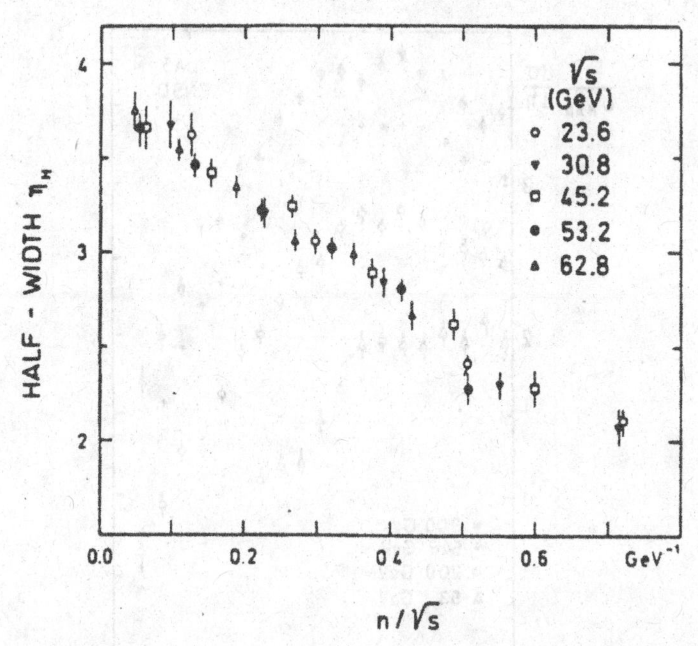

Figure 2.6 Half-width at half maximum ($\frac{1}{2}\Delta\eta|_{FWHM}$) of normalized charged particle densities (Figure 2.3) as a function of $n/\sqrt{s}$, where n is the observed charged multiplicity in the range $|\eta|$ <4 [112].

the η distributions decrease; and for fixed multiplicity, n, the widths of the distributions increase with increasing $\sqrt{s}$. In general, the distributions look much more trapezoidal than Gaussian and the distributions are typically characterized by their full width at half maximum, $\Delta\eta|_{FWHM}$, and the height of the central plateau, $dn/dy|_{y_{cm}^{NN}}$. The widths of the **semi-inclusive** η distributions at fixed n, $\Delta\eta|_{FWHM;n}$, obey an interesting scaling law, as a function of n and $\sqrt{s}$ as shown in Figure 2.6, where they are seen to depend on the single variable $n/\sqrt{s}$ for all the data. The latest data for the fully inclusive $dn/d\eta$ distribution [121] indicate that the width of the inclusive η distribution, $\Delta\eta|_{FWHM}$, increases less fast with $\sqrt{s}$ than the available rapidity range (see Figure 2.7).

2.2.7 Center-of-mass energy dependence of the height of the central plateau, $dn/dy|_{y_{cm}^{NN}}$

The variation in $dn/dy|_{y_{cm}^{NN}}$ with $\sqrt{s}$ – which is so clearly visible in Figure 2.7 at CERN collider energies – was not so apparent at lower energies. The first measurement [122] to show unambiguously that the height of the plateau, $dn/dy|_{y_{cm}^{NN}}$, was not a constant but rose steadily with increasing $\sqrt{s}$ was from the

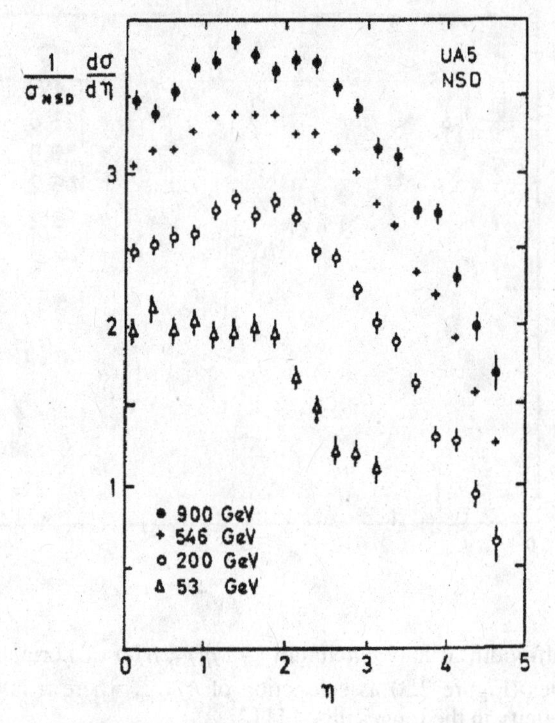

Figure 2.7 Pseudorapidity distributions for non-single diffractive events, UA5 data at four c.m. energies (pp, $p\bar{p}$) [121].

CERN-ISR. The best present data on the $\sqrt{s}$ dependence of the inclusive central (mid-rapidity) density [123],

$$\rho(0) = \frac{1}{\sigma_I} \frac{d\sigma}{d\eta}\bigg|_{\eta^*=0}, \tag{2.23}$$

are shown in Figure 2.8. When the data from AGS energies [118] are included, it becomes clear that the preferred fit over the range $\sqrt{s}$ of 5–1000 GeV is[2]:

$$\rho(0) = 0.74 \left(\sqrt{s(\text{GeV})}\right)^{0.210}; \tag{2.24}$$

there is a clear but very slow increase of $\rho(0)$ with $\sqrt{s}$. Note that the increase in mid-rapidity multiplicity density in going from $\sqrt{s_{NN}} = 5.39$ GeV, which is the nucleon–nucleon c.m. energy for the BNL heavy ion program, to $\sqrt{s_{NN}} = 19.4$ GeV at CERN, is only a factor of 1.31.

The slow variation of the *inclusive* mid-rapidity density $\rho(0)$ as a function of $\sqrt{s}$ is considerably different from the variation of the *semi-inclusive* mid-rapidity

[2] The first results from the LHC for p–p collisions up to $\sqrt{s} = 7000$ GeV (see later chapter) confirm this trend.

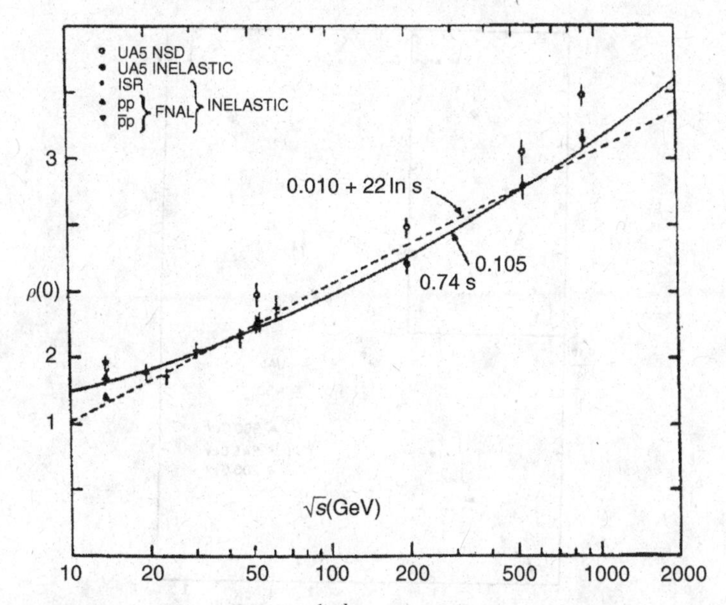

Figure 2.8 Central density $\rho(0) = \frac{1}{\sigma_I}\frac{d\sigma}{d\eta}|_{\eta^*=0}$ plotted as a function of c.m. energy for data from Fermilab fixed target to CERN collider energies. Fits to inelastic data using linear dependence in $\ln s$ or a power-law dependence on s are shown [123].

density as a function of the multiplicity n, $\rho(0)|_n$. The UA5 streamer chamber at the CERN-SPS collider [123] found a systematic relationship between the semi-inclusive mid-rapidity density at fixed multiplicity, $\rho(0)|_n$, divided by the fully inclusive mid-rapidity density of multiplicity, $\rho(0)$, which in their c.m. energy range (see Figure 2.9) was a function only of the scaled multiplicity, $z \equiv n/\langle n \rangle$, where $\langle n \rangle$ is the mean multiplicity at a given $\sqrt{s}$.

2.3 Semi-inclusive reactions – E_T and multiplicity distributions

In the previous section, the "longitudinal phase space" behavior of multiparticle production could be illustrated using single particle inclusive measurements – the $\langle p_T \rangle$ of pions produced is largely independent of rapidity and c.m. energy. At the end of the section, an example of a semi-inclusive measurement was given – the systematic behavior of the width and height of the multiplicity density in rapidity was shown as a function of the observed charged particle multiplicity n.

In the decade of the 1980s, multiparticle semi-inclusive measurements, in which many but not all of the particles from an interaction are measured, became one of the leading tools of both elementary particle physics and relativistic heavy ion physics. To quote Van Hove again [124], it is indicative of the "transformation of

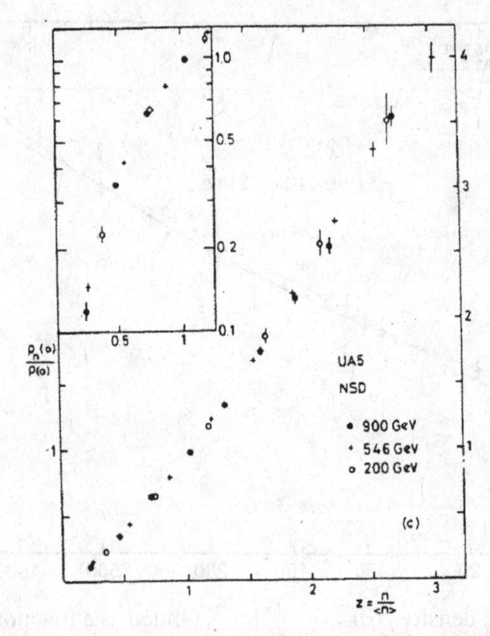

Figure 2.9 UA5 streamer chamber plot of the scaled central pseudorapidity density $\rho_n(0)/\rho_{\langle n\rangle}(0)$ against $z = n/\langle n\rangle$ from the CERN-SPS collider at $\sqrt{s}$= 200, 546, and 900 GeV. The inset shows on a logarithmic scale the data for small values of z [123].

elementary particle physics into many-body physics." The two principal multiparticle semi-inclusive variables are the charged particle multiplicity distribution, either over all phase space or in restricted intervals of rapidity, and the transverse energy flow, or E_T distribution, where

$$E_T = \sum E_i \sin\theta_i \tag{2.25}$$

and the sum is taken over all particles emitted on an event into a fixed but large solid angle. These variables are very closely related, but it took some time for this fact to be understood [125, 126].

2.3.1 Soft collisions and E_T distributions

The charged particle multiplicity, one of the classical observables in the study of high energy collisions, is based on high cross section, "soft," multiparticle physics. By contrast, the original measurements of transverse energy distributions [127] were stimulated by the desire to detect and study the jets from "hard" scattering, with an unbiased and theoretically more efficient trigger than the single particle inclusive probe through which the constituent scattering and jet phenomena were originally discovered (recall Figure 2.4). Contrary to early expectations, the transverse energy flow is dominated by "soft collisions" – the transverse energy is made up of a structureless cloud of average transverse momentum ($\langle p_T \rangle \sim$ 0.4 GeV/c) particles [128]. Jets are swamped [129, 130]. Thus, the transverse energy flow in rapidity, dE_T/dy, is simply related to dn/dy in soft collisions: $dE_T/dy \sim \langle p_T \rangle \times dn/dy$. See Chapter 11 for a full discussion.

3

Some experimental techniques

3.1 Relation of observables to experimental techniques

Every probe in the search for the QGP in relativistic heavy ion collisions tends to have a different experimental technique associated with it. In all cases the multiplicities in nuclear collisions are so large that all the detectors used are very highly segmented. For measuring the charged multiplicity or dn/dy, a segmented multiplicity detector, usually an array of proportional tubes with pad readout, or a silicon pad array were used in early experiments, while Time Projection drift Chambers (TPC) have become popular more recently. For measuring transverse energy flow, dE_T/dy, a hadron calorimeter is used. Some groups use an electromagnetic shower counter for this purpose. This has the advantage of being smaller, cheaper and higher in resolution than a full hadron calorimeter; but it has the disadvantage of being biased, since only π^0 and η^0 mesons are detected (via their two photon decay). Nuclear fragmentation products are detected by calorimeters in the projectile direction and by E, dE/dx scintillator arrays in the target fragmentation region. The particle composition and transverse momentum distributions are measured using magnetic spectrometers with particle identification. Typically, time-of-flight, gas and aerogel Cerenkov counters, and dE/dx are used to separate pions from kaons, protons, deuterons, etc. Drift chambers are generally utilized for charged particle tracking, although streamer chambers and TPCs are also in use. Lepton pair detectors are very specialized, and usually combine magnetic spectrometers with lepton identification (muons by penetration, and electrons by "gas" and "glass").

One of the specific problems in this field is how to detect, with minimum bias, when a nucleus–nucleus collision has taken place. For fixed target measurements, where the experiment takes place in an external beam, two techniques are used. The first is to put a calorimeter at zero degrees to determine whether the projectile has the full beam energy or has lost some energy. The second uses a so-called bullseye counter downstream of the target, sized just large enough to detect all the beam particles. The bullseye also measures the charge of the beam particles since the pulse height is proportional to Z^2. If a particle misses the bullseye, or the charge changes, this is taken as an indication of a nuclear interaction. Collider experiments are more complicated because charged spectator fragments at zero degrees to the collision axis remain captured in the beam so that only neutral fragments or free neutrons can be detected in a calorimeter at zero degrees.

3.2 Some details of experimental technique and analysis

Although it is beyond the scope of this book to give a full course in experimental physics, a few topics will be mentioned here which will be of use for the detailed discussions of the experimental results to follow. These will be focussed on how to analyze and understand the measurements rather than on how to build the detectors.

3.2.1 Passage of charged particles through matter

The electric charge of a particle passing through matter [84, 131–133] causes an electromagnetic force to be exerted on the atomic electrons of the material and causes them to be ejected from the atom, or **ionized**. The energy gained by the ions is lost by the incident particle, which slows down as it passes through matter, leaving a wake of ion-pairs in its path. Detection of these ion-pairs forms the basis of most charged particle detectors.

This is true for all particles except high energy electrons, which are so light that they essentially lose all their energy by radiation (Bremsstrahlung) due to interaction with the highly charged target nuclei. The probability for an electron of energy E to emit a photon of energy k, in range dk in passing through a thickness of material x is:

$$d\mathcal{P} = \frac{x}{X_o}\frac{dk}{k} \times F(k/E). \tag{3.1}$$

For thin radiators,

$$F(k/E) = [1 + (1 - k/E)^2 - \frac{2}{3}(1 - k/E)] \simeq 1, \tag{3.2}$$

so that the energy lost by electrons per unit length of material to photon emission can be written

$$-\frac{dE}{dx} \equiv k\frac{d\mathcal{P}}{dx} \simeq \frac{\int_0^E dk}{X_o} = \frac{E}{X_o}. \tag{3.3}$$

From this equation, it is clear that the **radiation length**, X_o, is the length over which the energy of an incident electron is degraded to a fraction $1/e$ of its original value. The radiation length is inversely proportional to Z^2, where Z is the atomic number (nuclear charge) of the medium, and X_o is usually tabulated in units of g cm^{-2} (strictly $\rho \times X_o$ where ρ is the density of the medium in g cm^{-3}). The **critical energy** of a medium is defined as the energy of electrons for which dE/dx by ionization loss and radiation are equal. Thus, for electrons above the critical energy, typically 10 to 40 MeV, the dominant source of energy loss is radiation.

For heavier particles (starting with the muon) ionization loss is predominant (until the TeV regime). The rate of this loss is given by the Bethe–Bloch formula, which takes the form of a kinetic factor which depends only on the velocity of the incident particle, $\beta = v/c$, times another factor with a slight β and $\log Z$ dependence:

$$\frac{-dE}{dx} = \frac{D\, n_e\, z_I^2}{\beta^2} \mathcal{F}(\beta, \ln Z)\,, \tag{3.4}$$

where D is a constant. The important features of the ionization loss are that it depends on the square of the charge of the incident particle, z_I (in units of the electron charge e), and linearly on n_e the number of electrons per unit volume in the target, because it is the result of Coulomb scattering of the incident particle by the atomic electrons in the target. Apart from the dominant $1/\beta^2$ dependence, the ionization loss is slowly varying, with a broad minimum; and it is usually sufficiently accurate to represent the ionization loss of an incident particle of charge z_I, invariant mass m and momentum P as:

$$\frac{-dE}{dx} = \left[1 + \left(\frac{mc}{P}\right)^2\right] \times z_I^2 \times \left.\frac{dE}{dx}\right|_{\min}. \tag{3.5}$$

Thus, for fixed momentum, heavy particles have a larger ionization loss. As the momentum becomes much larger than the mass of the particle, the ionization loss reduces to a constant value for all particles, represented as the charge of the particle squared times $dE/dx|_{\min}$, the **minimum ionization** for singly charged particles, which is typically 1 to 2 MeV cm^2 g^{-1}.

In contrast to the elastic scattering of the incident particle by the atomic electrons in a medium, which results in significant energy transfer due to the light mass of the atomic electrons, elastic scattering of the incident particle by the highly charged atomic nuclei results in negligible energy transfer, since the nuclei are heavy. However, the incident particle suffers a very large number of very small angle Coulomb elastic scatterings with the target nuclei, in passing through a medium, resulting in a smearing of the angles of the incident particles by a random walk process, **multiple scattering**. The angular distribution for small angle scattering is given by the famous Rutherford formula (see Chapter 4), from which the mean squared scattering angle can be derived. After passing through a thickness x of material,

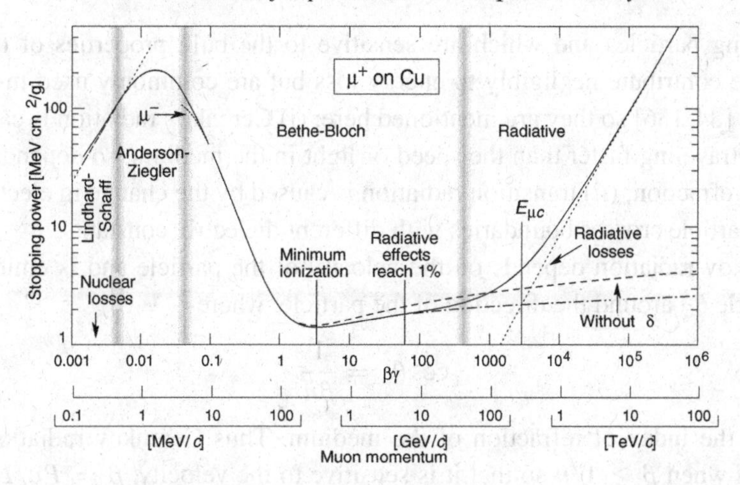

Figure 3.1 dE/dx of a μ^+ in copper as a function of muon momentum [84]. The large radiative energy loss at large momentum is dominated by direct pair production.

the root mean squared multiple scattering angle in space (polar angular deflection) of a particle of momentum P, velocity β is:

$$\theta|_{rms} = \sqrt{\langle\theta^2\rangle} = \frac{21.1\,\mathrm{MeV}/c}{(P\beta)} z_I \sqrt{\frac{x}{X_o}}. \tag{3.6}$$

Note that the multiple scattering increases as the square root of the number of radiation lengths (X_o) traversed. Although multiple scattering has nothing to do with radiation, the fact that the medium dependence of both processes scales as the number of radiation lengths is because both are Coulomb interactions with the target nuclei.

Another radiative process closely related to Bremsstrahlung is direct pair production. Basically any radiated photon can also convert internally into an e^+e^- pair. In fact direct pair production dominates over Bremsstrahlung for muons and heavier particles so that even the energy loss dE/dx for muons as a function of momentum is very complicated (Figure 3.1) [84].

3.2.1.1 Cerenkov and transition radiation

In addition to Bremsstrahlung, i.e. radiation by scattered particles in matter, which is determined by the nucleus and atomic electrons of individual atoms and is not generally sensitive to the bulk properties of the matter traversed,[1] there are also two different types of radiation that do not involve scattering or acceleration of

[1] There is also coherent Bremsstrahlung [134] which is sensitive to the crystal structure of the matter, as well as the LPM effect and dielectric suppression [135] which are sensitive to the bulk properties of the matter but will not be discussed here.

the passing particles and which are sensitive to the bulk properties of the matter. These contribute negligibly to energy loss but are commonly used in particle detectors [84, 136] so they are mentioned here: (i) Cerenkov radiation is caused by particles traveling faster than the speed of light in the medium, so depends on the index of refraction; (ii) transition radiation is caused by the change in electric field when a particle crosses boundaries with different dielectric constants.

Cerenkov radiation depends on the velocity of the particle and is emitted at a polar angle θ_C around the direction of the particle, where

$$\cos\theta_C = \frac{1}{\beta n} \tag{3.7}$$

and n is the index of refraction of the medium. Thus Cerenkov radiation has a threshold when $\beta \geq 1/n$ so that it is sensitive to the velocity, $\beta = Pc/E$, of the particle and is typically used to identify particles of different masses at a given momentum. The probability $\mathcal{P}$ for a particle with charge $z_I\, e$ to emit a Cerenkov photon of energy k in range dk in passing through a thickness of material x (or the number of photons, N_γ, emitted if $\mathcal{P} > 1$) is:

$$\frac{dN_\gamma}{dk} = x\,\frac{\alpha}{\hbar c} z_I^2 \sin^2\theta_C = 370\,\mathrm{eV}^{-1}\,\mathrm{cm}^{-1}\, x\, z_I^2 \sin^2\theta_C. \tag{3.8}$$

Unlike Bremsstrahlung, Cerenkov radiation is independent of k apart from the dispersion in the medium, i.e. the dependence of the index of refraction $n(k)$ on k. For Cerenkov counters [137] one also uses a figure of merit or response parameter for the number of photons detected, $N_0 = 370\,\mathrm{cm}^{-1} \int_{k_{min}}^{k_{max}} \varepsilon(k)dk$, where k_{min}, k_{max} are the lower and upper limits of photon energies k (in eV) detected with efficiency $\varepsilon(k)$.

Transition radiation does not have a threshold but has the property [84, 138] that the energy radiated by a particle with charge $z_I\, e$ entering a medium from the vacuum is linearly proportional to the relativistic γ factor of the particle. For transition radiation, the relevant bulk property of the medium is the plasma frequency, $\hbar\omega_p$, and the energy radiated is:

$$E = \alpha z_I^2 \gamma \hbar\omega_p/3 \tag{3.9}$$

where

$$\hbar\omega_p = \sqrt{4\pi N_e r_e^3}\; m_e c^2/\alpha = \sqrt{\rho(\mathrm{g\,cm}^{-2}) Z/A} \times 28.81\ \mathrm{eV}. \tag{3.10}$$

The photon spectrum is very complicated [138] and only for $\gamma > 1000$ it extends into the soft x-ray (keV) region practical for detectors [139]. Thus transition radiation detectors are useful for identifying electrons with $E > 5$ GeV, where atmospheric pressure gas Cerenkov counters can no longer separate electrons from pions.

3.2.2 Strongly interacting particles – hadrons

In addition to the electromagnetic interactions which predominate in the target, an incident hadron may also suffer an occasional nuclear interaction. This is usually represented by a **nuclear interaction length** λ_I. For an interaction cross section σ_I, the probability of suffering a nuclear interaction in a thin slab of material is just the interaction cross section (cm^2) times the number of target nuclei per cm^2 presented by the slab of material. If the thickness is x, measured in g cm^{-2}, where $x = t \times \rho$, t is the thickness of material in cm, and ρ is the density of the material in g cm^{-3}, then

$$\mathcal{P} = \sigma_I \times \frac{N_o \, \rho \, t}{A} \equiv \frac{x}{\lambda_I} \tag{3.11}$$

so that

$$\frac{1}{\lambda_I} = \frac{N_o \, \sigma_I}{A} \tag{3.12}$$

and λ_I is in g cm^{-2}. Here N_o is Avogadro's number, the number of atoms per A grams of material of atomic weight A, and $N_o \rho \, t / A$ is the number of atoms per cm^2 in the slab of material. Note that nuclear interaction lengths are usually tabulated for incident nucleons. When an incident nucleus is involved, experimenters should usually check the computation themselves.

3.2.3 Passage of photons through matter

Photons are uncharged, and so do not have any long range electromagnetic forces. Thus, photons do not suffer ionization loss or multiple scattering when passing through a material. However, photons couple directly to electric charge with a "point-like" coupling. Thus, when a photon does interact, it is usually catastrophic, resulting in the absorption or loss of the photon. The three major electromagnetic processes via which photons interact in matter are the photoelectric effect and Compton scattering, with the atomic electrons, and pair production from the highly charged nuclei. For energies below ~1 MeV, the atomic phenomena are dominant and photons are strongly absorbed in material, with the absorbtion being inversely proportional to the energy: the lower the energy, the stronger the absorption. Above a few MeV, photopair production becomes dominant, with a cross section that increases logarithmically with the photon energy until **complete screening** by the atomic electrons sets in and the cross section saturates. The interplay of these phenomena has the experimentally important consequence that photons of energies near the critical energy, typically 5 to 30 MeV, have the minimum absorption in matter, and thus can travel relatively long distances in a solid medium. Many of the troublesome mysterious backgrounds (or *albedo*) in experiments are caused by these photons near the critical energy, for which detectors are relatively transparent.

For high energy photons, the pair production process is very strongly related to the Bremsstrahlung process for high energy electrons. The probability of a high energy photon to pass through a thickness x of material without undergoing pair production [140] is

$$\mathcal{P}_{NC} = \exp\left(-\frac{7}{9}\frac{x}{X_o}[1-\zeta]\right) \tag{3.13}$$

where $\mathcal{P}_{NC}$ is the non-conversion probability, X_o is the radiation length, and $\zeta \lesssim 0.05$. Here the radiation length comes in because radiation and pair production are really just two aspects of the same process.

3.2.4 Electromagnetic showers, electromagnetic calorimeters

In actual fact the concept of single Bremsstrahlung or pair production is only valid for very thin radiators or converters, $x \ll X_o$. For thick radiators, $x \gtrsim 10X_o$, an electromagnetic cascade shower develops. An incident photon converts into an e^+e^- pair, then each member of the pair radiates photons, these photons in turn convert, making more electrons, etc. At first the number of electrons in the shower increases with the depth, and then decreases roughly exponentially. The total depth of the shower increases logarithmically with the energy. It is as if the incident electron or photon is converted by radiative processes into electrons and positrons at the critical energy, which then stop radiating and lose the rest of their energy by ionization. The measurement of electromagnetic showers forms the basis of high resolution electron and photon detectors, or in today's jargon, **electromagnetic calorimeters**.

There are two classes of shower counters: sampling or total absorption. In a sampling counter, layers of high Z plates such as Pb or U create the showers, and the ionization is detected or sampled in layers of active material such as liquid argon or plastic scintillator. In a Total Absorption Shower Counter (TASC) there is only one high Z medium which both creates the showers and detects them. The showers are detected either by scintillation or by Cerenkov light. The most popular scintillating TASCs are made of NaI or CsI. These counters have the best energy resolution and are used over the range from keV to TeV photons. A much cheaper, and hence very popular, TASC is the lead glass (PbGl) Cerenkov counter. Lead glass for particle detection contains ~55% PbO by weight, or about twice that of high quality crystal used in glassware and chandeliers. The radiation length of lead glass (SF-5) is about 2.4 cm, the critical energy is 15.8 MeV and the index of refraction is 1.67. The electrons in the shower are detected by Cerenkov radiation. The number of Cerenkov photons per unit length saturates very quickly for particles above the threshold $\beta_t = 1/n = 0.6$ so that electrons at the critical energy travel to the end of their range while emitting a constant amount of Cerenkov radiation per unit length.

An estimate of the energy resolution of a PbGl counter can be made by assuming that the incident electron or photon of energy E is converted into electrons at the critical energy E_c, which then each emit a constant amount of Cerenkov light, represented by n_C detected Cerenkov photons, in coming to the end of their range. In this simple model, all fluctuations are ignored except in the statistics of the total number N of detected Cerenkov photons:

$$N = n_C \times \frac{E}{E_c}. \tag{3.14}$$

Then,

$$\frac{\sigma_E}{E} = \frac{1}{\sqrt{N}} = \sqrt{\frac{E_c/n_C}{E}} = \frac{\sqrt{0.0158\,\text{GeV}/n_C}}{\sqrt{E(\text{GeV})}} = \frac{13\%/\sqrt{n_C}}{\sqrt{E(\text{GeV})}}. \tag{3.15}$$

The actual energy resolution of clear lead glass, where absorption of the Cerenkov photons is not important is (in r.m.s)

$$\frac{\sigma_E}{E} = \frac{4\%}{\sqrt{E(\text{GeV})}}, \tag{3.16}$$

implying a few detected Cerenkov photons per critical energy electron. In real life, lead glass becomes radiation damaged and turns brown.

3.2.5 Hadron calorimeters

At sufficiently high energies, $\gtrsim$10 GeV, hadron interactions in thick absorbers $x \gg \lambda_I$ create cascade showers. At these high energies, multiparticle production dominates inelastic hadron reactions. Thus, a shower develops as an incident hadron suffers an inelastic interaction, many new hadrons are produced, they in turn interact, etc. (It is interesting to point out here that multiple hadron interactions inside a nucleus do not in general produce cascade showers and this is an important difference in the microscopic and macroscopic physics of hadron interactions with nuclei.) Hadron showers are not nearly as elegant as electromagnetic showers and in general the fluctuations are much worse. The typical resolution for a hadron calorimeter [141] is:

$$\frac{\sigma_E}{E} = \frac{64\%}{\sqrt{E(\text{GeV})}}. \tag{3.17}$$

Also, since the nuclear interaction length of most materials is much longer than the radiation length, hadron calorimeters tend to be very large objects.

3.3 Measurement of the momentum of a charged particle

The previous discussion on the interactions of charged particles in matter led to the method of **energy** measurements of electrons, photons, and hadrons by total absorption counters or calorimeters. The **momentum** of charged particles is more conventionally measured by a magnetic spectrometer.

A particle with charge e, momentum P in a uniform magnetic field B undergoes uniform circular motion with radius:

$$r = \frac{P\,c}{e\,B}. \tag{3.18}$$

In practical units, $r = 3$ meters for $P = 1$ GeV/c and $B = 1.111...$ Tesla, for unit charged particles like those encountered by particle physicists. This is the principle of particle accelerators and solenoidal magnetic spectrometers: once the field B is given, the momentum determines the radius (or vice versa). In fixed target experiments, a magnetic spectrometer usually takes the form of a magnet with a rectangular shaped field region (see Figure 3.2).

The magnetic field deflects the charged particle, and the momentum is determined from the measured deflection. Taking the z direction normal to the face of the magnet, and the magnetic field in the y direction, the particle deflection is in the x–z plane, the plane of the drawing of Figure 3.2. The vector equation:

$$\mathbf{F} = \frac{d\mathbf{P}}{dt} = \frac{e}{c}\,\mathbf{v} \times \mathbf{B} \tag{3.19}$$

leads to no change in the y component of momentum since $\mathbf{B}$ is in the y direction but leads to the coupled equations for P_x and P_z, the x and z components of momentum:

$$\frac{d\,P_x}{dt} = -\frac{e}{c} v_z\,B_y \qquad \frac{d\,P_z}{dt} = \frac{e}{c} v_x\,B_y. \tag{3.20}$$

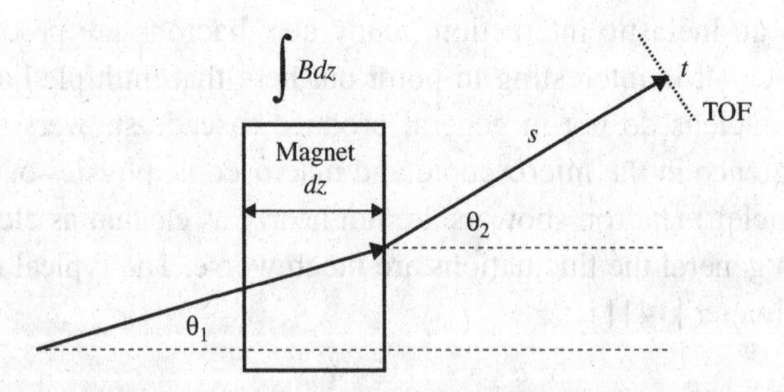

Figure 3.2 The trajectory of a charged particle entering and exiting a magnetic spectrometer. The field is normal to the page and the measurements of the entrance and exit angles are assumed to take place in a field free region. The particle travels a distance s over a time t from a defined starting point.

Since **B** acts perpendicular to the momentum, the magnitude of the momentum is conserved, so it is sufficient to integrate the x component equation:

$$\Delta P_x = -\frac{e}{c}\int_{traj} B_y\, v_z dt = -\frac{e}{c}\int B\, dz. \tag{3.21}$$

This leads to the usual description of a magnetic spectrometer as having a transverse momentum kick $\int Bdz = 300$ MeV/c per Tesla meter, for unit charged particles. For highly charged particles, this quantity is simply multiplied by z_I (i.e. $z_I \times 300$ MeV/c per T m).

3.3.1 Momentum resolution of a magnetic spectrometer

In the limit of small y component of momentum, Eq. 3.21 reduces to

$$\Delta \sin\theta = \sin\theta_2 - \sin\theta_1 = \frac{-(e/c)\int B\, dz}{P} \tag{3.22}$$

from which it should be apparent that $1/P$ is the quantity determined from a measurement of the difference in $\sin\theta$. Thus, the error in the quantity $1/P$ is related to the error in measurement of $\Delta \sin\theta$, which we may assume is normally distributed with standard deviation σ_{meas}:

$$\Delta(1/P) = \frac{\Delta P}{P^2} = \frac{\sigma_{meas}}{\int B\, dz}. \tag{3.23}$$

Measurement error usually dominates at high momenta, when the bending angle is small. At low momenta, multiple scattering usually dominates. For relativistic particles, the angular uncertainty due to multiple scattering (Eq. 3.6) is inversely proportional to P, leading to the neat relation when plugged into Eq. 3.23:

$$\left.\frac{\Delta P}{P}\right|_{ms} = \frac{15\,\text{MeV}/c\,\sqrt{x/X_o}}{\int B\, dz} \tag{3.24}$$

where the 21.1 MeV/c from Eq. 3.6 is divided by $\sqrt{2}$ to obtain the 15 MeV/c in Eq. 3.24, since only the component of multiple scattering in the bend plane affects the momentum resolution. Finally, since the uncertainties of multiple scattering and measurement error are uncorrelated so that they add in quadrature, the uncertainty in momentum measurement, the **momentum resolution**, is usually quoted in the form:

$$\frac{\Delta P}{P} = \sqrt{(a)^2 + (b\,P)^2} \tag{3.25}$$

where the first term is due to multiple scattering in the spectrometer and the second term is from measurement error.

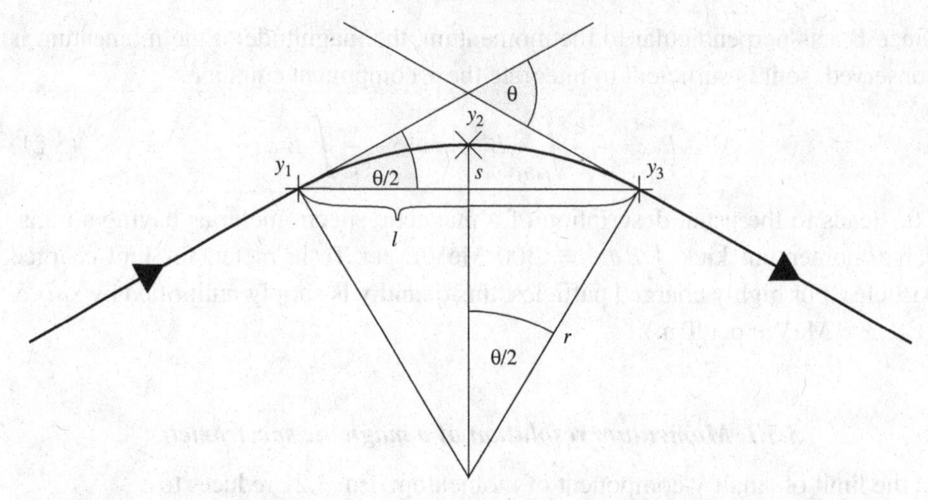

Figure 3.3 Trajectory of a charged particle in a solenoid with magnetic field along the collision axis which is normal to the page. The trajectory is measured over a length l given by the points y_1 and y_3 through which the particle bends by an angle θ. The momentum of the particle is determined by measuring the sagitta s at point y_2.

3.3.2 Momentum measurement in a solenoid

Most detectors at colliders are based on a solenoid to measure the momentum of charged particle tracks. The nicest feature of a well designed solenoid is that the magnetic field **B** is constant and along the beam axis (z) so that the particle orbits are helices. These orbits are circles in the x–y plane perpendicular to the axis of the colliding beams and provide a direct measurement of the transverse momentum, P_T. Figure 3.3 shows a view of a particle trajectory in the plane normal to the beam axis.

$$r(\text{cm}) = \frac{P_T(\text{MeV}/c)}{3B(\text{Tesla})}, \tag{3.26}$$

where the 3 is really $2.9979 = c\ (\text{cm s}^{-1})/10^{10}$. For the case of small bending angles the relation between the radius, the bend angle and the sagitta is simply estimated:

$$\sin\theta/2 = \frac{l}{2r} \tag{3.27}$$

$$s = r(1 - \cos\theta/2) = 2r\sin^2(\theta/4) \simeq r\theta^2/8 \tag{3.28}$$

$$s \simeq \frac{l^2}{8r} \tag{3.29}$$

$$r = \frac{l^2}{8s}. \tag{3.30}$$

Assume that the momentum is determined by measuring the sagitta with r.m.s. precision Δs over a known distance l, then the resolution is:

$$\frac{\Delta r}{r} = \frac{\Delta s}{s} = \frac{8r}{l^2}\Delta s$$

or

$$\frac{\Delta r}{r} = \frac{\Delta P_T}{P_T} = \frac{8r}{l^2}\Delta s = \frac{8P_T(\mathrm{GeV}/c)\times 10^3}{3B(\mathrm{T})}\frac{\Delta s(\mathrm{cm})}{l(\mathrm{cm})^2}. \tag{3.31}$$

The minimum measurement of the sagitta comes from the three points y_1, y_2 and y_3 all with assumed equal r.m.s. spatial resolution σ_y. Then

$$s = y_2 - \left(\frac{y_1 + y_3}{2}\right) \tag{3.32}$$

$$(\Delta s)^2 = \sigma_y^2 + \frac{\sigma_y^2}{4} + \frac{\sigma_y^2}{4} = \frac{6}{4}\sigma_y^2. \tag{3.33}$$

The effect of multiple scattering in solenoids is nicely given by Gluckstern [142].

3.3.3 Resolution smearing of a momentum spectrum

Suppose that x_o is a quantity to be measured, for example momentum, which is distributed with a steeply falling distribution, exponential for example:

$$d\mathcal{P}(x_o) = f(x_o)\,dx_o = e^{-bx_o}dx_o. \tag{3.34}$$

Further suppose that the true quantity x_o is measured with a Gaussian resolution function so that the result of the measurement is the quantity x, where

$$\mathcal{R}(x, x_o)\,dx = \mathrm{Prob}(x)|_{x_o}\,dx = \frac{1}{\sqrt{2\pi\sigma^2}}\exp\left[-\frac{(x-x_o)^2}{2\sigma^2}\right]dx. \tag{3.35}$$

The result for the measured spectrum is simply

$$f(x)\,dx = \int_{x_o=x-\infty}^{x_o=x+\infty} dx_o\ f(x_o)\ \mathrm{Prob}(x)|_{x_o}\,dx \tag{3.36}$$

$$f(x) = \frac{1}{\sqrt{2\pi\sigma^2}}\int dx_o\ \exp(-bx_o)\ \exp\left[-\frac{(x-x_o)^2}{2\sigma^2}\right]. \tag{3.37}$$

Complete the square:

$$f(x) = e^{-bx}\,e^{b^2\sigma^2/2}\times\frac{1}{\sqrt{2\pi\sigma^2}}\int_{x_o=x-\infty}^{x_o=x+\infty} dx_o\ \exp\left[-\frac{(x-b\sigma^2-x_o)^2}{2\sigma^2}\right]. \tag{3.38}$$

The result, since the Gaussian is normalized over $(-\infty, +\infty)$, is simply

$$f(x) = e^{b^2\sigma^2/2} \times e^{-bx} = e^{-b(x-b\sigma^2/2)}. \tag{3.39}$$

There are some important implications of this deceptively simple formula. It can be interpreted in two ways: the measured spectrum is shifted higher than the true spectrum by $\Delta x = b\sigma^2/2$; or equivalently, the measured spectrum at a true quantity x_o is higher than the true spectrum by a factor $\exp(b^2\sigma^2/2)$. Also, the steeper is the spectrum (larger b), the larger is the effect of the resolution smearing. This is a consequence of the fact that as the spectrum becomes steeper, it is relatively less probable to obtain larger values of the quantity of interest from the distribution itself, compared to the fluctuations due to resolution.

As in most of these discussions, real life is considerably more complicated than the simple examples. Clearly, the momentum resolution is not Gaussian, with a constant σ for all momenta, as assumed above, but depends on the true momentum x_o as in Eq. 3.25. Thus, Eq. 3.38 does not integrate so neatly. In general a non-exponential tail is produced at higher (measured) momenta. Also, the momenta spectra of particles are not in general exponential. However, it remains true that the steeper is the spectrum, at any value of momentum, the larger is the effect of the momentum resolution.

3.3.4 *Shift in a spectrum due to the effects of binning*

In addition to the effect of momentum resolution in smearing a measured spectrum from the true distribution, unintended shifts may be introduced into measured (momentum) spectra by experimenters if the effects of binning are neglected. To clarify this point, again consider a steeply falling spectrum in a quantity x_o

$$d\mathcal{P}(x_o) = f(x_o)\, dx_o. \tag{3.40}$$

One would like to measure the differential distribution $f(x_o)$. However, the real world intrudes and what is actually measured is the integral of the distribution in small intervals (bins): for example for a bin of width Δ centered at the value x_o

$$F(x_o, \Delta) = \int_{x_o-\Delta/2}^{x_o+\Delta/2} f(x)\, dx. \tag{3.41}$$

Typically, the differential distribution in a bin is estimated by dividing by the bin width:

$$\overline{f(x_o)} \equiv \frac{F(x_o, \Delta)}{\Delta}. \tag{3.42}$$

Clearly, in the limit of zero bin width, $\Delta \to 0$, the measured distribution approaches the true distribution. For the case of wide bins, there is a reasonable procedure (not always followed) of plotting the value $\overline{f(x_o)}$ at a value of $\overline{x_o}$ shifted from the center of the bin, typically to a slightly lower value for a falling spectrum. **If this procedure is not followed, then the experimental spectrum will be higher than the true spectrum.** Especially, in the case of steeply falling spectra, not correcting for the effect of bin width can cause an apparently upward shift in the normalization.

To calculate the correct plotting point, we have to know the shape of the true spectrum, although an exponential is usually a good approximation in any individual bin. The plotting point $\overline{x_o}$ is chosen so that it lies on the true curve, or in mathematics:

$$\overline{f(x_o)} \equiv \frac{1}{\Delta} \int_{x_o-\Delta/2}^{x_o+\Delta/2} f(x)\,dx \equiv f(\overline{x_o}). \tag{3.43}$$

For an exponential, $f(x) = A\exp(-bx)$, the solution is straightforward:

$$\overline{f(x_o)} = \frac{A\,e^{-bx_0}}{b\Delta}\,[e^{b\Delta/2} - e^{-b\Delta/2}]\,. \tag{3.44}$$

To obtain the correct solution, the exponentials must be expanded to second order in the small quantity:

$$e^z - e^{-z} = 2\,[z + z^3/3!] \simeq 2\,z\,e^{z^2/3!}, \tag{3.45}$$

whereupon

$$\overline{f(x_o)} = A\,e^{-bx_0}\,e^{b^2\Delta^2/24} = f(\overline{x_o}) \tag{3.46}$$

and the place to plot the point is shifted slightly from the center of the bin

$$\overline{x_o} = x_o - \frac{b\Delta^2}{24}. \tag{3.47}$$

As a general rule, beware of any experimental measurement of a falling spectrum where the points are plotted at the centers of the bins unless the experiment describes the following method for correcting the data points to the center of the bin.

3.3.5 Bin-shift correction

For recent experiments in p–p and $A+A$ collisions at RHIC and the LHC, precise ratios of the cross sections in p–p and $A+A$ collisions are necessary. Thus it is desirable to plot the points at the centers of the bins correctly so that taking ratios of different data sets is straightforward. This is called the bin-shift correction. Instead of plotting the value $\overline{f(x_o)}$ at the point $\overline{x_o}$ given by Eq. 3.43, we shift the point to the center of the bin x_o by plotting:

$$Y(x_o) \simeq \overline{f(x_o)} \times \frac{f(x_o)}{f(\overline{x_o})}. \tag{3.48}$$

Clearly a good estimation of the true functional form of the spectrum $f(x)$ is required in order to calculate the ratio of the function at the center of the bin to the value of the function at the correct average point. This is usually obtained by fitting the spectrum to a functional form and then iterating the bin-shift procedure [143].

3.4 Lorentz transformations, kinematics, spectra of decay products

In addition to measuring the spectra of primary particles, experimentalists are forced quite often to deal with the spectra of unstable particles which can only be inferred from their decay products. Two examples are

$$\pi^0 \to \gamma + \gamma \qquad \Lambda^0 \to p + \pi^-. \tag{3.49}$$

Lorentz transformations and kinematics also play a real role in the daily life of an experimentalist, since we cannot always position our equipment in the desirable rest frame.

3.4.1 Lorentz transformations

Consider a particle of vector momentum $\mathbf{P}$, energy E, in a particular rest frame, and invariant mass m. The **four-vector** momentum p of this particle is denoted such that the fourth (time) component $P_t \equiv iE$:

$$p = (\mathbf{P}, iE) \qquad \text{or} \qquad p = (P_x, P_y, P_z, P_t) \tag{3.50}$$

in units where the speed of light c is taken as unity. The four-dot product of two four-momentum vectors p_1 and p_2 is denoted:

$$\begin{aligned} p_1 \cdot p_2 &\equiv \mathbf{P}_1 \cdot \mathbf{P}_2 - E_1 E_2 \\ &= P_1 P_2 \cos\theta - E_1 E_2 \end{aligned} \tag{3.51}$$

where $\mathbf{P}_1 \cdot \mathbf{P}_2$ is the dot product of the three-momentum vectors, P_1 and P_2 are the moduli of the three-vectors and θ is the angle between them. The squared modulus of a four-vector is denoted

$$p^2 = p \cdot p = P^2 - E^2 = -m^2 \tag{3.52}$$

and is invariant under a Lorentz transformation.

We are often obliged to deal with particles in different reference frames. Let a particle of invariant mass m have a four-momentum

$$p^* = (\mathbf{P}^*, iE^*) \tag{3.53}$$

as measured in a coordinate system moving with a velocity β relative to the reference system in which the particle four-momentum is

$$p = (\mathbf{P}, iE). \tag{3.54}$$

The **Lorentz transformation** relates the momentum components in the reference frame to those measured in the moving coordinate system:

$$\begin{aligned} E &= \gamma\,(E^* + \beta P_L^*) \\ P_L &= \gamma\,(P_L^* + \beta E^*) \\ P_T &= P_T^* \end{aligned} \tag{3.55}$$

where $P_L = P\cos\theta$, $P_T = P\sin\theta$, $P_L^* = P^*\cos\theta^*$, $P_T^* = P^*\sin\theta^*$ are the components of momentum, longitudinal and transverse to the direction of motion, in the respective frames; θ and θ^* are the angles of the particle relative to the direction of motion in the two frames, and

$$\gamma = \frac{1}{\sqrt{1-\beta^2}}. \tag{3.56}$$

Note that the transverse momentum, P_T, is conserved between the two frames, as is the **transverse mass**, $m_T = \sqrt{P_T^2 + m^2}$, while the energy and longitudinal momentum are **boosted**. The transformation of the angle relative to the direction of motion in the two frames is obtained immediately from Eq. 3.55:

$$\tan\theta = \frac{P_T}{P_L} = \frac{\sin\theta^*}{\gamma\,(\cos\theta^* + \beta/\beta^*)} \tag{3.57}$$

where $\beta^* = P^*/E^*$ is the velocity of the particle in the moving frame.

The convention used in this book [144,145], in which the spatial components of a momentum three-vector are real while the fourth component, representing energy, is taken as imaginary, is called Cartesian or Euclidean space. This is because the modulus of a four-vector extends straightforwardly the modulus of a three-vector to a fourth dimension, allowing us to keep our familiar vector and matrix algebra [146] for the four-vector p and the vector-product:

$$p \cdot q = \sum_{i=1,\ldots 4} p_i q_i = \tilde{\mathsf{p}}\mathsf{q}. \tag{3.58}$$

The modulus of a four-vector is called the metric, which is the distance squared between two points in space, $\mathbf{x}_1 - \mathbf{x}_2 = \Delta\mathbf{r}$, or space-time, $x_1 - x_2 = \Delta r$, where

$$r = (\mathbf{x}, it) \quad \text{or} \quad r = (x, y, z, it) \tag{3.59}$$

and

$$d^2 = (\Delta r)^2 = \sum_{i=1,\ldots,4} (\Delta x_i)^2 = (\Delta x)^2 + (\Delta y)^2 + (\Delta z)^2 - (\Delta t)^2. \tag{3.60}$$

Thus, as in Eq. 3.52 for a real particle on its mass shell, where the magnitude of the time component is larger than the magnitude of the space component, it is better terminology to say that the four-vector is time-like, which is generally true, rather than negative which depends on the convention. Other conventions used in field theory and general relativity add a metric tensor in the vector product (Eq. 3.58) and keep both the space and time coordinates real. The advantage of the present metric is that standard vector and matrix algebra [146] is used without the need to introduce a metric tensor. The Lorentz transformation in Eq. 3.55 can be written simply in matrix notation as $\mathsf{p} = \mathsf{L}\mathsf{p}^*$, or

$$\begin{pmatrix} P_L \\ iE \end{pmatrix} = \begin{pmatrix} \gamma & -i\gamma\beta \\ i\gamma\beta & \gamma \end{pmatrix} \begin{pmatrix} P_L^* \\ iE^* \end{pmatrix} \tag{3.61}$$

where only the P_L (or z) component along the direction of motion is indicated in the matrix p because the components P_x and P_y transverse to the direction of motion are not affected by the Lorentz transformation. From Eq. 3.61 it is easy to see that the modulus of a four-vector remains invariant under a Lorentz transformation:

$$p \cdot p - P_T^2 = \tilde{\mathsf{p}}\mathsf{p} = \tilde{\mathsf{p}}^* \tilde{\mathsf{L}}\mathsf{L}\mathsf{p}^* = p^* \cdot p^* - P_T^2 \tag{3.62}$$

because

$$\tilde{\mathsf{L}}\mathsf{L} = \gamma^2 \begin{pmatrix} 1-\beta^2 & 0 \\ 0 & 1-\beta^2 \end{pmatrix} = \begin{pmatrix} 1 & 0 \\ 0 & 1 \end{pmatrix}. \tag{3.63}$$

As a note of terminology, the modulus of a four-vector is Lorentz invariant, it is the same in any frame; while four-vectors are said to be Lorentz covariant, they transform according to Eqs. 3.55, 3.61.

The simplicity of the rapidity variable Eq. 2.2 becomes apparent when the Lorentz transformation between the two frames is expressed in this variable. Let

$$y = \ln\left(\frac{E + P_L}{m_T}\right) = \frac{1}{2}\ln\left(\frac{E + P_L}{E - P_L}\right) \tag{3.64}$$

denote the rapidity of the particle in the reference frame, y^* be the rapidity of the particle measured in the moving frame and Y be the rapidity of the moving system:

$$Y = \frac{1}{2} \ln \left(\frac{1+\beta}{1-\beta} \right). \tag{3.65}$$

Then it follows from substituting Eq. 3.55 in Eq. 3.64 that

$$y = Y + y^*. \tag{3.66}$$

The details are left as an exercise for the reader.

3.4.2 Transformation to the rest system of a particle

The rest system of a particle is defined as the system in which the particle is at rest, i.e $P^* = 0$, $E^* = m$. It is then easy to see that the rest system of the particle moves with the particle velocity (as seen in the laboratory system in which the particle has momentum P and energy E):

$$\begin{aligned} E &= \gamma\, m \\ P &= \gamma\, \beta\, m. \end{aligned} \tag{3.67}$$

3.4.3 Two particle collisions – the laboratory and c.m. systems

The description of a two particle collision is a useful exercise in relativistic kinematics. In the laboratory system, an incident particle with momentum $\mathbf{P}_1$, energy E_1 and mass m_1 collides with a particle with mass m_2, at rest. The four-vectors are:

$$p_1 = (\mathbf{P}_1, iE_1) \qquad p_2 = (0, im_2). \tag{3.68}$$

In the center-of-mass (c.m.) system the momenta of the particles are equal and opposite, and the four-vectors are:

$$p_1^* = (\mathbf{P}_1^*, iE_1^*) \qquad p_2^* = (-\mathbf{P}_1^*, iE_2^*). \tag{3.69}$$

The transformation between the laboratory and c.m. systems is given in terms of the four-vector total momentum of the system $p_1 + p_2$, which is conserved in a collision. The modulus of the total four-momentum is a Lorentz invariant quantity, which is the same in all reference systems (also before and after the collision because of four-momentum conservation):

$$-s \equiv (p_1 + p_2)^2 = (p_1^* + p_2^*)^2 = -(E_1^* + E_2^*)^2. \tag{3.70}$$

Thus, it is clear that $\sqrt{s}$ is the total energy in the c.m. system, which is the same as the invariant mass of the c.m. system.

We also use the fact that while the modulus of the four-momentum is conserved in all reference frames, the four-momenta in different reference frames are related by Lorentz transformations. We can then write the four-vector total momentum of the system, $p_1 + p_2$, as well as the modulus, $-s$, in three different ways: in terms of the momentum and energy of the c.m. system, $\mathbf{P}^{cm}$, E^{cm}, as observed in the laboratory system (Eq. 3.71); in terms of the original laboratory system momenta and energy of the colliding particles (Eq. 3.72); and also as observed in the c.m. system (Eq. 3.73)

$$
\begin{aligned}
-s \equiv (p_1 + p_2)^2 &= -(\mathbf{P}^{cm}, iE^{cm})^2 && (3.71)\\
&= -(\mathbf{P}_1, i(E_1 + m_2))^2 && (3.72)\\
&= -(0, i\sqrt{s})^2. && (3.73)
\end{aligned}
$$

Then, using the Lorentz transformation between the c.m. reference frame and the laboratory reference frame in which the c.m. system is moving with velocity β^{cm}, γ^{cm} (Eq. 3.55), it is immediately apparent that the c.m. rest frame with invariant mass $\sqrt{s}$ moves in the laboratory system (along the direction of $\mathbf{P}_1$) with momentum $\mathbf{P}^{cm}$ and energy E^{cm} in the laboratory reference frame

$$
\begin{aligned}
E^{cm} &= E_1 + m_2 = \gamma^{cm}\sqrt{s}\\
\mathbf{P}^{cm} &= \mathbf{P}_1 \qquad = \gamma^{cm}\vec{\beta}^{cm}\sqrt{s}\\
\vec{\beta}^{cm} &= \mathbf{P}^{cm}/E^{cm} = \frac{\mathbf{P}_1}{E_1 + m_2},
\end{aligned}
\qquad (3.74)
$$

where the velocity β^{cm} corresponds to

$$\gamma^{cm} = \frac{E_1 + m_2}{\sqrt{s}} \quad \text{and} \quad Y^{cm} = \cosh^{-1}\gamma^{cm}, \qquad (3.75)$$

and that the c.m. energy $\sqrt{s}$ is given in terms of the laboratory quantities of the incident particles as

$$s = m_1^2 + m_2^2 + 2E_1 m_2. \qquad (3.76)$$

Another useful quantity is Y^{beam}, the rapidity of the incident particle in the laboratory system

$$Y^{beam} = \cosh^{-1}\frac{E_1}{m_1}, \qquad (3.77)$$

where for equal mass projectile and target:

$$Y^{cm} = Y^{beam}/2. \qquad (3.78)$$

3.4.4 Kinematics of two-to-two scattering

Let us now consider the kinematics of a two particle collision where only two particles emerge in the final state, e.g.

$$p_1 + p_2 \rightarrow p_3 + p_4. \tag{3.79}$$

In the laboratory system, the four-vectors of the outgoing particles are

$$p_3 = (\mathbf{P}_3, iE_3) \qquad p_4 = (\mathbf{P}_4, iE_4), \tag{3.80}$$

and in the center-of-mass (c.m.) system they are

$$p_3^* = (\mathbf{P}_3^*, iE_3^*) \qquad p_4^* = (\mathbf{P}_4^*, iE_4^*). \tag{3.81}$$

If the two outgoing particles are the same as the incident particles, e.g. $m_3 = m_1$ and $m_4 = m_2$, then this is called elastic scattering.

In certain types of two-to-two particle production, one or both of the outgoing particles may be different from the incident particles, hence with different masses, for example:

- photoproduction of a ρ^0 meson in hydrogen

$$\gamma + p \rightarrow \rho^0 + p, \tag{3.82}$$

- excitation of a nucleon resonance by an electron

$$e + N \rightarrow e + N^*, \tag{3.83}$$

- associated production

$$\pi^- + p \rightarrow \Lambda^0 + K^0. \tag{3.84}$$

However, in all cases, since both momentum and energy are conserved, the three-momenta of the outgoing particles are equal and opposite, but the energies will not be equal unless the masses are equal.

The four-momentum, i.e. momentum and energy, is conserved in all reactions so that in any reference frame:

$$p_1 + p_2 = p_3 + p_4. \tag{3.85}$$

It is convenient to use the Lorentz invariant Mandelstam variables [147] to describe two-to-two kinematics, where it is assumed for elastic scattering that p_3 is the outgoing momentum of particle p_1 and p_4 is the outgoing momentum of particle p_2:

$$\begin{aligned} -s &= (p_1 + p_2)^2 = (p_3 + p_4)^2 \\ -t &= (p_1 - p_3)^2 = (p_4 - p_2)^2 \\ -u &= (p_1 - p_4)^2 = (p_3 - p_2)^2. \end{aligned} \tag{3.86}$$

As noted above (Eq. 3.70), $\sqrt{s}$ is the total energy in the c.m. system, the c.m. energy of the collision. For the other two invariants, t is the negative of the four-momentum transfer squared of the collision (also defined as $t = -Q^2$, since t is space-like for elastic scattering), and u represents the invariant of the "crossed channel," when the momenta p_3 and p_4 are switched. The three Mandelstam invariants obey the relation:

$$s + t + u = m_1^2 + m_2^2 + m_3^2 + m_4^2. \tag{3.87}$$

This is again left as an exercise for the student.

3.4.4.1 Elastic scattering

We will deal with the detailed kinematics of elastic scattering in a later chapter. For the present, we derive a deceptively simple result for elastic scattering in the laboratory system which is important for understanding the parton model, knock-on electron production in material, etc.

Assume that the collision takes place in the system in which particle p_2, "the target," is at rest. As in Eqs. 3.68, 3.80 above, the four vectors are:

$$\begin{aligned} p_1 &= (\mathbf{P}_1, iE_1) \qquad p_2 = (0, im_2) \\ p_3 &= (\mathbf{P}_3, iE_3) \qquad p_4 = (\mathbf{P}_4, iE_4). \end{aligned} \tag{3.88}$$

The four-momentum transfer squared is

$$\begin{aligned} Q^2 = -t = (p_1 - p_3)^2 &= (p_4 - p_2)^2 \\ &= (\mathbf{P}_4, i\,[E_4 - m_2])^2 \\ &= P_4^2 - E_4^2 - m_2^2 + 2E_4 m_2 \\ &= 2E_4 m_2 - m_4^2 - m_2^2. \end{aligned} \tag{3.89}$$

If we rewrite this in terms of the kinetic energy T_4 of the recoil target particle p_4, $T_4 = E_4 - m_4$, we obtain

$$Q^2 = 2T_4 m_2 - (m_4 - m_2)^2. \tag{3.90}$$

Note that we did not assume $m_4 = m_2$ in this derivation which means that it could apply, for example, to the excitation of a nucleon resonance in electron proton scattering (Eq. 3.83). However, the elastic scattering case in which $m_1 = m_3$ and $m_2 = m_4 \equiv M$, $T_4 \equiv T$ is most instructive. In this case Eq. 3.90 reduces to the simple and fundamental relation:

$$Q^2 = 2MT \tag{3.91}$$

$$T = \frac{Q^2}{2M} = E_1 - E_3. \tag{3.92}$$

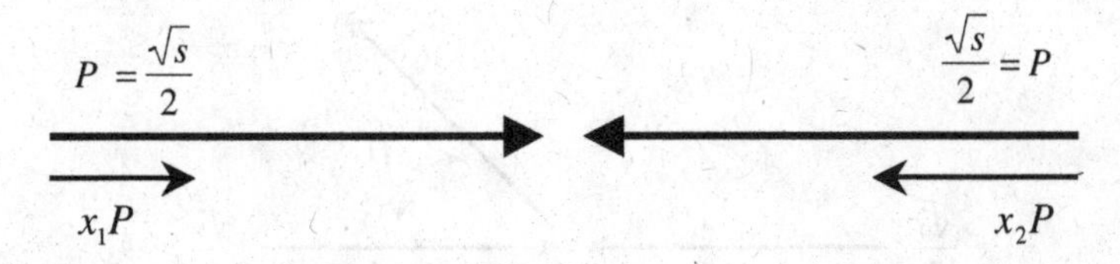

Figure 3.4 Initial parton kinematics in the p–p scattering c.m. system.

3.4.5 Kinematics of the parton model

Another simple but instructive example of elastic scattering is the parton model of particle production at large transverse momentum, high p_T. Suppose there is a p–p collision at c.m. energy $\sqrt{s}$. In the p–p c.m. system, the protons (with mass M) approach each other with equal energies $\sqrt{s}/2$ and opposite momenta

$$P = \frac{\sqrt{s}}{2}\sqrt{1 - \frac{4M^2}{s}}$$

which we take as $\mathbf{P}_1 = -\mathbf{P}_2 \approx \sqrt{s}/2$, since these collisions occur at very large $\sqrt{s} \gg 2M$. Each proton has a parton with momentum fraction x_i, $i = 1, 2$ which generally results in the c.m. system of the parton–parton collision moving in the p–p c.m. system (Figure 3.4). Since the partons are taken to be massless and the initial kinematics are along the collision axis with no transverse momentum, the c.m. energy squared ($\hat{s}$) of the parton–parton collision is:

$$\begin{aligned}\hat{s} &= -(p_1 + p_2)^2 = -([x_1 P - x_2 P], i\,[x_1 P + x_2 P])^2 \\ &= 4P^2 x_1 x_2 \\ &= s x_1 x_2.\end{aligned} \tag{3.93}$$

The parton–parton c.m. system has invariant mass $\sqrt{\hat{s}}$, longitudinal momentum $(x_1 - x_2)P$, total energy $(x_1 + x_2)P$, in the p–p c.m. system and thus has rapidity, $\hat{y}$:

$$\begin{aligned}\hat{y} &= \ln\left(\frac{E + P_L}{\sqrt{\hat{s}}}\right) \\ &= \ln\left(\frac{2Px_1}{2P\sqrt{x_1 x_2}}\right) \\ &= \ln\sqrt{\frac{x_1}{x_2}}.\end{aligned} \tag{3.94}$$

Also, Eqs. 3.94, 3.93 can be solved for x_1 and x_2:

$$x_1 = \sqrt{\frac{\hat{s}}{s}}\,e^{\hat{y}} \qquad x_2 = \sqrt{\frac{\hat{s}}{s}}\,e^{-\hat{y}}. \tag{3.95}$$

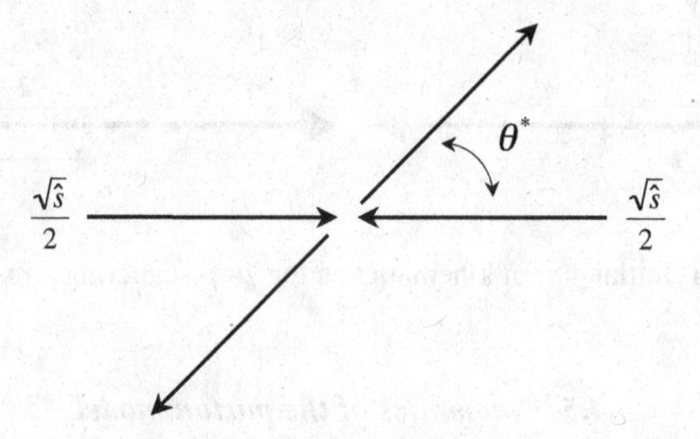

Figure 3.5 Elastic scattering in the parton–parton c.m. system.

The parton–parton c.m. system moves in the p–p c.m. system with:

$$\beta_{\hat{s}} = \frac{P_L}{E} = \frac{x_1 - x_2}{x_1 + x_2}$$
$$\gamma_{\hat{s}} = \frac{E}{\sqrt{\hat{s}}} = \frac{\sqrt{s}}{2\sqrt{s x_1 x_2}}(x_1 + x_2) = \frac{x_1 + x_2}{2\sqrt{x_1 x_2}} \tag{3.96}$$
$$\gamma_{\hat{s}} \beta_{\hat{s}} = \frac{x_1 - x_2}{2\sqrt{x_1 x_2}}.$$

In the parton–parton c.m. system, the two colliding partons have equal and opposite momenta, $\sqrt{\hat{s}}/2$.

Now, suppose that the incident partons undergo elastic scattering through scattering angle θ^* in the parton–parton c.m. system, where we require $0 \leq \theta^* \leq \pi$, $0 \leq \sin\theta^* \leq 1$ since we can rotate the collision plane 180° about the incident parton–parton axis (Figure 3.5). The scattered parton three-momenta are equal and opposite, $\mathbf{P}_3^* = -\mathbf{P}_4^*$, and their energies are equal, and equal to the magnitude of their three-momenta, $E_3^* = E_4^* = P_3^* = P_4^* = \sqrt{\hat{s}}/2$, because both outgoing partons are assumed to be massless, $m_3 = m_4 = 0$. If we take the z axis along the direction of the initial partons and the y axis perpendicular to the z axis in the scattering plane, then we can write the parton four-momenta in the parton–parton c.m. system as:

$$p_1^* = \frac{\sqrt{\hat{s}}}{2}(0, 0, 1, i) \quad p_2^* = \frac{\sqrt{\hat{s}}}{2}(0, 0, -1, i)$$
$$p_3^* = \frac{\sqrt{\hat{s}}}{2}(0, \sin\theta^*, \cos\theta^*, i) \quad p_4^* = \frac{\sqrt{\hat{s}}}{2}(0, -\sin\theta^*, -\cos\theta^*, i), \tag{3.97}$$

where

$$P_{L_3}^* = -P_{L_4}^* = \frac{\sqrt{\hat{s}}}{2}\cos\theta^* \qquad P_{T_3}^* = -P_{T_4}^* \equiv p_T = \frac{\sqrt{\hat{s}}}{2}\sin\theta^*. \tag{3.98}$$

The Mandelstam invariants of the parton–parton scattering are $\hat{s}$, $\hat{t}$, $\hat{u}$, where $\sqrt{\hat{s}}$ (Eq. 3.93) is the c.m. energy of the parton–parton scattering and $\hat{t}$ is the invariant four-momentum transfer squared,

$$\hat{s} + \hat{t} + \hat{u} = 0. \tag{3.99}$$

The invariants are related by the c.m. scattering angle:

$$\begin{aligned}\hat{Q}^2 = -\hat{t} = (p_1^* - p_3^*)^2 &= \frac{\sqrt{\hat{s}}}{2}[(0, -\sin\theta^*, 1 - \cos\theta^*, 0)]^2 \\ &= \frac{\hat{s}}{4}[\sin^2\theta^* + (1 - \cos\theta^*)^2] \\ &= \hat{s}\frac{(1 - \cos\theta^*)}{2};\end{aligned} \tag{3.100}$$

and likewise

$$-\hat{u} = (p_1^* - p_4^*)^2 = \hat{s}\frac{(1 + \cos\theta^*)}{2}. \tag{3.101}$$

For the outgoing partons, the rapidities are equal and opposite in the parton–parton c.m. system:

$$y_3^* = \frac{1}{2}\ln\left(\frac{E_3^* + P_{L_3}^*}{E_3^* - P_{L_3}^*}\right) \equiv -\frac{1}{2}\ln\left(\frac{E_3^* - P_{L_3}^*}{E_3^* + P_{L_3}^*}\right) = -y_4^* \tag{3.102}$$

because $P_{L_4}^* = -P_{L_3}^*$, and for zero (or equal) mass partons, $E_4^* = E_3^*$. Equivalently, we can write:

$$y_3^* = \ln\left(\frac{E_3^* + P_{L_3}^*}{m_T}\right) = \ln\left(\frac{m_T}{E_3^* - P_{L_3}^*}\right) = \ln\left(\frac{m_T}{E_4^* + P_{L_4}^*}\right) = -y_4^* \tag{3.103}$$

where $m_{T_3} = m_{T_4} \equiv m_T = p_T$.

Since we usually observe the parton–parton scattering in the p–p c.m. system in collider detectors, the Lorentz transformation from the parton–parton to the p–p c.m. systems is important. The four-vectors of the outgoing partons in the p–p c.m. system are:

$$p_3 = (\mathbf{P}_3, iE_3) \qquad p_4 = (\mathbf{P}_4, iE_4) \tag{3.104}$$

where $P_{T_3} = P_{T_3}^* \equiv p_T$, $P_{T_4} = P_{T_4}^* = -p_T$, and $m_{T_3} = m_{T_4} \equiv m_T = p_T$, which is invariant to the Lorentz transformation. The rapidities of the scattered partons in the p–p c.m. system are given by the additive rule:

$$y_3 = \hat{y} + y_3^* \tag{3.105}$$

$$y_4 = \hat{y} + y_4^* \tag{3.106}$$

$$y_3 + y_4 = 2\,\hat{y}$$

$$\frac{y_3 + y_4}{2} = \hat{y} \tag{3.107}$$

$$y_3 - y_4 = y_3^* - y_4^*, \tag{3.108}$$

where Eq. 3.107 relates the rapidity of the parton–parton c.m. system in the p–p c.m. system to the rapidities, y_3, y_4, of the outgoing scattered partons in this reference frame. Also, Eq. 3.108, together with $y_3^* = -y_4^*$ (Eq. 3.102) for zero (or equal) mass outgoing partons, allow the c.m. scattering angle $\cos\theta^*$ to be determined from y_3 and y_4:

$$y_3 - y_4 = y_3^* - y_4^* = 2y_3^* = 2\eta_3^*. \tag{3.109}$$

It then follows from Eq. 2.6 that:

$$\tanh\eta_3^* = \cos\theta^* = \tanh\frac{(y_3 - y_4)}{2}. \tag{3.110}$$

The initial parton momentum fractions x_1 and x_2 can also be calculated from y_3 and y_4 of the outgoing scattered partons in the p–p c.m. system. Since this is simply two-to-two scattering, we start with four-momentum conservation in the p–p c.m. system:

$$p_1 + p_2 = p_3 + p_4. \tag{3.111}$$

Then, from Eqs. 3.93:

$$\begin{aligned}
E_1 + E_2 = E = \frac{\sqrt{s}}{2}(x_1 + x_2) &= E_3 + E_4 = m_{T_3}\cosh y_3 + m_{T_4}\cosh y_4 \\
P_1 + P_2 = P_L = \frac{\sqrt{s}}{2}(x_1 - x_2) &= P_3 + P_4 = m_{T_3}\sinh y_3 + m_{T_4}\sinh y_4 \\
E + P_L = x_1\sqrt{s} &= m_{T_3}(\cosh y_3 + \sinh y_3) + m_{T_4}(\cosh y_4 + \sinh y_4) \\
E + P_L = x_1\sqrt{s} &= m_{T_3}e^{+y_3} + m_{T_4}e^{+y_4}.
\end{aligned} \tag{3.112}$$

Similarly

$$\begin{aligned}
E - P_L = x_2\sqrt{s} &= m_{T_3}(\cosh y_3 - \sinh y_3) + m_{T_4}(\cosh y_4 - \sinh y_4) \\
E - P_L = x_2\sqrt{s} &= m_{T_3}e^{-y_3} + m_{T_4}e^{-y_4},
\end{aligned} \tag{3.113}$$

or

$$x_1 = \frac{m_{T_3}}{\sqrt{s}} e^{+y_3} + \frac{m_{T_4}}{\sqrt{s}} e^{+y_4} \tag{3.114}$$

$$x_2 = \frac{m_{T_3}}{\sqrt{s}} e^{-y_3} + \frac{m_{T_4}}{\sqrt{s}} e^{-y_4}. \tag{3.115}$$

If both partons 3 and 4 are massless then $m_{T_3} = p_T = m_{T_4}$ and we get the simpler formula, where $x_T = 2p_T/\sqrt{s}$:

$$x_1 = x_T \frac{e^{+y_3} + e^{+y_4}}{2} \tag{3.116}$$

$$x_2 = x_T \frac{e^{-y_3} + e^{-y_4}}{2}. \tag{3.117}$$

Equations 3.116 and 3.117 are useful if you want to find the x_1 and x_2 range for a parton–parton collision with outgoing partons of a given p_T in a specific region of rapidity, or conversely if you know what range of x_1, x_2 you want to explore, what range of rapidity you need.

3.4.6 Two body decay of a heavy particle into light particles

This is a more advanced example of the transformation to a rest system, which has some interesting experimental applications. Consider a heavy particle which decays to two light particles. Examples would include:

$$J/\Psi \to e^+ e^- \qquad \pi^0 \to \gamma\gamma \qquad Z^0 \to \mu^+ \mu^-. \tag{3.118}$$

To be specific, a particle of momentum $\mathbf{P}$, energy E, rest mass m, decays into two particles with $\mathbf{P}_3$, E_3, m_3 and $\mathbf{P}_4$, E_4, m_4. There are many questions we might wish to ask with this starting point. However, we start out with a simple example on the use of rapidity and then follow with more complicated questions near and dear to the heart of somebody with a photon detector.

3.4.6.1 Rapidity acceptance of $H \to \gamma + \gamma$

Suppose we have a heavy particle, H with $p_T = 0$ at rapidity $\hat{y}$, i.e. moving along the axis of the beams in a p–p collider. Where do the two decay photons go? The kinematics are identical to Eq. 3.107 with $\sqrt{\hat{s}} = M_H$,

$$\hat{y} = y_H = \frac{y_3 + y_4}{2} = \frac{\eta_3 + \eta_4}{2} \tag{3.119}$$

where η_3 and η_4 are the pseudorapidities of the decay photons which are the same as their rapidities since they are massless. The same kinematics apply to the detection of Drell–Yan pairs or $J/\Psi \to e^+ + e^-$, since the $e^\pm$ or $\mu^\pm$ are effectively massless at the energies considered.

Suppose we consider a forward detector which covers an angular range $\eta \geq \eta_0$. In order for both leptons and photons to be detected, $\eta_3 \geq \eta_0$ and $\eta_4 \geq \eta_0$. Substituting into Eq. 3.119, we find:

$$\hat{y} = y_H \geq \eta_0. \tag{3.120}$$

This provides a simple and elegant relationship between the angular aperture of a detector ($\eta \geq \eta_0$) and its acceptance in rapidity, y_H, for a heavy particle (with $p_T = 0$) which decays to two light particles which are both detected in this angular aperture.

3.4.6.2 *More complicated rapidity acceptance*

Now suppose we have a photon detector which covers the angular range from η_{min} to η_{max}, what will be the rapidity range of the detected particle $H \to \gamma + \gamma$?

Since both γ must be detected, y_H falls in the range $\eta_{min} \leq y_H \leq \eta_{max}$. The simplest case is when $\eta_3 = \eta_4$ which occurs for 90° emission in the H rest frame, $\sin\theta^* = 1$, $p_T = E_1^* = M_H/2$. As y_H becomes larger, it is limited by the available energy in the p–p c.m. system ($\sqrt{s}$). Assuming formation via, for example, $g + g \to H \to \gamma+\gamma, q+\bar{q} \to H \to \gamma+\gamma$, then $M_H^2 = \hat{s} = s x_1 x_2$, and the kinematic limits occur due to the requirement $0 \leq x_1, x_2 \leq 1$. We can use Eqs. 3.116, to find the kinematic limits for a given p_T of the decay photons, where $p_T \leq M_H/2$, $x_T = p_T/(\sqrt{s}/2)$

$$0 \leq x_1 = x_T \frac{e^{+\eta_3} + e^{+\eta_4}}{2} \leq 1 \tag{3.121}$$

$$0 \leq x_2 = x_T \frac{e^{-\eta_3} + e^{-\eta_4}}{2} \leq 1; \tag{3.122}$$

and the c.m. decay angle θ^* is limited by the condition that for a given y_H, the two decay photons with η_3 and η_4 be within the acceptance

$$\eta_3 - y_H = \eta_3^* = -\ln\tan\theta^*/2 = -\eta_4^* = y_H - \eta_4 \tag{3.123}$$

where $\eta_{min} \leq \eta_3, \eta_4 \leq \eta_{max}$.

The same kinematics applies to the detection of Drell–Yan pairs or $J/\Psi \to e^+ + e^-$. Similar considerations apply to the detection of the W boson via the decay

$$W \to e + \nu$$

except that only one of the outgoing particles, the electron, is detected.

3.4.6.3 *Detecting both decay particles in a typical geometry – a PbGl wall*

Suppose the heavy particle (or π^0) were traveling in a direction normal to the surface of a PbGl wall, heading toward the detector, when it decayed a distance L

from the front surface. What is the distribution in the distance d between the two decay particles when they strike the front surface of the wall?

The two decay particles, here denoted 1 and 2, arrive at the wall at lateral distances d_1 and d_2 from where the point of impact of the parent would have been. Since it is a two body decay, all three particles lie in the same plane, so that the distance d between the two decay particles is simply

$$d = d_1 + d_2 = L(\tan\theta_1 + \tan\theta_2) \tag{3.124}$$

where θ_1, θ_2 are the decay angles of particles 1 and 2 in the laboratory system and L is the distance of the decay from the detector. In the rest system of the parent, the two particles have equal and opposite momenta, so that

$$x \equiv \cos\theta_1^* = -\cos\theta_2^* \tag{3.125}$$

and $\sin\theta_1^* = \sin\theta_2^* = \sqrt{1-x^2}$. This causes a few neat cancelations when Eq. 3.57 is substituted in Eq. 3.124 for both particles. In the case of equal masses for the two decay particles ($m_1 = m_2$), the exact result for d has the simple form:

$$\frac{d}{2L} = \beta\beta^* \frac{\sqrt{1-x^2}}{\gamma(\beta^2 - \beta^{*2}x^2)} \tag{3.126}$$

where β^* is the velocity of the decay particles in the rest system of the parent. The case $\beta^* = 1$ is particularly interesting since it involves massless particles in the final state (photons), for instance from $\pi^0 \to \gamma + \gamma$:

$$\frac{d}{2L} = \frac{\beta\sqrt{1-x^2}}{\gamma(\beta^2 - x^2)}. \tag{3.127}$$

For $|x| \geq \beta$, $d \to \infty$ or is negative (impossible for an inherently positive quantity), which means that one of the photons goes backwards in the laboratory and therefore cannot hit the detector. This is an important reminder that a massless particle cannot be turned around by a Lorentz transformation.

We now consider the distribution in d for two cases: either the total energy of the parent ($E = E_1 + E_2$) is held constant, or the energy of one of the photons (E_1) is held constant. This will be a good example in the use of conditional probability. The energies of the two photons have the same constraint on $x = \cos\theta^*$ as used above, so that the ratio of the energies is easily computed, with the result:

$$r \equiv \frac{E_2}{E_1} = \frac{1-\beta x}{1+\beta x} \quad \text{and} \quad x = \frac{1}{\beta}\frac{1-r}{1+r}. \tag{3.128}$$

For E fixed, we can ignore the case when one of the photons misses the detector, and the relativistic limit ($\beta \to 1$) of Eq. 3.127 may be taken simply:

$$\frac{d}{2L} = \frac{1}{\gamma\sqrt{1-x^2}}. \tag{3.129}$$

For the case E_1 fixed, the variables to use are E_1 and $r = E_2/E_1$. In this case, care must be taken about the divergences, so that terms $\sim 1-\beta^2$ cannot be ignored. We solve Eq. 3.128 for $\sqrt{1-x^2}$ in terms of r, in the limit $\beta \to 1$:

$$\sqrt{1-x^2} = \frac{2\sqrt{r}}{1+r} \tag{3.130}$$

and substitute into Eq. 3.129, using the relation

$$E = E_1 + E_2 = E_1(1+r) \qquad \gamma = \frac{E}{m} = \frac{E_1}{m}(1+r) \tag{3.131}$$

to obtain

$$\frac{d}{2L} = \frac{m}{2E_1\sqrt{r - m^2/2E_1^2}}. \tag{3.132}$$

Note that the subtraction constant in the square root is due to a term $\sim 1-\beta^2$, and corresponds to the value of $r_\infty = m^2/2E_1^2$, for which d diverges when E_1 is held fixed.

To complete the kinematics at fixed E_1, we note that the minimum value of r occurs when $x = 1$ and both decay photons are collinear with the parent: one going forward in the laboratory on the same trajectory as the parent and hitting the detector, while the other photon goes exactly backwards on the same trajectory, heading away from the detector. In this case, from Eq. 3.128,

$$r_{min} = \frac{1-\beta}{1+\beta} = \frac{1-\beta^2}{(1+\beta)^2} \to \frac{1}{4\gamma^2} = \frac{m^2}{4E_1^2}. \tag{3.133}$$

With the kinematics out of the way, we can now concentrate on finding the distribution in d.

The case E fixed is easy since it only depends on the angular distribution of decay particles in the parent rest frame, the distribution in $x = \cos\theta^*$. For fixed E there is a minimum separation of the two photons at the wall, which occurs for the symmetric decay, $x = 0$

$$\frac{d_{min}}{L} = \frac{2m}{E} = \frac{2}{\gamma}. \tag{3.134}$$

Use d_{min} in Eq. 3.129 to eliminate γ and obtain

$$\sqrt{1-x^2} = \frac{d_{min}}{d} \tag{3.135}$$

or

$$x = \sqrt{1 - d_{min}^2/d^2}. \tag{3.136}$$

The integral probability that two photons land on the wall separated by a distance $s \le d$ is just given by the integral probability that x lies between $\pm x(d)$, where $x(d)$ is given by Eq. 3.136. For the decay $\pi^0 \to \gamma + \gamma$, the decay is isotropic,

$$\frac{d\mathcal{P}}{dx} = \frac{1}{2} \tag{3.137}$$

so that

$$\mathcal{P}(s \le d) = \mathcal{P}(-x(d) \le x \le x(d)) = x(d) = \sqrt{1 - d_{min}^2/d^2}. \tag{3.138}$$

For the case in which one of the decay photon energies, E_1, is held constant, the relationship between the spectra of the parent and the decay particles is required. This will be discussed in the following section, after which the distribution in d will be presented.

3.4.7 The spectra of decay particles – the parent–daughter factor

A parent particle of momentum **P**, energy E, rest mass m, decays into two particles, as above. Furthermore, the differential probability to produce a parent particle with momentum P in range dP is given by the function (recall Eq. 2.15):

$$d\mathcal{P}(P) = f(P)\, P\, dP \tag{3.139}$$

where for example we take $f(P)$ to be of the form

$$f(P) = A\, P^{-n} \tag{3.140}$$

so that

$$d\mathcal{P}(P) = A\, P\, P^{-n}\, dP. \tag{3.141}$$

The conditional probability of finding a daughter with energy E_1, given a parent with momentum P, depends on the $x = \cos\theta^*$ distribution of the decay:

$$\left.\frac{\partial\mathcal{P}}{\partial E_1}\right|_P = \left.\frac{\partial\mathcal{P}/\partial x}{\partial E_1/\partial x}\right|_P. \tag{3.142}$$

For equal mass decay particles,

$$E_1 = \frac{E}{2}(1 + \beta\,\beta^*\, x), \tag{3.143}$$

while the other particle in the pair has:

$$E_2 = \frac{E}{2}(1 - \beta\,\beta^*\, x) \tag{3.144}$$

so that

$$\frac{\partial E_1}{\partial x} = \frac{E\beta\beta^*}{2} = \frac{P\beta^*}{2} \tag{3.145}$$

and

$$\left.\frac{\partial \mathcal{P}}{\partial E_1}\right|_P = \frac{2}{P\beta^*}\frac{\partial \mathcal{P}}{\partial x}. \tag{3.146}$$

For a uniform decay distribution,

$$\frac{\partial \mathcal{P}}{\partial x} = \frac{1}{2} \tag{3.147}$$

over the range $-1 \leq x \leq +1$ so we get

$$\left.\frac{\partial \mathcal{P}}{\partial E_1}\right|_P = \frac{1}{P\,\beta^*}. \tag{3.148}$$

The joint probability of finding a parent of momentum P and a daughter of energy E_1 is given by the rules of conditional probability:

$$\partial^2\mathcal{P}(P, E_1) = \partial\mathcal{P}(P)\,\partial\mathcal{P}(E_1)\,\big|_P = \frac{f(P)}{\beta^*} dP\,dE_1. \tag{3.149}$$

The marginal probability distribution for E_1 is found by integrating over all values of P consistent with E_1. We evaluate the case for massless particles in the final state, $\beta^* = 1$, and only consider the relativistic limit, ignoring the difference between P and E. A more exact treatment has been given by Sternheimer [148]:

$$\partial\mathcal{P}(E_1) = \int_{E_{min}=E_1(1+r_{min})}^{\infty} A P^{-n} dP\,dE_1 = (1 + r_{min})^{-n+1}\frac{1}{n-1} A\,E_1\,E_1^{-n}\,dE_1. \tag{3.150}$$

When Eq. 3.150, the daughter spectrum, is compared to Eq. 3.141, the parent spectrum, we see that they are precisely the same form (for a power-law) but the daughter spectrum is suppressed by a factor of $1/(n-1)$, for a spectrum falling with the $(n-1)$ power. This is called the **parent–daughter suppression factor**.

3.4.8 Two photon decay – the π^0

Equation 3.150 refers to the spectrum of one of two massless particles, assumed distinguishable, say an e^+ from an e^+, e^- pair in the relativistic limit. For a $\pi^0 \to \gamma + \gamma$ decay, Eq. 3.150 also applies for the E_2 spectrum of the second photon. Since each decay of the π^0 gives two photons, one with E_1 and one with E_2, the total spectrum of photons with E_γ from a π^0 with momentum spectrum given by Eq. 3.141 is

$$\partial\mathcal{P}(E_\gamma) = \frac{2}{n-1} A\,E_\gamma\,E_\gamma^{-n}\,dE_\gamma. \tag{3.151}$$

Thus the ratio of the spectrum of photons from π^0 decay to the π^0 spectrum of Eq. 3.139 at any given energy is a constant:

$$\frac{\gamma}{\pi^0}\bigg|_{\pi^0} = \frac{2}{n-1}. \tag{3.152}$$

It is also useful to define the energy asymmetry of the two photons from the decay of a π^0 with momentum P

$$\alpha \equiv \left| \frac{E_1 - E_2}{E_1 + E_2} \right| = \beta \cos\theta^* \tag{3.153}$$

which has a uniform distribution over the range $-\beta \leq \alpha \leq \beta$,

$$\frac{\partial \mathcal{P}}{\partial \alpha}\bigg|_P = \frac{1}{2\beta}, \tag{3.154}$$

where $\beta = P/E$ is the velocity (divided by c) of the π^0.

The π^0 are detected by forming the invariant mass of all pairs of identified photons on an event, typically restricting the energy asymmetry $\alpha < 0.8$. This may vary according to the background and the π^0 energy depending on the ability of the detector to resolve two distinct photons separated by a distance d (Eq. 3.134). The measured invariant mass distribution of photon pairs in the π^0 region from the PHENIX experiment is shown in Figure 3.6 for both p–p [149] and Au+Au collisions [150] at $\sqrt{s_{NN}} = 200$ GeV. For p–p, the peak is fit with a Gaussian and the signal is taken from the region within 2σ of the mean, as indicated,

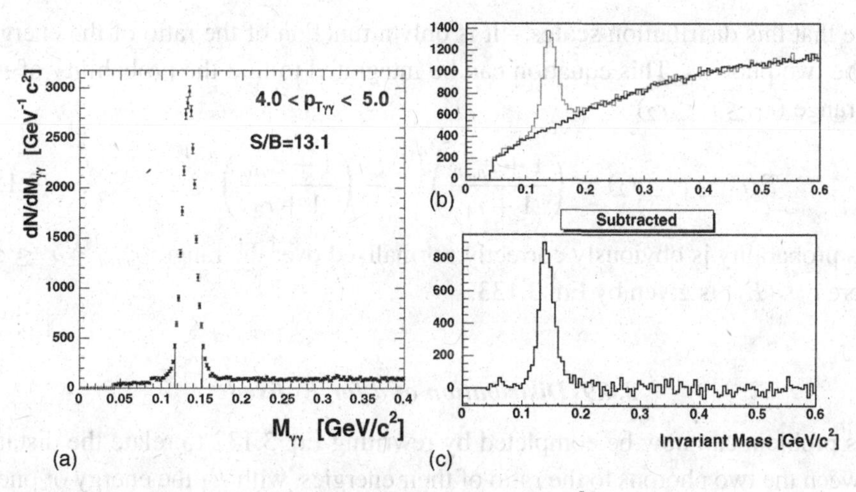

Figure 3.6 The 2γ invariant mass distribution in the π^0 region for $4.0 < p_{T_{\gamma\gamma}} <$ 5.0 GeV/c: (a) in p–p collisions [149]; and in Au+Au collisions [150], (b) observed mass distribution (gray) with mixed event background (black), (c) mass distribution with background subtracted.

with a first or second order polynomial background subtracted, with excellent signal/background ratio. For Au+Au, the peak sits on a large combinatoric background (Figure 3.6b) which is estimated by mixed events (taking the second photon of the pair from different events with the same centrality and vertex position) and subtracted.

To continue the analysis, we now wish to find the spectrum of the second photon, E_2, for the energy of the other photon, E_1, held constant. The joint probability distribution of E_2 and E_1 is trivially related to the joint probability distribution of E and E_1, since

$$E = E_1 + E_2 \qquad \text{so that} \qquad \partial E_2|_{E_1} = \partial E|_{E_1}. \tag{3.155}$$

The joint probability distribution comes directly from Eq. 3.149

$$\partial^2 \mathcal{P}(E_1, E_2) = f(E_1 + E_2)\, dE_1\, dE_2. \tag{3.156}$$

We now use a famous rule of conditional probability to find the distribution of E_2 given E_1

$$\mathcal{P}(A, B) = \mathcal{P}(A) \times \mathcal{P}(B)|_A. \tag{3.157}$$

Thus the conditional probability for E_2, given E_1, is just the joint probability for both E_1 and E_2 divided by the marginal probability for E_1, or Eq. 3.156 divided by Eq. 3.150:

$$\partial \mathcal{P}(E_2)|_{E_1} = \frac{\partial^2 \mathcal{P}(E_1, E_2)}{\partial \mathcal{P}(E_1)} = (1 + r_{min})^{n-1} \frac{n-1}{(1+r)^n} dr. \tag{3.158}$$

Note that this distribution **scales** – it is only a function of the ratio of the energies of the two photons. This equation can be integrated to find the probability of r in the range $(r_1 \leq r \leq r_2)$:

$$\mathcal{P}(r_1 \leq r \leq r_2) = \left(\frac{1 + r_{min}}{1 + r_1}\right)^{n-1} - \left(\frac{1 + r_{min}}{1 + r_2}\right)^{n-1}. \tag{3.159}$$

This probability is obviously correctly normalized over the range $(r_{min} \leq r \leq \infty)$ where $r_{min}(E_1)$ is given by Eq. 3.133.

3.4.9 Distribution of d for E_1 fixed

This problem can now be completed by rewriting Eq. 3.132 to relate the distance between the two photons to the ratio of their energies, with E_1 the energy of one of the photons being fixed:

$$r - \frac{m^2}{2E_1^2} = \frac{m^2/E_1^2}{d^2/L^2}. \tag{3.160}$$

The probability for d in any range is found by using Eq. 3.160 to find the corresponding values of r for the range and then using Eq. 3.159 to find the probability.

3.4.10 *Why spend such effort on minute kinematic details?*

These derivations and problems may seem a bit long-winded; but experimentalists spend most of their time dealing with such details rather than making great discoveries. Of course, if you do not spend enough effort coping with the details, you sometimes find a "discovery" which is nothing but a detail that you did not work out correctly.

3.5 Methods of direct single photon measurements

Here we discuss in detail experimental methods of π^0 and photon measurements including isolation and statistical subtraction methods.

3.5.1 *Exact versus approximate background estimates*

In Section 3.4.8 the ratio of the spectrum of photons from π^0 decay to the π^0 spectrum of Eq. 3.139 at any given energy is a constant (Eq. 3.152) repeated here for convenience:

$$\left.\frac{\gamma}{\pi^0}\right|_{\pi^0} = \frac{2}{n-1}.$$

It is important to be aware that this simple calculation is only valid near mid-rapidity where $E_\pi = p_{T_\pi}$. For the more general case where π^0 are not emitted perpendicular to the beam axis, the simple scaling does not hold. The exact background must be calculated in a Monte Carlo calculation which includes all the holes in the detector as well as the exact kinematics at all angles and not just simple approximations which work at mid-rapidity. Nevertheless, in the case of mid-rapidity collider detectors such as CCOR, PHENIX, etc., with $|y| < 0.8$, Eq. 3.152 serves a good approximation and is very useful as a check on the detailed Monte Carlo calculations.

In order to compute the total γ ray background in an experiment, one must also include the γ rays from other decaying mesons, the most prominent being $\eta \to \gamma + \gamma$ with 39% branching ratio. A Monte Carlo calculation for the PHENIX experiment which includes all relevant decays to γ rays is shown in Figure 3.7a for Au+Au collisions at $\sqrt{s_{NN}} = 200$ GeV [151]. Figure 3.7b shows only the total $R_\gamma = \gamma/\pi^0$ for all decays from the left panel, in comparison to a calculation

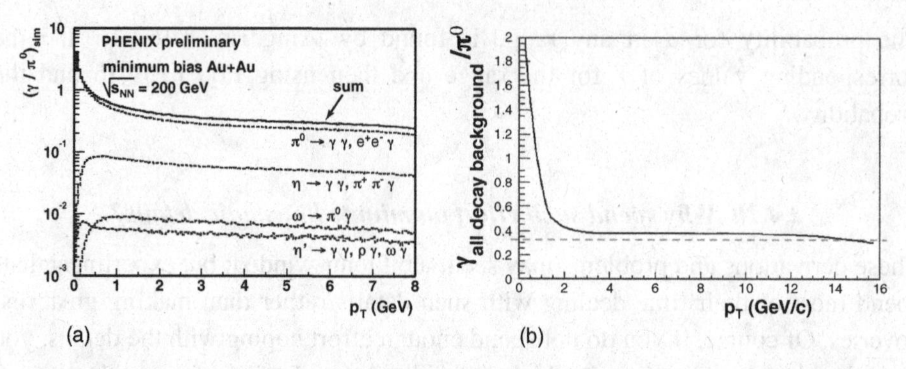

Figure 3.7 (a) Monte Carlo simulation of γ spectrum from known meson decays expressed as the ratio $R_\gamma = \gamma/\pi^0$ [151]; (b) the total background from (a) (solid line) together with calculation from Eq. 3.161 (dashed line).

with Eq. 3.152 for $n = 8.10 \pm 0.05$ which includes only the π^0 and the η where $\eta/\pi^0 = 0.48 \pm 0.03$ is a constant for $p_T \geq 2$ GeV/c [152]:

$$\gamma_{background}/\pi^0 = (1 + 0.48 \times 0.39) \times 2/7.1 = 1.19 \times 2/7.1 = 0.334. \quad (3.161)$$

The agreement is excellent.

This example illustrates the importance of simple estimates in experimental physics: a simple analytical calculation provides a check of an often complicated Monte Carlo program.

3.5.2 Direct photons, tagging cut, isolation cut, fake rate due to π^0 decay

In p–p collisions, if one is searching for direct single photon production, i.e. where the single γ does not come from the decay of another particle, one does not have

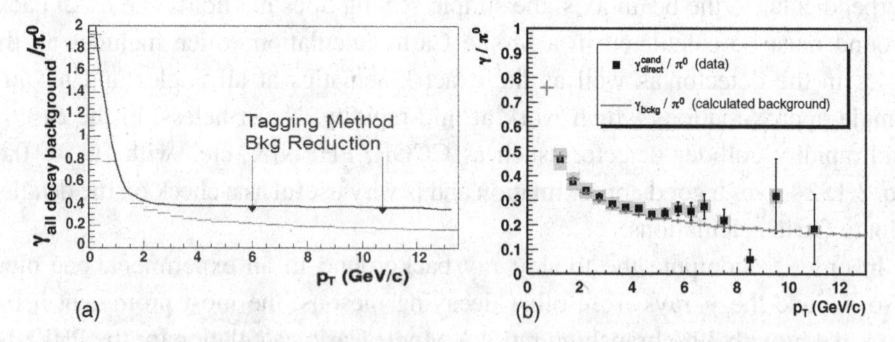

Figure 3.8 (a) Reduction in the total decay γ background from Figure 3.7 with the addition of a tagging cut (gray dashes); (b) the background from (a) (dashes) together with the measured ratio of direct photon candidates to the π^0 spectrum (data points) in p–p collisions at $\sqrt{s} = 200$ GeV [154].

to accept all the background from $\pi^0 \rightarrow \gamma + \gamma$ decays (decay photons). There are two ways to reduce this background

(i) Tagging can be used, i.e. rejecting direct γ candidates which form an invariant mass in the π^0 mass window with another photon in the event;
(ii) an isolation cut can be used, selecting direct γ candidates with for example less than 10% additional energy within a cone of radius $\Delta r = \sqrt{(\Delta\eta)^2 + (\Delta\phi)^2} = 0.5$ around the candidate [153].

The effect of a tagging cut on the Monte Carlo total decay γ background from Figure 3.7 is shown in Figure 3.8a. Figure 3.8b shows the measured ratio of direct γ candidates to the π^0 spectrum from an early PHENIX publication in p–p collisions [154] together with the background from the left panel.

4

The search for structure

4.1 Rutherford scattering

Shortly after the discovery of the radioactivity of uranium by Becquerel in 1896 [155] and its ability to ionize gases, Rutherford [156] began a study of the rate of discharge of a parallel plate capacitor in gas by placing successive layers of thin aluminum foil over the surface of a layer of uranium oxide on one plate. He concluded that "the uranium radiation is complex, and that there are present at least two distinct types of radiation: one that is readily absorbed which will be termed for convenience the α radiation, and the other of a more penetrative character, which will be termed the β radiation." In 1906, Rutherford [157] observed that α particles from the decay of radium scattered, i.e. deviated from their original direction of motion, when passing through a thin sheet of mica, but did not scatter in vacuum. He made this observation by passing α particles through narrow slits and making an image on a photographic plate. In vacuum, the edges of the image were sharp while the image of α particles that passed through the mica was broadened and showed diffuse edges. This observation was controversial because it was not expected that α particles would scatter [158]: "Since the atom is the seat of intense electrical forces, the β particle in passing through matter should be much more easily deflected from its path than the massive α particle."

Rutherford had Geiger [159] follow up on these measurements using scintillations on a phosphorescent screen with convincing results that the deflections are on the average small, on the order of a few degrees, but, as Geiger noted "some of the α particles after passing through the thin leaves – the stopping power of one leaf corresponded to about 1mm. of air – were deflected through quite an appreciable angle." This was then followed up by the much more striking observation by Geiger and Marsden [160] that about 1/8000 α particles incident on a layer of Au 6×10^{-5} cm thick (2 mm air) "can be turned through an angle of 90° and even more." It is reported [161] that Rutherford remarked about this observation,"It was as if one had fired a large naval shell at a piece of tissue paper and it had bounced back."

At this time, the theory of the structure of matter was J. J. Thomson's "plum pudding" model [162], in which [9], "The atom is supposed to consist of a number N of negatively charged corpuscles, accompanied by an equal quantity of positive electricity uniformly distributed throughout a sphere." (Like the "raisins in plum pudding," as told by Maurice Goldhaber [163].) In this case, the scattering of β particles traversing matter would be via a large number of small deflections, as in multiple Coulomb scattering (Eq. 3.6). This was apparently verified by Crowther [164], who measured the scattering of β particles in various substances, by observing β particles scattered through angles $\phi < 18°$ as defined by an aperture stop. The prediction was that the average scattering angle $\langle\theta\rangle$ for a particle passing through a thickness t of material with n atoms per unit volume was equal to $\langle\theta\rangle = \theta_0\sqrt{n\pi b^2 t}$, where b is the radius of the atom and θ_0 is the average deflection due to an encounter with a single atom. The angular distribution of scattering would be:

$$\frac{d\sigma}{d\Omega} = \frac{d\sigma}{2\pi \sin\theta d\theta} \approx \frac{d\sigma}{2\pi\theta d\theta} \propto e^{-\theta^2/\langle\theta\rangle^2}, \tag{4.1}$$

which is Gaussian. Thus, in the small angle approximation of Eq. 4.1, the probability for scattering through an angle less than ϕ is:

$$\frac{I}{I_0} = 1 - e^{-\phi^2/\langle\theta\rangle^2} = 1 - e^{-k/t}, \tag{4.2}$$

where $k = \phi^2/(\theta_0^2 n\pi b^2)$ is a constant for any given value of ϕ. Equation 4.2 was verified by Crowther [164] by scattering through various thicknesses of aluminum for $\phi < 18°$ but he never checked the angular distribution, Eq. 4.1.

Rutherford, in 1911 [9], then compared the multiple scattering (or as he described it compound scattering) for an α or β particle passing through the distributed charge of Thomson's model to a single scattering from the positive charge of an atom all located at a central point surrounded by the negative electrons uniformly distributed throughout a sphere of radius R of the order of the radius of the atom $\sim 10^{-8}$ cm. For close collisions with the central charge $\sim 10^{-11}$ cm, the effect of the electrons would be negligible. Rutherford concluded that single scattering in his model would always dominate multiple/compound scattering and that agreement of his theory and Geiger and Marsden's measurement [160] was "reasonably good" "Considering the difficulty of the experiments."

In a beautiful derivation, Rutherford [9] used the hyperbolic orbits of a central $1/r^2$ repulsive force for which the eccentricity, $\varepsilon = \sec\frac{\pi-\theta}{2}$ (Figure 4.1). He used conservation of energy to relate the initial kinetic energy of the projectile to the kinetic energy at the distance of closest approach, and conservation of angular momentum to relate the impact parameter to the distance of closest approach, leading to the relation between the impact parameter b and the scattering angle θ (Figure 4.1):

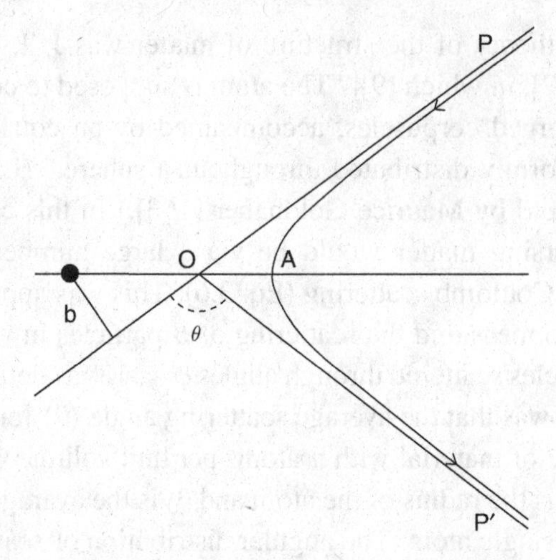

Figure 4.1 Scattering diagram (after Rutherford [165]). The impact parameter b is the perpendicular distance from the nucleus (solid circle) to the initial line of motion (P–O) of the α particle.

$$2\tan\frac{\theta}{2} = \frac{2ZZ'e^2}{pvb}, \tag{4.3}$$

where Ze is the charge of the α particle with initial momentum $p = Mv$, mass M, and initial velocity v, and $Z'e$ is the charge of the nucleus. For a given scattering center, the probability of a collision with impact parameter between b and $b + db$ is $dP = 2\pi b db$, which with a few changes of variables gives the Rutherford scattering cross section per unit solid angle per atom:

$$\frac{d\sigma}{d\Omega} = \frac{K^2}{16}\frac{1}{E^2 \sin^4\theta/2} \tag{4.4}$$

where E is the initial kinetic energy of the α particle and $K = ZZ'e^2$.

The experiment of Geiger and Marsden in 1913 [166] then verified Rutherford scattering (Figure 4.2) which clearly showed a dramatic large angle power-law tail compared to the Thomson compound scattering Gaussian model. In the words of Rutherford and collaborators [165], "These experiments thus afford abundant proof of the law of scattering with angle deduced by Rutherford from the nuclear theory of the atom." This was both the birth of the nucleus and the birth of high p_T physics.

4.1.1 Rutherford scattering and high p_T physics

Rutherford scattering clearly established the fundamental relationship between large transverse momentum scattering and small distances probed (Eq. 4.3). This

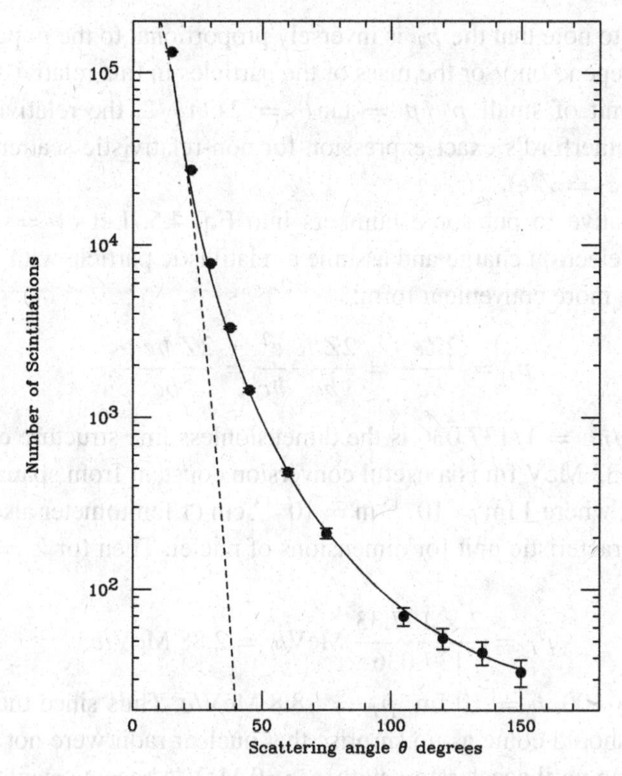

Figure 4.2 Geiger and Marsden measurement of α scattering in Au [166] with Eq. 4.4 (solid curve). Dashes represent Eq. 4.1 with $\langle\theta\rangle = 12.6°$ chosen arbitrarily for purpose of illustration. (After March [167].)

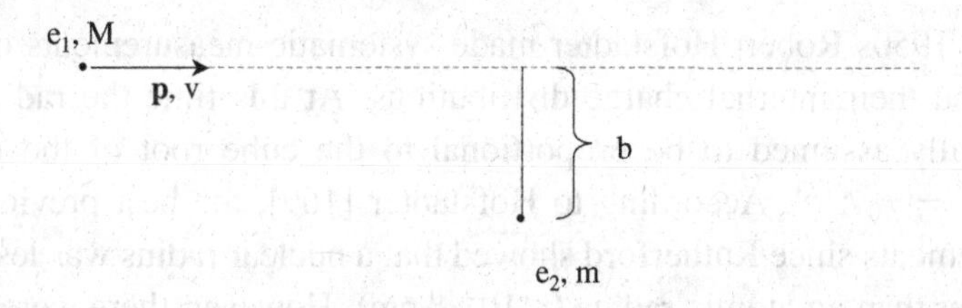

Figure 4.3 After Jackson [138].

is also nicely illustrated in an example from Jackson's textbook [138] (Figure 4.3) which calculates the change in momentum Δp of an incident particle with charge e_1, velocity v and energy $E = \gamma Mc^2$, which passes an electron with charge e_2 and mass $m \ll M$, at rest, with impact parameter b. The change in momentum of the heavy particle is transverse to the motion, $\Delta p = p_T$, and the electron recoils with the opposite p_T which is given by:

$$\Delta p = p_T = \frac{2e_1e_2}{bv}. \tag{4.5}$$

It is important to note that the p_T is inversely proportional to the impact parameter and does not depend on γ or the mass of the particles in this relativistic treatment. Also in the limit of small $p_T/p = \tan\theta = 2\tan\theta/2$, the relativistic result is identical to Rutherford's exact expression for non-relativistic scattering (Eq. 4.3, with $e_1 = Ze$, $e_2 = Z'e$).

It is informative to put some numbers into Eq. 4.5. Let $e_1 = Ze$, $e_1 = e$ where e is the electron charge and assume a relativistic particle with $v \simeq c$. Then, converting to a more convenient form:

$$p_T = \frac{2Ze^2}{bc} = \frac{2Z\hbar c}{bc}\frac{e^2}{\hbar c} = \frac{2Z\,\hbar c\,\alpha}{bc}, \tag{4.6}$$

where $\alpha = e^2/\hbar c = 1/137.036$ is the dimensionless fine structure constant [84], and $\hbar c = 197.33$ MeV fm is a useful conversion constant from spatial to momentum units [84], where 1 fm $= 10^{-15}$ m $= 10^{-13}$ cm (1 femtometer also known as 1 fermi) is a characteristic unit for dimensions of nuclei. Then for $Z = 1$ (a proton) and $b = 1$ fm,

$$p_T = \frac{2 \times 197.33}{137.036}\ \text{MeV}/c = 2.88\ \text{MeV}/c, \tag{4.7}$$

while for $Z = 100$, $b = 10$ fm, $p_T = 28.8$ MeV/c. Thus since the p_T must be less than p it should come as no surprise that nuclear radii were not measured by elastic scattering until accelerators with $p \gtrsim 30$ MeV/c became available [168].

4.2 Hofstadter – measurement of radii of nuclei and the proton

In the early 1950s Robert Hofstadter made systematic measurements of the radii of nuclei and their internal charge distributions. At this time the radii of nuclei were generally assumed to be proportional to the cube root of the number of nucleons, $r = r_0 A^{1/3}$. According to Hofstadter [169], the best previous scattering measurements since Rutherford showed that a nuclear radius was less than 105 times smaller than an atomic radius ($\lesssim 10^{-10}$ cm). However, there were also other methods, as, for instance, discussed by Sherr [170].

Regarding the charge distribution inside a nucleus, it was not known whether the protons were distributed uniformly inside the nucleus or, for example, whether they were all collected at the outer periphery [171]. The sensitivity of Coulomb scattering to the charge distribution is simply related to Gauss' law – the electric field depends only on the charge inside a closed surface, e.g. of radius b, the impact parameter, which can be determined from p_T, or more formally from Q^2, the four-momentum transfer squared. Once $Q > \hbar c/R$, where R is the nuclear radius, the scattering will be reduced due to the reduced effective charge at radii equal to $\hbar c/Q < R$ [172]. Another possibility for a reduction in the scattering could be a breakdown of electrodynamics at small distances. This was a great topic of research in the period 1950–1970 which will not be discussed further here.

4.2.1 Form factors and nuclear structure

Hofstadter [173] and all following electron (and muon) elastic scattering measurements use the Rosenbluth formula [174] which is the relativistic quantum electrodynamic extension of Rutherford scattering for e–p scattering. However, a more general illustration of how to include the effect of a charge distribution in this formula was given using the earlier Mott scattering formula [175, 176] which was the first relativistic treatment of scattering Dirac particles such as electrons from point nuclei. In the relativistic limit for the electron, $\beta \to 1$, the Mott scattering formula for an electron, with spin, scattering with angle θ from a point nucleus of charge Ze is:

$$\left(\frac{d\sigma}{d\Omega}\right)_{Mott} = \frac{Z^2\alpha^2 \cos^2\theta/2}{4E^2 \sin^4\theta/2}. \tag{4.8}$$

For a nucleus with structure represented by a charge density $\rho(\mathbf{r})$, the Z in the Mott formula 4.8 is multiplied by a "form factor" [172, 173]:

$$F(q) = \frac{4\pi}{q}\int_0^\infty \rho(r)\sin(qr)\, r\, dr \tag{4.9}$$

where $\rho(\mathbf{r})$ is normalized to 1, $\int \rho(\mathbf{r})d^3\mathbf{r} = 1$ and q^2 is the invariant four-momentum transfer squared of the scattering. Thus the cross section is multiplied by $F^2(q)$. Hofstadter [173] gives many examples of possible nuclear charge distributions and their corresponding form factors.

4.2.2 Kinematics of e–p elastic scattering

Except for the HERA e–p collider at DESY from 1991–2007, all the lepton–hadron scattering experiments have been performed in the fixed target geometry. The kinematics are crucial for understanding the physics (Figure 4.4). An electron of initial energy E scatters through angle θ from a proton at rest, which recoils with kinetic energy T leaving the scattered electron with energy $E' = E - T$.

To be slightly formal, from Eq. 3.88

$$\begin{aligned} p_1 &= (0, 0, E, iE) & p_2 &= (0, 0, 0, iM) \\ p_3 &= (0, E'\sin\theta, E'\cos\theta, iE') & p_4 &= (\mathbf{P_p}, i\,[T + M]) \end{aligned} \tag{4.10}$$

so that

$$s = -(p_1 + p_2)^2 = (E + M)^2 - E^2 = 2ME + M^2 \tag{4.11}$$

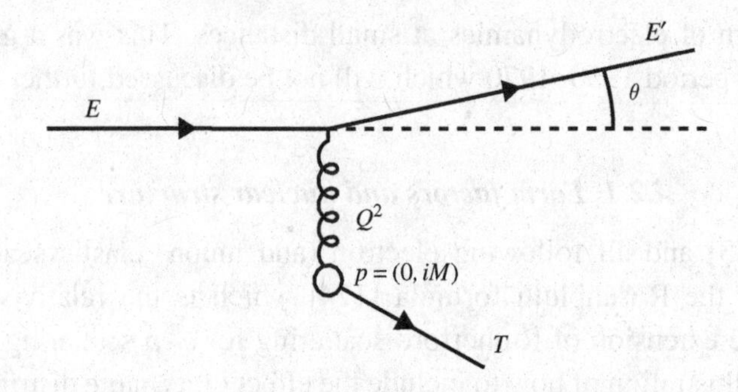

Figure 4.4 Electron–proton elastic scattering.

$$
\begin{aligned}
Q^2 &= (p_1 - p_3)^2 = (0, -E' \sin\theta, E - E' \cos\theta, i\,[E - E'])^2 \\
&= (E' \sin\theta)^2 + E^2 - 2EE' \cos\theta + (E' \cos\theta)^2 - E^2 - E'^2 + 2EE' \\
&= 2EE'(1 - \cos\theta) \\
&= 4EE' \sin^2\theta/2.
\end{aligned} \tag{4.12}
$$

From Eqs. 3.89–3.92, we know that

$$
Q^2 = 2MT
$$

$$
T = \frac{Q^2}{2M} = E - E', \tag{4.13}
$$

so that

$$
\frac{E'}{E} = 1 - \frac{Q^2}{2ME}. \tag{4.14}
$$

Substituting in Eq. 4.12 and solving for Q^2 we obtain the two important relations:

$$
Q^2 = \frac{4E^2 \sin^2\theta/2}{1 + (2E/M)\sin^2\theta/2} \tag{4.15}
$$

$$
\frac{E'}{E} = \frac{1}{1 + (2E/M)\sin^2\theta/2}. \tag{4.16}
$$

We can also solve Eq. 4.15 for $\sin^2\theta/2$ and then find $\cos^2\theta/2 = 1 - \sin^2\theta/2$ with result:

$$
\left(1 - \frac{Q^2}{2ME}\right)\sin^2\theta/2 = \frac{E'}{E}\sin^2\theta/2 = \frac{Q^2}{4E^2} \tag{4.17}
$$

$$
\left(1 - \frac{Q^2}{2ME}\right)\cos^2\theta/2 = \frac{E'}{E}\cos^2\theta/2 = 1 - \frac{Q^2}{2ME} - \frac{Q^2}{4E^2}. \tag{4.18}
$$

4.2.3 Determination of nuclear and proton form factors

With the kinematics out of the way, we can finally write the Rosenbluth formula [174] for e–p scattering with form factors:

$$\left(\frac{d\sigma}{d\Omega}\right) = \left(\frac{d\sigma}{d\Omega}\right)_{Mott} \frac{E'}{E} \left\{ \frac{G_E^2(Q^2) + \tau G_M^2(Q^2)}{1+\tau} + 2\tau G_M^2(Q^2) \tan^2 \theta/2 \right\} \tag{4.19}$$

where $\tau = Q^2/4M^2$. Here, $G_E(Q^2)$ and $G_M(Q^2)$ represent the form factors of the electric charge (normalized to $G_E(0) = 1$) and magnetic moment (normalized to $G_M(0) = \mu$), where $\mu = 2.79$ is the magnetic moment of the proton in nuclear magnetons ($e\hbar/2M$). For small angles and small $Q^2 \ll 4M^2$, electric scattering (G_E) dominates, while for large momentum transfers, $Q^2 \gtrsim 4M^2$ $G_M(Q^2)$ dominates. $G_E(Q^2)$ and $G_M(Q^2)$ can be determined separately by measuring elastic scattering for several different angles keeping Q^2 constant by varying the incident energy E (Eq. 4.17).

Hofstadter and collaborators in 1953–1956 [173] made extensive measurements of elastic electron scattering in many nuclei at the Stanford High Energy Physics Laboratory (HEPL) $E \lesssim 1$ GeV electron linac, in which the angle and energy of the scattered electrons were measured in a high precision moveable single arm spectrometer, with particle identification being provided by Cerenkov counters, and later with added electromagnetic shower counter calorimeters. A typical illustration of the same technique [177] is shown from the 30 GeV Stanford electron Linear Accelerator Laboratory (SLAC) in Figure 4.5.

In addition to the beautiful measurements of the radii and structure functions of nuclei, Hofstadter made the astounding discovery that the proton is not a point-like elementary particle [178] but is an extended object with a radius and a structure (Figure 4.6a). A major problem with rigid body structure is that it is incompatible with relativity – if a rigid body were pushed to move with relativistic velocity the information would have to be transmitted across the body faster than the speed of light. The only way structure can be accommodated in relativity is by the exchange

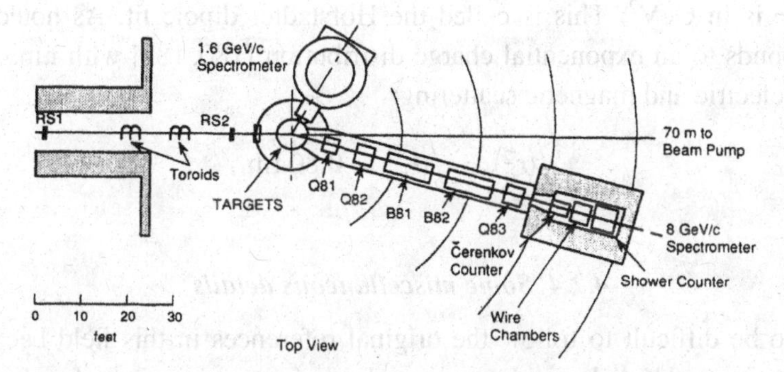

Figure 4.5 Electron spectrometers at SLAC. The 8 GeV spectrometer contains quadrupole (Q) focussing and dipole (B) bending magnets as well as a final detector consisting of a Cerenkov counter, wire chambers and a shower counter [177].

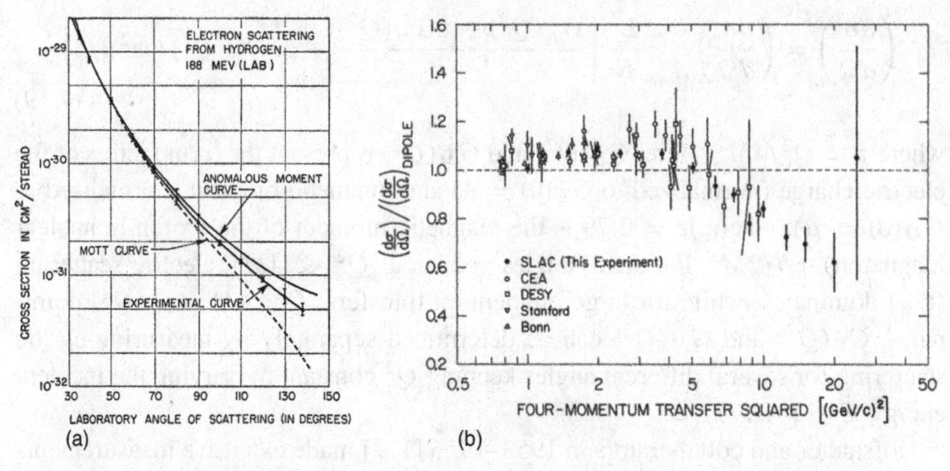

Figure 4.6 (a) Discovery of proton structure [178]. Curves shown are the experimental curve, the Mott curve (Eq. 4.8) and the Rosenbluth curve (Eq. 4.19) with point charge, point magnetic moment. (b) Ratios of measured e–p elastic scattering cross sections to the dipole fit (Eq. 4.20) for $0.8 \leq Q^2 \leq 25$ GeV^2 [179].

of particles, like in a pion cloud around the proton [174]. In purely phenomenological terms, a nice discussion of the different models of the proton charge density consistent with the data is given in reference [173], where the exponential model (IV) [180] will turn out to be preferred by later data [181] including measurements from SLAC at much larger Q^2 [179] (Figure 4.6b).

The result of all these measurements, generally accepted until quite recently (see below), is that

$$G_E(Q^2) = G_M(Q^2)/\mu = \frac{1}{(1+Q^2/0.71)^2}, \tag{4.20}$$

where Q^2 is in GeV^2. This is called the Hofstadter dipole fit. As noted above, it corresponds to an exponential charge distribution [180, 181] with r.m.s. proton radiii for electric and magnetic scattering:

$$\sqrt{\langle r_e^2 \rangle} = \sqrt{\langle r_m^2 \rangle} = 0.80 \text{ fm}. \tag{4.21}$$

4.2.4 Some miscellaneous details

It tends to be difficult to follow the original references in this field because the notation varies and has changed over time. For reference we give a few formulas.

The no-spin, no-structure cross section σ_{NS} is

$$\left(\frac{d\sigma}{d\Omega}\right)_{NS} = \left(\frac{d\sigma}{d\Omega}\right)_{Mott} \frac{E'}{E}. \tag{4.22}$$

The Mott cross section in terms of Q^2 is

$$\left(\frac{d\sigma}{dQ^2}\right)_{Mott} = \frac{4\pi\alpha^2}{Q^4}\cos^2\theta/2. \tag{4.23}$$

The Rosenbluth formula in invariant quantities [182] is

$$\frac{d\sigma}{dQ^2} = \frac{4\pi\alpha^2}{Q^4}\left\{\frac{G_E^2(Q^2)+\tau G_M^2(Q^2)}{1+\tau}\left[1-\frac{Q^2}{2ME}-\frac{Q^2}{4E^2}\right]+\frac{Q^2}{2E^2}\tau G_M^2(Q^2)\right\} \tag{4.24}$$

where $2ME = s - M^2$ (Eq. 4.11).

4.2.5 *A recent controversy*

Elastic e–p scattering still seems to be a lively topic of research at present [183]. Results from the Rosenbluth extraction of $G_E(Q^2)$ [184,185] which have given us the equality $G_E(Q^2) = G_M(Q^2)/\mu$ are at present contradicted by direct measurements of $G_E(Q^2)/G_M(Q^2)$ from the ratio of transverse to longitudinal polarization of recoil protons in the scattering plane [183, 186] from longitudinally polarized electrons scattering on unpolarized protons. Similar issues were discussed many years ago for the scattering of polarized muons [187] but no such measurement was made. A possible explanation is that two photon effects play a larger role in one or both of these measurements than previously thought [188]. This can be determined by experimental measurements of the two photon effect from the difference between e^-–p and e^+–p scattering which have not yet, at this writing, been performed.

4.3 DIS – deeply inelastic electron scattering

Until the advent of the Stanford Two-Mile Accelerator (SLAC) [189] which provided beams of electrons with energies up to 20 GeV in 1966, inelastic electron scattering was concerned mainly with photoproduction using the virtual photons radiated from the electron (Eq. 3.82) whose energy $\nu = E - E'$ could be measured precisely [190–192], or excitation of nucleon resonances (Eq. 3.83). However a sea-change occurred at SLAC where the energy ν and four-momentum transfer squared, Q^2 of the exchanged virtual photon could both be significantly larger than 1 GeV and 1 (GeV/c)^2, respectively – the region of "Deeply Inelastic Scattering" (DIS).

4.3.1 *Kinematics of e–p inelastic scattering*

As usual, it is important to start with the kinematics of a typical experiment from that period in which an electron of four-momentum k, initial energy E, scatters

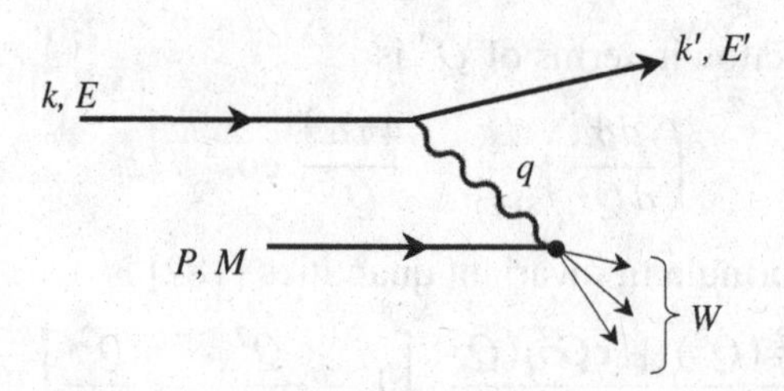

Figure 4.7 Electron–proton inelastic scattering.

through an angle θ from a proton at rest resulting in an outgoing electron with k', E' and an outgoing nucleon system in the final state with invariant mass $M_F \equiv W$, which could be a nucleon resonance as in Eq. 3.83, or a more complicated state (Figure 4.7).

To quote a paper from that period [193], "there are unfortunately as many notations as authors in this field." Thus, we try to present a consistent notation starting with the kinematics, and we must warn the reader that it is somewhat confusing to read the original papers. The initial and final four-vectors are:

$$\begin{aligned} p_1 &\equiv k = (0, 0, E, iE) & p_2 &\equiv P = (0, 0, 0, iM) \\ p_3 &\equiv k' = (0, E' \sin\theta, E' \cos\theta, iE') & p_4 &\equiv w = (\mathbf{P_F}, iE_F). \end{aligned} \tag{4.25}$$

There are several other important quantities.

The momentum transfer four-vector is:

$$q = (k - k') = (0, -E' \sin\theta, E - E' \cos\theta, i\,[E - E']) \tag{4.26}$$

$$q^2 = (k - k')^2 = Q^2 = 4EE' \sin^2 \theta/2. \tag{4.27}$$

There are two things worthy of note in Eq. 4.27: $q^2 = Q^2 = -t$ is space-like, or positive in this metric so there is no need to introduce Q^2 which is defined as positive (as done in the other metric); the formula for Q^2 is the same in Eq. 4.27 as in elastic scattering (Eq. 4.12) when expressed only in terms of the incident and scattered electron variables E, E', θ.

The invariant mass of the final hadron state is W, where:

$$W^2 = -w^2 = -(p_4)^2 = E_F^2 - P_F^2. \tag{4.28}$$

This is also the same as the c.m. energy squared of the virtual photon–proton collision:

$$s_{\gamma p} = -(q + P)^2 = -(p_4)^2 = W^2. \tag{4.29}$$

The c.m. energy of the electron–proton collision is slightly different but as an initial state quantity is obviously identical to that of e–p elastic scattering (Eq. 4.11):

$$s_{ep} = -(k+P)^2 = -k^2 - P^2 - 2k \cdot P = +M^2 + 2ME, \tag{4.30}$$

where we have used the relation

$$k \cdot P = -ME \tag{4.31}$$

which can be read off by inspection from Eqs. 4.25. A related quantity which can be read off by inspection from Eqs. 4.25 and Eq. 4.26 is

$$q \cdot P = -M\nu \tag{4.32}$$

and we can use this to obtain the relation between W^2, ν and Q^2:

$$W^2 = -w^2 = -(q+P)^2 = -q^2 - P^2 - 2q \cdot P = -Q^2 + M^2 + 2M\nu. \tag{4.33}$$

Another related quantity which is sometimes used is K:

$$K = \frac{W^2 - M^2}{2M} = \nu - \frac{Q^2}{2M}, \tag{4.34}$$

which is the energy of a real photon in the laboratory system which produces a final state with invariant mass W when it is absorbed by a proton at rest [191].

It is evident that the nomenclature of the 1960s was heavily influenced by photo- and electro-production of nucleon resonances with well defined invariant mass.

4.3.2 Inelastic scattering cross section

The Drell–Walecka formula [194] (Eq. 4.35) gives the inelastic electron scattering cross section, for the case when only the outgoing electron is detected, in terms of two form factors $W_2(Q^2, \nu)$ and $W_1(Q^2, \nu)$ which are functions of the independent variables Q^2 and ν, since in inelastic scattering, the constraint $Q^2 = 2M\nu$ (Eq. 4.13) no longer applies:

$$\frac{d^2\sigma}{dQ^2 d\nu} = \frac{4\pi\alpha^2}{Q^4}\left\{W_2(Q^2,\nu)\left(1 - \frac{\nu}{E} - \frac{Q^2}{4E^2}\right) + 2W_1(Q^2,\nu)\frac{Q^2}{4E^2}\right\}. \tag{4.35}$$

The similarity to the Rosenbluth formula for elastic scattering (Eq. 4.24) is striking. This becomes even more apparent by using the formula for Q^2 which holds for both elastic and inelastic scattering (Eq. 4.27), so that $Q^2/(4E^2) = (E'/E)\sin^2\theta/2$, as well as recognizing that the ν/E term in Eq. 4.35 is the same as the $Q^2/2ME$ term in Eq. 4.24 where the elastic constraint $\nu = Q^2/2M$ has been used.

We also recognize that by using Eqs. 4.17, 4.18, the Drell–Walecka formula [194] (Eq. 4.35) can be written in the typical form used in the SLAC papers of the late 1960s and early 1970s [195, 196]:

$$\frac{\pi}{EE'}\frac{d^2\sigma}{d\Omega dE'} = \frac{d^2\sigma}{dQ^2 d\nu} = \frac{4\pi\alpha^2}{Q^4}\frac{E'}{E}\left[W_2\cos^2(\theta/2) + 2W_1\sin^2(\theta/2)\right], \quad (4.36)$$

while the Rosenbluth formula (Eq. 4.24) can be written as

$$\frac{d\sigma}{dQ^2} = \frac{4\pi\alpha^2}{Q^4}\frac{E'}{E}\left\{\frac{G_E^2(Q^2) + \tau G_M^2(Q^2)}{1+\tau}\cos^2(\theta/2) + 2\tau G_M^2(Q^2)\sin^2(\theta/2)\right\}. \quad (4.37)$$

This shows the close connection between the magnetic form factor $G_M(Q^2)$ and the structure function $W_1(Q^2,\nu)$ which dominate the scattering at large angles [197].

The relationship of the structure functions $W_2(Q^2,\nu)$ and $W_1(Q^2,\nu)$ to the cross sections for virtual photon interactions with the target nucleon or nucleus [191,198] is also important:

$$W_2 = \frac{1}{4\pi^2\alpha}\frac{Q^2}{\sqrt{Q^2+\nu^2}}\left[\sigma_T(Q^2,\nu) + \sigma_L(Q^2,\nu)\right] \quad (4.38)$$

$$W_1 = \frac{1}{4\pi^2\alpha}\frac{Q^2}{\sqrt{Q^2+\nu^2}}\frac{Q^2+\nu^2}{Q^2}\sigma_T(Q^2,\nu) \quad (4.39)$$

so that

$$\frac{W_1}{W_2} = \left(1 + \frac{\nu^2}{Q^2}\right)\frac{\sigma_T(Q^2,\nu)}{\sigma_T(Q^2,\nu) + \sigma_L(Q^2,\nu)}, \quad (4.40)$$

where $\sigma_T(Q^2,\nu)$ and $\sigma_L(Q^2,\nu)$ are the cross sections for transversely and longitudinally polarized virtual photons, respectively. For real photons, $\sigma_L(0,\nu) = 0$, since massless photons can only have transverse polarization as defined by the direction of the electric field.

4.3.3 The first DIS measurements from SLAC

In 1968, the first results from the new SLAC accelerator by the SLAC-MIT group were shown by Panofsky [53] at the ICHEP in Vienna with publication taking place nearly a year later [2,51] and final results published in 1972 [199]. The results fell into two categories: typical [51] (Figure 4.8a,b,c) and spectacular [2] (Figure 4.8d). The typical results (Figure 4.8a,b,c) showed a spectrum of nucleon resonances on a continuum background. With increasing range of q^2 the resonances become suppressed relative to the continuum which becomes the dominant effect at large q^2. This is shown in a more spectacular manner in Figure 4.8d. Here the inelastic cross section divided by the Mott cross section, i.e. Eq. 4.36/Eq. 4.23, is equal to the elastic cross section at $q^2 = 1$ (GeV/c)^2 but remains nearly constant as a function of q^2 up to 6 (GeV/c)^2 while the elastic cross section drops by a factor of nearly 1000.

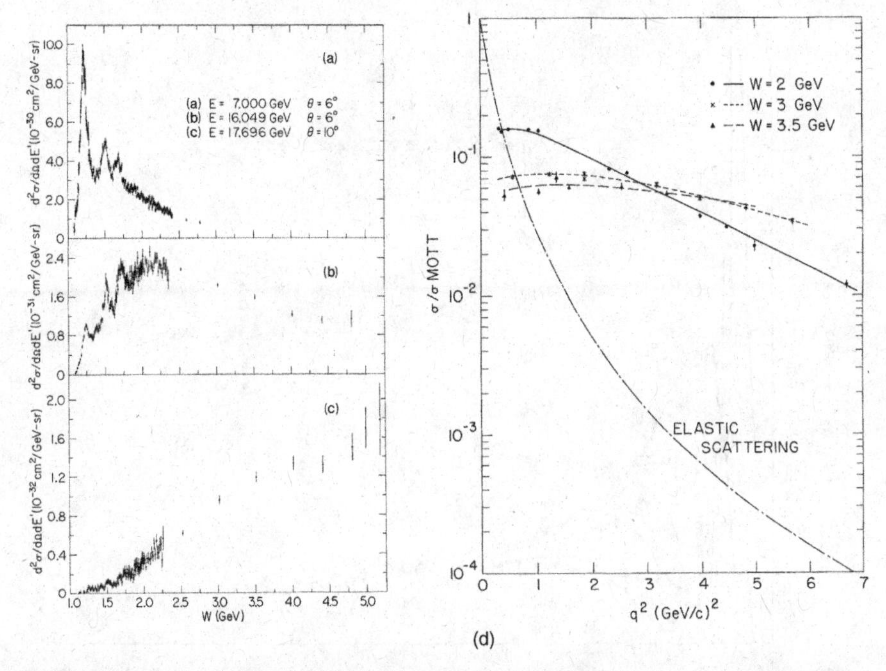

Figure 4.8 (a, b, c) Three radiatively corrected scattered electron spectra for incident electrons with E, θ as indicated. The ranges of q^2 covered are (a) $0.2 \leq q^2 \leq 0.5$ (GeV/c)^2, (b) $0.7 \leq q^2 \leq 2.6$ (GeV/c)^2, (c) $1.6 \leq q^2 \leq 7.3$ (GeV/c)^2. The elastic peaks are not shown [51]. (d) $(d^2\sigma/d\Omega dE')/\sigma_{Mott})$ in GeV^{-1} versus q^2 for three values of W indicated, compared to $(d\sigma/d\Omega)_{elastic}/\sigma_{Mott})$ calculated at 10° [2].

Since there are two structure functions, W_2 and W_1, which each depend on two variables Q^2 and ν, one could imagine pages of data plots of the form of Figure 4.9a, as shown by Panofsky in 1968 [53], in which the function $W_2(Q^2, \nu)$ is plotted as a function of ν for $\theta = 6°$ for several values of Q^2 indicated with the assumption that $2W_1 \sin^2\theta/2 \ll W_2 \cos^2\theta/2$. Admittedly there does seem to be a convergence of the data for $\nu > 5$ GeV.

However, the spectacular became revolutionary with a treatment of the data suggested by Bjorken [72, 200] (Figure 4.9b). When the data are plotted in the scaled variable[1] $\omega = \nu/Q^2$, the measurements of $F(\omega) = \nu W_2(Q^2, \nu)$ seem to collapse onto a universal curve for all the different values of Q^2 indicated (Figure 4.9b). There is only a small dependence of the universal curve on the assumed value of $R = \sigma_L/\sigma_T$ from 0 to ∞. The final results of this first experiment [199] showed impressive scaling of both MW_1 and νW_2 using the measured average value of $R = 0.18 \pm 0.10$ (Figure 4.10).

[1] In later notation, used at present, $\omega \equiv 2M\nu/Q^2$.

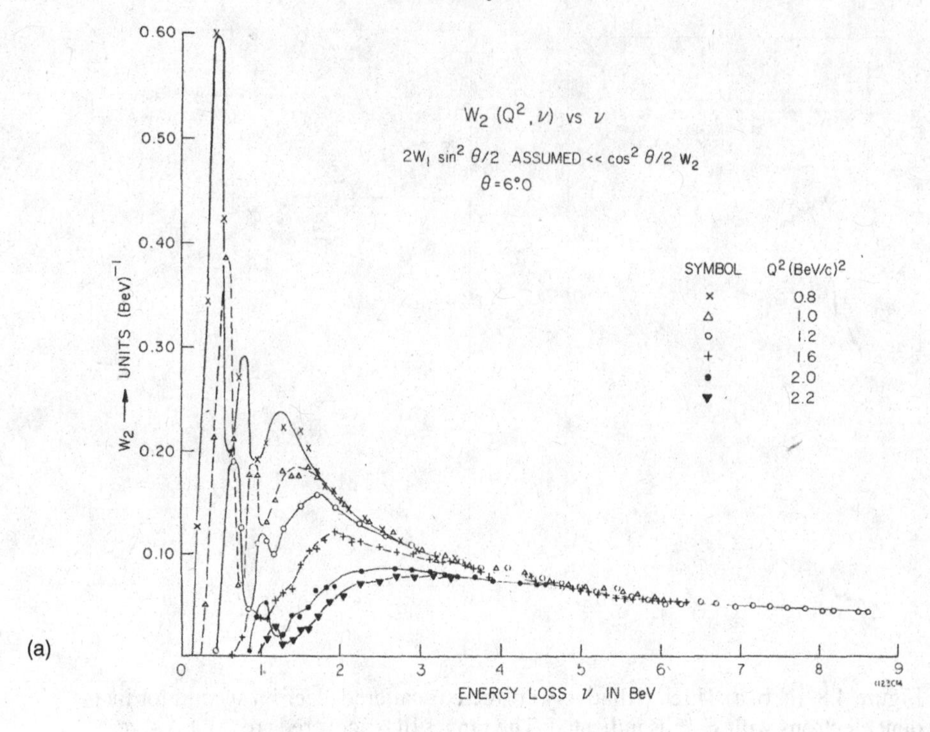

(a)

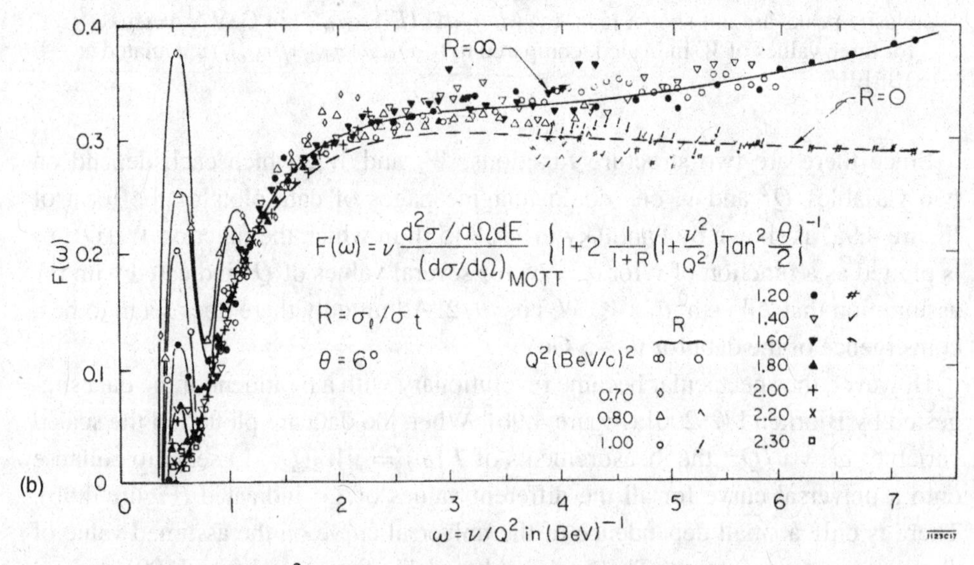

(b)

Figure 4.9 (a) $W_2(Q^2, \nu)$ as a function of ν at a fixed angle, $\theta = 6°$ for several values of Q^2 indicated. (b) $F(\nu/Q^2) = \nu W_2$ as a function of $\omega = \nu/Q^2$ for $\theta = 6°$, with the Q^2 values indicated, for two assumed values of $R = \sigma_L/\sigma_T$ [53].

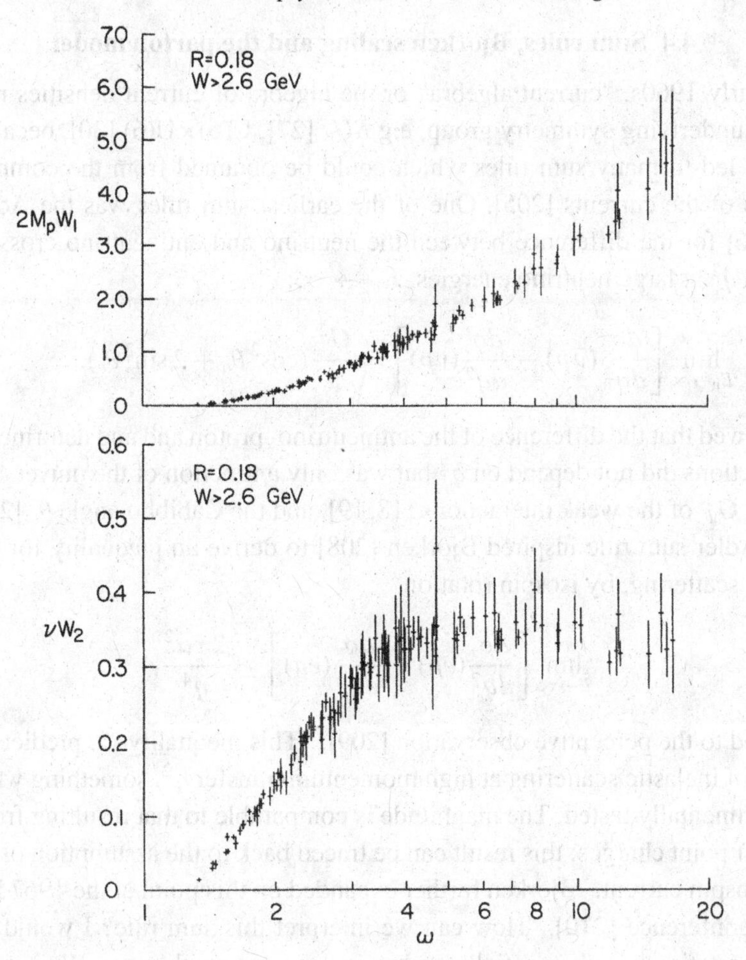

Figure 4.10 Dimensionless structure functions $2MW_1$ and νW_2 as a function of the dimensionless variable $\omega = 2M\nu/q^2$ for $R = 0.18$, $W > 2.6$ GeV, and $q^2 > 1$ (GeV/c)^2 [199].

The scaling of the structure functions had been predicted by Bjorken [72, 200], which led to a fundamental change in the concept of the structure of the proton. This was hinted at by Panofsky in 1968 when he noted that "theoretical speculations are focused on the possibility that these data might give evidence on the behaviour of point-like charged structures within the nucleon." He mentioned a sum rule by Gottfried [201] (who he called Godfrey) to support this view, but added a note of skepticism that "There is no visible quasi-elastic peak at a defined inelasticity $\nu = Q^2/2m$, where m is some characteristic mass" (recall Eq. 3.90) [202, 203]. The skepticism for point-like structures inside the proton was shared by most high energy physicists at that time [200, 204], with one notable exception being Bjorken.

4.4 Sum rules, Bjorken scaling and the parton model

In the early 1960s, "current algebra" or the algebra of current densities resulting from an underlying symmetry group, e.g SU_3 [27], U(6)×U(6) [30], became popular and led to many sum rules which could be obtained from the commutation relations of the currents [205]. One of the earliest sum rules was the Adler sum rule [206] for the difference between the neutrino and antineutrino cross section for fixed q^2 at large neutrino energies, $E_\nu \to \infty$:

$$\lim_{E_\nu \to \infty} \left[\frac{d\sigma}{dq^2}(\bar{\nu} p) - \frac{d\sigma}{dq^2}(\nu p) \right] = \frac{G_F^2}{\pi}(\cos^2\theta_c + 2\sin^2\theta_c). \tag{4.41}$$

This showed that the difference of the antineutrino–proton and and neutrino–proton cross sections did not depend on q^2 but was only a function of the universal Fermi constant G_F of the weak interactions [18, 19], and the Cabibbo angle θ_c [207].

The Adler sum rule inspired Bjorken [208] to derive an inequality for electron inelastic scattering, by isospin rotation,

$$\lim_{E \to \infty} \left[\frac{d\sigma}{dq^2}(ep) + \frac{d\sigma}{dq^2}(en) \right] \geq \frac{2\pi\alpha^2}{q^4}, \tag{4.42}$$

which led to the perceptive observation [209]: "This inequality ... predicts a large amount of inelastic scattering at high momentum transfer q^2, something which can be experimentally tested. The magnitude is comparable to that resulting from scattering off point charges; this result can be traced back to the assumption of locality of the isospin current." Bjorken further expanded on this point at the 1967 Lepton–Photon conference [210]. "How can we interpret this sum rule? I would suggest an interpretation based essentially on history and not much more. We assume that the nucleon is built out of some kind of point-like constituents which could be seen if you could really look at it instantaneously in time, rather than in processes where there is a time averaging and in which the charge distribution or matter distribution of these constituents is smeared out." Then, a bit later, "To make a guess we turn to history and suppose that the scattering from these constituents (if they exist) is quasi-free in the limit of large q^2 so that we can use free particle kinematics. That does not mean necessarily that the constituents have to escape the nucleon.... Using quasi-free kinematics in the relativistic case, one finds that the energy loss, just like the elastic scattering, goes like q^2 divided by twice the mass of the constituent:

$$\Delta E = \nu \sim q^2/2\overline{m}$$

In order for the picture to make any sense at all, this means that the mass of the constituent had better be less than the mass of the target..."

The next step was "Bjorken scaling" [72], a fundamental change in the concept of the structure of the proton and its application to both DIS and proton–proton scattering. Bjorken [72] related the structure functions $W_1(Q^2, \nu)$ and $W_2(Q^2, \nu)$ to "matrix elements of commutators of currents at almost equal times at infinite momentum" and found that in the limit $Q^2 \rightarrow \infty$, $\nu \rightarrow \infty$, with the ratio Q^2/ν held constant, the structure functions $W_1(Q^2, \nu)$ and $W_2(Q^2, \nu)$, if they remained finite, *scaled*, i.e. were only functions of the ratio Q^2/ν. Bjorken had suggested this to the SLAC-MIT group in 1968 [200]; and the scaling and the finiteness of the structure functions were both observed in the first presentation of the data [53], as noted above.

The fact that the structure functions *scaled* was of enormous importance and utility because it allowed cross sections measured at one c.m. energy to be related to those at another c.m. energy by scaling, without detailed knowledge of the underlying physics. Additionally, the observation of scaling and the definition of the new scaling structure functions [72]

$$\nu W_2(Q^2, \nu) = F_2(x) \tag{4.43}$$

$$M\, W_1 = F_1(x) \tag{4.44}$$

in the new variable, "Bjorken" x

$$x \equiv \frac{Q^2}{2M\nu} \tag{4.45}$$

strongly supported the idea that the inelastic scattering from the proton was simply due to quasi-elastic scattering of point-like constituents of the proton with effective mass Mx, i.e.

$$\nu = \frac{Q^2}{2Mx} \tag{4.46}$$

even without the observation of a quasi-elastic peak at some effective mass $\overline{m} = Mx$. This was shortly put on a firm basis in the "parton model" by Bjorken, strongly motivated by Feynman [196,211].

In the parton model, the proton is composed of point-like electrically charged constituents or parts (partons) which carry a certain fraction x of the proton's longitudinal momentum in the infinite-momentum frame. The infinite-momentum frame is used so that any transverse motion of the partons in the protons or any interactions among them can be neglected and the incident electron scatters instantaneously, incoherently and elastically from the individual partons, "the impulse approximation" [54] (Figure 4.11).

The individual partons are assumed to be massless in the infinite-momentum frame and neither the charge nor the spin of the partons is specified, although these could be determined experimentally. One interesting observation in this regard was

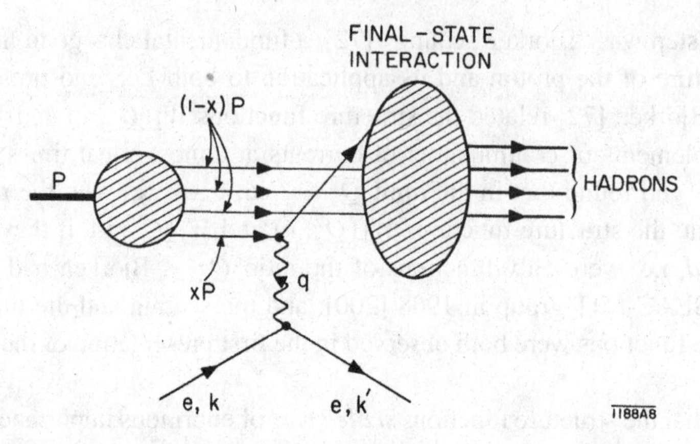

Figure 4.11 Kinematics of e–p scattering in the parton model [196].

by Callan and Gross [212], who showed that for spin 1/2 partons in the Bjorken limit q^2, $\nu \to \infty$, x finite,

$$F_2(x) = 2x F_1(x). \tag{4.47}$$

This can also be seen by setting $\sigma_L = 0$ in Eq. 4.40 using the definitions Eqs. 4.43, 4.44.

The parton model provided a clean and elegant explanation of scaling and the DIS structure functions $F_1(x)$ and $F_2(x)$, but it left many unresolved issues. Were there also neutral partons in the proton that did not scatter from electrons since they had no electric charge? Kuti and Weisskopf [67] made a convincing case that the partons were Gell-Mann and Zwieg's fractionally charged quarks [29, 30] and also included quark–antiquark pairs and neutral gluons among the constituents of the nucleon. However, no free quarks had yet been found in the seven years since they had been proposed. Also, experiments searched for quarks coming out along the q vector in DIS but only found soft particles [213, 214]. If the partons were really point-like, then a nucleus with A nucleons would just be a collection of A times as many point-like partons which would not shadow each other so that the DIS cross section should be A times that of a nucleon. In fact $A^{1.00}$ was observed [215–218] (Figure 4.12).

Another major issue which was not clearly addressed in the parton model was how or whether the partons interact with each other. Are the interactions between partons hard, falling off quickly with distance as in Coulomb scattering [219], or soft, varying more slowly with distance? There was no consensus on this issue [75, 76], although it was clear that Bjorken was on the side of hard scattering. For instance, the discussion cited in reference [76] was due to Bjorken having shown this prescient prediction (Figure 4.13a) for the transverse momentum distribution to

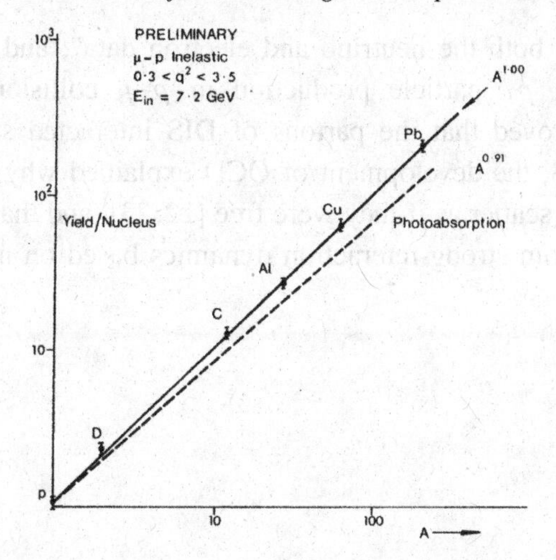

Figure 4.12 The A dependence of the inelastic muon cross section [217, 218].

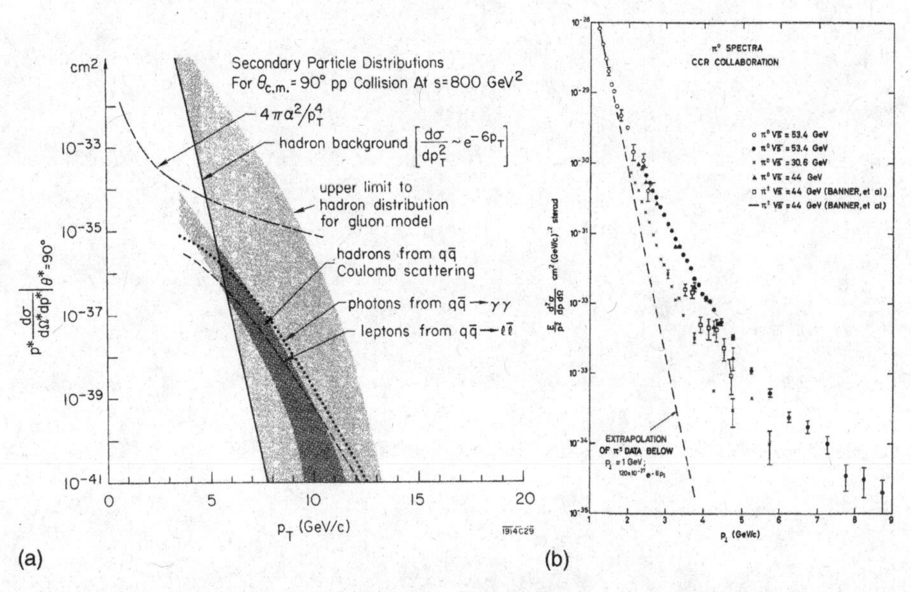

Figure 4.13 (a) Predicted typical transverse momentum distribution for pions in the reaction $pp \to \pi$ + anything from Bjorken [220]; the function $4\pi\alpha^2/p_T^4$ is also shown. (b) CCR p_T spectrum shown at ICHEP 1972 [55].

be expected from parton–parton scattering in proton–proton collisions [220] which was controversial.

These issues began to be resolved at the International Conference on High Energy Physics (ICHEP) in 1972: (i) with measurements of DIS in neutrino scattering presented by Perkins [221] who proclaimed that "In terms of constituent models, the fractionally charged (Gell-Mann/Zweig quark model) is the *only* one which fits both the neutrino and electron data"; and (ii) the discovery of very large high p_T particle production in p–p collisions at the CERN-ISR [55], which proved that the partons of DIS interacted strongly with each other. Then in 1973, the development of QCD explained why quarks were confined yet seemed to scatter as if they were free [22, 23], and that "Bjorken scaling may be obtained from strong-interaction dynamics based on non-Abelian gauge symmetry" [23].

5

Origins of high p_T physics – the search for the W boson

5.1 Why were some people studying "high p_T" physics in the 1960s?

The quick answer is that they were looking for a "left handed" intermediate boson $W^{\pm}$, the proposed carrier of the weak interaction [35]. Although the possibility of Fermi's point-like weak interaction of β decay [18, 19] being transmitted by a boson field was originally discussed by Yukawa [222] and other authors, the modern concept of the parity-violating intermediate vector bosons $W^{\pm}$ as the quanta that transmit the weak interaction was introduced by Lee and Yang in the year 1960 [35, 36] to avoid a breakdown of unitarity in neutrino scattering at high energies if the weak interaction remained point-like [35]; and experiments were proposed to detect the intermediate bosons with high energy neutrinos [37, 38]. Neutrino beams at the new BNL-AGS and CERN-PS accelerators provided the first opportunity to study weak interactions at high energy, whereas previously weak interactions had only been studied via radioactive decay.

The first high energy neutrino experiment at the BNL-AGS [39] set a limit on the mass of the intermediate boson of roughly less than the mass of the proton $M_W \lesssim M_p$. However, much more importantly, this experiment discovered that the neutrinos from charged pion decay

$$\pi^{\pm} \rightarrow \mu^{\pm} + (\nu/\bar{\nu})$$

produced only muons. This discovery of a second neutrino that coupled only to muons, in addition to the original (now-electron) neutrino from β decay which coupled to electrons was dramatic. It led to the concept of families of leptons with conserved lepton number (and later by analogy to generations or "flavors" of quarks).

The plan view of this pioneering experiment is shown in Figure 5.1. The detector was a spark chamber made of 2.54 cm thick aluminum plates, equal to $0.29X_o$, so they could detect electrons by their shower development and muons by passage all the way through the detector without interacting or showering (see Figure 5.2).

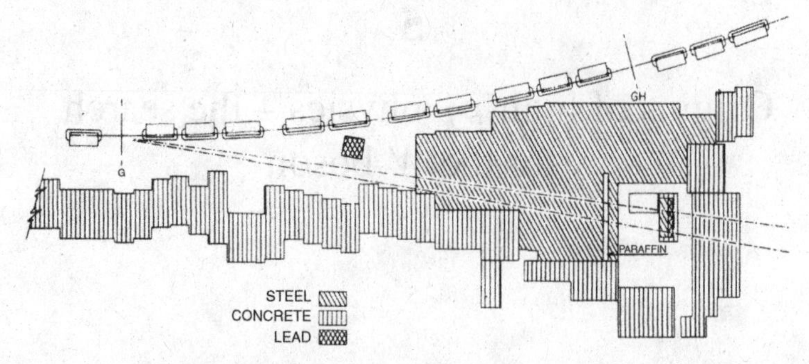

Figure 5.1 Plan view of the two neutrino experiment at the AGS [39]. $\pi^{\pm}$ produced by 15 GeV protons striking the target at G decay in flight before they hit a 13.5 m thick iron shield wall 21 m from the target. The 15 GeV energy is chosen to keep the rate of decay $\mu^{\pm}$ penetrating the shield tolerable.

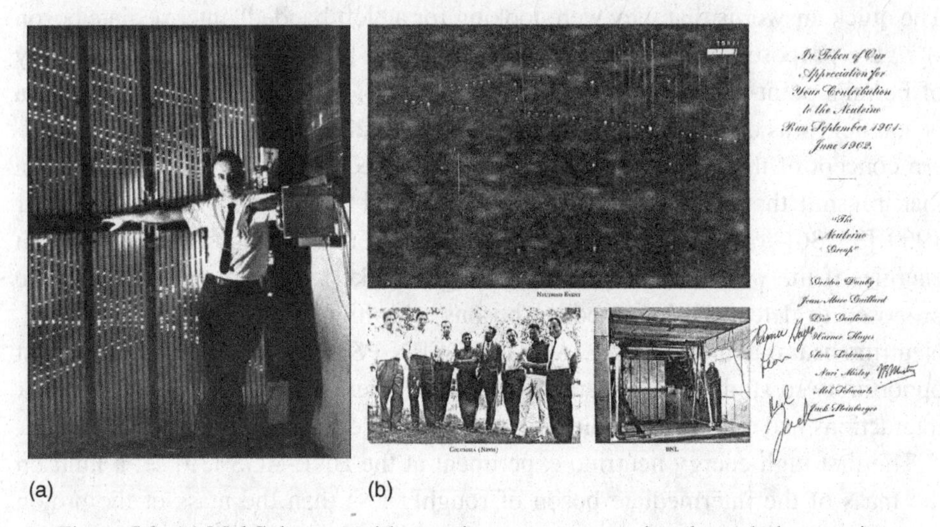

Figure 5.2 (a) Mel Schwartz with cosmic ray muons passing through the neutrino detector. (b) Actual μ track from ν_μ with an autographed photo of the entire group of experimenters and the detector. (BNL photos used with permission.)

The $W^{\pm}$ was not observed in the first few neutrino experiments at BNL [223] or CERN [224, 225] and only modest limits on the mass of the W (>2 GeV/c^2) were obtained due to the relatively low energies of the neutrino beams. However, it was soon realized that the intermediate bosons that mediate the weak interaction might be produced more favorably in nucleon–nucleon collisions [40–42] than in neutrino interactions. The signature of the heavy $W^{\pm}$ would be given by the two body semi-leptonic decay:

$$W^+ \rightarrow e^+ + \nu_e \qquad \text{or} \qquad W^+ \rightarrow \mu^+ + \nu_\mu. \tag{5.1}$$

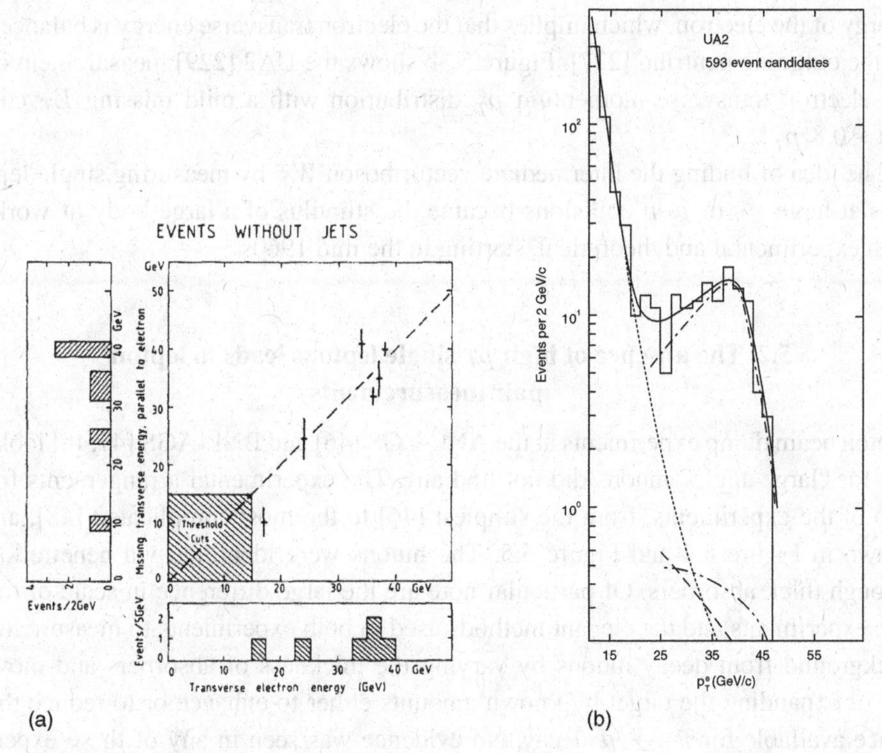

(a) (b)

Figure 5.3 (a) Missing energy opposite to the electron in the transverse plane versus the observed transverse energy of the electron from UA1 [227]; (b) p_T^e of $W \to e + \nu$ from UA2 [229].

The canonical method for discovering the W boson in p–p collisions was described by Nino Zichichi in a comment at the 1964 ICHEP [226], which deserves a verbatim quote because it was essentially how the W was discovered at CERN 19 years later (Figure 5.3) [227, 228]: "We would observe the μ's from W-decays. By measuring the angular and momentum distribution at large angles of K and π's, we can predict the corresponding μ-spectrum. We then see if the μ's found at large angles agree with or exceed the expected numbers." The $W^{\pm}$ would be visible above the background as a peak at lepton transverse momentum

$$p_T^e = \frac{1}{2} M_W \tag{5.2}$$

for the assumed isotropic decay, where M_W is the mass of the intermediate boson. The only addition to this method for the actual W discovery was a requirement of unbalanced or missing transverse energy p_T^{ν} from the undetected ν in a "4π 'hermetic' calorimeter" [230]. Figure 5.3a shows that the missing energy parallel to the outgoing electron in the transverse plane balances the observed transverse energy of the electron, which implies that the electron transverse energy is balanced by the outgoing neutrino [227]. Figure 5.3b shows the UA2 [229] measurement of the electron transverse momentum p_T^e distribution with a mild missing E_T cut, $p_T^\nu > 0.8|p_T^e|$.

The idea of finding the intermediate vector boson $W^\pm$ by measuring single leptons at large p_T in p–p collisions became the stimulus of a large body of work, both experimental and theoretical, starting in the mid 1960s.

5.2 The absence of high p_T single leptons leads to lepton pair measurements

Proton beam dump experiments at the ANL-ZGS [46] and BNL-AGS [47,48] looking for "large angle" muons did not find any. The experimental arrangements for two of the experiments, from the simplest [46] to the most complicated [48], are shown in Figure 5.4 and Figure 5.5. The muons were identified via penetration through thick absorbers. Of particular note are the large difference in scale of the two experiments and the elegant methods used in both experiments to measure the background from decay muons by varying the thickness of absorbers and moving or expanding the target by known amounts either to enhance or to reduce the space available for $\pi \to \mu$ decay. No evidence was seen in any of these experiments for prompt muon production from the $W \to \mu + \nu$ reaction. This was expressed quantitatively by reference [48] as $B_{\mu\nu}\sigma_W \lesssim 6 \times 10^{-36}$ cm^2 for an incident proton beam of $E_p = 28.5$ GeV, where σ_W is the cross section and $B_{\mu\nu}$ the branching ratio; with an interesting limit of $\mu/\pi < 10^{-6}$ for the ratio of prompt muons to the pion flux at the same energy and angle, corresponding to $0.7 < p_T < 1.5$ GeV/c.

After these null results, the big question became, "How do you know how many Ws should have been produced?" It was emphasized by Chilton, Saperstein and Shrauner [45] that the time-like form factor of the proton must be known in order

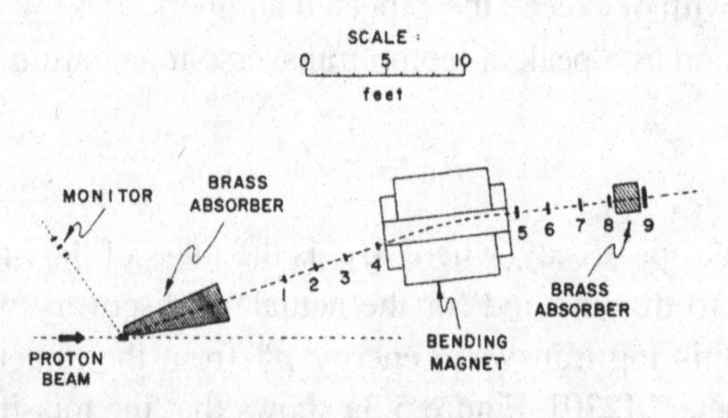

Figure 5.4 Layout of experiment [46]. Numbers 1–9 indicate plastic scintillation counters.

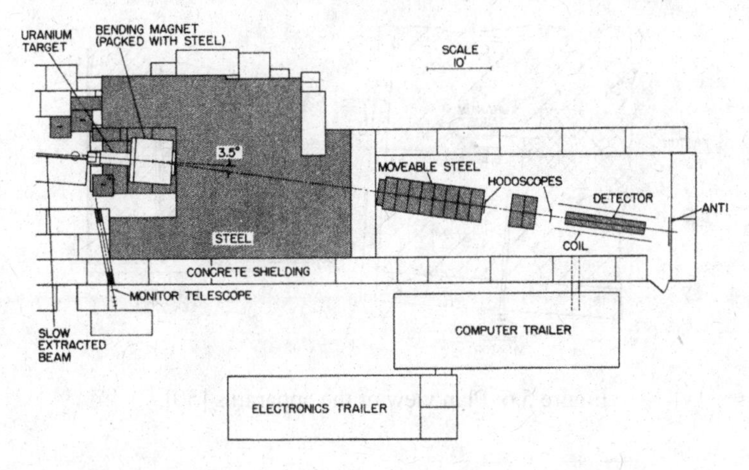

Figure 5.5 Sketch of the experimental apparatus [48].

to calculate the W production rate, and that this had been neglected in previous calculations, notably for the beautiful reaction $p+p \to d+W^+$ [231–233], which resulted in a factor of $\sim$1000 overestimate of the rate.[1] Then, Yamaguchi [49] proposed that the time-like form factor could be found by measuring the number of lepton pairs (e^+e^- or $\mu^+\mu^-$), "massive virtual photons," of the same invariant mass as the W via the reaction

$$p+p \to \gamma^* + \text{anything} \tag{5.3}$$

(but also noted that the individual leptons from these electromagnetically produced pairs might mask the leptons from the W). This set off a spate of single and di-lepton experiments,[2] notably the discovery by Lederman *et al.* of "Drell–Yan" pair production at the BNL-AGS [50, 54] followed by proposals for experiments E70 at the new National Accelerator Laboratory (now Fermilab) and CCR at the CERN-ISR.

The discovery of "Drell–Yan" pairs at the AGS proved to be seminal in future relativistic heavy ion physics as well as providing an interesting lesson. The experiment measured the reaction $p + \text{U} \to \mu^+\mu^- + X$, with the full proton beam striking a uranium target or "beam dump" (Figure 5.6). The large beam intensity resulted in a large flux of muons from pion and kaon decay penetrating the thick shielding wall. These were suppressed by the tapered iron absorber which required

[1] Such an experiment had been proposed [234] and approved at BNL and a special magnet built before the error in the rate calculation became known.

[2] Note these experiments were quite different in intent than a previous di-lepton measurement in the elastic annihilation channel, $\bar{p}+p \to e^+ + e^-$, which placed upper limits on the elastic time-like form factor of the proton [235, 236].

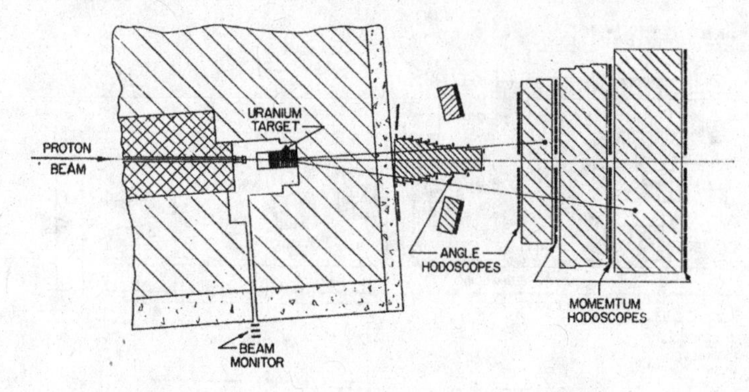

Figure 5.6 Plan view of the apparatus [50].

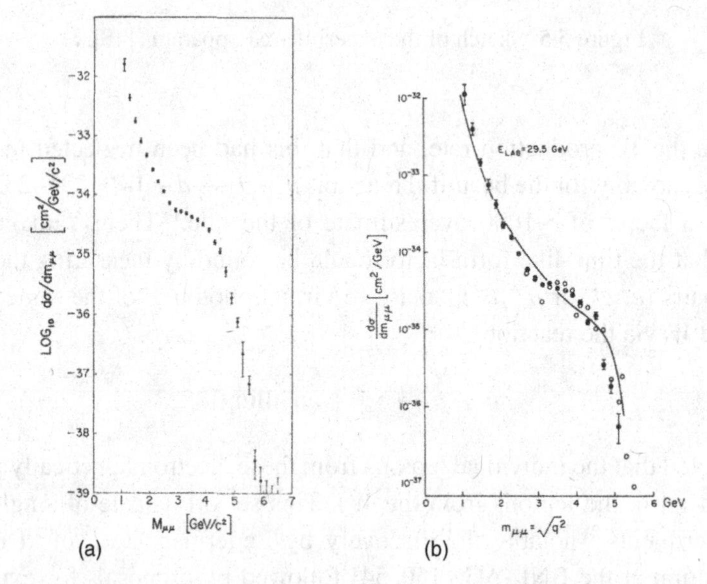

Figure 5.7 (a) Di-muon invariant mass spectrum $d\sigma/dm_{\mu\mu}$ [50]; (b) theoretical prediction [237] for $d\sigma/dm_{\mu\mu}$.

$p_T > 0.5$ GeV/c for detected muons whose angles were measured by scintillator hodoscopes and whose energy was measured by their range.

Figure 5.7a shows the di-muon invariant mass spectrum $d\sigma/dm_{\mu\mu}$ from the collisions of 29.5 GeV protons in a thick uranium target. There was definitely a dispute in the group about the meaning of the shoulder or "bump" or "??" for $2.5 < m_{\mu\mu} < 4.0$ GeV/c^2, which was apparently resolved adequately by the long forgotten theory paper [237] which produced the curve that agreed beautifully with the data (Figure 5.7b) with the explanation,"The origin of the shoulder comes from an interplay between the phase-space control of the integration region and the q^2 dependence of the coefficient." The important lesson to be learned from Figure 5.7

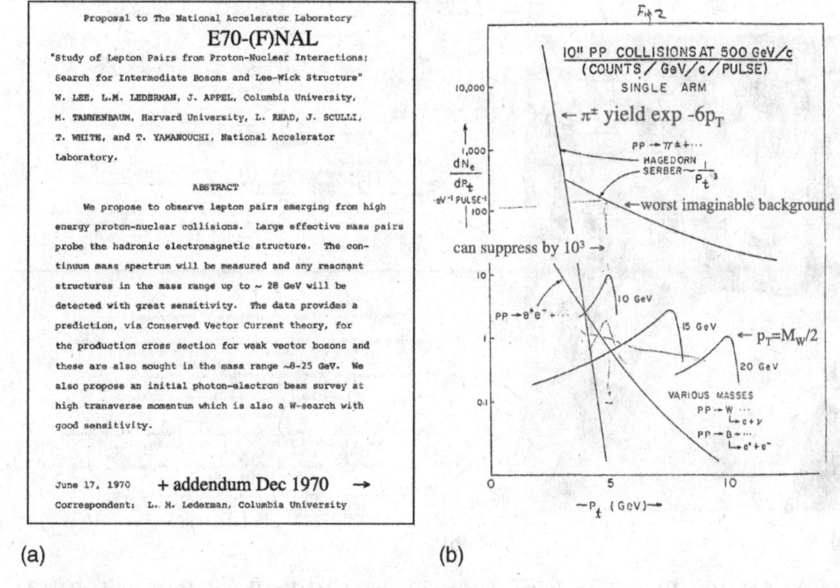

Proposal to The National Accelerator Laboratory

E70-(F)NAL

"Study of Lepton Pairs from Proton-Nuclear Interactions: Search for Intermediate Bosons and Lee-Wick Structure"

W. LEE, L.M. LEDERMAN, J. APPEL, Columbia University, M. TANNENBAUM, Harvard University, L. READ, J. SCULLI, T. WHITE, and T. YAMANOUCHI, National Accelerator Laboratory.

ABSTRACT

We propose to observe lepton pairs emerging from high energy proton-nuclear collisions. Large effective mass pairs probe the hadronic electromagnetic structure. The continuum mass spectrum will be measured and any resonant structures in the mass range up to ~ 28 GeV will be detected with great sensitivity. The data provides a prediction, via Conserved Vector Current theory, for the production cross section for weak vector bosons and these are also sought in the mass range ~8-25 GeV. We also propose an initial photon-electron beam survey at high transverse momentum which is also a W-search with good sensitivity.

June 17, 1970 + addendum Dec 1970 →

Correspondent: L. M. Lederman, Columbia University

(a) (b)

Figure 5.8 (a) Proposal for E70 Fermilab, June 17, 1970; (b) cross section and background calculation for $W^{\pm} \to e^{\pm} + X$ from Addendum [238].

is NEVER to be influenced by theoretical curves which "explain" your data. It is only when the curves fail to explain the data that you learn something definitive: the theory is wrong.

Leon Lederman was very excited in 1970 to be in the possession of the di-muon continuum mass spectrum, $d\sigma/dm_{\mu\mu}$, because by combining this result at $\sqrt{s} = 7.4$ GeV with the newly found Bjorken scaling [72] as used by Drell and Yan [54], he could calculate the W cross section at any $\sqrt{s}$, and hence the sensitivities of his two proposals E70 [238] and CCR [239] (see Figure 5.8). Details worthy of note from Figure 5.8b are the e^{-6p_T} pion yield, the line with the worst imaginable background and the Jacobian peaks at $p_T = M_W/2$ from $W \to e + X$ for various W masses. It is also important to note that the proposal for E70 (Figure 5.8a) was for measurement of the lepton pair continuum but did in fact mention that "any resonant structures in the (lepton-pair) mass range up to ~28 GeVc^{-2} will be detected with great sensitivity." It also proposed "an initial photon-electron beam survey at high transverse momentum which is also a W-search with good sensitivity."

5.3 The November revolution

Lederman's shoulder was explained in November 1974 by the discovery of the J/ψ at the BNL-AGS [60] and at SLAC [61] (Figure 5.9). This discovery revolutionized high energy physics since it was a heavy vector meson with a very narrow width

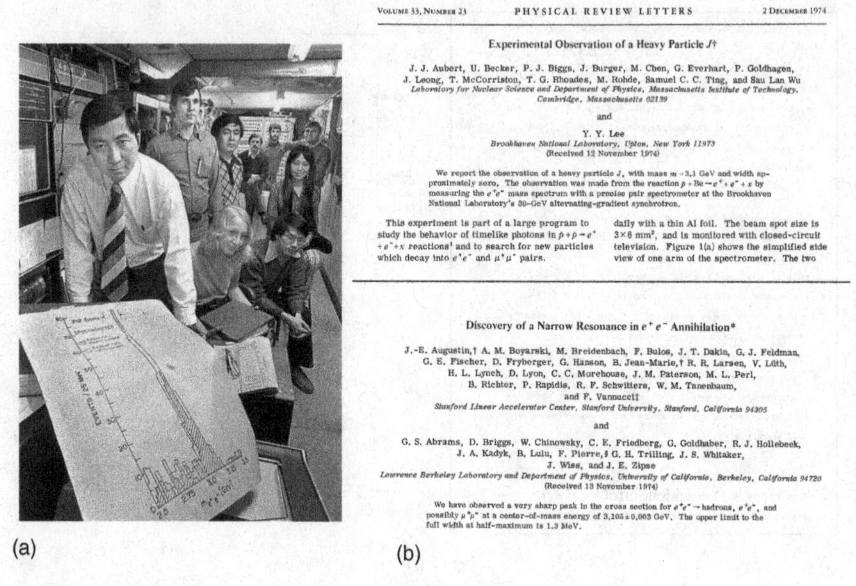

(a)

VOLUME 33, NUMBER 23 PHYSICAL REVIEW LETTERS 2 DECEMBER 1974

Experimental Observation of a Heavy Particle J†

J. J. Aubert, U. Becker, P. J. Biggs, J. Burger, M. Chen, G. Everhart, P. Goldhagen, J. Leong, T. McCorriston, T. G. Rhoades, M. Rohde, Samuel C. C. Ting, and Sau Lan Wu
Laboratory for Nuclear Science and Department of Physics, Massachusetts Institute of Technology, Cambridge, Massachusetts 02139

and

Y. Y. Lee
Brookhaven National Laboratory, Upton, New York 11973
(Received 12 November 1974)

We report the observation of a heavy particle J, with mass $m = 3.1$ GeV and width approximately zero. The observation was made from the reaction $p + \mathrm{Be} \to e^+ + e^- + x$ by measuring the e^+e^- mass spectrum with a precise pair spectrometer at the Brookhaven National Laboratory's 30-GeV alternating-gradient synchrotron.

This experiment is part of a large program to study the behavior of timelike photons in $p + p \to e^+ + e^- + x$ reactions[1] and to search for new particles which decay into e^+e^- and $\mu^+\mu^-$ pairs.

daily with a thin Al foil. The beam spot size is 3×6 mm^2, and is monitored with closed-circuit television. Figure 1(a) shows the simplified side view of one arm of the spectrometer. The two

Discovery of a Narrow Resonance in e^+e^- Annihilation*

J.-E. Augustin,† A. M. Boyarski, M. Breidenbach, F. Bulos, J. T. Dakin, G. J. Feldman, G. E. Fischer, D. Fryberger, G. Hanson, B. Jean-Marie,† R. R. Larsen, V. Lüth, H. L. Lynch, D. Lyon, C. C. Morehouse, J. M. Paterson, M. L. Perl, B. Richter, P. Rapidis, R. F. Schwitters, W. M. Tanenbaum, and F. Vannucci‡
Stanford Linear Accelerator Center, Stanford University, Stanford, California 94305

and

G. S. Abrams, D. Briggs, W. Chinowsky, C. E. Friedberg, G. Goldhaber, R. J. Hollebeek, J. A. Kadyk, B. Lulu, F. Pierre,§ G. H. Trilling, J. S. Whitaker, J. Wiss, and J. E. Zipse
Lawrence Berkeley Laboratory and Department of Physics, University of California, Berkeley, California 94720
(Received 13 November 1974)

We have observed a very sharp peak in the cross section for $e^+e^- \to$ hadrons, e^+e^-, and possibly $\mu^+\mu^-$ at a center-of-mass energy of 3.105 ± 0.003 GeV. The upper limit to the full width at half-maximum is 1.3 MeV.

(b)

Figure 5.9 (a) Sam Ting with J discovery at BNL; (b) J [60] and ψ [61] discoveries in PRL.

(it decayed slowly) which implied a new conservation law, similar to the discovery of strange particles in cosmic rays [25]. The J/ψ was quickly understood to be a bound state of heavy c–$\bar{c}$ quarks (charmonium) [63, 64] – the hydrogen atom of QCD. This was clear evidence for a "second" generation of quarks, and made all physicists believe in quarks and QCD.

This discovery also changed the paradigm of di-lepton production from the measurement of background for W production to the search for new resonances. This will be further discussed in later chapters.

6

Discovery of hard scattering in p–p collisions

The late 1960s were very exciting for experimental high energy physicists. Two new accelerators, the CERN-ISR, in Europe [240], and the National Accelerator Laboratory (NAL, now Fermilab), in the USA [241,242], were under construction. The call for proposals went out in January 1969 for the CERN-ISR [243], with eventual first collisions in January 1971. For NAL, proposals were requested in March 1970 [244] with first operation in March 1972. Everybody who was anybody in experimental high energy physics at that time was involved in proposals at one or both laboratories.

Fermilab's accelerator was a traditional fixed target machine which provided 200 to 400 GeV primary proton beams and a large variety of secondary beams, while the CERN-ISR was the first proton–proton collider.[1] The size and scope of the machines was quite different although both were destined to make important contributions to high p_T physics, a subject that did not exist before these machines operated. Also, Fermilab was a brand new laboratory totally dedicated to the new accelerator while CERN had been in existence since 1954 with several operating accelerators [248].

6.0.1 The Fermilab program, circa 1970

Fermilab had held summer studies in 1968 and 1969 for users to help design and specify the various beams and facilities at the new laboratory. The layout of the accelerator [249] together with a more expanded view of the initial (circa 1973–1975) beam lines and facilities [250] is shown in Figure 6.1. Note the 1 km radius of the main accelerator.

To get a feeling for the status and thinking in high energy physics at the startup of Fermilab, the initial complement of accepted proposals (February 1971) is shown

[1] It was also the first collider to run with d–d, α–α, α–p and $\overline{p}$–p collisions [245–247].

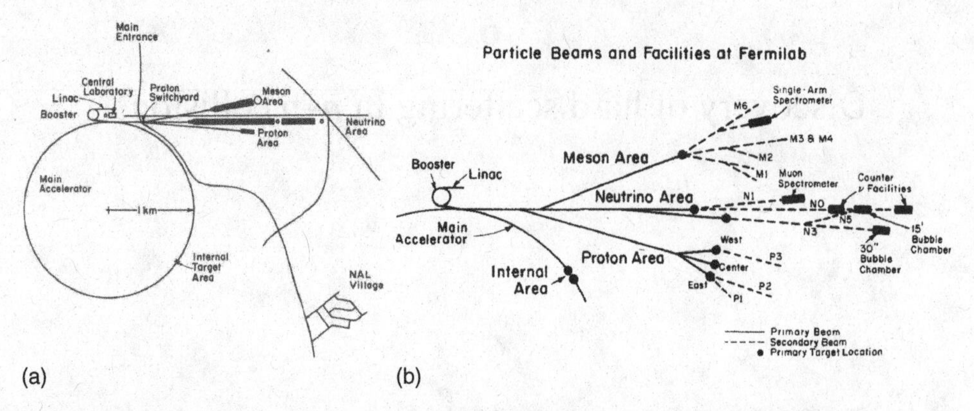

Figure 6.1 (a) Schematic diagram of the NAL accelerator and experimental areas [249]. (b) Expanded view of the various particle beams and facilities [250].

in Figure 6.2 [251]. The only experiment looking at large angles (for the W boson) is Lederman's E70, which will be described in detail later. E48 was also a W boson search but in the forward (beam) direction, not at wide angles.

Some history is worth mentioning here. In the 1968 summer study, there were serious discussions about whether the W boson would be more copiously produced with muon (μ) beams, via coherent production on a nucleus [252, 253], than in p–p collisions. This was largely based on a theoretical calculation in 1968 [254] of a huge cross section, $\sigma_W \sim 10^{-30}$ cm^2, for μ induced single W production on a proton. This calculation was eventually shown to be wrong in late 1969 [255] because it was not gauge invariant. Another historical note is the absence from Figure 6.2 of Jim Cronin's single-arm high p_T spectrometer experiment, E100 [256], which was submitted in December 1970, well after the 1970 summer study and PAC meeting which reviewed the first round of proposals. This experiment has had immense influence in the field of relativistic heavy ion physics, although perhaps not in the way the proponents originally envisioned or desired. Further discussion is in order.

The layout of E100 [257] is shown in Figure 6.3 (note the scale of the experiment). The apparatus was a single-arm spectrometer located at a fixed angle of 77 mrad with respect to the incident proton beam. The proposal was to study the particle composition of a beam produced at approximately 90° in the p–p c.m. system by 200–400 GeV protons striking a fixed target ($\sqrt{s}$ of 19.4–27.4 GeV). The purpose was an exploratory investigation to provide information on: (1) hadron production at high p_T; (2) the possible existence of the W boson, heavy photons and heavy leptons by searching for leptons with high p_T; (3) the possible existence of long-lived particles (with or without fractional charge); (4) with slight modification of the apparatus, direct-photon production.

List of Proposals Accepted.

Proposal Number	Title	Scientific Spokesman	Institutions
1-A	NAL Neutrino Proposal	D. Cline	Wisconsin Pennsylvania Harvard
3	Proposal for a Search for Magnetic Monopoles at NAL	P. H. Eberhard	LRL, SLAC
4-A	Neutron-Proton Diffraction Scattering and Neutron	M. J. Longo	Michigan, ANL
4-B	Total Cross Sections up to 200 GeV		
7	A Proposal to Measure $\pi^{\pm}p$ and p-p Differential Elastic Scattering Cross Sections from 50 to 170 GeV/c	D. Meyer	Michigan, ANL NAL
8	Experiments in a Neutral Hyperon Beam	L. G. Pondrom	Wisconsin Michigan
12	A Study of Neutron-Proton Charge-Exchange Scattering in the Momentum Range 50-200 GeV/c	N. W. Reay	Ohio State Michigan State Carleton
21	Neutrino Physics at Very High Energies	B. Barish	NAL Caltech
22	Experimental Proposal to the NAL for a Search for Multigamma Events from Magnetic Monopole Pairs	G. B. Collins	
26	High Momentum Transfer Inelastic Muon Scattering and Test of Scale Invariance at NAL	K. W. Chen	Princeton Cornell
32	Test and Calibrate a Large NaI (Tl) Tanc Detector and to Measure Neutral Hadron Total Cross Sections	R. Hofstadter	Stanford
34	Nuclear-Electromagnetic Cascade Development Study (Ionization Spectrometer Development)	R. W. Huggett	Louisiana State Max-Planck Inst.
48	A Measurement of the Intensity and Polarization of Muons Produced Directly by the Interactions of Protons with Nuclei	R. K. Adair	Yale, BNL Princeton
54	Quasi-Two-Body Reactions at 50-200 GeV	J. Pine	Caltech UCLA, NAL
55	Proposal to Study $\pi^- p \to \pi^0 n$ and $\pi^- p \to \eta n$ at High Energy	A. V. Tollestrup	Caltech
63	Survey of Particle Production in Proton Collisions at NAL	J. K. Walker	NAL
69-A	Elastic Scattering of the Hadrons	J. Lach	NAL, Yale
70	Study of Lepton Pairs from Proton-Nuclear Interactions; Search for Intermediate Bosons and Lee-Wick Structure	L. M. Lederman	Columbia Harvard, NAL
72	Experimental Proposal to NAL Quark Search	R. K. Adair	Yale, BNL
74	Proposal to National Accelerator Laboratory for a Search for Magnetic Monopoles	R. L. Fleischer	General Electric NAL
75	A Proposal to Search for Fractionally Charged Quarks	T. Yamanouchi	NAL
76	Search for Magnetic Monopoles Produced at NAL	R. A. Carrigan	NAL
81	Preliminary Survey of 200-GeV Proton Interactions with Complex Nuclei	G. Butler	ANL, BNL Chicago Carnegie-Mellon Purdue SUNY(Buffalo)
82	Proposal to Investigate Regeneration of Neutral K-Mesons at Very High Energies	V. L. Telegedi	Chicago, SLAC U. of Calif. (San Diego)
96	Focusing Spectrometer Facility (Replaces 64 and 73)	D. M. Ritson	ANL, Cornell U. of Bari Brown, CERN MIT, NAL Northeastern Stanford
97	Elastic Scattering of the Hyperons (Replaces 69-Y)	J. Lach	NAL, Yale
98	Muon-Proton Inelastic Scattering Experiment at the National Accelerator Laboratory (Replaces 29 and 33)	L. W. Mo	Chicago Harvard
104	Measurement of Total Cross Sections on Hydrogen and Deuterium (Replaces 10 and 56)	W. F. Baker	NAL, BNL Rockefeller Univ.

Figure 6.2 The initial complement of accepted proposals at Fermilab in February 1971 [251] following reviews by the Program Advisory Committee (PAC) at meetings in August 1–7, and December 11–12, 1970. These were followed by further rounds of proposal submission and reviews.

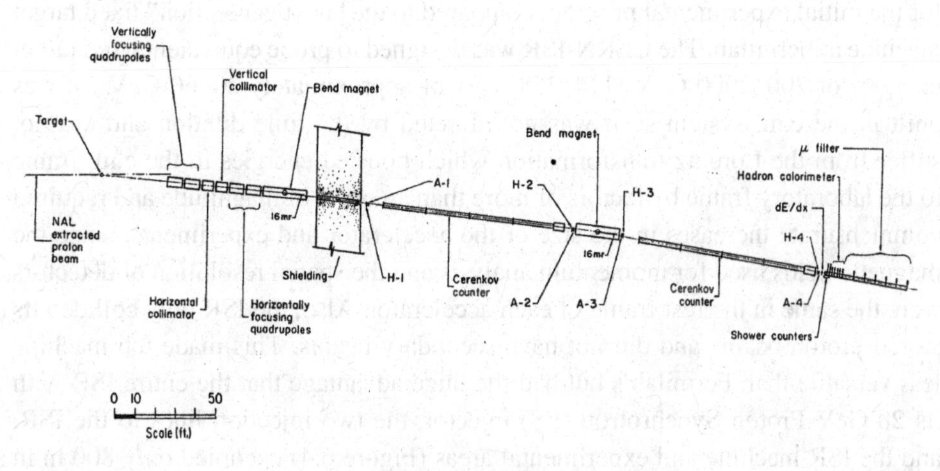

Figure 6.3 The E100 spectrometer [257].

The angle of 90° in the p–p c.m. system, or $y^* = 0$ (mid-rapidity), corresponds (recall Eqs. 3.78, 3.77) to laboratory rapidity:

$$y = Y^{cm} = \frac{Y^{beam}}{2} = \frac{1}{2}\cosh^{-1}\frac{E_1}{m_1} \tag{6.1}$$

where E_1 and m_1 are the energy and mass of the incident proton. This requirement results in laboratory angles of 97, 79, 68 mrad for 200, 300, 400 GeV protons, respectively, or conversely, for the fixed laboratory angle of 77 mrad, results in c.m. angles of 77°, 88° and 97°.

To get an idea of what the proponents were thinking (and with the benefits of hindsight), a few comments from the proposal are worth repeating here [256]:

(i) We have suggested a number of investigations that can be carried out with a single beam at a large fixed angle. The basic principle in mind is the following: it is always worthwhile to look in a region where one is supposed to find nothing. Anything found is therefore very likely of some interest.

(ii) In most cases we have suggested the use of targets of complex nuclei. We would guess that most of the events at high transverse momentum would be independent of whether the nucleon is in a complex nucleus or not. The only significant effect is that of Fermi momentum which smears the available energy in the center of mass. It would be valuable to have the option of installation of a hydrogen target.

6.0.2 The CERN-ISR program, circa 1971

The CERN-ISR, being the first hadron–hadron collider, had a different set of issues for the initial experimental program compared to the "next generation" fixed target machine at Fermilab. The CERN-ISR was designed to probe equivalent fixed target energies of 200–2000 GeV [114, 258] ($\sqrt{s}$ of approximately 20–60 GeV). It was built in the c.m. system so it was not affected by the time dilation and did not suffer from the Lorentz transformation which boosted energies in the c.m. frame to the laboratory frame by factors of more than an order of magnitude and required commensurate increases in the size of the accelerator and experiments, since the magnetic fields used for momentum analysis and the spatial resolution of detectors were the same in the rest frame of each accelerator. Also, the ISR only collided its stored proton beams and did not have secondary beams. This made the machine less versatile than Fermilab's but had the huge advantage that the entire ISR with its 28 GeV Proton Synchrotron (PS) injector, the two injection lines to the ISR, and the ISR machine and experimental areas (Figure 6.4) occupied only 800 m in length, less than the radius of the Fermilab main accelerator (compare Figure 6.1a), although it spanned two countries.

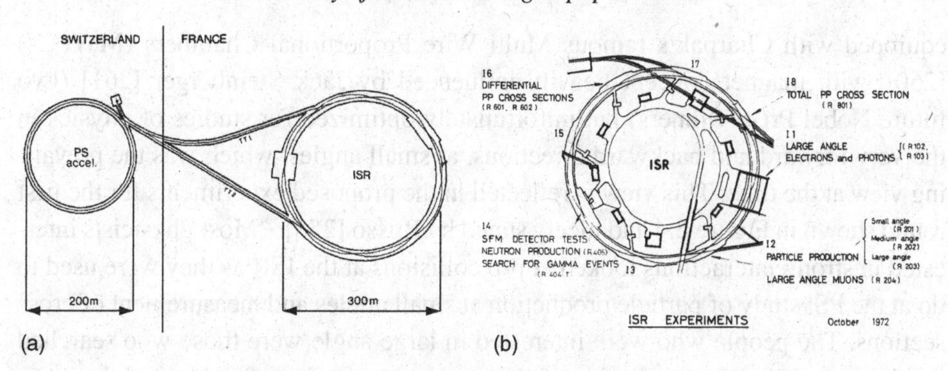

Figure 6.4 (a) Schematic diagram [240] of the CERN-ISR together with the PS injector which supplied the accelerated proton beams via the transfer lines TT1 and TT2 to the ISR collider. (b) Expanded view of the ISR building with the eight interaction regions, I1 to I8, and the first round experiments indicated.

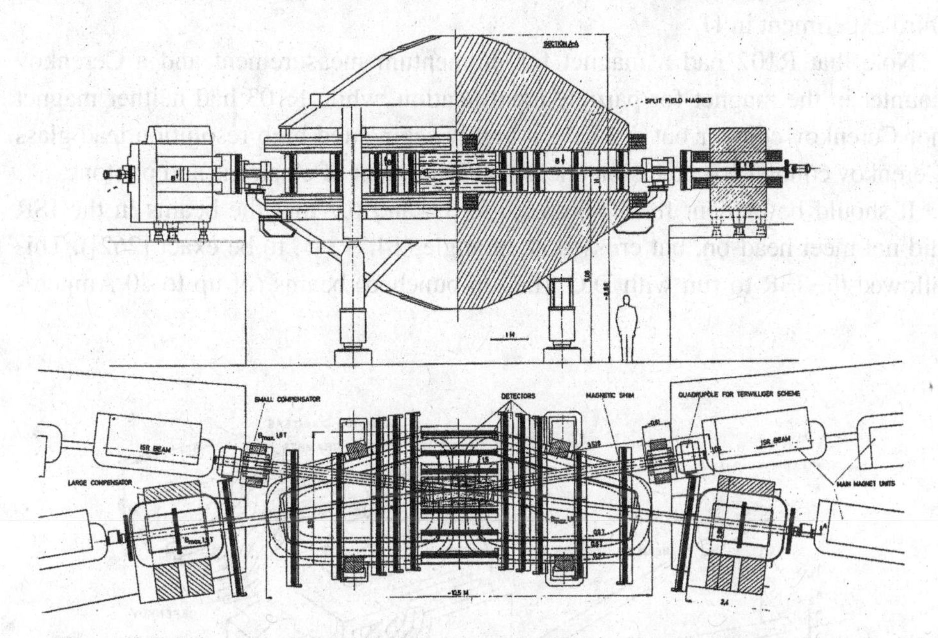

Figure 6.5 Schematic layout [259] of the split field magnet facility on the intersection region I4. The MWPCs for the detection of the collision products are also outlined.

Because of the much smaller number of interaction points at a collider than at a fixed-target facility, and the smaller size of the experiments, the initial ISR physics program was geared to many relatively small experiments which either could occupy the same interaction region at the same time or could be moved in and out relatively easily to take turns. There was also one major facility built, the Split Field Magnet (SFM) (Figure 6.5) [259], which was huge and very high-tech for its time, equipped with Charpak's famous Multi Wire Proportional Chambers (MWPCs) [260], with magnet concept heavily influenced by Jack Steinberger [261] (two future Nobel Prize winners) but unfortunately optimized for studies of physics in the very forward and backward directions, at small angles, which was the prevailing view at the time. This view is reflected in the proposed experiments for the first round shown in Figure 6.4 and nicely stated by Russo [258]: "Most physicists interested in strong interactions looked at p-p collisions at the ISR as they were used to do at the PS: study of particle production at small angles and measurement of cross sections. The people who were interested in large angle were those who searched for rare events like the production of electrons or muons from the decay of the intermediate vector bosons..." The most influential of these experiments were both in I1, Saclay–Strasbourg, R102 [56] (Figure 6.6), and CERN–Columbia–Rockefeller (CCR), R103 [3], (Figure 6.7), where the ISR nomenclature, e.g. R103, means the third experiment in I1.

Note that R102 had a magnet for momentum measurement and a Cerenkov counter in the magnet for particle identification, while R103 had neither magnet nor Cerenkov counter but did have a highly segmented high resolution lead glass Cerenkov counter for precision energy measurement of electrons and photons.

It should be evident from Figures 6.5, 6.6 and 6.7 that the beams in the ISR did not meet head-on, but crossed at an angle (14.7734°, to be exact [262]). This allowed the ISR to run with D.C. (i.e. un-bunched) beams (of up to 40 Amperes

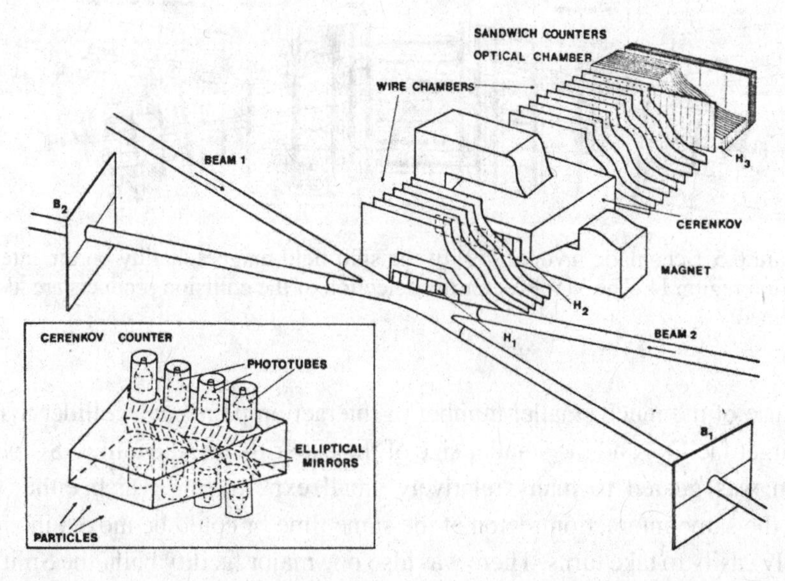

Figure 6.6 Isometric view of Saclay–Strasbourg experiment R102 [56].

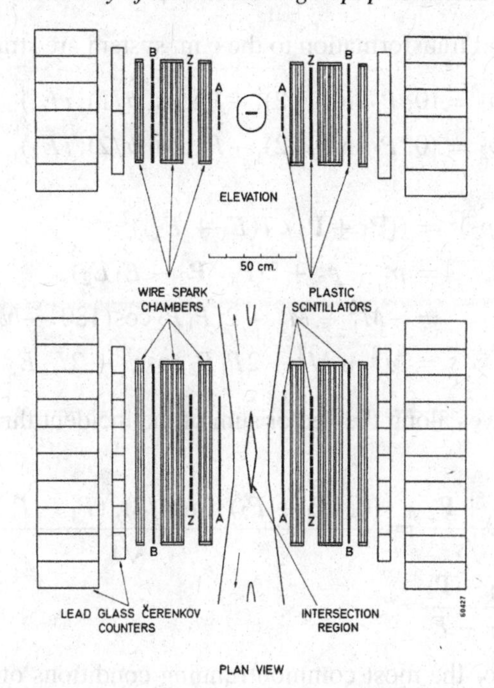

Figure 6.7 Elevation and plan views of CERN–Columbia–Rockefeller experiment R103 [3].

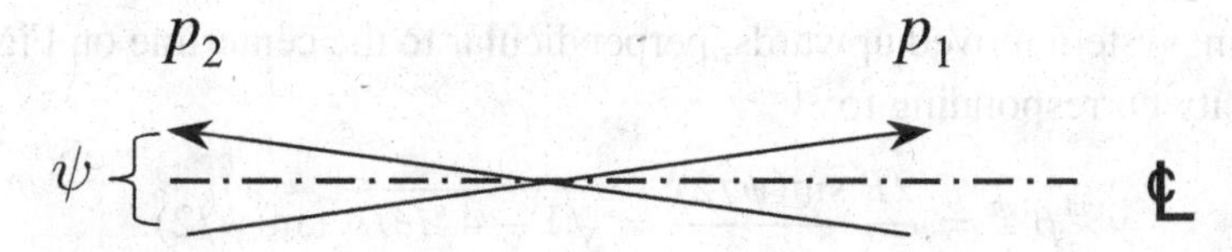

Figure 6.8 Beams of four-momentum p_1 and p_2 colliding with angle ψ.

of protons) which gave a continuous interaction rate similar to fixed-target experiments, so that the instantaneous and average interaction rates were identical. This is in distinction to all future colliders up to the present era which have bunched beams and much larger instantaneous rate of collisions (proportional to the difficulty of triggering and of separating individual collisions) compared to the average interaction rate (proportional to the number of events collected).

6.0.2.1 Complication due to the crossing angle at the CERN-ISR

The crossed beams implied that the laboratory frame and the c.m. frame were not identical at the CERN-ISR, but that the c.m. system had a small velocity $\vec{\beta}^{cm}$ perpendicular to the centerline of the crossing beams (Figure 6.8).

The kinematics and transformation to the c.m. system are straightforward.

$$p_1 = (0, P_1 \sin(\psi/2), +P_1 \cos(\psi/2), iE_1)$$
$$p_2 = (0, P_2 \sin(\psi/2), -P_2 \cos(\psi/2), iE_2), \quad (6.2)$$

$$\begin{aligned} -s = (p_1 + p_2)^2 &= ((\mathbf{P}_1 + \mathbf{P}_2), i(E_1 + E_2))^2 \\ &= p_1^2 + p_2^2 + 2(\mathbf{P}_1 \cdot \mathbf{P}_2 - E_1 E_2) \\ &= -M_1^2 - M_2^2 + 2(P_1 P_2 \cos(180° - \psi) - E_1 E_2), \\ s &= M_1^2 + M_2^2 + 2P_1 P_2 \cos\psi + 2E_1 E_2. \end{aligned} \quad (6.3)$$

The c.m. system moves along the vector sum of the incident three-momenta, $\mathbf{P}_1 + \mathbf{P}_2$ with velocity:

$$\begin{aligned} \gamma^{cm} \vec{\beta}^{cm} &= \frac{\mathbf{P}_1 + \mathbf{P}_2}{\sqrt{s}} = \frac{(0, (P_1 + P_2)\sin(\psi/2), (P_1 - P_2)\cos(\psi/2))}{\sqrt{s}} \\ \vec{\beta}^{cm} &= \frac{\mathbf{P}_1 + \mathbf{P}_2}{E_1 + E_2}. \end{aligned} \quad (6.4)$$

This meant that for the most common running conditions of equal momentum beams, $P_1 = P_2 = P$, $E_1 = E_2 = E$, $M_1 = M_2 = M$,

$$\sqrt{s - 4M^2} = 2P \cos(\psi/2), \quad (6.5)$$

and the c.m. system moved upwards, perpendicular to the centerline on Figure 6.8, with velocity corresponding to:

$$\gamma^{cm} \beta^{cm} = \frac{2P \sin(\psi/2)}{\sqrt{s}} = \sqrt{1 - 4M^2/s}\, \tan(\psi/2). \quad (6.6)$$

This was roughly constant at $\gamma^{cm} \beta^{cm} \approx \tan(\psi/2) = 0.12964$ as a function of $\sqrt{s}$; but the factor of $\sqrt{1 - 4M^2/s}$ was generally included in the correction for the moving c.m. system. This meant that, for example for the CCR experiment, which was symmetric about the centerline, the relations between the actual transverse momentum p_T^* in the c.m. system and the measured p_T were different in the detectors to the left and right of the beam pipe in Figure 6.7 (which correspond to up and down in Figure 6.8). This may explain the occasional emphasis on p_T^* rather than simply p_T in some publications from the ISR [263]. Also, since the c.m. system moved to the left in Figure 6.7, the acceptances of the left and right detectors were different in the c.m. system. Thus, the constraint of requiring the same result in both the left and right detectors provided an excellent systematic check of both the detectors and the Lorentz transformations to the c.m. system.

The lesson from this long discussion is that small details can lead to inconsistent or wrong experimental results if they are not properly taken into account.

6.1 Bjorken scaling and the parton model in *p–p* collisions

As discussed in Section 4.3, the idea of hard scattering in $p + p$ collisions dates from the first studies at SLAC of deeply inelastic electron–proton scattering, i.e. scattering with large values of four-momentum transfer squared, Q^2, and energy loss, ν. The discovery that the Deeply Inelastic Scattering (DIS) structure function

$$F_2(Q^2, \nu) = F_2\left(\frac{Q^2}{\nu}\right) \tag{6.7}$$

"scaled," i.e just depended on the ratio

$$x = \frac{Q^2}{2M\nu} \tag{6.8}$$

independently of Q^2 [2], as originally suggested by Bjorken [72], led to the concept of a proton composed of point-like "partons." The deeply inelastic scattering of an electron from a proton is simply quasi-elastic scattering of the electron from point-like partons of effective mass Mx, with quasi-elastic energy loss, $\nu = Q^2/2Mx$. The probability for a parton to carry a fraction x of the proton's momentum is measured by $F_2(x)/x$.

Since the partons of DIS are electrically charged, and hence must scatter electromagnetically from each other in a $p+p$ collision (Figure 6.9), Berman, Bjorken and Kogut (BBK) [264] calculated the cross section of the inclusive reaction

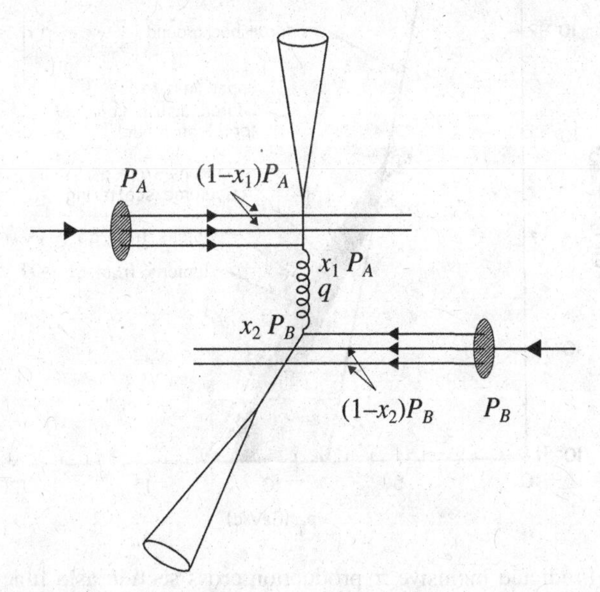

Figure 6.9 Schematic parton–parton scattering in a *p–p* collision (after Fig. 4.11 [196]).

$$A + B \rightarrow C + X \tag{6.9}$$

based on the elastic electromagnetic parton–parton scattering:

$$a + b \rightarrow c + d \tag{6.10}$$

where a and b are the incident partons, c, d, the scattered partons, A and B are the incident hadrons and particle C has $p_T \gg 1$ GeV/c, "which may be viewed as a *lower bound* on the real cross section at large p_T" (Figure 6.10).

BBK proposed a general form for high p_T cross sections, for the electromagnetic (EM) scattering:

$$E\frac{d^3\sigma}{dp^3} = \frac{4\pi\alpha^2}{p_T^4}\mathcal{F}\left(x_1 = \frac{-\hat{u}}{\hat{s}}, x_2 = \frac{-\hat{t}}{\hat{s}}\right). \tag{6.11}$$

The two factors are a $1/p_T^4$ term, characteristic of single photon exchange, and a form factor $\mathcal{F}$, where $\hat{s}$, $\hat{t}$ and $\hat{u}$ are the constituent-scattering invariants. Note that $x_{1,2}$ in Eq. 6.11 are not Bjorken x (Eq. 6.8). The point is that $\mathcal{F}$ **scales**, i.e. is only a function of the ratio of momenta. Vector ($J = 1$) gluon exchange gives the same form as Eq. 6.11 but would be much larger.

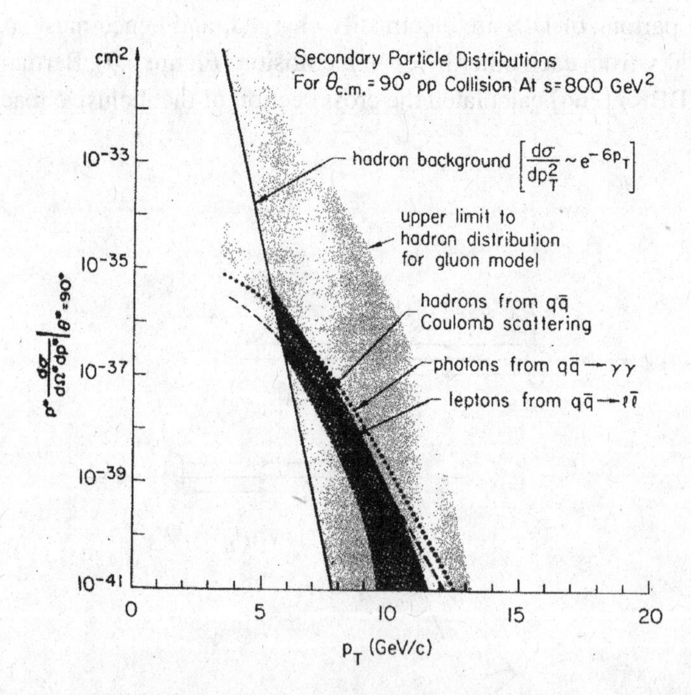

Figure 6.10 Predicted inclusive π production cross section as a function of p_T in $p + p$ collisions [264]. Note the difference compared to the left panel of Figure 4.13 [220].

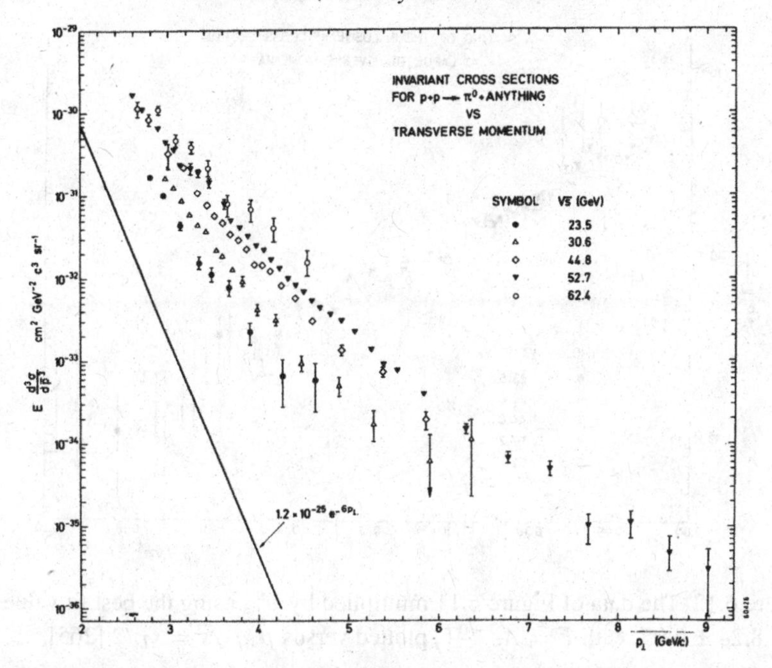

Figure 6.11 The transverse momentum dependence of invariant cross sections of five center-of-mass energies. The errors are statistical only [265].

6.2 ISR data, notably CCR 1972–1973

The CERN–Columbia–Rockefeller (CCR) collaboration [3] (and also the Saclay–Strasbourg [56] and British Scandinavian [57] collaborations) measured high p_T pion production at the CERN-ISR (Figure 6.11). The e^{-6p_T} breaks to a power-law at high p_T with characteristic $\sqrt{s}$ dependence. The large rate represented the discovery that partons interact strongly ($\gg$ EM) with each other, *but* [3], "Indeed, the possibility of a break in the steep exponential slope observed at low p_T was anticipated by Berman, Bjorken and Kogut. However, the electromagnetic form they predict, $p_\perp^{-4} F(p_\perp/\sqrt{s})$, is not observed in our experiment. On the other hand, a constituent exchange model proposed by Blankenbecler, Brodsky and Gunion, and extended by others, does give an excellent account of the data" [3]. The data fit $p_\perp^{-n} F(x_T)$, with $n \simeq 8$ (Figure 6.12), where $x_T \equiv 2p_T/\sqrt{s}$.

6.3 Constituent interchange model (CIM), 1972

Inspired by the "dramatic features of pion inclusive reactions" revealed by "the recent measurements at CERN-ISR of single-particle inclusive scattering at 90° and large transverse momentum," Blankenbecler, Brodsky and Gunion [266] proposed a new general scaling form:

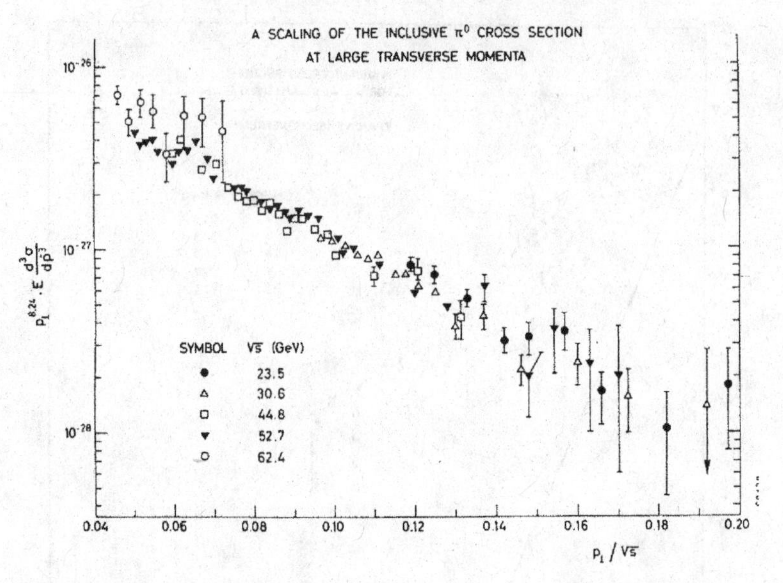

Figure 6.12 The data of Figure 6.11 multiplied by $p_{\perp}^{n}$, using the best fit value of $n = 8.24 \pm 0.05$, with $F = Ae^{-bx_T}$, plotted versus $p_{\perp}/\sqrt{s} = x_T/2$ [265].

$$E\frac{d^3\sigma}{dp^3} = \frac{1}{p_T^{n_{eff}}} F\left(\frac{p_T}{\sqrt{s}}\right) \tag{6.12}$$

where n_{eff} relates to the form of the force law between constituents, i.e. the quantum exchange governing the reaction. More precisely, in the inclusive hadron scattering of Eq. 6.9, the hadrons are assumed to be composite systems consisting of elementary constituents (a, b, c, d) and the basic CIM mechanism [267] is the rearrangement of the elementary constituents a and b to c and d in the reaction. In the CIM framework, $n_{eff} = 2(n_{active} - 2)$, where n_{active} is the number of elementary fields (lepton–photon–quark) participating in the large-angle subprocess, $n_{active} = n_a + n_b + n_c + n_d$. "Thus for electron-quark or photon-quark or quark-quark scattering ($n_{active} = 4$) one obtains the standard scale invariant $p_{\perp}^{-4}$ predictions of the parton model" [267]. In other words, for QED or vector gluon exchange, $n = 4$, as in Eq. 6.11. Perhaps more importantly, the CIM prediction for the case of quark–meson scattering by the exchange of a quark ($n_{active} = 6$) gives $p_{\perp}^{-8}$, as apparently observed.

6.4 First prediction using "QCD" 1975 – WRONG!

R. F. Cahalan, K. A. Geer, J. Kogut and Leonard Susskind [268] generalized, in their own words (but with emphasis added here):

the naive, point-like parton model of Berman, Bjorken and Kogut to scale-invariant and asymptotically free field theories. The asymptotically free field generalization is studied in detail. Although such theories contain vector fields, **single vector-gluon exchange contributes insignificantly to wide-angle hadronic collisions.** This follows from (1) the smallness of the invariant charge at small distances and (2) the *breakdown of naive scaling* in these theories. These effects should explain the apparent absence of vector exchange in inclusive and exclusive hadronic collisions at large momentum transfers observed at Fermilab and at the CERN-ISR.[2]

Nobody is perfect, they got *one* thing right! They introduced the "effective index" $n_{eff}(x_T, \sqrt{s})$ to account for "scale breaking":

$$E\frac{d^3\sigma}{dp^3} = \frac{1}{p_T^{n_{eff}(x_T,\sqrt{s})}} F\left(\frac{p_T}{\sqrt{s}}\right) = \frac{1}{\sqrt{s}^{n_{eff}(x_T,\sqrt{s})}} G\left(\frac{p_T}{\sqrt{s}}\right). \quad (6.13)$$

6.5 Experimental improvements, theoretical improvements

An upgrade of CCR (to be described in more detail later (Figure 9.2b in Chapter 9)), the CCOR experiment [269] (Figure 6.13) with a larger apparatus and much increased integrated luminosity, extended their previous π^0 measurement [3, 270] to much higher p_T. The p_T^{-8} scaling fit which worked at lower p_T extrapolated below the higher p_T measurements for $\sqrt{s} > 30.7$ GeV and $p_T \geq 7$ GeV/c (Figure 6.13). The new fit [269] (not shown in Figure 6.13) is $Ed^3\sigma/dp^3 \simeq p_T^{-5.1\pm0.4}(1 - x_T)^{12.1\pm0.6}$, for $7.5 \leq p_T \leq 14.0$ GeV/c, $53.1 \leq \sqrt{s} \leq 62.4$ GeV (including *all* systematic errors).

An important feature of the scaling analysis (Eq. 6.13) for determining $n_{eff}(x_T, \sqrt{s})$ is that *the absolute p_T scale uncertainty cancels!* In Figure 6.14a, the CCOR data of Figure 6.13 for the three values of $\sqrt{s}$ are plotted versus x_T on a log–log scale [269]. $n_{eff}(x_T, \sqrt{s})$ is determined for any two values of $\sqrt{s}$ by taking the ratio as a function of x_T as shown in Figure 6.14b. $n_{eff}(x_T, \sqrt{s})$ clearly varies with both $\sqrt{s}$ and x_T, it is not a constant. For $\sqrt{s} = 53.1$ and 62.4 GeV, $n_{eff}(x_T, \sqrt{s})$ varies from ~8 at low x_T to ~5 at high x_T. The effect of the absolute scale uncertainty, which is the main systematic error in these experiments, can be gauged from Figure 6.14c [271] which shows the π^0 cross sections from several experiments. The absolute cross sections disagree by factors of ~3 for different experiments but the values of $n_{eff}(x_T, \sqrt{s})$ for the CCOR [269] (Figure 6.14b) and ABCS [271] experiment (Figure 6.14d) are in excellent agreement due to the cancelation of the error in the absolute p_T scale.

[2] There is an acknowledgement in this paper that is worthy of note: "Two of us (J. K. and L. S.) also thank S. Brodsky for *emphasizing to us repeatedly* that the present data on wide-angle hadron scattering *show no evidence for vector exchange.*"

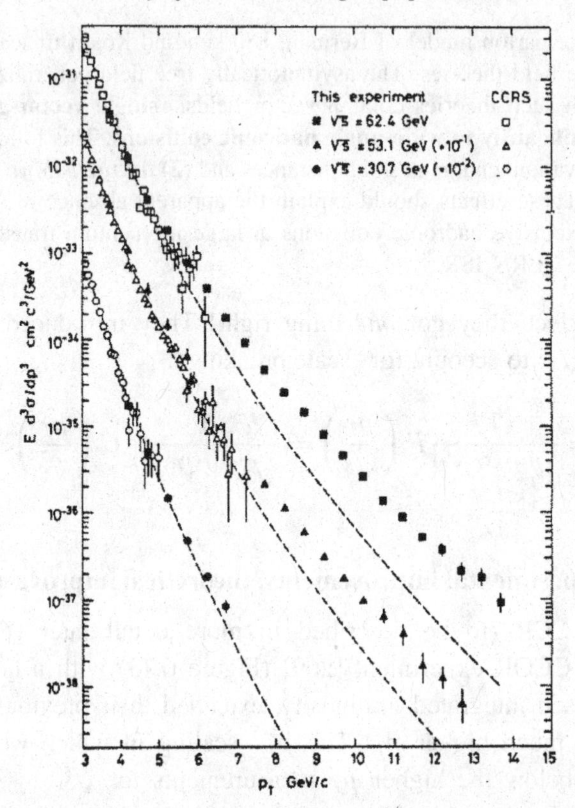

Figure 6.13 CCOR [269] transverse momentum dependence of the invariant cross section for $p + p \to \pi^0 + X$ at three center-of-mass energies. Cross sections are offset by the factors noted. Open points and dashed fit are from a previous experiment, CCRS [270].

The variation of $n_{eff}(x_T, \sqrt{s})$ with x_T and $\sqrt{s}$, tending to a value $\approx$5 at the highest measured p_T, 7–14 GeV/c at the ISR, were strongly suggestive of the $n = 4$ for QED or vector gluon exchange of Eq. 6.11, modified by the scale breaking of QCD as in the "effective index" of Eq. 6.13. In fact, it was shown on a roughly contemporaneous timescale by Jeff Owens and collaborators [69,70], closely followed and corroborated by Feynman and collaborators [71], that the high p_T single particle inclusive cross sections, including non-scaling, could be described by QCD once the effect of the smearing of the steeply falling p_T spectrum by the initial state transverse momentum of partons inside the proton, the k_T effect, was taken into account [272]. This will be discussed more extensively in Chapter 9.

6.6 State of the art at Fermilab 1977 – but misleading!

The best data at Fermilab in 1977 (Figure 6.15) [273] appeared to show beautifully the CIM scaling with $n_{eff} \sim 8$ over the range $0.2 \leq x_T \leq 0.6$ for p–p collisions

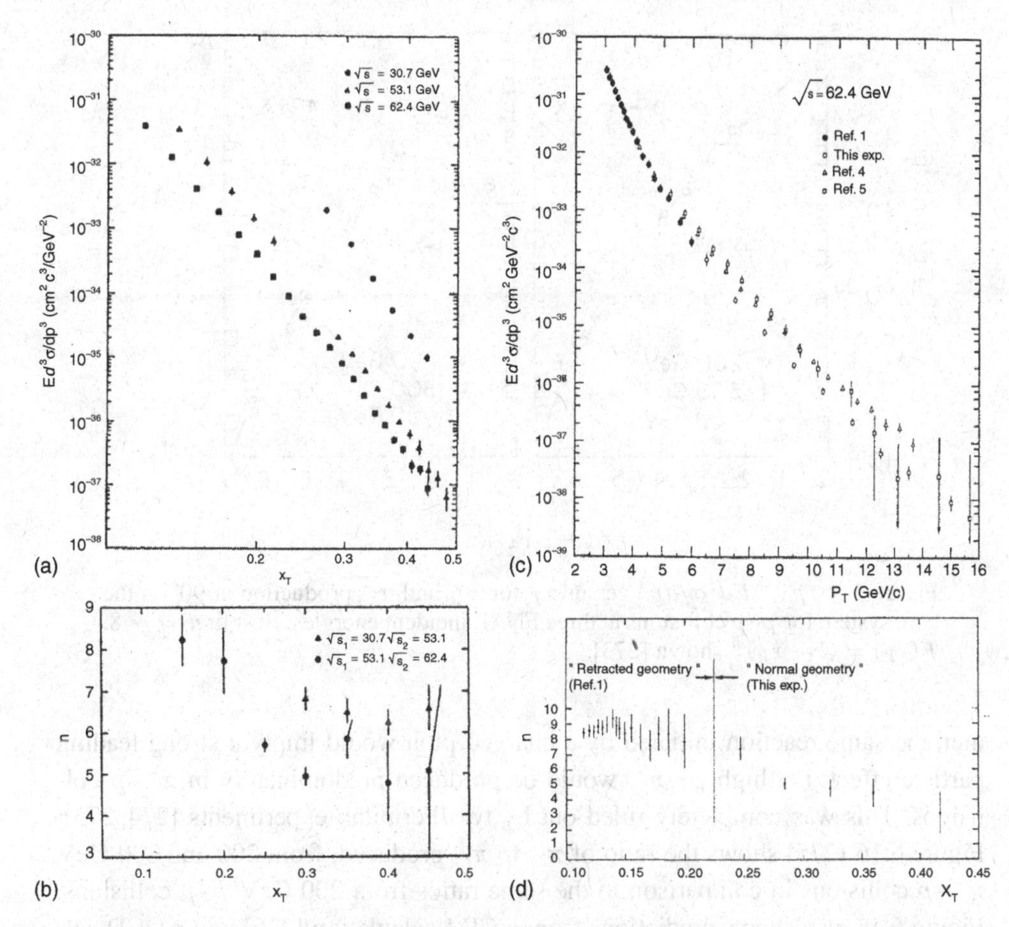

Figure 6.14 (a) Log–log plot of CCOR [269] invariant cross sections versus $x_T = 2p_T/\sqrt{s}$. (b) CCOR [269] $n_{eff}(x_T, \sqrt{s})$ derived from the combinations indicated. The systematic normalization error at $\sqrt{s} = 30.6$ GeV has been added in quadrature. There is an additional common systematic error of ± 0.33 in n_{eff}. (c) Invariant cross section for inclusive π^0 from several ISR experiments, compiled by the ABCS collaboration [271]. (d) $n_{eff}(x_T, \sqrt{s})$ from ABCS 52.7, 62.4 GeV data only. There is an additional common systematic error of ± 0.7 in n.

with 200, 300 and 400 GeV incident energies, where $x_T = 2p_T/\sqrt{s}$. However, this effect turned out not to be due to CIM, but to the "broadening" by initial state transverse momentum, the "k_T effect" [272].

In addition, further measurements at Fermilab, where there were secondary beams of $\pi^{\pm}$, allowed a more stringent test of the CIM, which it totally failed. If the CIM were in fact the explanation of $n_{eff} \sim 8$ in p–p collisions (Figure 6.15) via quark–meson scattering by the exchange of a quark,

$$q + \pi \leftrightarrow \pi + q,$$

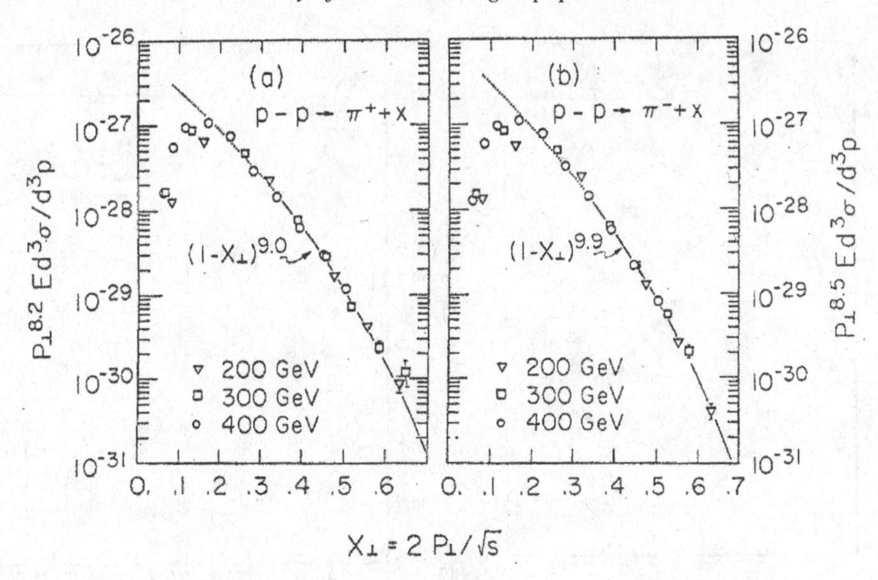

Figure 6.15 $p_\perp^{n_{eff}} Ed^3\sigma/dp^3$ versus x_T for π^+ and π^- production at 90° in the c.m. system for p–p collisions at three FNAL incident energies. Best fit $n_{eff} \sim 8$, $F(x_T) = (1 - x_T)^m$ shown [273].

then the same reaction initiated by a charged pion would imply a strong leading particle effect, i.e. high p_T π^- would be produced predominantly in π^-–p collisions. This was completely ruled out by two Fermilab experiments [274, 275]. Figure 6.16 [275] shows the ratio of π^- to π^+ produced, from 200 and 300 GeV π^-–p collisions in comparison to the same ratios from 200 GeV p–p collisions. Figure 6.16 also shows predictions from a CIM calculation [276] and a QCD calculation [277]. Although the CIM calculations fit the production of $\pi^\pm$ in p–p collisions, they "disagree with the pion-induced data by a large factor" [275] while the QCD calculations [277] agree in both cases. The same conclusion was reached for 200 GeV π^+–p scattering by a different experiment [274] where a CIM calculation [278] was shown "to disagree completely with the π^+–p result" [274]. Both these measurements also contradict the much later claim [279] that "high-p_T hadrons are produced by different mechanisms at fixed-target and collider energies. For pions, higher-twist subprocesses where the pion is produced directly dominate at fixed target energy..."

Another major result from Fermilab deserves mention here: the famous "Cronin effect" [280]. In deeply inelastic lepton scattering, where hard scattering was discovered, the cross section for DIS in lepton–nucleus collisions was found to be proportional to $A^{1.00}$, indicating that the partons were indeed point-like in electromagnetic scattering from a nucleus of A nucleons and did not shadow each other. In hadron collisions, the A dependence in p–A collisions was also found to be

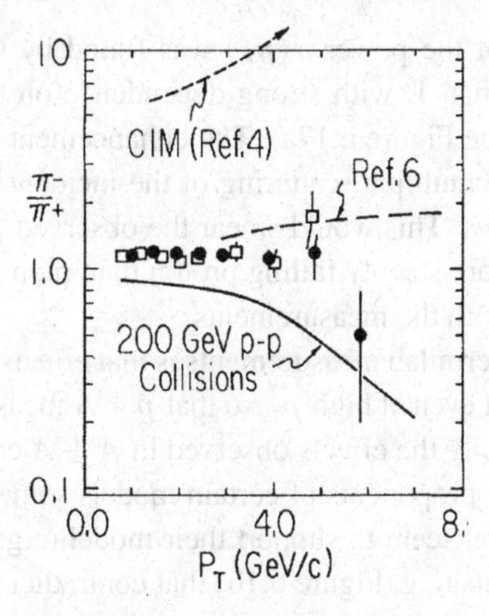

Figure 6.16 The ratio of π^-/π^+ produced in 200 GeV π^-–p collisions [275]. Also shown is the corresponding ratio in p–p collisions, and the theoretical predictions of the CIM (Ref.4) [276] and QCD (Ref.6) [277].

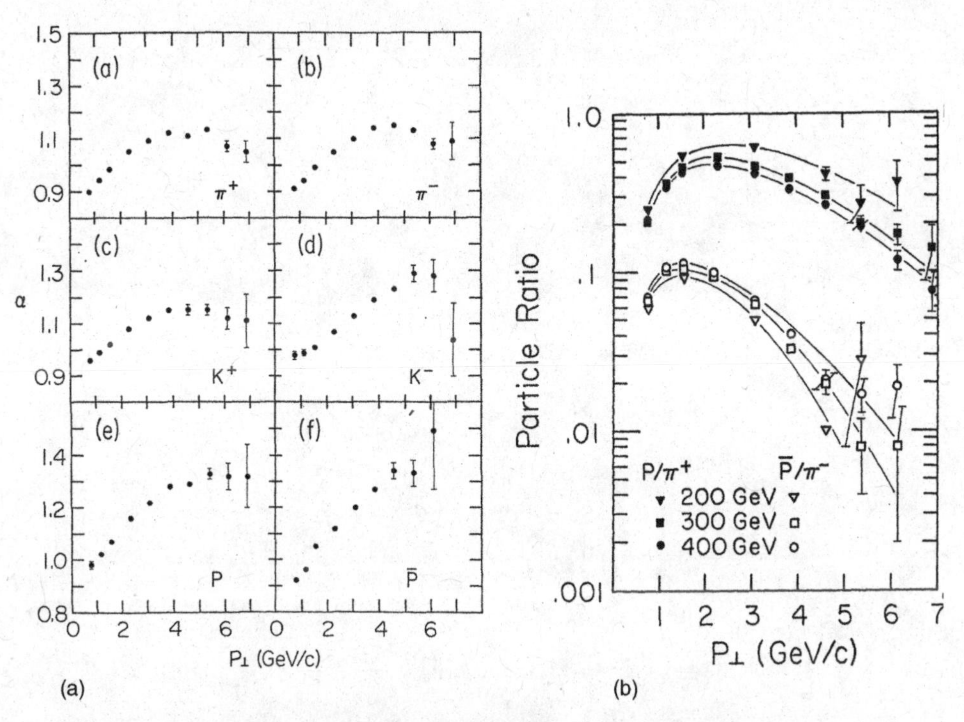

Figure 6.17 (a) The power α of the A dependence of the invariant cross sections versus $p_\perp$ for the production of identified hadrons $\pi^\pm$, $K^\pm$, $p^\pm$, as labeled (a)–(f), by 400 GeV protons [257]. (b) The particle ratios p/π versus $p_\perp$ for 200, 300 and 400 GeV p–p collisions [257].

a power-law, $A^{\alpha(p_T)}$, but the power $\alpha(p_T)$ was found by Cronin and collaborators [280] to be larger than 1, with strong dependence on the p_T and identity of the produced particle (see Figure 6.17a). The enhancement (relative to $A^{1.00}$) was thought to be due to the multiple scattering of the incident partons while passing through the target nucleus. This would smear the observed p_T spectrum, which is a larger effect for the more steeply falling proton than pion spectra (Figure 6.17b) in apparent agreement with the measurements.

The lesson from the Fermilab measurements is that effects in cold nuclear matter are tricky and significant even at high p_T so that $p + A$ measurements are required in order to understand fully the effects observed in $A + A$ collisions.

Another lesson is that proponents of certain models sometimes pay more attention to measurements that seem to support their model (e.g. Figure 6.15) and less attention to measurements (e.g. Figure 6.16) that contradict it.

7

Direct single lepton production and the discovery of charm

7.1 The CCRS experiment at the CERN-ISR

A very interesting thing happened in the second round of ISR experiments, two first round experiments decided to combine their detectors and join forces. The Saclay–Strasbourg experiment (R102) had proposed a second spectrometer arm (Arm 2) to study what was produced opposite in azimuth to balance the transverse momentum of the high p_T hadrons [281]. Then, together with the CCR experiment (R103), it was proposed [282] to add the CCR lead glass counters (PbGl) behind the Arm 2 to enable improved detection of single electrons and gamma rays at high p_T as well as to continue the search for e^+e^- pairs. This became R105, the CERN–Columbia–Rockefeller–Saclay experiment, CCRS (Figure 7.1). This detector turned out to be very powerful and was rewarded with one major discovery, one near-miss and several excellent measurements.

The key features of this detector were the following:

(i) $\geq 10^5$ charged hadron rejection from electron identification in the Cerenkov counter combined with matching the momentum and energy of an electron candidate in the magnetic spectrometer and the PbGl;

(ii) minimum of material in the aperture to avoid external conversions;

(iii) zero magnetic field on the axis to avoid de-correlating conversion pairs;

(iv) rejection of conversions in the vacuum pipe (and small opening angle internal conversions) by requiring single ionization in a hodoscope of scintillation counters H' close to the vacuum pipe, preceded by a thin track chamber to avoid conversions in the H' counters;

(v) precision measurement of π^0 and η, the predominant background source;

(vi) precision background determination in the direct single $e^\pm$ signal channel by adding an external converter, to distinguish direct single $e^\pm$ from $e^\pm$ from photon conversion.

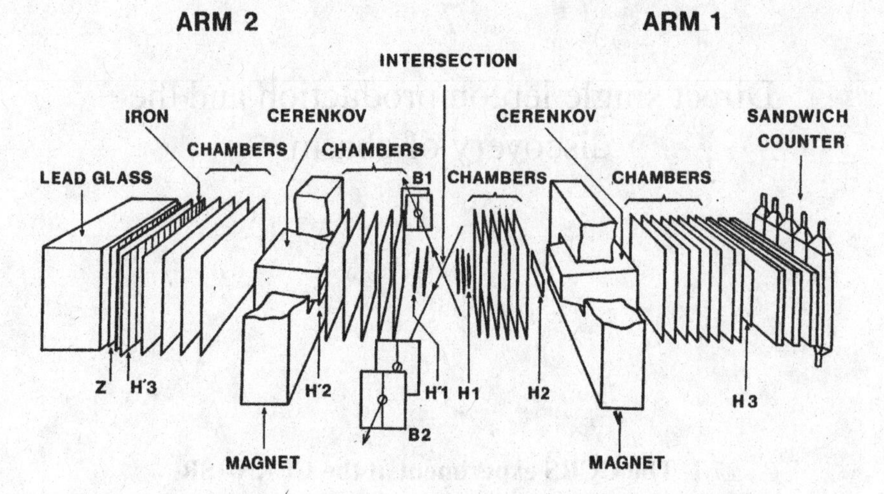

Figure 7.1 Isometric view of CCRS experiment R105 [282].

The solid angle acceptance of the two-arm spectrometer was $\Delta\Omega^* = 0.168$ sr for Arm 2, with $\Delta\phi^* = 7°$, and $\theta^* = 90 \pm 20°$. The c.m. system at the ISR moved towards Arm 2 and away from Arm 1 with velocity $\beta = 0.1285$ so that transverse momenta measured in Arm 2 were 13.8% above their c.m. value, while those in Arm 1 were 12.1% below their c.m. value. The Arm 1 acceptance was $\Delta\Omega^* = 0.068$ sr.

7.2 Experimental issues in direct single lepton production

7.2.1 Detecting electrons, photons, muons

Electron and photon detection are intimately connected and require an open geometry, which also allows the measurement of all hadrons whose decays constitute the major background to direct single $e^\pm$ production. Muons are identified by passage through a thick absorber, a closed geometry in which no other particles penetrate. In the CCRS experiment, the PbGl electromagnetic calorimeter measures the energy of γ and $e^\pm$ and reconstructs π^0 from two photons as previously discussed (Section 3.4.8). Electrons are identified principally by a count in the Cerenkov counter, with matching energy and momentum, p/E, used to clean up any residual hadron background. In Figure 7.2a [283], curve a shows single electron candidates with all cuts except the p/E matching as a function of p/E, while curve b shows the measured p/E response of the PbGl to charged hadrons and curve c the measured p/E response to electrons from photon conversions measured in situ. The residual hadron background to curve a is calculated by fitting to a linear combination of curves b and c which are normalized to the fitted values in the figure.

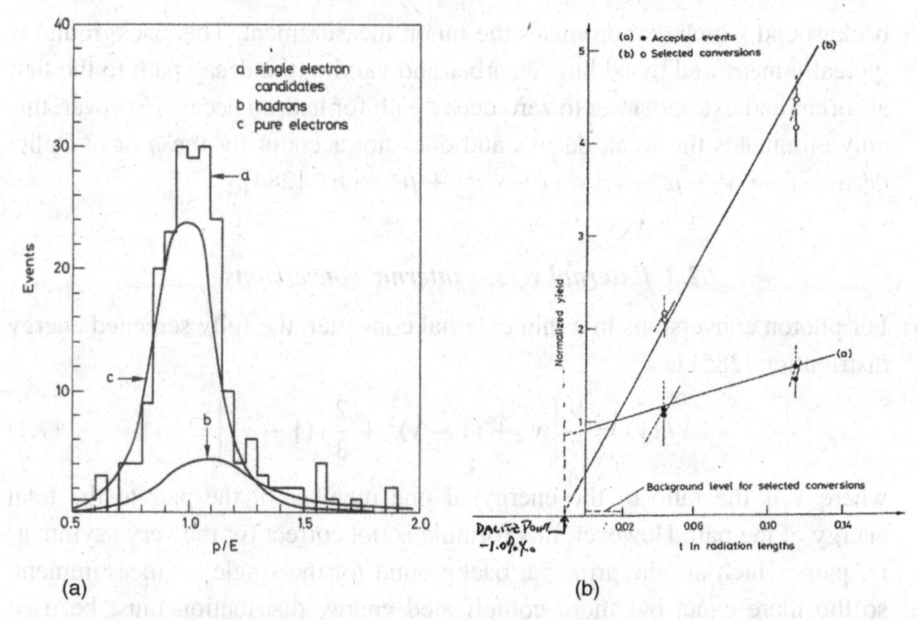

Figure 7.2 (a) p/E distribution for (b) single electron candidates (curve a), identified charged hadrons (curve b), and identified electrons (curve c), as described in the text. (b) CCRS [283] yield of inclusive electrons versus total external radiation length t/X_0 for accepted events with prompt electrons (non-zero intercept at the Dalitz point) (curve a) and selected conversions (extrapolates to zero at the Dalitz point) (curve b). The yields are both normalized to unity at $t/X_o = 0.016$, the normal thickness for data collection.

7.2.2 Background

The main experimental problem for detecting direct single $e^{\pm}$ is the fierce background from internal (Dalitz) and external conversions of photons from the decays $\pi^0 \to \gamma + \gamma$, $\eta \to \gamma + \gamma$, $\omega^0 \to \pi^0 + \gamma$. It is nearly impossible to calculate this background correctly to the accuracy required. Thus, the conversion and Dalitz pairs must be rejected as strongly as possible and any remaining background must be **measured** in order for experiments to give reliable results. Muons are detected by their penetration through meters of absorber, typically iron or beryllium (to minimize multiple scattering). This poses two problems:

(i) If the absorber is close to the interaction point, all other particles are absorbed, killing everything except muon measurements: this precludes direct measurement of any hadrons that may cause background from their decay to a muon or di-muon channel.

(ii) If the absorber is far enough from the interaction point to allow a hadron detector, then hadron decays in the open space (mostly $\pi, K \to \mu + X$) make a huge background which compromises the muon measurement. This background is typically measured by adding absorber and varying the decay path to the first absorber and extrapolating to zero decay path for hadron decay. However, this only eliminates the weak decays and does not account for the η or ω Dalitz decays, $\eta \to \gamma + \mu^+ + \mu^-$, $\omega^0 \to \pi^0 + \mu^+ + \mu^-$ [284].

7.2.3 External versus internal conversions

(i) For photon conversions in a thin external converter, the fully screened energy distribution [285] is

$$f(y) = \frac{9}{7}\left[y^2 + (1-y)^2 + \frac{2}{3}y(1-y)\right] \tag{7.1}$$

where y is the ratio of the energy of one member of the pair to the total energy of the pair. However, this formula is not correct for the very asymmetric pairs which are the principal background for the single $e^\pm$ measurement, so the more exact but more complicated energy distribution must be used (Figure 7.3) [286]. If packaged Monte Carlo simulations are used for these calculations, it is important to verify that the curves of Figure 7.3 are reproduced accurately.

(ii) For Dalitz decays [287–289], e.g. $\pi^0 \to \gamma + e^+ + e^-$:

$$\frac{1}{\Gamma_{\gamma\gamma}}\frac{d\Gamma_{ee\gamma}}{dm} = \frac{4\alpha}{3\pi}\frac{1}{m}(1-x)^3\left(1-\frac{x_n}{x}\right)^{1/2}\left(1+\frac{x_n}{2x}\right) \tag{7.2}$$

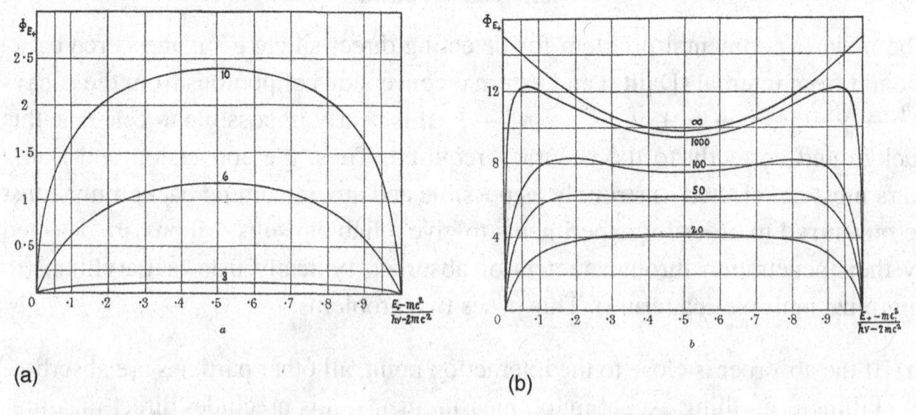

(a) (b)

Figure 7.3 Energy distribution [286] y of e^+e^- pairs from external conversion of a photon. Φ_{E_+} is the cross section (units $Z^2 r_0^2/137$) for the creation of a positron with kinetic energy $E_+ - mc^2$. The numbers on the curves refer to the energy of the photon in units of the positron mc^2 ($E/0.511$ MeV): (a) is valid for any element (screening neglected); (b) refers to Pb.

where m equals the mass of the e^+e^- pair, M is the mass of the π^0 or η, m_e is the mass of an electron, $x = m^2/M^2$, x_n is the minimum value of x (when $m = 2m_e$), and the maximum value of x is 1.0. The total Dalitz (internal conversion) probability is

$$\delta = \frac{\Gamma_{ee\gamma}}{\Gamma_{\gamma\gamma}} = \frac{2\alpha}{3\pi}\left(\ln\frac{M^2}{m_e^2} - \frac{7}{2}\right) = \begin{cases} 1.19\% & \text{for } \pi^0 \\ 1.62\% & \text{for } \eta, \end{cases} \tag{7.3}$$

and the decay angular distribution of the Dalitz pair in its rest frame is

$$\frac{d\mathcal{P}}{d\cos\theta} = \frac{3}{8}\,\frac{(1+\cos^2\theta + (x_n/x)\sin^2\theta)}{1 + x_n/2x}, \tag{7.4}$$

where θ is the polar angle of the Dalitz pair in its rest frame, with respect to the total momentum of the pair.

7.2.4 The converter method

Fortunately, for the case of direct single $e^\pm$, the $e^\pm$ from internal and external conversions of photons can be separated from the direct or prompt $e^\pm$ from non-photonic sources by measurement. The classical method to determine the conversion and Dalitz background from two photon decays of π^0 and η is to increase the external converter artificially (typically a very thin vacuum pipe) by adding material of a few percent of a radiation length (X_o). The probability of internal and external conversion per γ is

$$\frac{e^-|_\gamma}{\gamma} = \frac{e^+|_\gamma}{\gamma} = \frac{\delta_2}{2} + \frac{t}{\frac{9}{7}X_0} \equiv \delta_{eff} \tag{7.5}$$

where t/X_o, is the total external converter thickness in radiation lengths; $\delta_2/2$ is the Dalitz (internal conversion) branching ratio per photon, equal to 0.6% for $\pi^0 \to \gamma\gamma$, 0.8% for $\eta \to \gamma\gamma$; and the factor 9/7 comes from the ratio of conversion length to radiation length. For π^0 and η decays, the yield extrapolates to zero at the "Dalitz point" $-\frac{9}{14}\delta_2 \sim -1\%$, actually -0.8% for π^0 and -1.0% for η, so the method has the added advantage that it depends very little on the η/π^0 ratio. In Figure 7.2b, the extrapolation for the signal [283], curve a where Dalitz and conversions are rejected, shows only a small decrease from the normal thickness used for data collecting ($t/X_o = 1.6\%$) to the Dalitz point, while curve b where conversions are selected (instead of being rejected) extrapolates nicely to the Dalitz point, indicating a photonic source.

7.2.5 Effect of the falling spectrum

Still using the p_T^{-n} power-law for the π^0 and thus the decay γ spectra (Eq. 3.141) one finds:

$$\left.\frac{e^-}{\pi^0}\right|_{\pi^0}(p_T) = \left.\frac{(e^- + e^+)}{2\pi^0}\right|_{\pi^0}(p_T) = \delta_{eff} \times \frac{2}{(n-1)^2} \tag{7.6}$$

which for $n = 11$ at ISR energies gives $e^-/\pi^0|_{\pi^0} > 1.2\%/10^2 = 1.2 \times 10^{-4}$. Thus one needed $\sim 10^4$ rejection against $\pi^\pm$ at the ISR just to be able to see the $e^\pm$ background from π^0 Dalitz; while at RHIC, where $n = 8$, the numbers are $e^-/\pi^0|_{\pi^0} > 1.2\%/7^2 = 2.4 \times 10^{-4}$, or $\sim 5 \times 10^3$ rejection.

7.3 The discovery of direct single lepton production

At the ICHEP in London in July 1974, results from six experiments were presented on "the observation of leptons at large transverse momentum" [290]: four in the $\mu^\pm$ channel including E100 from Fermilab [291], one in the $e^\pm$ channel (CCRS) [292, 293] and one in both $e^\pm$ and $\mu^\pm$ channels, Lederman's E70 [294]. The results were presented in the form of the lepton/π ratio, for example $e/\pi \equiv (e^+ + e^-)/(\pi^+ + \pi^-)$, which was roughly independent of p_T in the ranges measured. The ratios reported were $\mu/\pi = 2.5 \times 10^{-5}$ at $\sqrt{s} = 12$ GeV at Serpukhov [295], $e, \mu/\pi \sim (0.8 - 1) \times 10^{-4}$ at $\sqrt{s} = 24.5$ GeV and $e/\pi = (1.2 \pm 0.2) \times 10^{-4}$ at $\sqrt{s} = 52.7$ GeV [293]. It is important to emphasize that these discoveries of direct single leptons were made before the J/Ψ or charm particles were known. Also it was shown by CCRS [293], who made a specific search for $\phi \rightarrow K^+ + K^-$, that the possible $\phi \rightarrow e^+ + e^-$ contribution was less than 1/2 to 1/4 of the observed $e^\pm$ signal.

Many further measurements of direct single leptons were made following up the initial discovery [283, 296]. The most comprehensive of these measurements was by CCRS [283] who found that direct single $e^\pm$ in the range $1.3 < p_T^* < 3.2$ GeV/c was roughly constant at a level of 10^{-4} of charged pion production at all five c.m. energies measured (Figure 7.4) [283]. Actually, in detail, a rise of e/π with $\sqrt{s}$ was indicated by the CCRS data, where the e/π ratio varied systematically by a factor of ~ 1.8 from $\sqrt{s} = 30$ to 60 GeV [283] (see Figure 7.5).

7.4 The direct single electrons are the first observation of charm

It took about two years, from 1974–1976 (which included the discovery of the J/Ψ) to understand that the direct single electrons were due to the semi-leptonic decay of charm particles. The first interpretation of the CCRS measurements was by Farrar and Frautschi [298] who proposed that the direct single $e^\pm$ were due to the internal conversion of direct photons with a ratio $\gamma/\pi^0 \sim 10$–20%. CCRS was

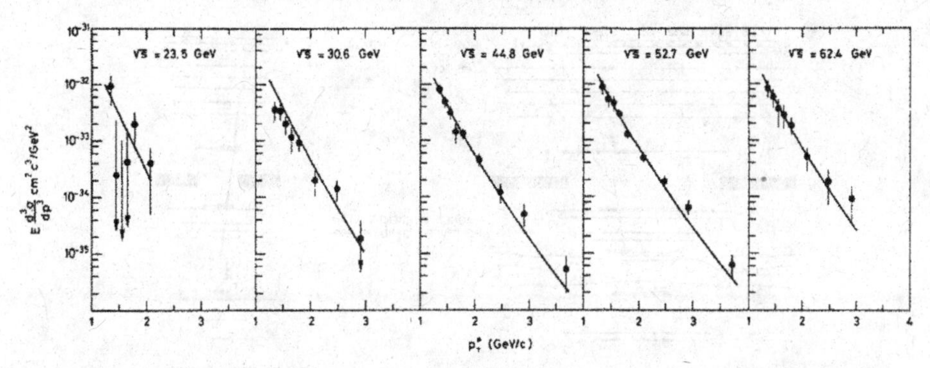

Figure 7.4 Invariant cross sections at mid-rapidity for five values of $\sqrt{s}$ in p–p collisions at the CERN-ISR: $(e^+ + e^-)/2$ (points) [283]. Lines represent a fit of the $(\pi^+ + \pi^-)/2$ data [297], multiplied by 10^{-4}.

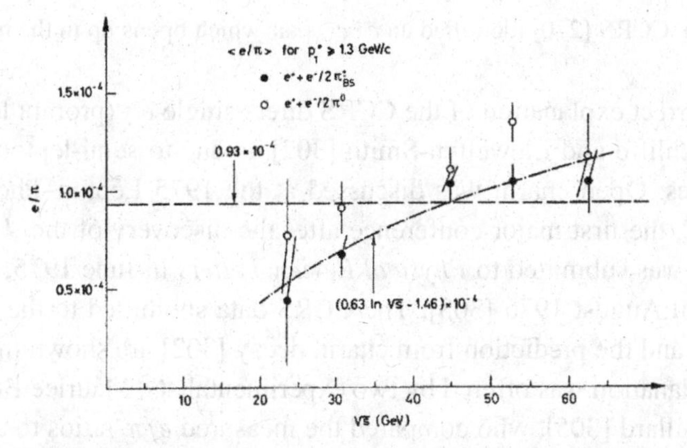

Figure 7.5 The ratio e/π for $p_T \geq 1.3$ GeV/c as a function of c.m. energy $\sqrt{s}$ from CCRS [283]. The pion data were obtained from the work of the British Scandinavan collaboration (solid points) and from the spectrum of selected conversions measured in this experiment [283]. The two curves shown are fits to the solid points.

able to detect cleanly (and reject) both external and internal conversions since there was zero magnetic field on the axis (Figure 7.6) [283]. Based on the observation of only two same-side e^+e^- pairs with $m_{ee} \geq 0.3$ GeV/c^2, for a sample of 2806 direct single $e^\pm$ candidates, CCRS was able to set limits excluding this explanation [283]: an internal conversion spectrum of the form $d\mathcal{P}/dm \propto 1/m$ could not account for more than 6.4% of the observed signal with 95% confidence. Although rejected by the measurements, this suggestion by Farrar and Frautchi [298], among others [299, 300], initiated the study of direct photons in hadron collisions well before the advent of QCD and the famous "inverse QCD Compton effect" [301], $g + q \rightarrow \gamma + q$ (see Chapter 10).

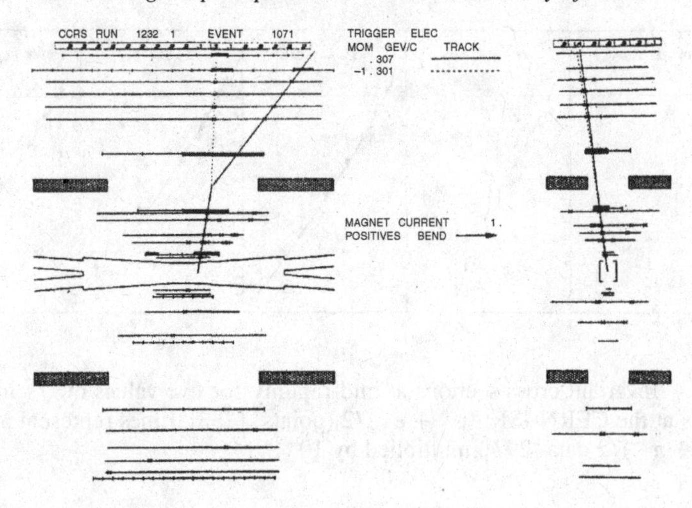

Figure 7.6 CCRS [270] identified an e^+e^- pair which opens up in the magnet.

The first correct explanation of the CCRS direct single $e^\pm$ (prompt leptons) was given by Hinchliffe and Llewellyn-Smith [302] as due to semi-leptonic decay of charm particles. Open charm was discussed at the 1975 Lepton–Photon conference at SLAC, the first major conference after the discovery of the J/Ψ and Ψ', and the paper was submitted to *Physical Review Letters* in June 1975, but was not published until August 1976 [303]. The CCRS data submitted to the SLAC conference [304] and the prediction from charm decay [302] are shown in Figure 7.7. A similar explanation was offered by two experimentalists, Maurice Bourquin and Jean-Marc Gaillard [305], who compared the measured e/π ratios to a cocktail of all known leptonic decays including *"Possible contributions from the conjectured charm meson..."* (Figure 7.8). The usage of "conjectured" charm is notable due to the delayed publication of reference [303]. Another notable point about both papers [302, 305] is that neither could fit the data points for $p_T < 1$ GeV/c at 30° from the CHORMN measurement [306] (Figure 7.9).

The important point here is that many experiments not designed for the purpose wanted to get into the prompt-lepton act, resulting in questionable data, and this is only one example, see reference [305] for more details. This led to the unfortunate situation as exemplified in Figure 7.10 from a typical paper [307] about charm circa 1990 which did not even give citations for the many disagreeing ISR publications of the direct single $e^\pm$ from charm, which were essentially ridiculed or, in the case of the original CCRS discovery [283], ignored [309] because it was published before either charm or the J/Ψ were discovered so there is no reference to the word "charm" in the publication. A fairer comparison of the ISR and fixed target measurements [310] of open charm is given in Figure 7.11 from the very much later first measurement of direct single $e^\pm$ at RHIC by PHENIX [308].

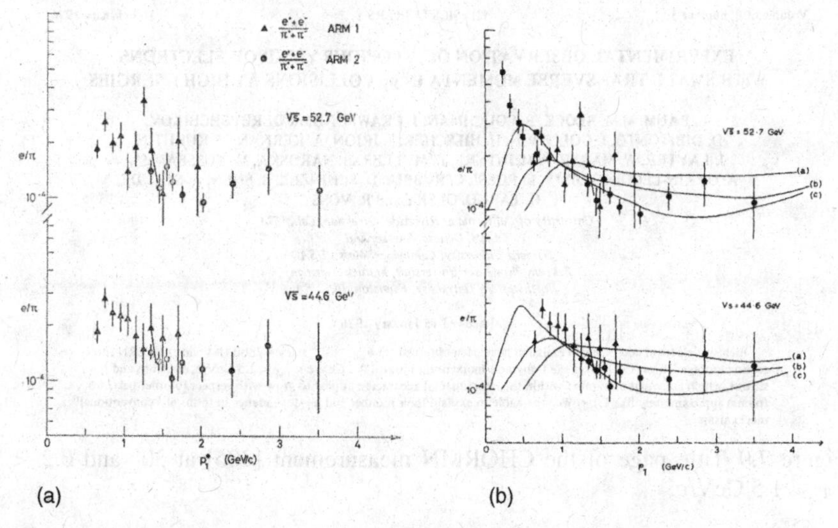

Figure 7.7 (a) CCRS e/π in two spectrometer arms at two values of $\sqrt{s}$ [304]; (b) same data with prediction, see reference [302] for details.

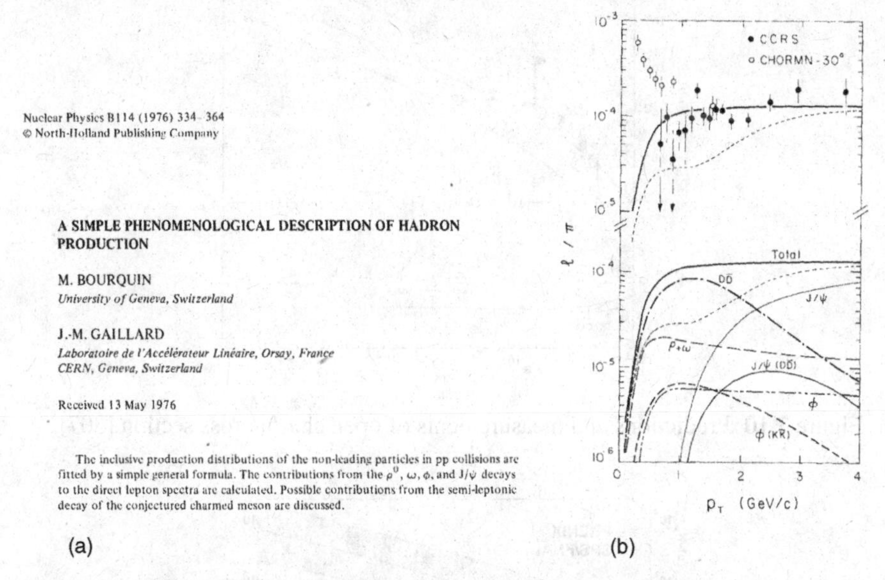

Nuclear Physics B114 (1976) 334–364
© North-Holland Publishing Company

A SIMPLE PHENOMENOLOGICAL DESCRIPTION OF HADRON PRODUCTION

M. BOURQUIN
University of Geneva, Switzerland

J.-M. GAILLARD
Laboratoire de l'Accélérateur Linéaire, Orsay, France
CERN, Geneva, Switzerland

Received 13 May 1976

The inclusive production distributions of the non-leading particles in pp collisions are fitted by a simple general formula. The contributions from the ρ^0, ω, ϕ, and J/ψ decays to the direct lepton spectra are calculated. Possible contributions from the semi-leptonic decay of the conjectured charmed meson are discussed.

Figure 7.8 (a) Bourquin–Gaillard title page [305]; (b) contribution of a cocktail of decays to the e/π ratio.

The "Basile" measurement mentioned in Figure 7.11 is another example of a second (more likely third) generation experiment at the CERN-ISR, this time using the split field magnet (Figure 7.12) [311]. It is well instrumented with many Cerenkov counters and includes a dE/dx chamber next to the vacuum pipe to veto on conversions. Although clearly stated in the introduction and conclusion

Volume 60B, number 5 PHYSICS LETTERS 16 February 1976

EXPERIMENTAL OBSERVATION OF A COPIOUS YIELD OF ELECTRONS WITH SMALL TRANSVERSE MOMENTA IN pp COLLISIONS AT HIGH ENERGIES

L. BAUM, M.M. BLOCK, B. COUCHMAN, J. CRAWFORD, A. DEREVSHCHIKOV, D. DIBITONTO, I. GOLUTVIN, H. HILSCHER, J. IRION, A. KERNAN, V. KUKHTIN, J. LAYTER, W. MARSH, P. McINTYRE, F. MULLER, B. NAROSKA, M. NUSSBAUM*, A. ORKIN-LECOURTOIS**, L. ROSSI, C. RUBBIA, D. SCHINZEL, B. SHEN, A. STAUDE, G. TARNOPOLSKY and R. VOSS

University of California at Riverside, Riverside, Cal., USA
CERN, Geneva, Switzerland
Harvard University, Cambridge, Mass., USA
Sektion Physik der Universität, Munich, Germany
Northwestern University, Evanston, Ill., USA

Received 15 January 1976

Inclusive electron and positron emission have been observed for $\theta_{cm} = 30°$ and $s = 2800$ GeV^2 at the CERN Intersecting Storage Rings (ISR). Over the transverse momentum interval 0.2 GeV/c < p_T < 1.5 GeV/c, electrons and positrons, which are equal in number within the experimental accuracies, appear to grow with respect to other particles (pions) approximately like $1/p_T$. We are unable to explain their number and p_T-dependence in terms of "conventional" mechanisms.

Figure 7.9 Title page of the CHORMN measurement [306] at 30° and $0.2 < p_T < 1.5$ GeV/c.

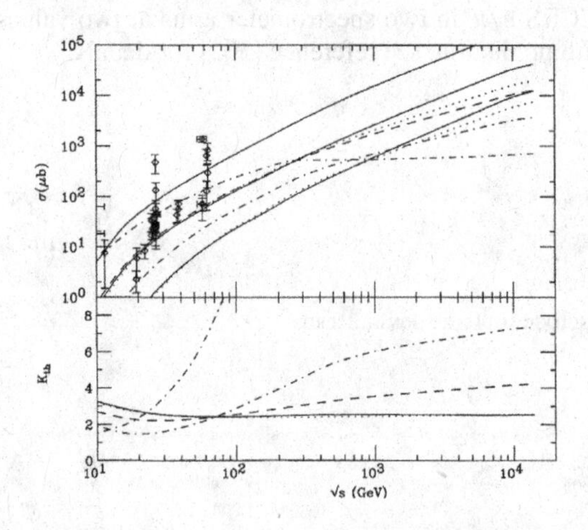

Figure 7.10 Predictions and measurements of open charm cross section [307].

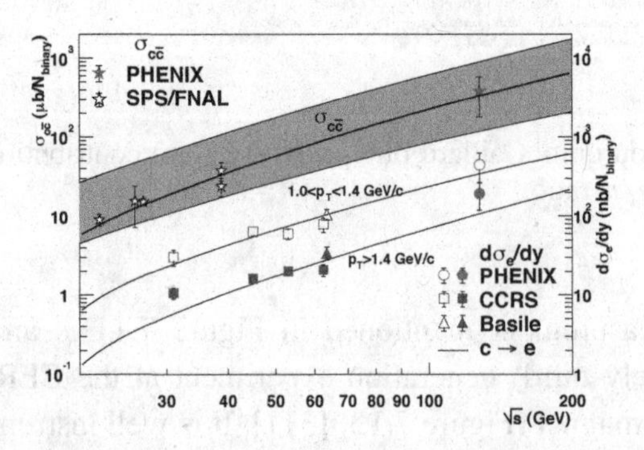

Figure 7.11 Predictions and measurements [308] of charm cross section $\sigma_{c\bar{c}}$.

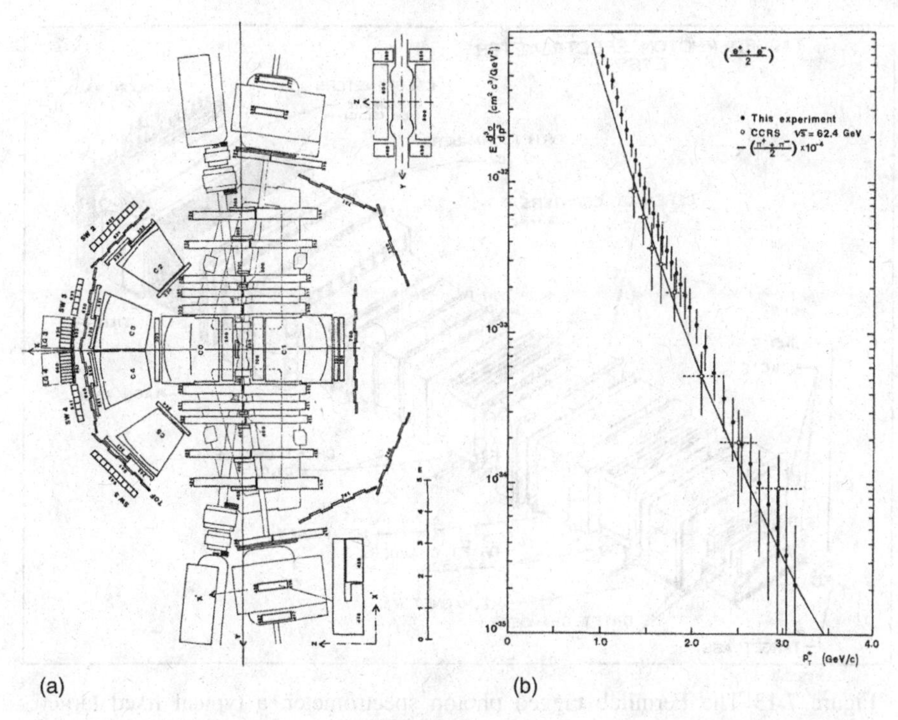

Figure 7.12 (a) Top view of the SFM detector, showing the MWPCs and the external apparatus for particle identification [311]: (i) gas threshold Cerenkov counters, C0 to C5; (ii) lead/scintillator sandwiches, SW2–SW5; (iii) lead glass arrays, LG3 and LG4; (iv) a time-of-flight system, TOF; (v) the dE/dx chamber labeled "209." (b) Invariant cross section of prompt $(e^+ + e^-)/2$ as a function of p_T^* in "this experiment" [311] (filled circles) compared to the CCRS measurement [283] (open circles). The solid line is the fit [312] to the charged averaged pion cross section at 90° multiplied by 10^{-4}.

that "the quantity e/π is strictly correlated, especially in the high-p_T range, with the semi-leptonic decay of the heavy flavours produced at the ISR," there is no mention of "charm" in either the title or abstract of this 1981 publication. In fact the first mention of the word "charm" (actually "charmed") in an experimental publication from the CERN-ISR was by the CCHK collaboration [313] who found "the first generally accepted evidence for charmed particle production in hadron–hadron interactions" [309], the observation of the charmed $D^+ \rightarrow K^+\pi^+\pi^-$, using the SFM in the forward direction (low p_T) for which it was originally designed.

The fixed target measurements, which like to claim the discovery of hadroproduction of charm, all used silicon vertex detectors, which is the first tiny element (SMD), actually an upgrade, in an otherwise giant experiment [314] (Figure 7.13). At least, Jeff Appel in his review article [310] did acknowledge the much earlier prompt lepton results:

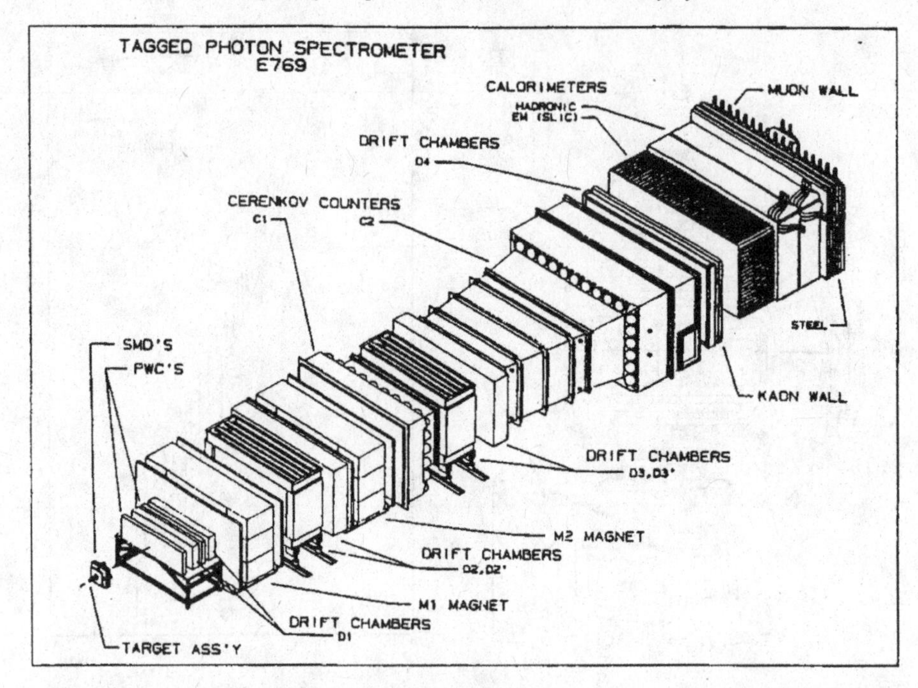

Figure 7.13 The Fermilab tagged photon spectrometer, a typical fixed target multi-particle apparatus [314]. Note the tiny SMD.

The early years of open charm hadroproduction were limited by the capability of detectors in the face of difficult experimental conditions. Among the difficulties (still faced today) are (a) the small fractional charm production cross section (one $c\bar{c}$ pair event per 103 interactions, typically), (b) the high multiplicity of particles in the charm events, and (c) the small branching ratios to specific final states (typically 1–10%). As it turned out, many of the more reliable early measurements were indirect. Among these were the observations of prompt leptons resulting from the semileptonic decays of charm particles. Most of these early experiments had goals other than charm production and decay as their primary motivation. Nevertheless, leptons with intermediate transverse momentum have been interpreted to come from charm decay. Electrons and muons were observed at rates of 10^{-4} to 10^{-3} of the charged pion at fixed-target and collider energies. Muons and neutrinos were also measured in beam dump experiments. The physics results, charm cross sections times average branching ratios, were extrapolated from total observed lepton rates under varied experimental conditions. These observed rates typically included much larger numbers of leptons from photon conversions or decays of particles containing strange, not charm, quarks.

7.4.1 1977 era controversies

To be fair, the direct single $e^{\pm}$ measurements as observations of semi-leptonic decays of open charm were not universally accepted by the experimental community in the late 1970s; and there were heavy hitters on both sides of the argument. First of all, after the discovery of the J/Ψ in November 1974 and, in particular, with the near miss of this discovery at the CERN-ISR, there was concern that the $J/\Psi \rightarrow e^+ + e^-$ decay could be the source of the single $e^{\pm}$. Fortunately, CCRS quickly demonstrated that the J/Ψ was not the source of the single $e^{\pm}$ [283] (see Chapter 8).

Sam Ting, in his Nobel Lecture [315] (Figure 7.14) also noted that the J meson could not explain the prompt leptons and indicated that he actually delayed announcing the J discovery at Viki Weisskopf's retirement ceremony in mid-October 1974 in order to investigate the prompt leptons in his AGS experiment.

On the other side of this argument was Jim Cronin, another Nobel Laureate (but incorrect on this issue), who prominently claimed in his plenary talk at the 1977 Lepton Photon Symposium in Hamburg [316] that "The origin of direct single leptons is principally due to the production of lepton pairs" (Figure 7.15). Also, as one can see from the discussion, a young (at the time) Associate Professor from the Rockefeller University vehemently disagreed with Cronin's explanation.

The lesson from all direct single electron measurements, past and present, is that "people who try to measure prompt leptons become the world's experts on η Dalitz decay" [317].[1]

[1] For instance, compare reference [318] to reference [319] to reference [320].

V. I was considering announcing our results during the retirement ceremony for V. F. Weisskopf, who had helped us a great deal during the course of many of our experiments. This ceremony was to be held on 17 and 18 October 1974. I postponed the announcement. for two reasons. First, there were speculations on high mass e^+e^-pair production from proton-proton collisions as coming from a two-step process : $p+N\rightarrow\pi+\ldots$, where the pion undergoes a second collision $\pi+N\rightarrow e^++e^-+\ldots$. This could be checked by a measurement based on target thickness. The yield from a two-step process would

S.C.C.Ting 335

increase quadratically with target thickness, whereas for a one-step process the yield increases linearly. This was quickly done, as described in point (iv) above.

Most important, we realized that there were earlier Brookhaven measurements [24] of direct production of muons and pions in nucleon-nucleon collisions which gave the μ/π ratio as 10^{-4}, a mysterious ratio that seemed not to change from 2000 GeV at the ISR down to 30 GeV. This value was an order of magnitude larger than theoretically expected in terms of the three known vector mesons, p, ω, φ, which at that time were the only possible "intermediaries" between the strong and electromagnetic interactions. We then added the J meson to the three and found that the linear combination of the four vector mesons could not explain the μ^-/π^- ratio either. This I took as an indication that something exciting might be just around the corner, so I decided that we should make a direct measurement of this number. Since we could not measure the μ/π ratio with our spectrometer, we decided to look into the possibility of investigating the e^-/x^- ratio.

We began various test runs to understand the problems involved in doing the e/π experiment. The most important tests were runs of different e^- momenta as a function of incident proton intensities to check the single-arm backgrounds and the data-recording capability of the computer.

On Thursday, 7 November, we made a major change in the spectrometer (see Fig. 13) to start the new experiment to search for more particles. We began by measuring the mysterious e/π ourselves. We changed the electronic logic and the target, and reduced the incident proton beam intensity by almost two orders of magnitude. To identify the e^-background due to the decay of π^0 mesons, we inserted thin aluminium converters in front of the spectrometer to increase the $\gamma\rightarrow e^++e^-$ conversion. This, together with the C_B counter which measures the $\pi\rightarrow\gamma+e^++e^-$ directly, enabled us to control the major e^- background contribution.

Figure 7.14 Excerpt from Ting's Nobel Lecture [315] © The Nobel Foundation 1976.

HADRON INDUCED LEPTONS AND PHOTONS

James W. Cronin
Enrico Fermi Institute, University of Chicago
Chicago, Illinois, 60637 U.S.A.

ABSTRACT

A review of direct production of leptons and photons in hadron-hadron collisions is presented. Production of lepton pairs with large mass is well accounted for by the Drell-Yan process. The origin of direct single leptons is principally due to the production of lepton pairs. A dominant source of lepton pairs is at low effective mass, m < 600 MeV/c.

III. SINGLE LEPTON PRODUCTION

Since 1974 direct single lepton production has been the subject of intensive experimental investigation. The important qualitative observation has been that direct electron or muon production for $p_\perp$ > 1 GeV/c and large c.m. angle is approximately 10^{-4} of pion production. We shall argue that a very large fraction of the observed single muons have their source as one member of a lepton pair. These sources are best studied in experiments in which both members of the pair can be observed, or if a neutrino accompanies the lepton that the neutrino is detected indirectly through an absence of total energy balance. We believe that in the future, experiments which observe only a single lepton are going to yield little new information. We have come to understand the single lepton problem only through the study of lepton pairs.

V. CONCLUSIONS

1. The Drell-Yan mechanism provides a qualitative and nearly quantitative description for lepton pair production with mass > 4 GeV/c^2.

2. Direct single lepton production is consistent with a source which is lepton pair production. An important ingredient in this discussion is the existence of a significant low mass dilepton continuum which is an order of magnitude larger than expected from the Drell-Yan process.

3. A dedicated experiment will be required to establish definitively a significant direct photon yield induced by hadrons.

M.J. Tannenbaum, Rockefeller University: With reference to explaining ISR single electrons from Fermilab data at x_F > .15: First of all, the CCRS experiment made a great effort to eliminate low mass electron pairs (M_{ee} < 0.50 GeV/c^2) but allowed a continuum of $d\sigma/dm$ 1/m or flatter to explain all the single electrons observed; secondly, the crucial parameters in the lepton pair cross section relevant to explaining the ISR(CCRS) single electrons are the width of the rapidity plateau and $d^2\sigma/dmdy|_{y=0}$, the value of the cross section on the plateau. The Fermilab measurements at x_F > .15 do not give these values and thus are irrelevant in trying to explain the ISR data. Thus, while it might seem reasonable that the dilepton continuum could explain all of the single electron signal, it is by no means proved.

J.W. Cronin: While technically it is true that the measurements of the CP(II) group do not correspond to the kinematic region observed in the CCRS experiment (Ref. 24), we do have some experience with continuity. The extrapolation of the CP(II) group was based on a constant cross section in rapidity between x_F = 0.15 and x_F = o. I insist on my main point, however, that in fact we learn little from the single lepton measurements alone. To "prove" that pairs explain the CCRS signal, pair measurements will have to be made at $x_F \sim$ o.

Figure 7.15 Excerpt from Cronin's talk and discussion at Lepton–Photon 1977 [316].

8

J/Ψ, Υ and Drell–Yan pair production

8.1 Di-lepton production in the parton model: Drell–Yan pairs

Looking to test Bjorken scaling and the parton model in a reaction other than DIS, Drell and Yan in 1970 [54], evidently influenced by preliminary releases of Lederman's di-muon measurement [321], proposed that in the limit $s \to \infty$ with m^2/s finite, the reaction .

$$p + p \to (\mu^+ + \mu^-) + X$$

would be produced by the partonic reaction shown in Figure 8.1,

$$\overline{q} + q \to \mu^+ + \mu^-$$

and would follow the scaling law:

$$\frac{d\sigma}{dm^2_{\mu\mu}} = \left(\frac{4\pi\alpha^2}{3m^2_{\mu\mu}}\right)\left(\frac{1}{m^2_{\mu\mu}}\right)\mathcal{F}(\tau) \tag{8.1}$$

where $\tau \equiv m^2_{\mu\mu}/s$, and the factor $4\pi\alpha^2/3m^2_{\mu\mu}$ is just the total cross section for e^+e^- annihilation into muon pairs in the relativistic limit. Also since the lepton pair is from a virtual photon, its polar angular distribution in the $\overline{q}$–q c.m. system is

$$\frac{dP}{d\cos\theta^*} = \frac{3}{8}(1 + \cos^2\theta^*). \tag{8.2}$$

As noted previously (Section 5.2) Leon Lederman was very excited about the Drell–Yan scaling for his $\sqrt{s} = 7.4$ GeV di-muon spectrum [50] (Figure 5.7) because this enabled him to calculate the cross section for W production at any $\sqrt{s}$, the main objective of his single-lepton and di-lepton proposals at Fermilab [238] and at CERN [239]. However, the development of di-lepton physics, and in fact of the whole of high energy particle physics, from 1970 until the W was discovered in 1983 [227, 228], proved to be revolutionary rather than evolutionary.

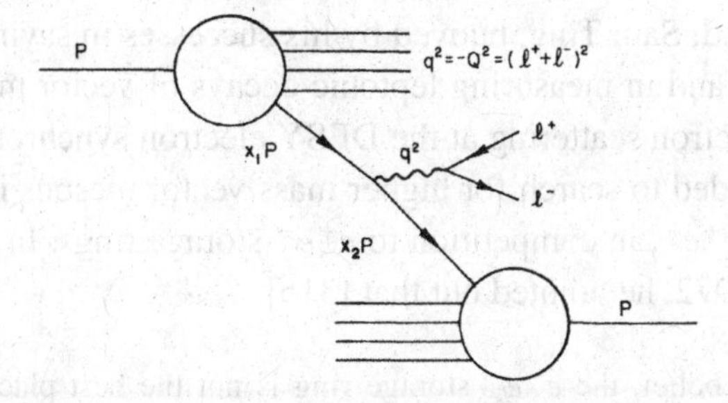

Figure 8.1 Kinematics of Drell–Yan production of a lepton pair in p–p collisions [322].

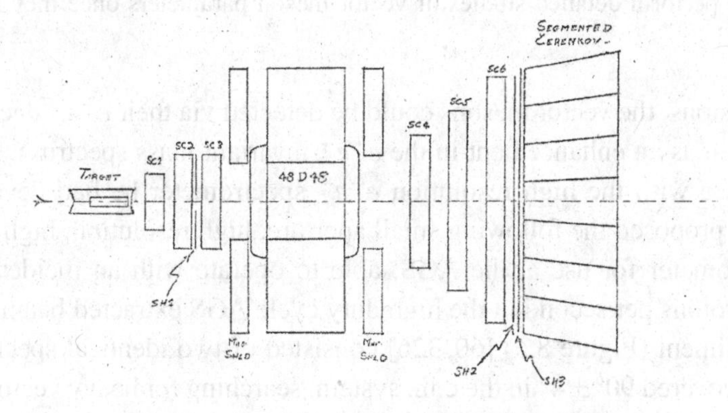

Figure 8.2 Proposed layout of P549 [323].

8.2 J/Ψ production

First, let us discuss an interesting experiment that did not happen. As also noted in Section 5.2, there were discussions within Lederman's group about the meaning of the shoulder in the data. Peter Limon and a few other young colleagues actually proposed in September 1970 [323] a followup experiment at the AGS, "a high resolution study of the production of electron pairs in hadronic interactions covering the mass range $0.7 \leq M_{e^+e^-} \leq 4.5$ GeV/c^2," to check this (Figure 8.2). The experiment used the 48D48 magnet from an existing spectrometer, plus spark chambers (SC) a segmented Cerenkov counter and Scintillator Hodoscopes (SH) for triggering. The request was to run in a 25 GeV/c positive beam (mostly protons) and a 20 GeV/c negative beam, mostly π^-. With (or maybe even without) an added EM calorimeter, for triggering and improved $\pi^\pm$ rejection, the proposed experiment probably would have discovered the J/Ψ. However, with Fermilab and the ISR starting up and with an AGS program of DIS of muons underway, the potential collaborators were all too busy to follow through!!

On the other hand, Sam Ting, buoyed by his successes in saving quantum electrodynamics [324] and in measuring leptonic decays of vector mesons [325], e.g. $\rho^0 \to e^+e^-$, in electron scattering at the DESY electron synchrotron in Hamburg, Germany, had decided to search for higher mass vector mesons in p–p collisions via their decay to e^+e^-, in competition to e^+e^- storage rings. In proposal 598 for the AGS in early 1972, he pointed out that [315],

> Contrary to popular belief, the e^+e^- storage ring is not the best place to look for vector mesons. In the e^+e^- storage ring, the energy is well-defined. A systematic search for heavier mesons requires a continuous variation and monitoring of the energy of the two colliding beams – a difficult task requiring almost infinite machine time. Storage ring is best suited to perform detailed studies of vector meson parameters once they have been found.

In p–p collisions, the vector mesons could be detected via their e^+e^- decay which would be seen as an enhancement in the e^+e^- invariant mass spectrum. Based on the experience with the high resolution e^+e^- spectrometer he had developed at DESY, Ting proposed the following small aperture, high resolution, high rejection e^+e^- spectrometer for use at the AGS, able to operate with an incident flux of 10^{11}–10^{12} protons per second in the high duty cycle AGS extracted beam.

The experiment (Figure 8.3) [60, 326] consisted of two identical spectrometers which each covered $90° \pm 4°$ in the c.m. system, searching for heavy vector mesons produced at rest in the c.m. system, decaying to symmetric e^+e^- pairs. M_0, M_1 and M_2 are dipole magnets. Bending is vertical to decouple the angle θ and momentum. C_B, C_0 and C_e are gas threshold Cerenkov counters. A_0, A, B and C are proportional wire chambers. Behind chambers A and B are situated two planes of hodoscopes for improving timing resolution. Behind the C chamber are two orthogonal banks of PbGl counters, $3X_o$ thick, followed by one horizontal bank of lead lucite shower counters, each $10X_o$ thick. The Cerenkov counters C_0 and C_e are filled with hydrogen with thin mylar windows to reduce knock-on electrons, multiple scattering and photon conversions. The Cerenkov counter C_B, is placed close to the target below a specially constructed magnet M_0, to detect the soft electrons from asymmetric conversions (from $\pi^0 \to \gamma e^+e^-$) in coincidence with the other electron detected in the spectrometer, and such events are rejected. Conversely one could accept these events as the source of a "pure electron beam" to calibrate the spectrometers. The Cerenkov counters C_0 and C_e are decoupled by magnets to avoid C_e triggering on knock-on electrons from C_0. The e/π rejection in each arm was $\gg 10^4$ for a $\gg 10^8$ rejection against hadron pairs. The experiment took place in an external beam of 28.5 GeV (kinetic energy) protons [327] with intensity $10^{10} - 2 \times 10^{12}$ per pulse which struck an extended target composed of nine pieces of beryllium 1.8 mm thick, each separated by 7.5 cm, so that particles created in one piece and accepted

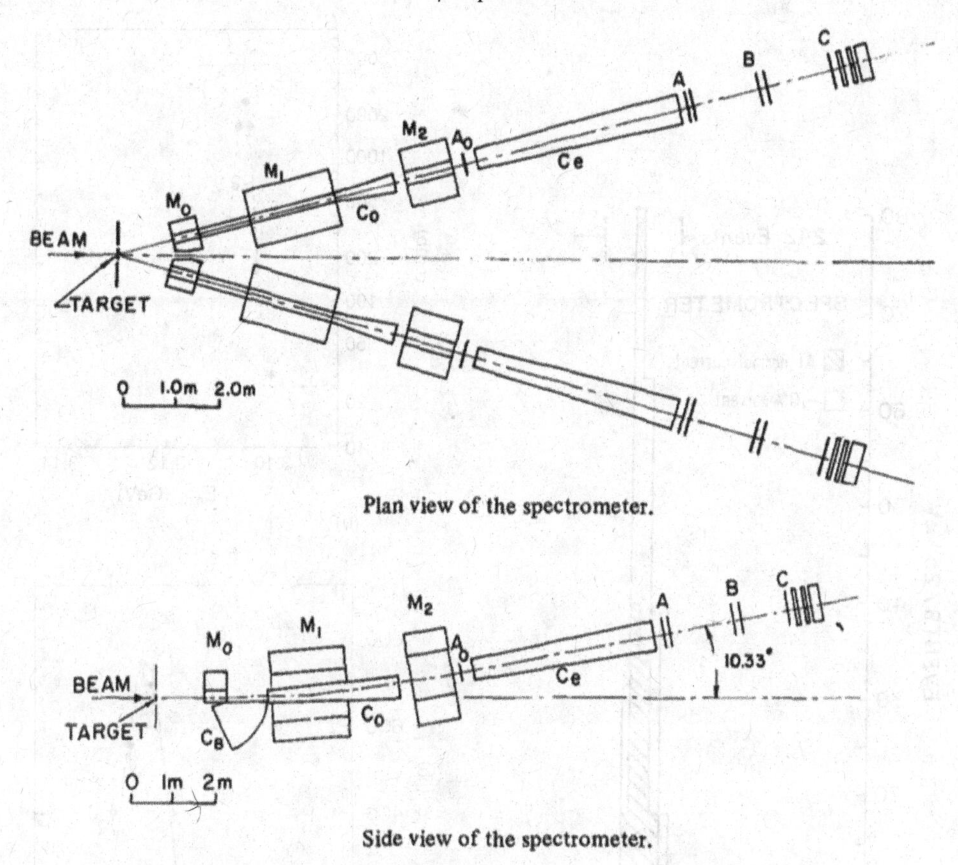

Figure 8.3 Plan (top) and side (bottom) views of the spectrometer [326].

by the spectrometer do not pass through the next piece. This reduces both multiple scattering and accidental pairs from different pieces of the target.

Many experimental checks and calibrations were performed [60,326] before the discovery "of a heavy particle J with mass $m = 3.1$ GeV/c^2 and width approximately zero" (Figure 8.4) was announced on November 12, 1974, incredibly, simultaneously with the observation of a narrow resonance in e^+e^- collisions at the same mass, at SLAC [61], which they called the ψ. There was some suspicion that the mass of the J may have been leaked to SLAC by Mel Schwartz, at the time a Professor at Stanford University, or his graduate students, who were running at the AGS just next to Ting. However, all doubts were dispelled with the discovery of a second narrow resonance at SLAC, just 12 days later, the ψ' with mass ~600 MeV/c^2 higher than the J/ψ [62].

As previously noted (Section 5.3), this discovery revolutionized high energy physics since it was a heavy vector meson with a very narrow width (it decayed

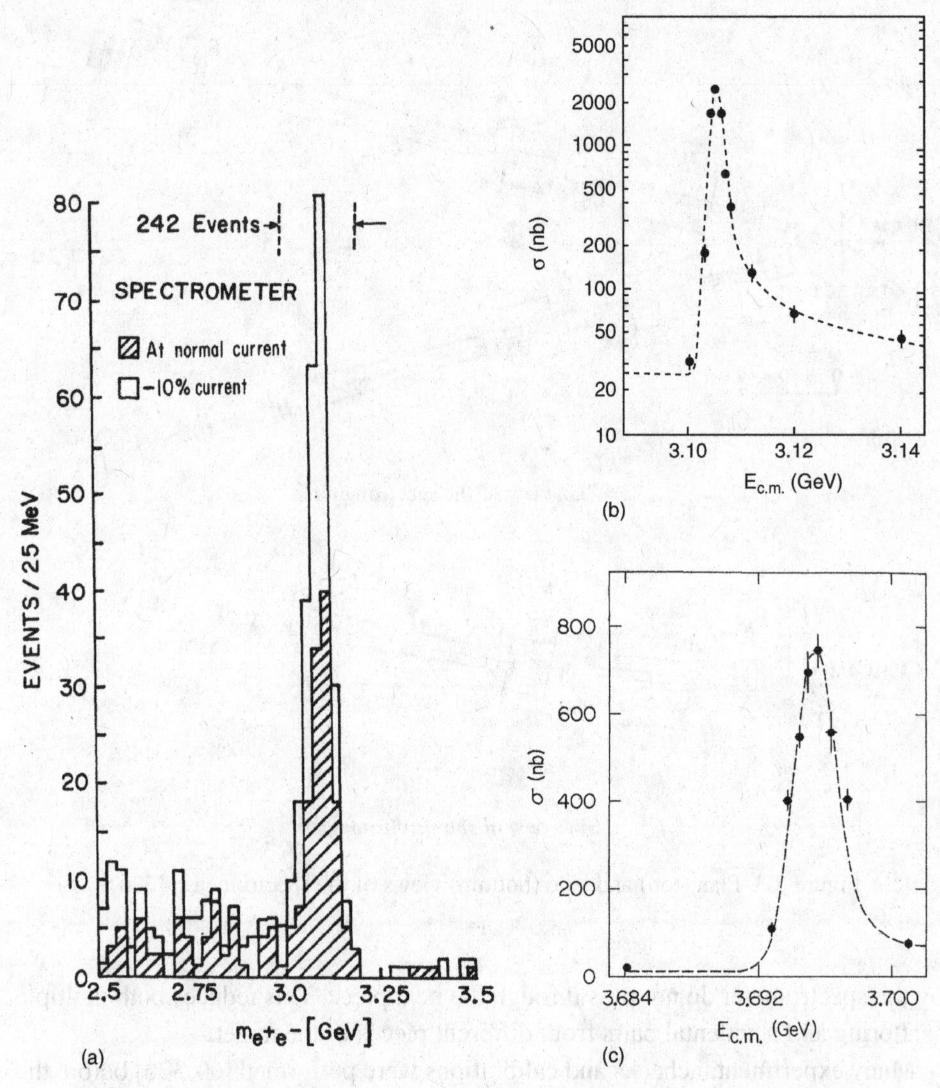

Figure 8.4 (a) Mass spectrum [60] for $p + p \to e^+ + e^- + X$ showing the existence of the J. Results from two spectrometer settings are plotted showing that the peak is independent of spectrometer currents. (b) Cross section versus c.m. energy for $e^+ + e^- \to$ multihadron final states [61] showing the ψ. The curve is the expected shape of a δ-function resonance folded with the Gaussian spread of the beams and including radiative processes. (c) Total cross section for $e^+e^- \to$ hadrons corrected for detection efficiency showing the ψ' at larger c.m. energy, following a search [62]. The dashed curve is the expected resolution folded with the radiative corrections.

slowly) which implied a new conservation law, similar to the discovery of strange particles in cosmic rays [25]. The J/ψ was quickly understood to be a bound state of heavy c–$\bar{c}$ quarks (charmonium) [63, 64] – the hydrogen atom of QCD [65, 66]. This was clear evidence for a "second" generation of quarks, and made all physicists believe in quarks and QCD. This discovery also changed the paradigm of di-lepton production from the measurement of background for W production to the search for new resonances.

The discovery of the $J \rightarrow e^+ + e^-$ di-lepton resonance in p–p collisions at $\sqrt{s} = 7.55$ GeV with $\sqrt{\tau} = 0.41$ was an incredible tour de force because many of the experiments that had discovered direct single electrons at much higher c.m. energies were also using two-arm spectrometers that originally searched in vain for the $e^{\mp}$ opposite in azimuth which would balance the p_T of the detected direct single $e^{\pm}$. At the CERN-ISR, the situation was soon rectified, first by CCRS [283, 328] (Figure 8.5a) who also showed that the J/Ψ could not explain their direct single electron data (Figure 8.5b), and a few years later by other experiments, e.g. Figure 8.5c [329]. If the J/Ψ were the source of the direct single $e^{\pm}$, then the ratio of observed e^+e^- pairs to direct single $e^{\pm}$ in CCRS would have been ~10 times larger than the value of $(0.5 \pm 0.2)\%$ observed [283]. Also, the $\langle p_T \rangle$ of the $J/\psi = 1.10 \pm 0.05$ GeV/c, from the data in Figure 8.5c [329], indicates a background single $e^{\pm}$ spectrum in Figure 8.5b well below the measurement.

Lederman's group at Fermilab was a bit overanxious on this issue and in early 1976 when they completed their double-arm e^+e^- spectrometer (Figure 8.6a) – essentially a second arm of the single-arm spectrometer that discovered the direct single $e^{\pm}$ [294], both at 90°±3 mrad in the c.m. system with vertical bending – they

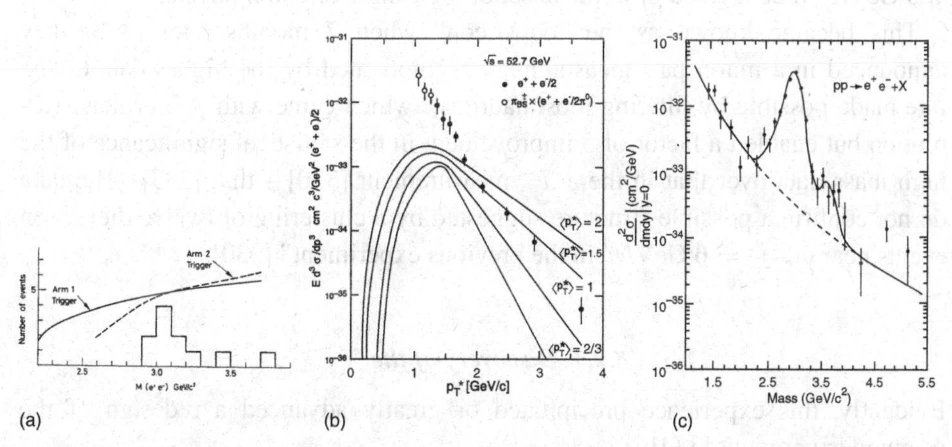

Figure 8.5 (a) First J/Ψ at ISR [328]; (b) direct $e^{\pm}$ data at $\sqrt{s} = 52.7$ GeV (Figure 7.4) with calculated $e^{\pm}$ spectrum for J/Ψ for several values of $\langle p_T \rangle$ [283]; (c) best $d\sigma_{ee}/dm_{ee}dy|_{y=0}$ [329].

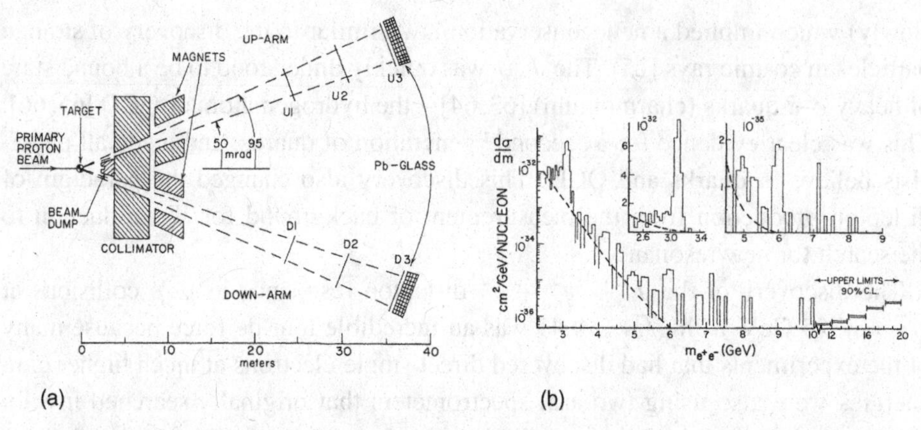

Figure 8.6 (a) Plan view of a di-lepton spectrometer for e^+e^- pairs [330]. One spectrometer bends upward and the other downward, with U1–3, D1–3 representing scintillation trigger counters, proportional wire chambers and scintillation hodoscopes. The Pb glass calorimeter in conjunction with the momentum measurement with resolution $\Delta p/p = 1.5\%$ (r.m.s.) identifies electrons with hadron rejection $\geq$2500 in each arm independently. All detection is done downstream of the magnet. (b) e^+e^- invariant mass spectrum $d\sigma/dm$ per nucleon assuming a linear A dependence [330]. Note bin-width changes.

published an e^+e^- invariant mass spectrum (Figure 8.6b) in which "a strong J/Ψ peak, a ψ' peak, and a higher-mass signal between ~4 and 10 GeV/c^2 are observed" [330, 331].[1] However, they also noted [330] "A statistically significant clustering near 6 GeV/c^2 suggests the existence of a narrow resonance" and also suggested in a footnote that "the name Υ (upsilon) be given either to the resonance at 6 GeV/c^2 if confirmed or to the onset of high-mass dilepton physics."

This became known as the "oopsLeon" when 7 months later [333] they announced in a muon pair measurement – "motivated by the higher data-taking rate made possible by filtering most hadrons," which came with poorer mass resolution but enabled a factor of 5 improvement in the statistical significance of the high mass data over that in the e^+e^- measurement [330] – that [333] "The data do not confirm a possible structure suggested by a clustering of twelve dielectron events near $m_{e^+e^-} = 6$ GeV/c^2 in the previous experiment" [330].

8.2.1 Discovery of the Υ

Evidently, this experience precipitated or greatly advanced a redesign of the di-muon experiment [334]:

[1] Reference [331] also confirmed that the J/Ψ was not the source of the direct single $e^{\pm}$, with the help of some friends, see for example reference [332] for comparison.

(i) to increase the data rate by at least a factor of 5;
(ii) while reconfiguring the spectrometers to improve the resolution;
(iii) adding an iron magnet behind the detectors to redetermine the muon momentum to $\lesssim$15%;
(iv) reconfiguring the target box to decrease the background;
(v) adding cast beryllium absorber blocks to minimize "inscattering of high hadron and muon flux" as well as to preserve the ~2% mass resolution (see Figure 8.7) [335].

This overhaul made precision di-muon measurements the class of the di-lepton spectra (Figure 8.8); and was rewarded by an equally fundamental discovery as the J/Ψ, the real Υ (upsilon), the first indication of a third family of quarks much heavier than the first two families [335], a bound state of b–$\bar{b}$ quarks. The width of the peak was wider than the spectrometer resolution and eventually was fit to two [336] and then three peaks (Figure 8.9a) [337]. Also, by the ratio of the $\Upsilon \rightarrow \mu^+\mu^-$ resonance signal to the continuum $\mu^+\mu^-$ cross section, it was

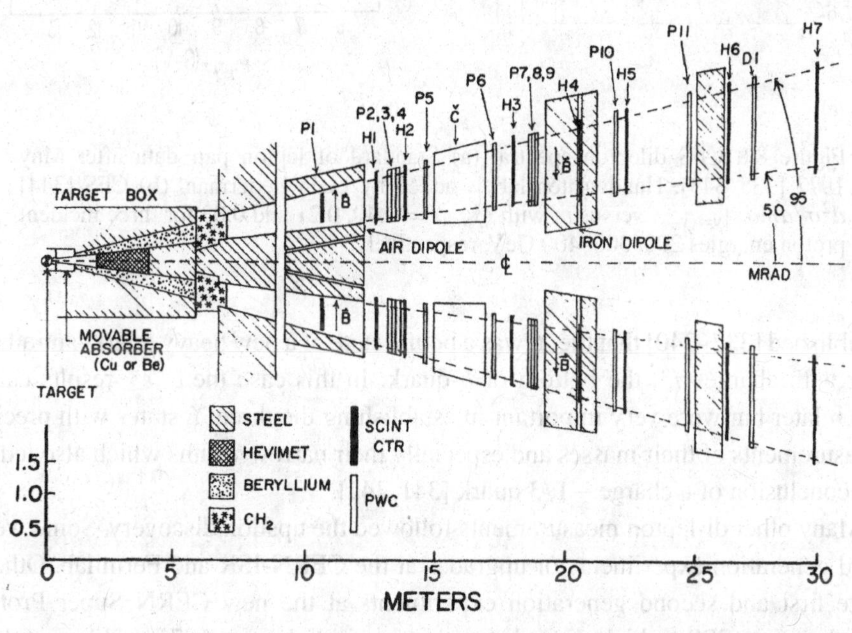

Figure 8.7 Plan view of redesigned di-muon apparatus [335]. Each spectrometer arm includes eleven PWCs, P1–P11, seven scintillation counter hodoscopes H1–H7, a drift chamber D1 and a gas-filled Cerenkov counter Č. Each arm is up/down symmetric and hence accepts both positive and negative muons. Note especially the muon filter containing low Z absorbers, beryllium and CH_2, in the spectrometer apertures, and the iron magnets to redetermine the muon momentum.

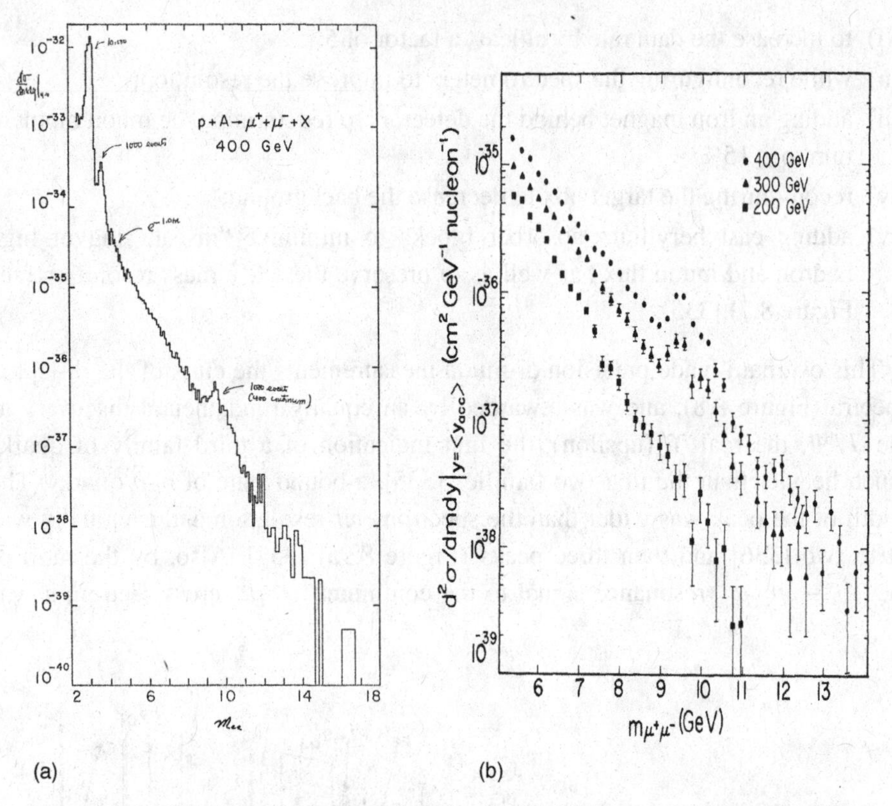

Figure 8.8 CFS dilepton spectra. (a) Standard of lepton pair data after May 1977 [335, 343]. Handwritten labels added by Leon Lederman. (b) CFS [344] $d^2\sigma/dmdy|_{y=\langle y_{acc}\rangle}$ versus m with $\langle y_{acc}\rangle = 0.40, 0.21$ and 0.03 for three incident proton energies 200, 300, 400 GeV, respectively.

established [337–340] that the Υ was a bound state of a new heavy quark–antiquark pair with charge 1/3, the bottom or b-quark. In this case the e^+e^- results came much later but were very important in establishing the three Υ states with precise measurements of their masses and especially their narrow widths which also led to the conclusion of a charge $-1/3$ quark [341, 342].

Many other di-lepton measurements followed the upsilon discovery. Some were third generation experiments or upgrades at the CERN-ISR and Fermilab. Others were first and second generation experiments at the new CERN Super Proton Synchrotron (SPS) which started construction in February 1971 and began data taking in January 1977 [346]. One of the SPS experiments, NA10, benefitted from the information of the Υ discovery. In March 1977, NA10 [347] had originally proposed a di-muon spectrometer based on magnetized solid iron toroids to be used in a high intensity $\pi^\pm$ beam to study Drell–Yan scaling and the parton distribution functions for pions. However, in November 1977 [348] the proposed

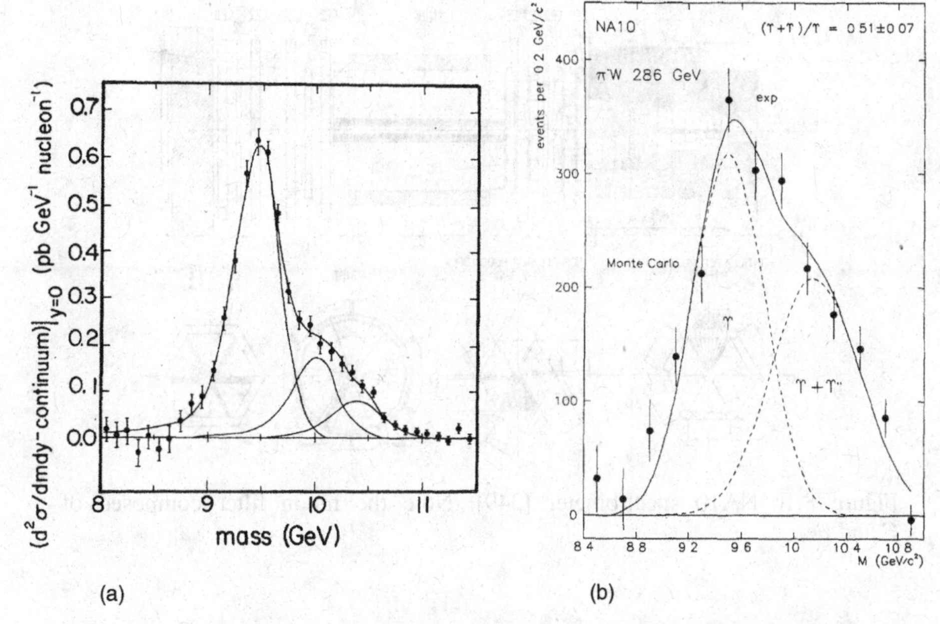

Figure 8.9 (a) CFS [337] mass spectrum for 400 GeV protons on a platinum target in the Υ region with continuum subtracted and with fits to three Υ. (b) NA10 [345] di-muon mass spectrum for 280 GeV π^- on a tungsten target after subtraction of the continuum, with fits to the $\Upsilon' + \Upsilon''$ and Υ shown.

experiment was turned into a high resolution device by eliminating the magnetized solid iron toroids and replacing them with a pulsed air-core toroidal magnet (Figure 8.10) [349], with a combined carbon and iron muon filter, capable of "mass resolution of the order of 2%" (Figure 8.9b). Of course Lederman's group had not been idle and had improved the mass resolution of their spectrometer (Figure 8.7) from 2.2% r.m.s. to 1.7% r.m.s., in order to separate the three Υ (Figure 8.9a), by lowering the beam intensity so that "a multiwire proportional chamber could be installed and operated halfway between the target and the analysis magnet" [337].

At the CERN-ISR, the CERN–Columbia–Rockefeller (CCR) collaboration, together with Oxford University, became the CCOR collaboration and designed a new detector [350] based on a thin-walled $(1X_o)$ aluminum stabilized superconducting solenoid magnet,[2] so that electrons and photons could exit and be detected in two EM calorimeters composed of arrays of lead glass. Four double-gap cylindrical drift chambers with three-dimensional readout inside the solenoid provided charged particle tracking over the full azimuth for $|y| < 0.7$ in the 1.4 T magnetic field [350] (see Figure 9.2b, in Chapter 9).

[2] This was the first such magnet at a collider, but soon became a favorite [351].

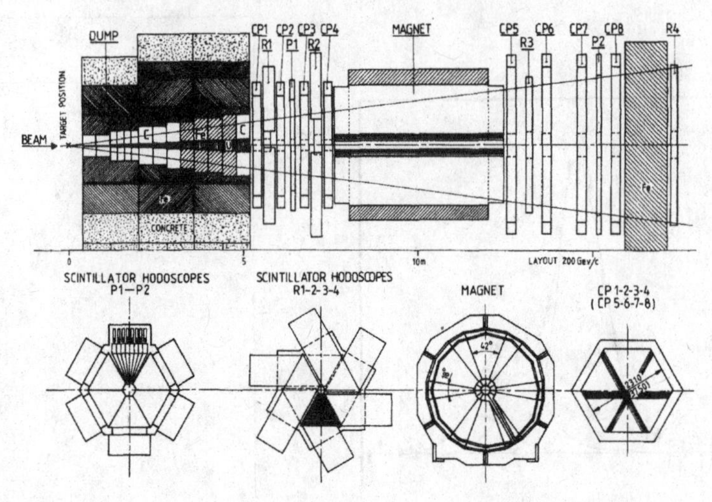

Figure 8.10 NA10 spectrometer [349]. Note the muon filter composed of C and Fe.

The principal disadvantage of this configuration, which lacked electron identification at the trigger level, was that π^0 pairs dominated the trigger rate, which forced a threshold of >2 GeV in each EM calorimeter, too high for J/Ψ detection but well suited for the Υ (Figure 8.11a) [352]. CCOR also found that the Υ and J/Ψ excitation curves were similar in shape as a function of $\sqrt{\tau} = M/\sqrt{s}$ as expected [339], with a ratio of $B_{ee}d\sigma/dy|_{y=0} \simeq 2 \times 10^{-3}$ which was smaller than expected (Figure 8.11b).

Several other J/Ψ and Υ measurements deserve to be mentioned, not so much for the influence of their results but rather to see the ingenuity of the various detectors used. One of the most influential in this regard was the Brookhaven–CERN–Syracuse–Yale collaboration [355] which was later joined by two groups from Athens [356]. In addition to the standard Proportional Wire Chambers (PWC) and plastic scintillator hodoscopes (Figure 8.12a), their detector pioneered the use of lithium foil transition radiators followed by xenon linear proportional chambers, which provided electron–hadron discrimination, as well as a lead/liquid-argon electromagnetic shower detector segmented laterally and longitudinally to provide measurement of the electron energy, and to achieve further rejection of hadrons. Since they had electron identification at the trigger level, they were able to obtain a nice e^+e^- invariant mass spectrum at mid-rapidity which included both the J/Ψ and the Υ as well as the continuum (Figure 8.12b) [356].

Another example is the di-muon experiment of Sam Ting and collaborators done at the CERN-ISR following the J/Ψ discovery at BNL. The detector, shown in Figure 8.13a [357], is a "large-acceptance spectrometer composed of seven

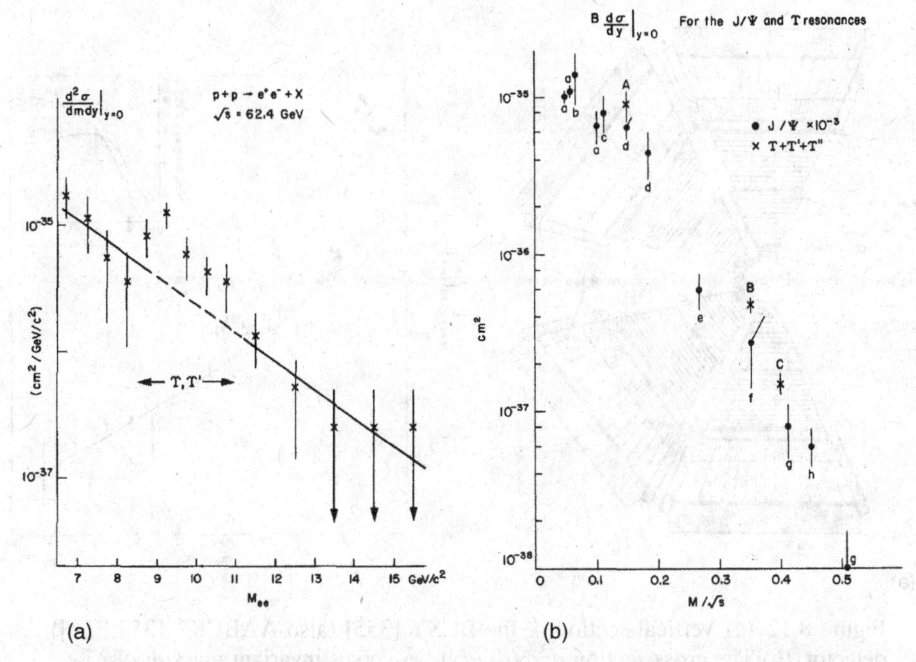

(a) (b)

Figure 8.11 (a) CCOR [352] measurement of the cross section $d^2\sigma/dmdy|_{y=0}$ for e^+e^- pairs with invariant mass in the range $6.5<m<15$ GeV/c^2. The solid line is an exponential fit to the continuum. (b) CCOR [352] compilation of $B_{ee}d\sigma/dy|_{y=0}$ for the J/Ψ and Υ, where the J/Ψ cross sections have been multiplied by 10^{-3}. See reference [352] for details. Also note that an earlier anomalously large ISR Υ measurement [353], which was omitted, was remeasured [354] and found to be in agreement with the measurements shown.

magnetized iron toroids, excited to 18 kG and totaling 450 tons, which provide both the hadron absorber and the magnetic field for the momentum analysis of muons. This method minimized background from hadron punch through, because there is more than 1.3 m of magnetized iron in the path of a penetrating track. To reduce background from hadron decays, the absorber starts about $\sim$ 40 cm from the interaction point. Muons are identified by penetration, requiring a minimum of 1.8 GeV/c momentum to traverse the absorber." Since the mass resolution is dominated by multiple scattering in the iron magnets, it is roughly constant at $\sigma_m/m \approx 11\%$ independent of mass. This is shown by the di-muon invariant mass spectrum (Figure 8.13b) [357], which well illustrates the positives and negatives of this style detector. Due to the large acceptance, the mass range covers from 2.5 to 30 GeV/c^2, but at the expense of mass resolution. It may be surprising even to see the wide J/Ψ peak since the muons from mid-rapidity J/Ψ cannot penetrate the iron; however it is for forward going J/Ψ with rapidity $0.9 < y < 2.0$.

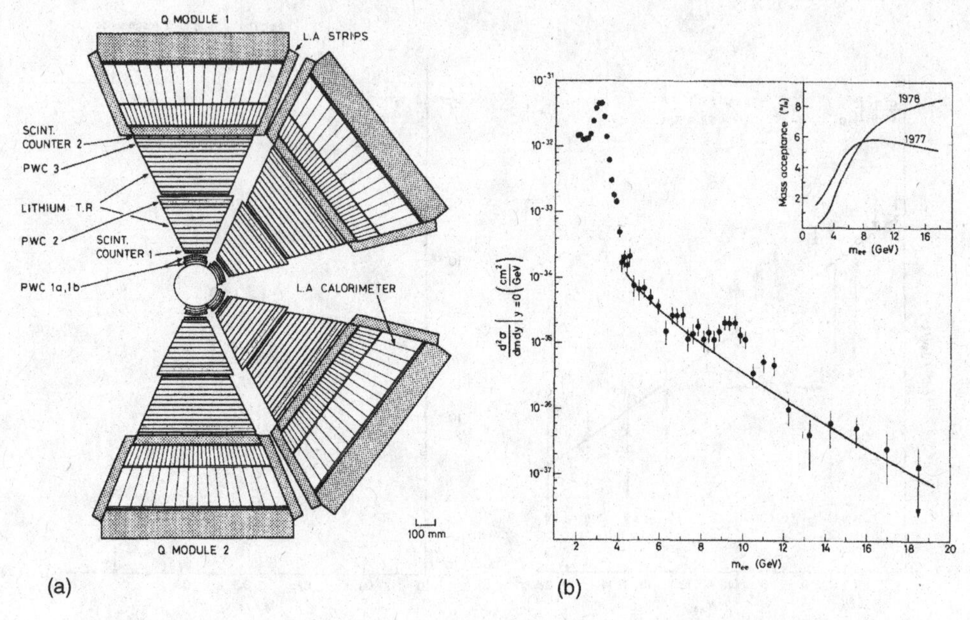

Figure 8.12 (a) Vertical section of the BCSY [355] (also AABCSY [354, 356]) detector. (b) The cross section $d^2\sigma/dmdy|_{y=0}$ versus invariant mass m of e^+e^- pairs in p–p collisions at $\sqrt{s}$ = 53 and 63 GeV combined [356]. A fit to the continuum is shown, as well as the different acceptances for the two years of data taking.

A nice illustration of a first round SPS experiment is the NA3 spectrometer (Figure 8.14) [358], proposed in October 1974 [359, 360] as a versatile detector for the study of high p_T hadrons, leptons, di-leptons produced by means of various hadron–hadron configurations (p–p, π–p, K–p), capable of detecting e^+e^- or $\mu^+\mu^-$ pairs by removing or inserting the absorber. Since the beam intensities are relatively low, e.g. (1–3) $\times$ 10^7 per pulse, compared to an external proton beam, the absorber was made of a 1.5 m long block of stainless steel with an embedded heavy (tungsten–uranium) conical plug of $\pm$30 mrad aperture inserted in the center. This gives a charged particle flux of $\sim$20% of the incident beam intensity at the exit of the dump, with a broad 35 cm fwhm before any sweeping by the magnet, which is acceptable, and a reduction of background μ from π, K decays to a level of 10^{-3} for a single particle trigger and 10^{-6} for a pair trigger. The target assembly and two PWCs are located inside a superconducting dipole magnet with a vertical field in a cylindrically shaped air gap 1.6 m in diameter, with other PWCs PC3–PC6 for tracking located downstream. Triggering is by scintillation counter hodoscopes, T1, T2, T3, M1, M2 with additional Fe muon absorber before the last trigger counter.

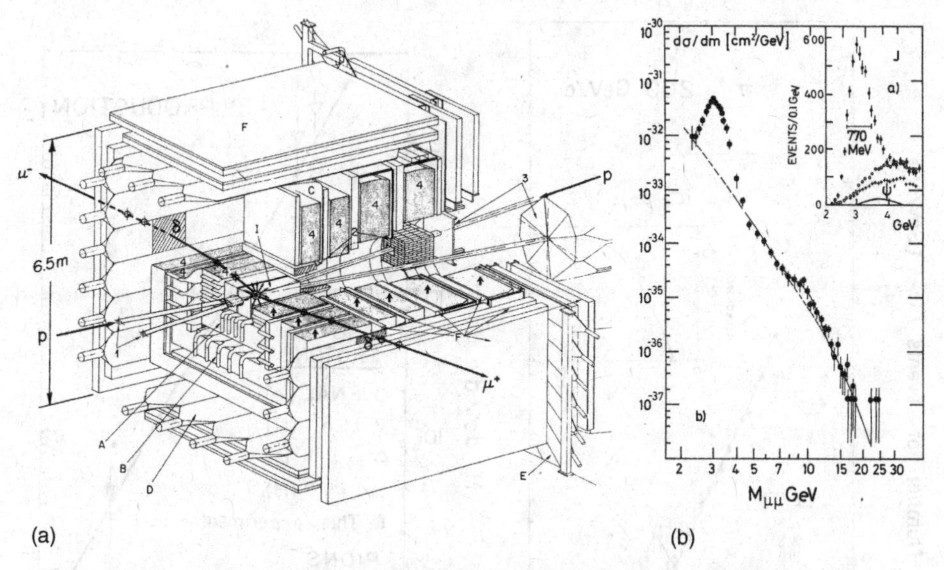

Figure 8.13 (a) Cutaway view of detector at the CERN-ISR [357]. Shown are beampipes (1), Pb absorbers (2) in front of luminosity monitors (3), magnetized iron toroids (4), trigger counter hodoscopes A–E and muon drift chambers (F). A di-muon event with $m = 24.5$ GeV/c^2 is sketched. (b) Measured cross section $d\sigma/dm$ as a function of di-muon mass. The dashed line is a fit to the continuum while the solid line includes the resonances. The inset shows the events in the $J/\Psi^{'}$ region.

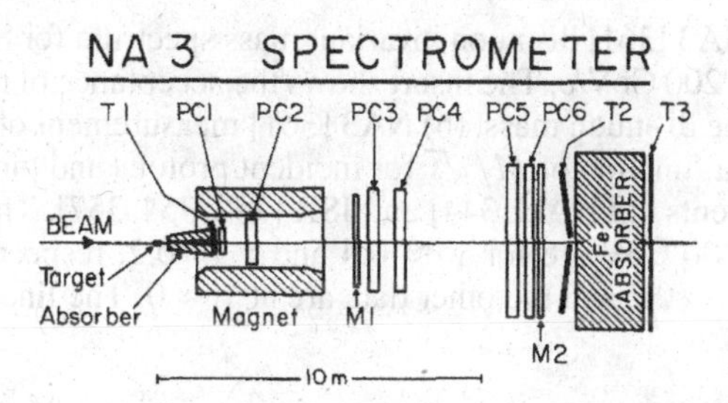

Figure 8.14 NA3 detector [358, 361] as described in the text.

The di-muon invariant mass spectrum from NA3 (Figure 8.15a) [361] is quite clean in this configuration as can be seen by the relatively small like-sign background. This spectrum represents the first measurement of Υ production by pions. Figure 8.15b [361] shows the excitation curve for Υ production, $Bd\sigma/dy|_{y=0}$ as a function of $M/\sqrt{s}$, as in Figure 8.11b for p–p collisions, in comparison to the NA3 measurements of Υ production by protons and $\pi^{\pm}$. The NA3 proton data are in agreement with the trend of the measurements at the other c.m. energies; while the production of Υ by pions, which is roughly the same for

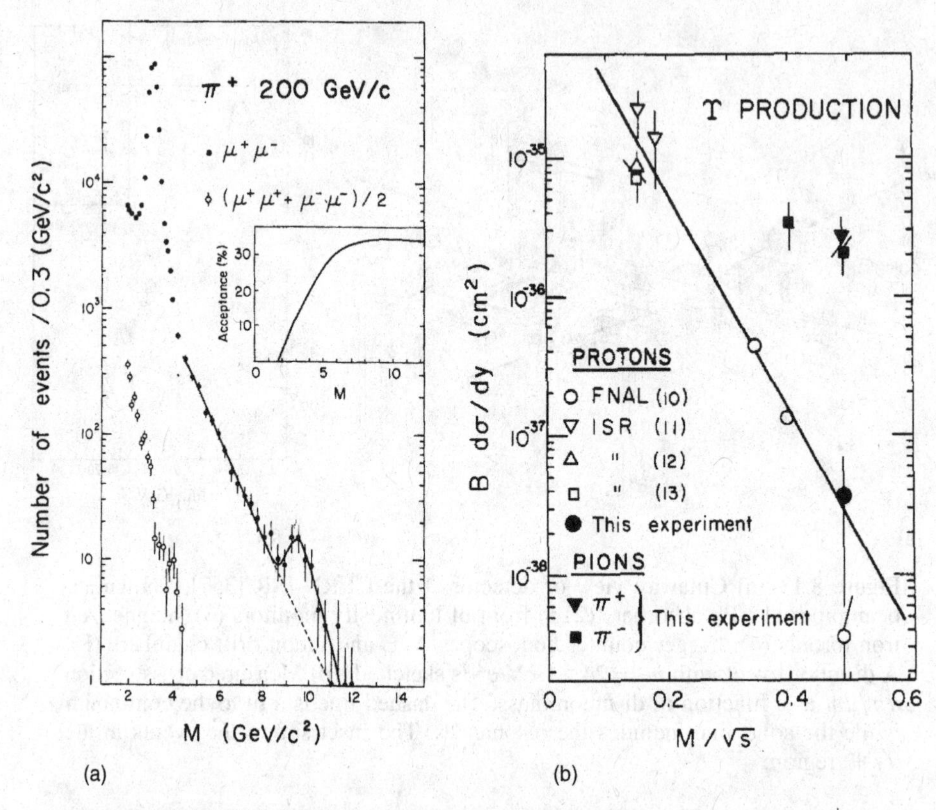

Figure 8.15 (a) NA3 [361] di-muon invariant mass spectrum for incident π^+ on a platinum target at 200 GeV/c. The insert shows the acceptance of the spectrometer as a function of the di-muon mass. (b) NA3 [361] measurement of $Bd\sigma/dy$ of the three Υ states as a function of $M/\sqrt{s}$ for incident protons and pions compared to proton measurements at FNAL [344] and ISR [352, 354, 357]. The FNAL proton data at 200 and 300 GeV are for $y \approx 0.4$ and $y \approx 0.2$, respectively. The NA3 pion data are at $y \approx 0.2$ and the other data are at $y \approx 0$. The line is an eyeball fit to the proton data.

π^+ and π^-, is about 30 times larger than by protons at 200 GeV incident energy, corresponding to $M/\sqrt{s} = \sqrt{\tau} = 0.49$.

8.3 Are J/Ψ and Υ production due to hard scattering?

Although somewhat diverted by the discoveries of the second and third generations of quarks evidenced by narrow resonances in the di-lepton channel, experiments also made thorough studies of the continuum of di-leptons, known as Drell–Yan pairs. As noted above, Drell–Yan pair production is closely related to both DIS in e–p collisions and hard scattering in p–p collisions. In fact, while it is clear that Drell–Yan pair production is really due to hard scattering of point-like partons because of its measured $A^{1.0}$ dependence in $p+A$ collisions [322] the same cannot be said of either J/Ψ or Υ which exhibit shadowing in $p+A$ collisions, corresponding to $A^{0.92\pm0.008}$ [362] for both J/Ψ and Ψ' and $A^{0.96}$ [363] for both the Υ (0.96 ± 0.01) and $\Upsilon' + \Upsilon''$ (0.95 ± 0.01). Also the values of the power α for the J/Ψ and Υ depend on both p_T and x_F (Figure 8.16).

These measurements of the A-dependence of J/Ψ and Υ production were made by E772 [366] using the E605 [367] apparatus which was a third generation spectrometer at Fermilab, a major redesign of the spectrometer that discovered the Υ (recall Figure 8.7) [335] to better explore the kinematic limits of lepton and hadron production, which is worth a comment here (Figure 8.17) [368]. There was a major change in philosophy (more like NA3) with strong magnetic sweeping replacing shielding as the main method to eliminate the lower p_T particles from the target and decays-in-flight, which caused large counting rates (background) in the detectors. The spectrometer bent vertically, with the three main magnets, SM0, SM12 and SM3 providing transverse momentum kicks of 1.3, 7.5, and 0.9 (0, 7.5, and −0.9) GeV/c, respectively for 400 (800) GeV/c incident proton beams on liquid hydrogen, liquid deuterium or heavy targets. The two large magnets SM12 and SM3

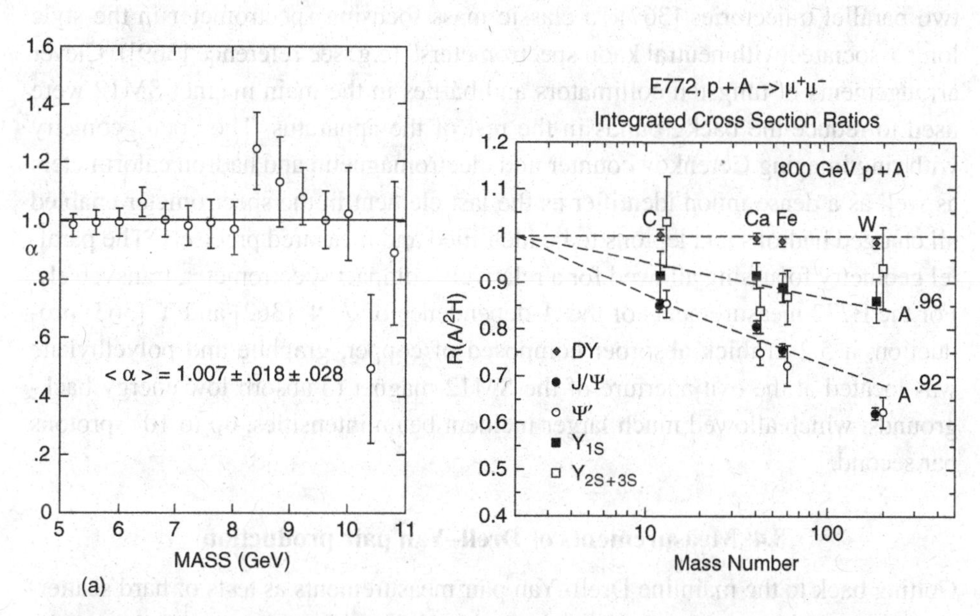

Figure 8.16 (a) A-dependence power α, for Drell–Yan pairs derived from Pt and Be targets, as a function of the di-muon mass integrated over all p_T [322]. (b) The ratio of heavy nucleus to deuterium yields per nucleon [362, 363] for Drell–Yan, J/Ψ, Υ integrated over the ranges $0 \leq x_F \leq 0.6$ and $0 \leq p_T \leq 4$ GeV/c [364, 365].

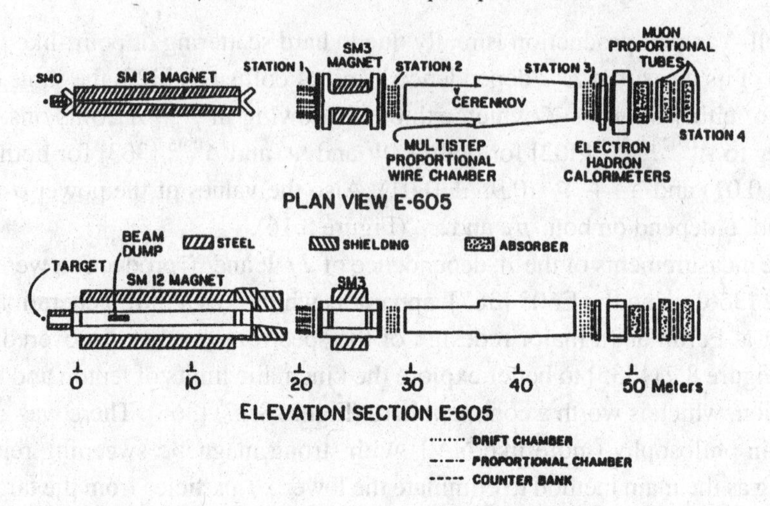

Figure 8.17 E605 Spectrometer [368].

were built with 2000 tons of steel from Columbia University's Nevis Cyclotron. In this geometry, particles of appropriate charge with $p_T = 1.3 + 7.5 + 0.9 = 9.7$ GeV/c exit the magnets approximately parallel to the initial beam direction; and oppositely charged particles of invariant mass $2 \times 9.7 = 19.4$ GeV/c^2 exit with two parallel trajectories [367] "a classic mass focusing spectrometer in the style long associated with neutral kaon spectrometers" (e.g. see reference [369]). Clever arrangements of tungsten collimators and baffles in the main magnet SM12 were used to reduce the backgrounds in the rest of the apparatus. The open geometry with ring imaging Cerenkov counter and electromagnetic and hadron calorimeters as well as a dense muon identifier as the last element in the spectrometer enabled all charged hadrons and leptons to be identified and measured precisely. The parallel geometry focussing allowed for a relatively compact spectrometer, transversely. For the E772 measurements of the A-dependence of J/Ψ [362] and Υ [363] production, a 5.2 m thick absorber composed of copper, graphite and polyethylene was located at the exit aperture of the SM12 magnet to absorb low energy backgrounds, which allowed much larger incident beam intensities, up to 10^{11} protons per second.

8.4 Measurements of Drell–Yan pair production

Getting back to the mainline Drell–Yan pair measurements as tests of hard scattering, the first feature of Drell–Yan pair production to be tested successfully was the scaling prediction (Eq. 8.1) written in slightly different form:

$$\frac{s\, d^2\sigma}{d\sqrt{\tau} dy} = G(\sqrt{\tau}). \tag{8.3}$$

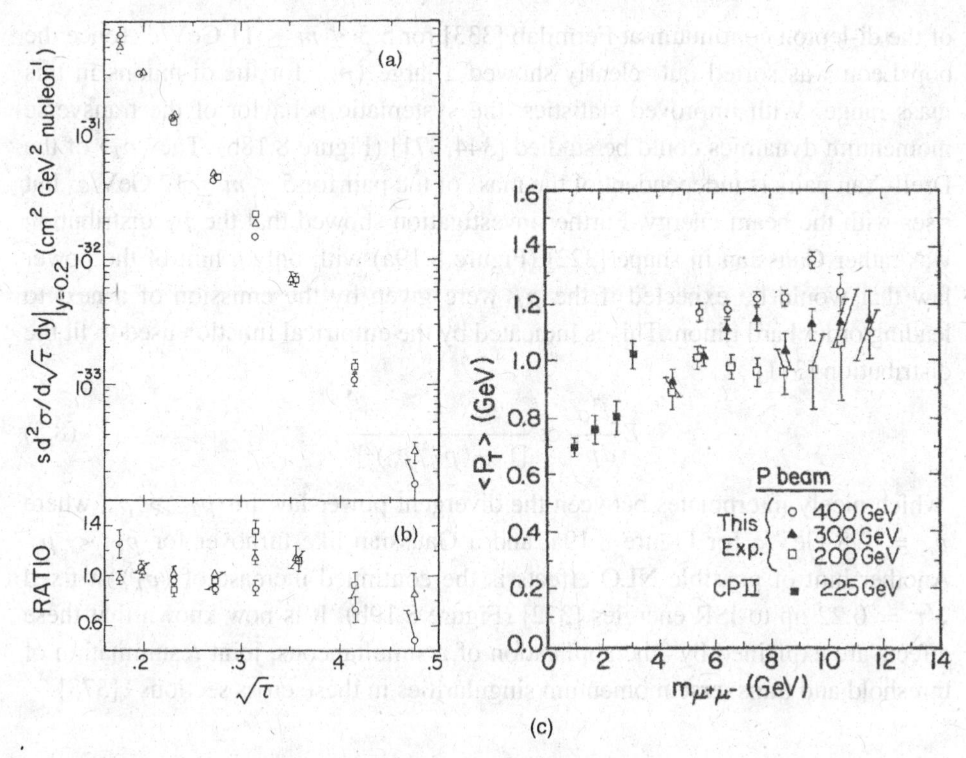

Figure 8.18 (a) CFS measurement of $s\,d^2\sigma/d\sqrt{\tau}dy|_{y=0.2}$ versus $\sqrt{\tau}$. Circles, triangles and squares correspond to 400, 300 and 200 GeV beam energy respectively [344]. (b) Same data divided by the overall fit (Eq. 8.4). (c) $\langle p_T \rangle$ versus m for Drell–Yan di-muons at three beam energies [344]. The data below $m_{\mu^+\mu^-} = 4$ GeV/c^2 are from another experiment, Chicago–Princeton II [370].

Figure 8.18a shows the CFS [344] data from Figure 8.8b for the continuum plotted in this manner. The scaled data were consistent with a global fit:

$$\left.\frac{s\,d^2\sigma}{d\sqrt{\tau}dy}\right|_{y=0.2} = (44 \pm 0.7 \pm 12.0)\exp - [(25.3 \pm 0.2 \pm 0.6)\sqrt{\tau}]\ \mu\text{b GeV}^2 \quad (8.4)$$

with a χ^2/dof of 173/145 (confidence level 10%). The ratios of the scaled cross sections of Figure 8.18a to this fit are shown in Figure 8.18b. The scaling works very well.

Other features gave more surprising results, notably the $\langle p_T \rangle$ of the lepton pair. In the naive parton model, the Drell–Yan pair should have zero net p_T with respect to the parton–parton collision axis. This also means that the pair should have zero net p_T with respect to the proton–proton collision axis, which is assumed to be the same as the parton–parton collision axis because the partons are assumed to be collinear with their proton, i.e. have no intrinsic p_T. However, the first observations of the di-lepton continuum at Fermilab [333] for $5.5 \leq m \leq 11$ GeV/c^2 (once the oopsLeon was sorted out) clearly showed a large $\langle p_T \rangle$ for the di-muons in this mass range. With improved statistics, the systematic behavior of the transverse momentum dynamics could be studied [344, 371] (Figure 8.18b). The $\langle p_T \rangle$ of the Drell–Yan pairs is independent of the mass of the pair for $5 \leq m \leq 12$ GeV/c^2 but rises with the beam energy. Further investigation showed that the p_T distribution was rather Gaussian in shape [322] (Figure 8.19a) with only a hint of the power law that would be expected if the p_T were given by the emission of a next to leading order hard gluon. This is indicated by the empirical function used to fit the distribution [371]

$$E\frac{d^3\sigma}{dp^3} \propto \frac{1}{[1 + (p_T/p_o)^2]^6}, \tag{8.5}$$

which nicely interpolates between the divergent power law for $p_T \gg p_o$, where $p_o = 2.8$ GeV/c for Figure 8.19a, and a Gaussian-like turnover for $p_T < p_o$. Another hint of possible NLO effects is the continued increase of $\langle p_T \rangle$ at fixed $\sqrt{\tau} = 0.22$ up to ISR energies [372] (Figure 8.19b). It is now known that these effects are explained by "the application of a simultaneous, joint resummation of threshold and transverse momentum singularities in these cross sections" [373].

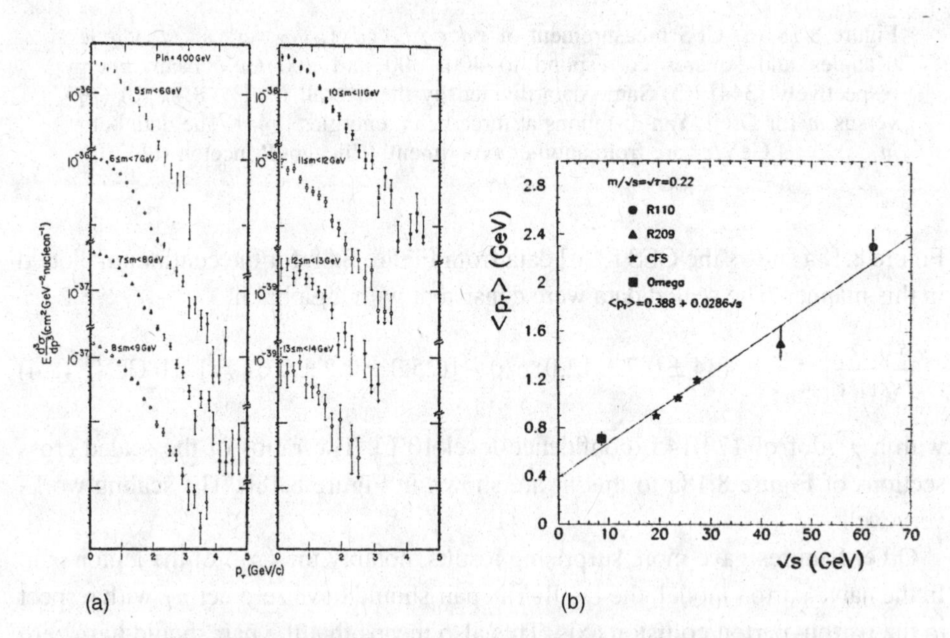

Figure 8.19 (a) Invariant yield of di-muons as a function of transverse momentum of the muon pair for 400 GeV protons [322]. (b) CMOR (R110) [372] plot of $\langle p_T \rangle$ of Drell–Yan lepton pairs as a function of $\sqrt{s}$ for fixed $\sqrt{\tau} = 0.22$.

Having been beaten to both the J/ψ and Υ discoveries by lower c.m. energy fixed target experiments with much higher luminosity, the CCOR collaboration, after being denied the use of the superconducting low β insertion at the ISR [374] which would have increased their luminosity from 5×10^{31} to 2×10^{32} cm^{-2} s^{-1} found a way [375] to increase the acceptance for high mass e^+e^- pairs by a factor of 5 by adding $14X_o$ thick lead scintillator shower counters inside the solenoid in the azimuthal range not covered by the existing lead glass detectors (Figure 8.20). The first $3.5X_o$ of the shower counters were read out separately for pre-shower hadron rejection. Similarly, a front wall of lead glass, $4X_o$ thick, followed by a Multiwire Proportional Chamber (MWPC) with cathode strip read-out was added as a pre-convertor in front of the previously existing lead glass arrays. A minor rearrangement of the existing drift chambers was also required. The collaboration also underwent a minor readjustment, substituting Michigan State University for Columbia so that it became the COR [376], then CMOR collaboration (R110).

The greatly increased acceptance allowed CMOR (Figure 8.21a) [372] to improve on the previous ISR Drell–Yan measurements, namely: R108 (CCOR) [352] (Figure 8.11a); R806 [356] (Figure 8.12b); R209 [357] (Figure 8.13b). The CMOR data [372] (Figure 8.21b) also showed that the Drell–Yan scaling observed by the Fermilab measurements, CFS [322,344] (recall Figure 8.18a), continued to hold for $\sqrt{s}$ up to 62.4 GeV.

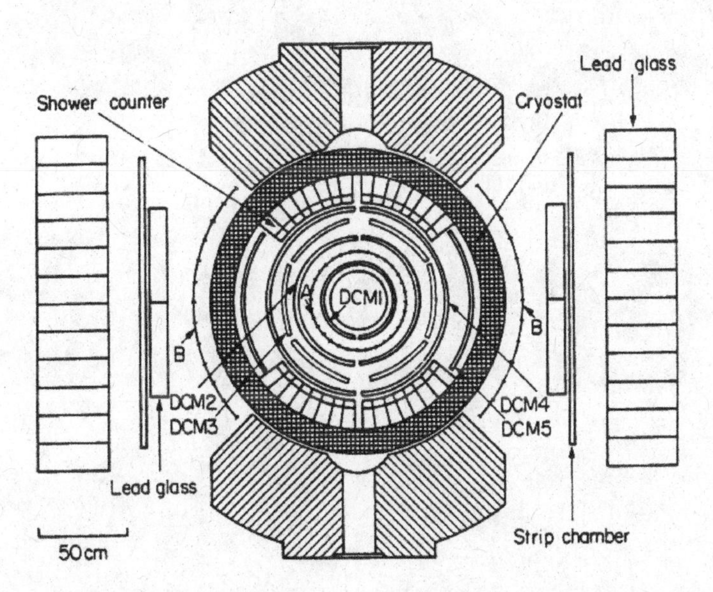

Figure 8.20 CMOR (R110) apparatus viewed along the beam axis [372, 376].

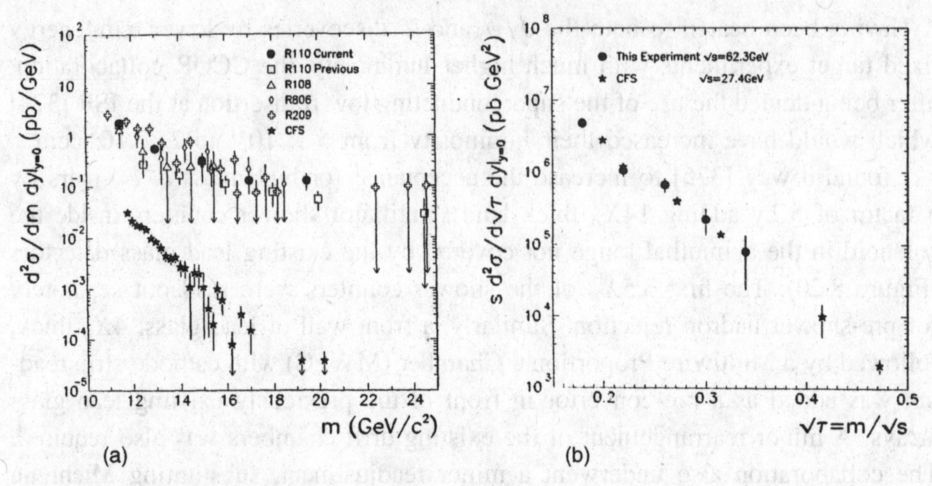

Figure 8.21 (a) CMOR [372] measurement of $d^2\sigma/dmdy|_{y=0}$ (filled circles) together with other data as indicated in the text. (b) Drell–Yan scaling of CMOR [372] data in the form $sd^2\sigma/d\sqrt{\tau}dy|_{y=0}$ at $\sqrt{s} = 62.3$ GeV compared to the CFS measurement [322] at $\sqrt{s} = 27.4$ GeV.

The di-lepton channel, which was full of exciting discoveries, also supported the Drell–Yan parton model of the continuum, with the observation of its predicted scaling. However, di-leptons did not provide the key tests of QCD which would be given instead by two particle correlation measurements in the di-hadron channel.

9

Two particle correlations

Following the discovery of hard scattering in p–p collisions at the CERN-ISR by the observation of the large yield of high p_T hadrons, the attention of experimentalists turned to the search for the predicted di-jet structure [377] of the hard scattering events using two particle correlations. There were also searches for jets, i.e. collections of particles with limited momentum transverse to a common axis representing the "parton fragmentation" to hadrons (and originally called "cores" by Bjorken [377]), but these searches were much more difficult than initially contemplated, with a long learning curve which included several false alarms (see Chapter 11).

9.1 Hard scattering in the parton model

The overall p–p hard scattering cross section [377] for the inclusive reaction, where more generally A and B are distinct hadrons

$$A + B \to C + D + X \tag{9.1}$$

and where C and D are outgoing particles or jets, is the sum over parton reactions $a + b \to c + d$ (e.g. $g + q \to g + q$) at parton–parton center-of-mass (c.m.) energy $\sqrt{\hat{s}}$:

$$\frac{d^3\sigma}{dx_1 dx_2 d\cos\theta^*} = \frac{s d^3\sigma}{d\hat{s} d\hat{y} d\cos\theta^*} = \sum_{ab} f_a^A(x_1) f_b^B(x_2) \frac{d\sigma^{ab\to cd}}{d\cos\theta^*} \tag{9.2}$$

where $f_a^A(x_1)$, $f_b^B(x_2)$, are parton distribution functions, the differential probabilities for partons a and b to carry momentum fractions x_1 and x_2 of their respective hadrons (e.g. $u(x_2)$ for a proton), and $d\sigma^{ab\to cd}/d\cos\theta^*$ is the parton–parton scattering cross section as a function of θ^*, the scattering angle in the parton–parton c.m. system. The parton–parton c.m. energy squared is $\hat{s} = x_1 x_2 s$, where $\sqrt{s}$ is the c.m. energy of the p–p collision. The parton–parton c.m. system moves with rapidity $\hat{y} = \frac{1}{2}\ln(x_1/x_2)$ in the p–p c.m. system. The quantities $f_a^A(x_1)$ and $f_b^B(x_2)$, the "number" distributions of the constituents, are related (for the electrically charged quarks) to the structure functions measured in Deeply Inelastic lepton–hadron Scattering (DIS), for example in $e + A$ scattering

$$F_{1A}(x, Q^2) = \frac{1}{2}\sum_a e_a^2\, f_a^A(x, Q^2)$$
$$F_{2A}(x, Q^2) = x\sum_a e_a^2\, f_a^A(x, Q^2) \tag{9.3}$$

where e_a is the electric charge on a quark in units of the proton charge.

The Mandelstam invariants $\hat{s}$, $\hat{t}$ and $\hat{u}$ of the constituent scattering have a clear definition in terms of the scattering angle θ^* in the parton–parton c.m. system (recall Eqs. 3.100, 3.101):

$$\hat{t} = -\hat{s}\,\frac{(1-\cos\theta^*)}{2} \quad \text{and} \quad \hat{u} = -\hat{s}\,\frac{(1+\cos\theta^*)}{2}. \tag{9.4}$$

The transverse momentum of a scattered parton is:

$$p_T = p_T^* = \frac{\sqrt{\hat{s}}}{2}\sin\theta^*, \tag{9.5}$$

and the scattered constituents c and d, the outgoing parton-pair, have equal and opposite momenta in the parton–parton c.m. system.

Equation 9.2 gives the p_T spectrum of outgoing parton c, which then fragments into a jet of hadrons, which may include for example π^0. The parton fragmentation function $dP_c^{\pi^0}/dz = D_c^{\pi^0}(z)$ is the probability for a π^0 to carry a fraction $z = \boldsymbol{p}^{\pi^0}\cdot\boldsymbol{p}^c/|p^c|^2 \approx p^{\pi^0}/p^c$ of the momentum of outgoing parton c (and similarly $D_d(z')$ for parton d). For single inclusive reactions, Eq. 9.2 must be summed over all subprocesses leading to a π^0 in the final state weighted by their respective fragmentation functions, and additionally weighted by D_d for di-jet (or di-hadron) studies. In this formulation, $f_a(x_1)$, $f_b(x_2)$, $D_c^{\pi^0}(z)$ (and $D_d(z')$) represent the "long-distance phenomena" to be determined by experiment (primarily from DIS or e^-e^+ collisions). The parton–parton scattering cross section, $d\sigma^{ab\to cd}/d\cos\theta^*$, can then be determined from p–p hard scattering experiments, if f_a, f_b, $D_c(z)$ (and $D_d(z')$) are known [377].

Since the outgoing jet pairs of hard scattering obey the kinematics of elastic scattering (of partons) in a parton–parton c.m. frame which is moving longitudinally with rapidity $y = \frac{1}{2}\ln(x_1/x_2)$ in the p–p c.m. frame, the jet pair formed from the scattered partons should be co-planar with the beam axis, with equal and opposite transverse momenta, and thus be back-to-back in azimuth. Viewed in a plane perpendicular to the beam axis, the events from hard scattering should show strong azimuthal correlations (Figure 9.1). For experiments with inclusive high transverse

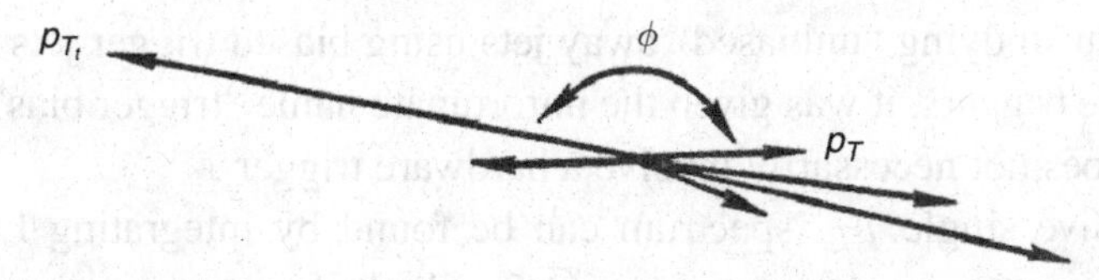

Figure 9.1 Schematic azimuthal projection of a parton–parton scattering event onto a plane perpendicular to the beam axis.

momentum triggers, one of the fragments with transverse momentum, denoted p_{T_t}, will satisfy the trigger.

It is not necessary to reconstruct the jets fully in order to measure their properties. In many cases two particle correlations are sufficient to measure the desired properties, and in some cases, such as the measurement of the net transverse momentum of a jet pair, may be superior, since the issue of the systematic error caused by missing some of the particles in the jet is not relevant. A helpful property in this regard is the "leading particle effect."

9.1.1 The leading particle effect: "trigger bias" and the "Bjorken parent–child relationship"

Due to the steeply falling power-law transverse momentum spectrum of the scattered partons, the inclusive single particle (e.g. π) spectrum from jet fragmentation for reactions near 90° in the p–p c.m. system is dominated by fragments with large z, where $z \approx p_{T\pi}/p_{T_q}$ is the fragmentation variable. The joint probability for a fragment pion, with momentum fraction z, from a parton with $p_{T_q} = p_{T\,jet}$ is:

$$\frac{d^2\sigma_\pi(p_{T_q}, z)}{p_{T_q} dp_{T_q} dz} = \frac{d\sigma_q}{p_{T_q} dp_{T_q}} \times D_\pi^q(z) = \frac{A}{p_{T_q}^n} \times D_\pi^q(z), \tag{9.6}$$

where $D_\pi^q(z) \sim e^{-bz}$ is the fragmentation function and n is the simple power fall-off of the parton invariant cross section (i.e. not the $n_{eff}(x_T, \sqrt{s})$ of Eq. 6.12 [378]). The change of variables, $p_{T_q} = p_{T_\pi}/z$, $dp_{T_q}/dp_{T_\pi}|_z = 1/z$, then gives the joint probability of a fragment π, with transverse momentum p_{T_π} and fragmentation fraction z:

$$\frac{d^2\sigma_\pi(p_{T_\pi}, z)}{p_{T_\pi} dp_{T_\pi} dz} = \frac{A}{p_{T_\pi}^n} \times z^{n-2} D_\pi^q(z). \tag{9.7}$$

Thus, the effective fragmentation function when a single fragment (with p_{T_π}) is detected (or selected), is weighted upward in z by a factor z^{n-2}. In other words, inclusive single particle production is dominated by fragments with large values of z. This is the "leading particle effect." Since this property, although general, is most useful in studying "unbiased" away jets using biased trigger jets selected by single particle triggers, it was given the unfortunate name "trigger bias" [379, 380] although it does not necessarily involve a hardware trigger .

The inclusive single p_{T_π} spectrum can be found by integrating Eq. 9.7 over all values of parton p_{T_q} from p_{T_π} to $\sqrt{s}/2$, which, for any p_{T_π}, corresponds to integrating over z from $2p_{T_\pi}/\sqrt{s}$ $(=x_T)$ to 1:

$$\frac{1}{p_{T_\pi}}\frac{d\sigma_\pi}{dp_{T_\pi}} = \frac{1}{p_{T_\pi}^n}\int_{x_T}^{1} A\, D_q^\pi(z)\, z^{n-2} dz. \tag{9.8}$$

Since the integral depends only weakly on p_{T_t}, due to the typically small values of x_T studied, one can see that the fragment π^0 invariant p_{T_π} spectrum is a power-law with the same power n as the original parton p_{T_q} spectrum. This is Bjorken's parent–child relationship [377]: the single particle cross section has the same power dependence as the parton cross section.

9.2 Two particle correlation measurements

Independently of any models, the next question asked by experimenters after the discovery of hard scattering was whether there were any other particles associated with a high p_{T_t} inclusive trigger and what their properties were. Given an inclusive trigger on a particle with transverse momentum p_{T_t}, it is straightforward to measure the conditional probability of observing an associated particle with transverse momentum p_T (recall Figure 9.1):

$$\left.\frac{d\mathcal{P}(p_T)}{dp_T}\right|_{p_{T_t}} = \frac{d^2\sigma(p_{T_t}, p_T)/dp_{T_t}dp_T}{d\sigma(p_{T_t})/dp_{T_t}}. \tag{9.9}$$

Many ISR experiments provided two particle correlation measurements [381, 382]. A large solid-angle device was desirable for such measurements in order to observe the full extent of the correlation in azimuth and rapidity, which was unknown. Typically, the most popular measurements involved either using the split field magnet in the self triggered mode (recall Figure 6.5), as done by the CERN–Collège de France–Heidelberg–Karlsruhe (CCHK) collaboration [383], or adding a triggering device, as done by the British–French–Scandavian collaboration (BFS) who added the British–Scandivian moveable "Cerenkov mode" Wide Angle Spectrometer (WAS) [384] to the SFM, which could provide identified $\pi^\pm$, $K^\pm$, $p^\pm$ triggers in the c.m. momentum range 0.5–4.5 GeV/c, and could move from 36° to 90° in the p–p c.m. system [385] (Figure 9.2a). Another CERN group (R412) [386] placed a 1 m^2 lead glass array 3.6 m from the SFM at 90° in the c.m. system to trigger on resolved $\pi^0 \to \gamma + \gamma$ for $2 < p_{T_t} < 4.1$ GeV/c.

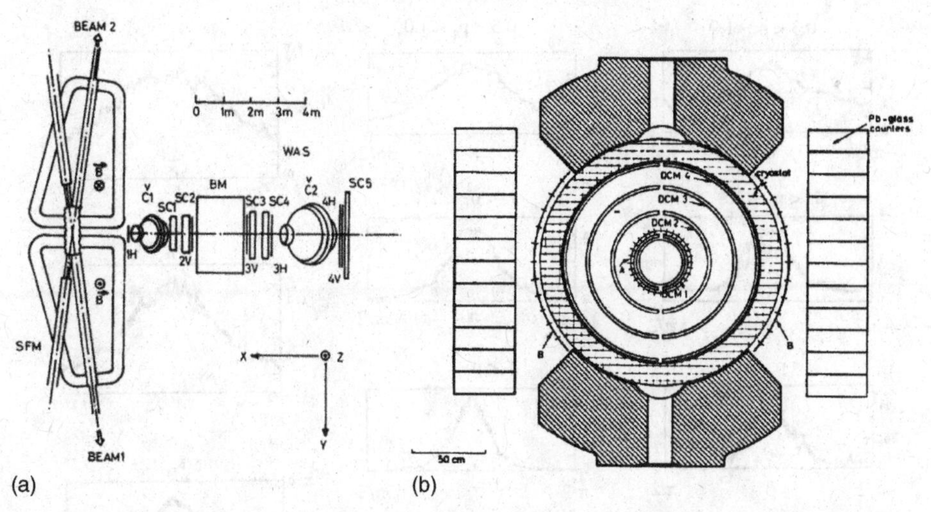

Figure 9.2 (a) Schematic plan view of BFS experiment [385] with spectrometer at 90° to the SFM (wire chambers not shown). The spectrometer consisted of a bending magnet (BM), spark chambers (SC1–SC5), horizontal and vertical scintillator hodoscopes (1H–4H, 1V–4V) and gas Cerenkov counters (C1–C2). (b) Azimuthal projection of CCOR experiment [387, 388]. Four layers of cylindrical drift chambers (DCM1–DCM4) with three-dimensional readout and an array of 32 scintillation counters (A) surround the beam pipe inside a thin-coil ($1X_o$) aluminum-stabilized superconducting solenoid with 1.4 T axial magnetic field uniform to $\pm1.5\%$ in a cylindrical volume 1.7 m long with diameter 1.4 m. Scintillation counters (B) just outside the cryostat were used to correct for energy lost in the solenoid coil and to detect conversions, while two PbGl arrays were used for triggering and measurement of π^0, γ and $e^{\pm}$.

A different tack was taken by the CCOR collaboration [387, 388] who developed a new third generation experiment which was the first to provide charged particle measurement with full and uniform acceptance over the entire azimuth, with pseudorapidity coverage $-0.7 \leq \eta \leq 0.7$, so that the jet structure of high p_T scattering could be easily seen and measured. Two electromagnetic calorimeters (one on either side of the intersection region) composed of lead glass Cerenkov counter arrays from the CCR and CCRS experiments were located outside a new thin-walled ($1X_o$) superconducting solenoid magnet containing magnetically compensated drift chambers (Figure 9.2b). The apparatus was triggered with adjustable thresholds on clusters of deposited energy from $e^{\pm}$, γ, π^0 in one or both EMCal arms. A minimum bias trigger consisting of a count in any of the 32 scintillation counters (A) which surround the beam pipe was also available. In both cases a vertex of two or more charged particles was required offline.

In Figure 9.3a,b the azimuthal distributions of associated charged particles relative to a π^0 trigger with transverse momentum $p_{T_t} > 7$ GeV/c, in the range

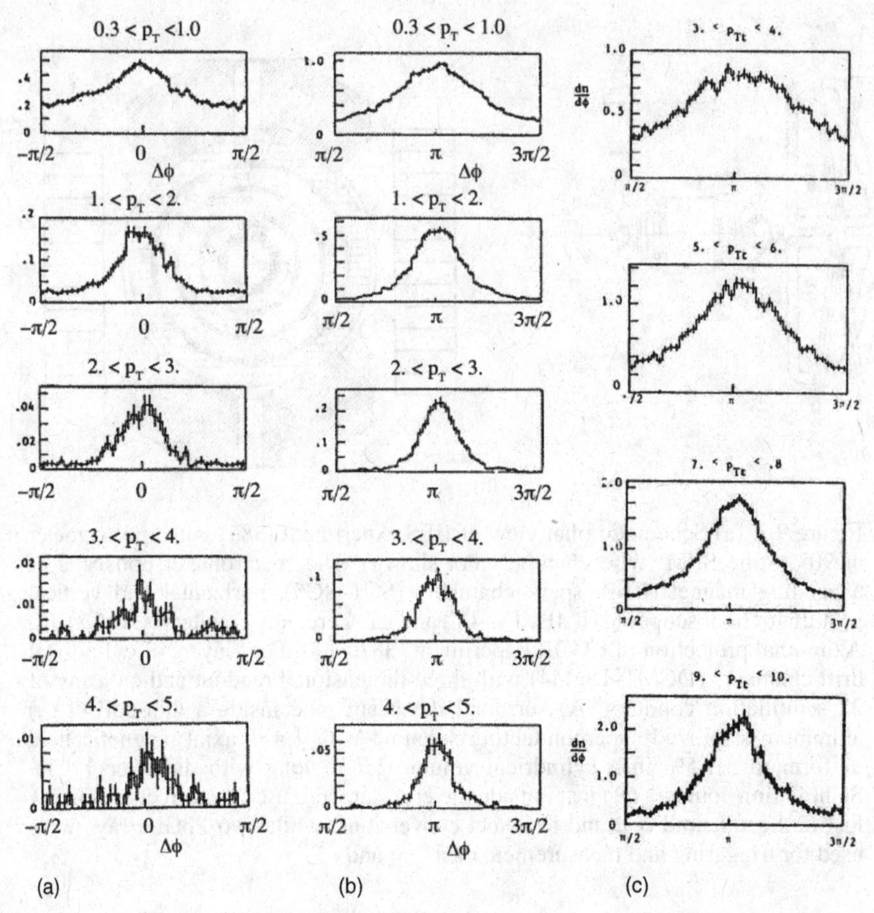

Figure 9.3 (a,b) Azimuthal distributions of charged particles of transverse momentum p_T, with respect to a trigger π^0 with $p_{Tt} \geq 7$ GeV/c, for five intervals of p_T: (a) for $\Delta\phi = \pm\pi/2$ rad about the trigger particle, and (b) for $\Delta\phi = \pm\pi/2$ about π radians (i.e. directly opposite in azimuth) to the trigger. The trigger particle is restricted to $|\eta| < 0.4$, while the associated charged particles are in the range $|\eta| \leq 0.7$. (c) Away-side azimuthal distributions integrated for $p_T > 0.30$ GeV/c for four intervals of trigger p_{T_t}: 3–4, 5–6, 7–8, 9–10 GeV/c [387].

$|\eta| < 0.4$, are shown for five intervals of associated particle transverse momentum p_T [387].

The quantity plotted

$$\left.\frac{dn(p_T)}{d\phi}\right|_{p_{T_t}} \tag{9.10}$$

is the average number of tracks, with pseudorapidity $|\eta| \leq 0.7$, per radian, per event with a π^0 trigger with $p_{T_t} \geq 7.0$ GeV/c. No correction is required for the azimuthal acceptance since the detector covers the full azimuth, $\Delta\phi = 2\pi$.

The plot is split into sections, the trigger side, $|\Delta\phi| \leq \pi/2$ (Figure 9.3a), and the away side, $\pi/2 < \Delta\phi < 3\pi/2$ (Figure 9.3b), each with different vertical scales. Also the vertical scales change for the five bands of associated transverse momentum.

In all cases, strong correlation peaks on top of a flat "background" level are observed on both the trigger and away sides, indicating a di-jet structure which is contained in an interval $\Delta\phi = \pm 60°$ about a direction towards and opposite to the trigger for all values of associated p_T (>0.3 GeV/c) shown. The small variation of the widths of the away-side peaks for $p_T > 1$ GeV/c (Figure 9.3b) indicates out-of-plane activity of the di-jet system beyond simple jet fragmentation (the k_T effect, see below). In each p_T band, the flat "background" level is approximately equal to the value measured with the minimum bias trigger. For associated particle transverse momenta $p_T > 1.0$ GeV/c, the flat "background" is negligible in comparison to the two peaks.

It is also informative to look at the away-side azimuthal distributions integrated over all associated particle transverse momenta $p_T > 0.30$ GeV/c, as a function of p_{T_t} of the trigger (Figure 9.3c). Compared to the flat "background," which remains constant in all cases, the height of the away peak increases with increasing trigger p_{T_t}. However, its width becomes narrower so that the overall away-side charged multiplicity changes only very slowly with increasing trigger transverse momentum.

This huge azimuthal peaking toward and away from a high p_{T_t} trigger corresponds to the naive di-jet picture of the parton model. It seems farfetched to invoke conservation of momentum to explain such strong peaking of 0.5 GeV/c particles opposite to a 3.0 to 9.0 GeV/c transverse momentum trigger particle in a collision with a total available energy of 62.4 GeV. Of course, the peaking on the trigger side cannot be explained by momentum conservation but could be due to resonances. However, as will be discussed below, resonances do not contribute significantly to the same-side correlation due to the "trigger-bias" and "parent–child" effects.

CCOR directly measured the trigger bias from these data by reconstructing the trigger jet from the associated charged particles with $p_T \geq 0.3$ GeV/c, within $\Delta\phi = \pm 60°$ from the trigger particle, using the algorithm $p_{T\,jet} = p_{Tt} + 1.5 \sum p_T \cos(\Delta\phi)$, where the factor 1.5 corrects the measured charged particles for missing neutrals. The measurements of $\langle z_{trig}\rangle = \langle p_{Tt}/p_{T\,jet}\rangle$ as a function of p_{Tt} for three values of $\sqrt{s}$ (Figure 9.4) show the property of x_T scaling, which was not expected [389].

9.2.1 x_E distributions and the fragmentation function

Historically [383, 386], in order to get some additional quantitative information from the away jet azimuthal angular distributions of Figure 9.3b, two variables

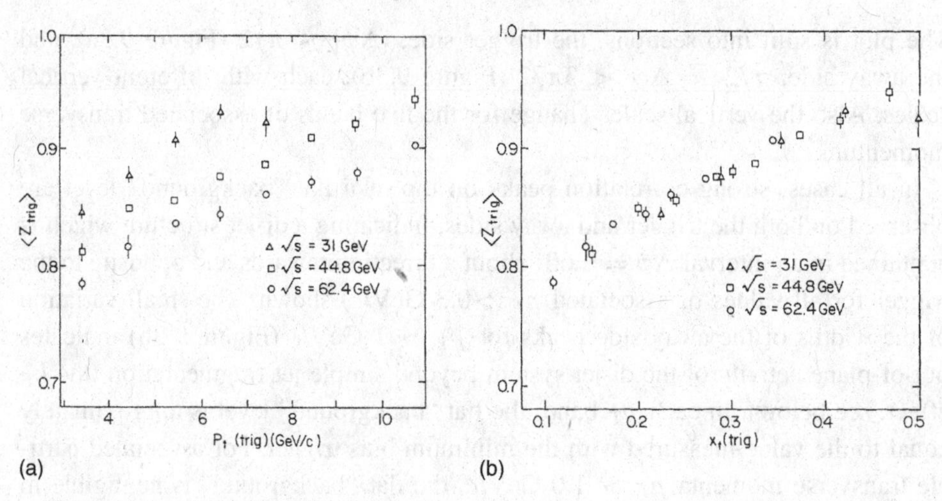

Figure 9.4 CCOR [390] measurement of $\langle z_{trig} \rangle$ as a function of (a) p_{Tt} and (b) $x_{Tt} = 2p_{Tt}/\sqrt{s}$.

were defined (Figure 9.5) [386]): x_E, the associated particle transverse momentum projected to the trigger transverse momentum axis and scaled by the trigger transverse momentum, $x_E \equiv \left| p_x/p_{T_t} \right|$; and $p_{out} = p_T \sin(\Delta\phi)$, the projected transverse momentum component of a track out of the "scattering" plane defined by the high p_{T_t} trigger and the beam direction. For any slice of associated particle transverse momentum p_T in Fig. 9.3b, p_{out} is proportional to the width of the away azimuthal distribution. The variable x_E was defined [386] in order to indicate how well individual away-side hadrons balanced the trigger π^0 transverse momentum. It also relates the fragmentation variable $z \simeq p_T/p_{T\,jet}$ of the away-side jet, which was thought to be unbiased, to that of the biased trigger jet, $z_{trig} \simeq p_{Tt}/p_{T\,jet}$, if the trigger and away jets had equal and opposite transverse momenta as expected:

$$x_E = \frac{-\mathbf{p}_T \cdot \mathbf{p}_{Tt}}{|\mathbf{p}_{Tt}|^2} = \frac{-p_T \cos(\Delta\phi)}{p_{Tt}} \simeq \frac{z}{z_{trig}}. \tag{9.11}$$

It was also expected that x_E would equal the fragmentation fraction z of the away jet, for highly biased trigger jets as $z_{trig} \to 1$.

Following the seminal article of Feynman, Field and Fox (FFF) [391], it was generally assumed that the p_T distribution of away-side hadrons from a single particle trigger with p_{T_t}, corrected for $\langle z_{trig} \rangle$, would be the same as that from a jet trigger and would follow the same fragmentation function as observed in e^+e^- or DIS. Because of the relatively large trigger bias at ISR energies and small range of $\langle z_{trig} \rangle$, $0.8 \lesssim \langle z_{trig} \rangle \lesssim 0.9$, it was also assumed that x_E scaling [380] would hold, i.e.

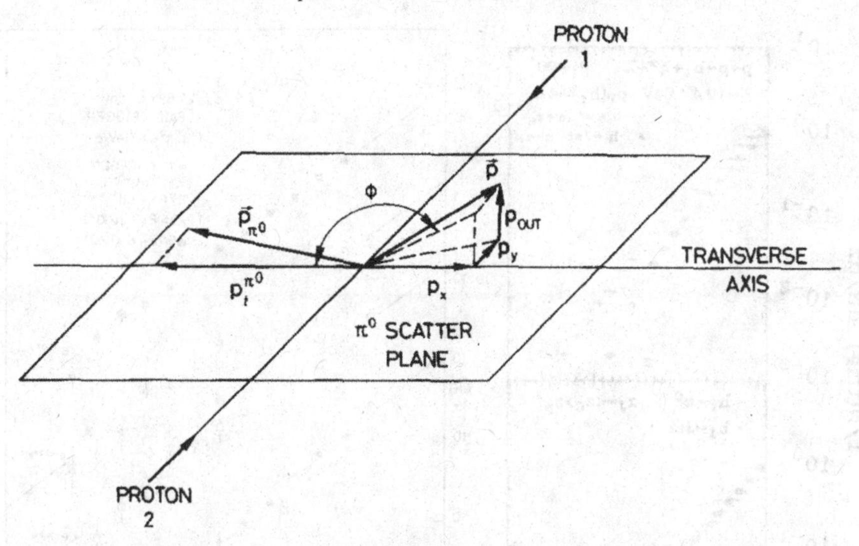

Figure 9.5 Diagram [386] of kinematic quantities for π^0–hadron correlations. The trigger π^0 has momentum $\boldsymbol{p}_{\pi^0}$ and transverse momentum $p_{T_t}^{\pi^0}$. The hadron momentum $\boldsymbol{p}$ is broken into three components: p_{out} perpendicular to the scattering plane formed by the colliding protons and $\boldsymbol{p}_{\pi^0}$, and $p_x = p_T \cos\phi$ opposite to the direction of $p_{T_t}^{\pi^0}$ in the scattering plane, where ϕ (which is called $\Delta\phi$ throughout the text) is the azimuthal angle between $p_{T_t}^{\pi^0}$ and p_T.

all x_E distributions measured at different p_{T_t} would be the same. To cite directly from reference [391], p. 25,[1]

There is a simple relationship between experiments done with single-particle triggers and those performed with jet triggers. The only difference in the opposite side correlation is due to the fact that the 'quark', from which a single-particle trigger came, always has a higher $p_\perp$ than the trigger (by factor $1/z_c$). The away-side correlations for a single-particle trigger at $p_\perp$ should be roughly the same as the away side correlations for a jet trigger at $p_\perp$(jet)=$p_\perp$(single particle)/ $\langle z_c \rangle$.

This point is reinforced in the conclusions (p. 59): "2. The distribution of away-side hadrons from a jet trigger should be the same as that from a single particle trigger except for a correction due to $\langle z_c \rangle$" (see Figure 9.6a). Another interesting point is, "8. Because the quarks scatter elastically (no quantum number exchange – except perhaps color), the away-side distribution of hadrons in pp collisions should be essentially independent of the quantum numbers of the trigger hadron." That is, the jets fragment independently.

[1] Note that in FFF the notation is $a + b \to c + d$ (as in Eq. 9.2) where a, b, c, d are called "quarks," so FFF call z_{trig}, z_c.

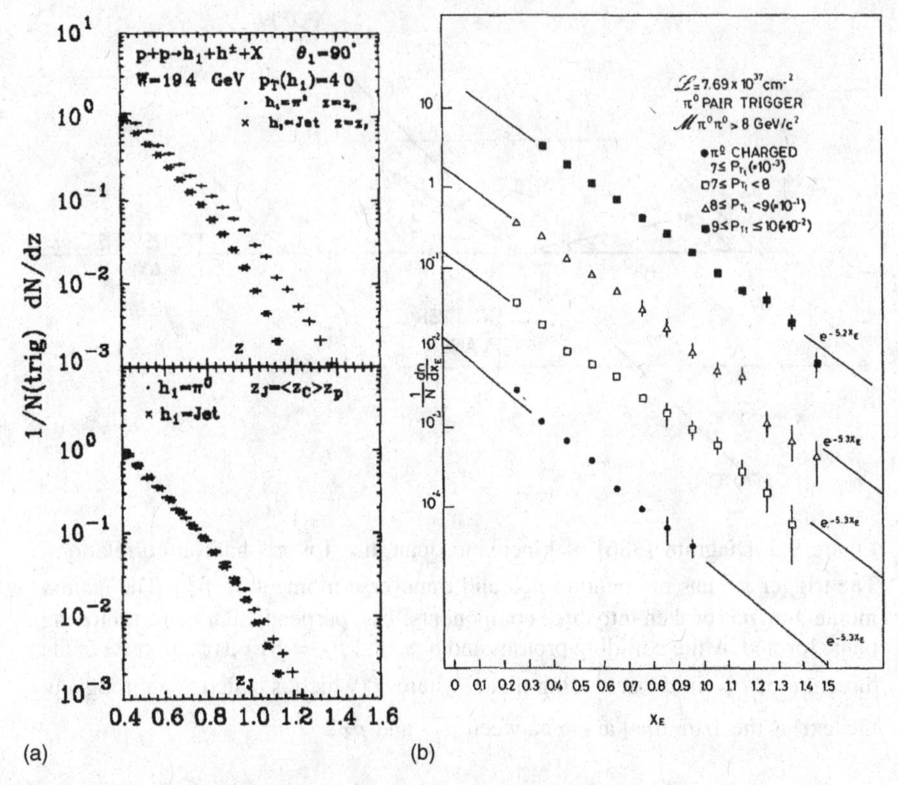

Figure 9.6 (a) Figure 23 from FFF [391] : (top) $dn/dz|_{p_{T_t}}$ plotted versus $z_p = x_E$ for a single particle trigger (dots) and $z_J = p_x/p_{T\,jet}$ for a jet trigger (crosses); (bottom) same plot except now for the single particle trigger the quantity $z_J = \langle z_c \rangle x_p$ is plotted for the single particle trigger instead of z_p. (b) CCOR [387, 392] x_E distributions, $dn/dx_E|_{p_{T_t}}$, from the data of Figure 9.3 for several values of p_{T_t}.

These predicted features were apparently observed in the CCOR x_E distributions [387, 392] from the data of Figure 9.3b as shown in Figure 9.6b. The x_E scaling is evident for all values of p_{T_t}, with the expected fragmentation behavior, $e^{-6z} \sim e^{-6x_E\langle z_{trig}\rangle}$. This figure also showed that there were no di-jets, each of a single particle, as claimed by another ISR experiment of that period [393], since there is no peak at $x_E = 1$.

The CCOR results (Figure 9.6b) [387, 392] and several others [394, 395] led to the strong belief at the time that jet fragmentation functions from ν–p, e^+e^- reactions and from p–p x_E distributions are the same, with the same dependence of the exponential slope b on p_{T_t} or $\sqrt{s}/2$ for e^+e^- (Figure 9.7 [396]).

It turns out that this belief was one of the very few results from the "high p_T discovery period" that did not stand the test of time. It was discovered at RHIC [149],

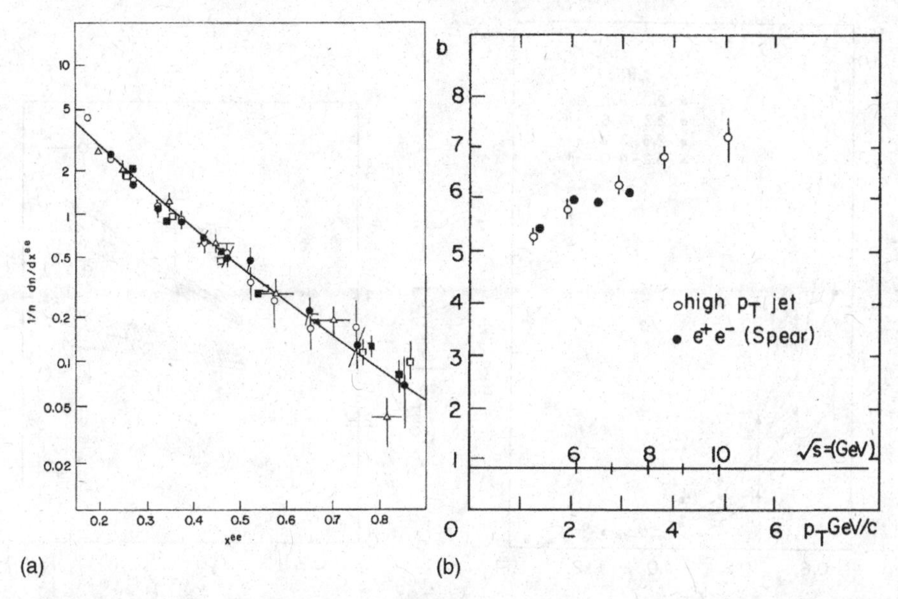

Figure 9.7 (a) Jet fragmentation functions from ν–p, e^+e^- compared to p–p x_E distributions [396]. (b) Exponential slopes b from these data as a function of p_T or $\sqrt{s}/2$ [396].

a quarter of a century later, that the shape of the x_E distribution triggered by the fragment of a jet, such as a π^0, was not sensitive to the shape of an exponential fragmentation function but instead depended only on the power n of the invariant single particle cross section. Details are given in Appendix E.

9.2.2 No x_E scaling for lower values of $p_{T_t} \lesssim 3\,\mathrm{GeV}/c$

Although the evidence for x_E scaling was "overwhelming" at the ISR for $p_{T_t} > 4$ GeV/c [395], the early measurements at the ISR did not find x_E scaling. Historically (in 1977) it was found by CCHK [383] that x_E scaling [380] did not work in the range $2.0 \leq p_{T_t} \leq 3.2$ GeV/c; i.e. different values of trigger p_{T_t} did not produce a universal x_E distribution (Figure 9.8a). CCHK also looked at the p_{out} variable and plotted $\langle |p_{out}| \rangle$ versus x_E for triggers in the range $2.0 \leq p_{T_t} \leq 3.2$ GeV/c. They found that the $\langle |p_{out}| \rangle$ increased with increasing x_E up to a maximum value of $\langle |p_{out}| \rangle \sim 0.65$ GeV/c (Figure 9.8b). This effect [391,397] coupled with the lack of x_E scaling for $p_{T_t} < 3$ GeV/c was taken by CCHK [383] as evidence for the transverse momentum of quarks inside the proton. Calculations from their parton scattering model with $k_T = 0$ and $\langle k_T \rangle = 610$ MeV/c are shown. This result inspired Feynman, Field and Fox (FFF) [391] to introduce formally $\boldsymbol{k_T}$, the transverse momentum of a parton in a nucleon, into their model of parton–parton scattering.

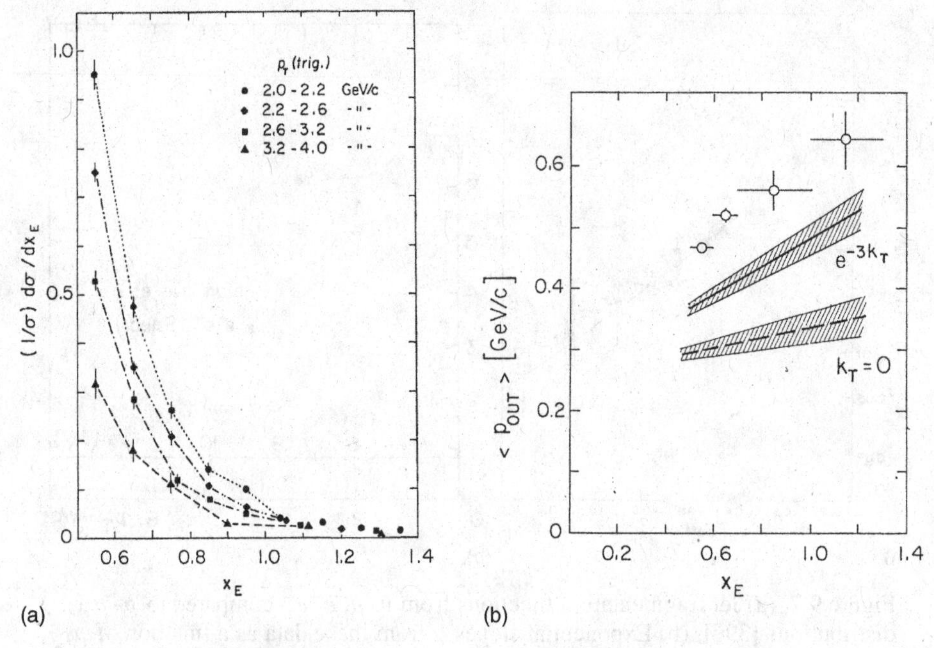

Figure 9.8 (a) CCHK [383] measurement of x_E distributions, $dn/dx_E|_{p_{T_t}}$, for intervals of p_{Tt} in the range 2.0–4.0 GeV/c. (b) CCHK [383] measurement of $\langle |p_{out}| \rangle$ versus x_E for triggers in the range $2.0 \leq p_{T_t} \leq 3.2$ GeV/c. Also shown are the predictions of their parton scattering Monte Carlo model with $k_T = 0$ and $dN/k_T dk_T \propto e^{-3k_T}$ (with the requirement $k_T < 5/3$ GeV/c).

Later results on x_E scaling were published in 1978 by the BFS collaboration [385] using an identified charged particle trigger at 90° c.m., with detection of the associated charged particles in the split field magnet at the CERN-ISR (recall Figure 9.2a). The acceptance for associated charged particles, which was the same as CCHK, covered a rapidity range $-4 \leq y \leq +4$, with an azimuthal aperture of $\pm 40°$ on the trigger side and $\pm 25°$ on the away side. BFS first used this large aperture to make a dramatic map of the two particle correlation function over nearly the full region in rapidity. Figure 9.9a shows the ratio of the density of tracks with $p_T > 0.5$ GeV/c, as a function of rapidity y and azimuthal angle ϕ, from a high p_{T_t} trigger with $3 < p_{T_t} < 4$ GeV/c located at $y = 0$, $\phi = 180°$, relative to minimum bias events. They call this ratio, which cancels the effects of acceptance variation, $R+1$:

$$R+1 = \frac{\text{mean track density for high } p_{T_t} \text{ events}}{\text{mean track density for minimum-bias events}}. \tag{9.12}$$

The main features of Figure 9.9a are an increase in the value of the correlation function in a small region in y and ϕ near the trigger particle and a much larger increase on the away side, mainly within $|\phi| < 45°$ (the limit of their acceptance), but extending in rapidity to $|y| \simeq 3$. This illustrates the di-jet coplanar structure

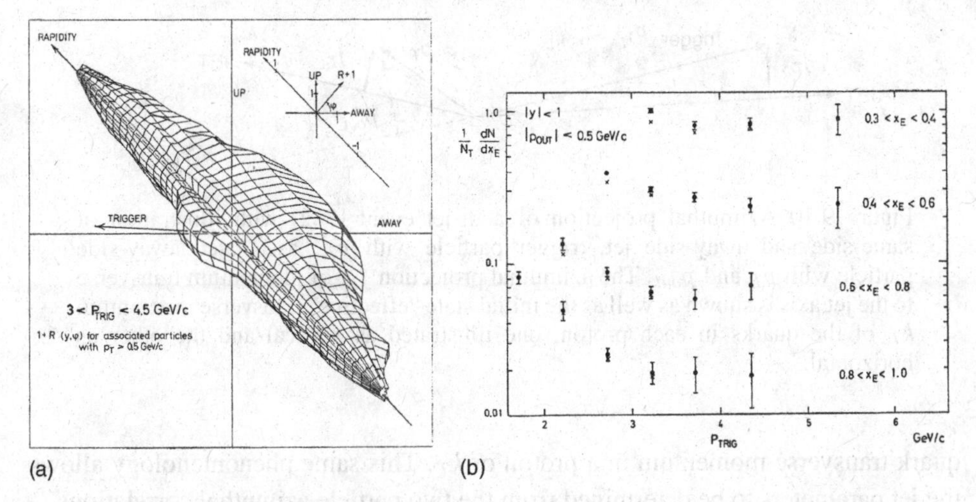

Figure 9.9 (a) BFS [385] measurement of $R + 1$ (Eq. 9.12). (b) BFS [385] measurement of x_E distributions.

of high p_{T_t} triggered events. One can also see a smaller same-side correlation extending out to $|y| \simeq 2$.

For the x_E distribution, in order to compare with the CCHK experiment, BFS plotted the values of dn/dx_E as a function of trigger p_{T_t} for several fixed values of x_E (Figure 9.9). In all cases, dn/dx_E decreases with increasing p_{T_t} until $p_{T_t} \simeq 3$ GeV/c whereupon dn/dx_E becomes independent of trigger p_{T_t} indicating x_E scaling for $p_{T_t} \gtrsim 4$ GeV/c. Together with the CCOR [387] and other measurements [398], at larger p_{T_t}, this provided Darriulat's "overwhelming" evidence for x_E scaling at the ISR for $p_{T_t} > 4$ GeV/c [395].

9.2.3 k_T, the transverse momentum of a quark in a nucleon

As noted above, the CCHK measurement [383] inspired Feynman, Field and Fox (FFF) [391] to introduce formally $\boldsymbol{k_T}$, the transverse momentum of a parton in a nucleon, into their model of parton–parton scattering. In this formulation, the net transverse momentum of an outgoing parton pair is $\sqrt{2}k_T$, which is composed of two orthogonal components, $\sqrt{2}k_{T_\phi}$, out of the scattering plane, which makes the jets acoplanar, i.e. not back-to-back in azimuth, and $\sqrt{2}k_{T_x}$, along the axis of the trigger jet, which makes the jets unequal in energy. In phenomenological terms [391], quark transverse momentum is thought to be composed of two parts, the "intrinsic" part due to the quark wave function, inside the proton, and an additional part due to QCD "radiative corrections" which combine to give an effective

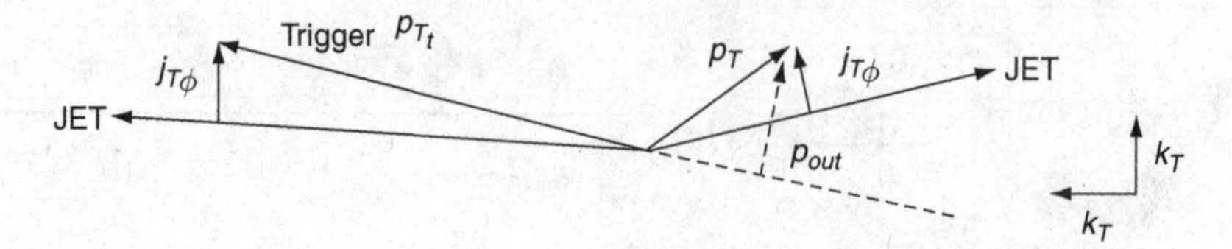

Figure 9.10 Azimuthal projection of a di-jet event [382] with illustration of same-side and away-side jet, trigger particle with p_{T_t}, associated away-side particle with p_T and p_{out}. The azimuthal projection j_{T_ϕ} of momentum transverse to the jet axis is shown as well as the initial state "effective" transverse momentum k_T of the quarks in each proton, one illustrated as vertical and the other as horizontal.

quark transverse momentum in a proton or k_T. This same phenomenology allows the jet parameters to be determined from the two particle azimuthal correlations.

Consider the azimuthal projection of a di-jet system, with each jet having a fragment with momentum $\boldsymbol{j}_T$ perpendicular to the jet axis (Figure 9.10) [382].

A fragment from one jet will be the trigger particle, and a fragment from the other jet will be the associated particle. The quarks in each proton have overall effective transverse momentum $\boldsymbol{k}_T$. Since there are two colliding protons, this gives to the di-jet system two independent vectors $\boldsymbol{k}_T$ which we may think of as acting horizontally (i.e. in the plane of the beam axis and one jet) and vertically (perpendicular to this plane). Combining the two $\boldsymbol{k}_T$ vectors randomly and taking components in the horizontal and vertical directions gives one component of magnitude $\sqrt{2}k_{T_x} = k_T$ acting horizontally and a component of magnitude $\sqrt{2}k_{T_\phi} = k_T$ acting vertically.[2] The horizontal component produces a smearing effect, analogous to resolution, which on the average increases the trigger jet transverse momentum relative to the away jet. This is the effect that supposedly spoils x_E scaling at low p_{T_t}. The vertical component contributes to p_{out}.

Feynman, Field and Fox [391,397], gave the approximate formula to derive both j_T and k_T from the measurement of p_{out} as a function of x_E:[3]

$$\langle|p_{out}|\rangle^2 = x_E^2\,[2\langle|k_{T_\phi}|\rangle^2 + \langle|j_{T_\phi}|\rangle^2] + \langle|j_{T_\phi}|\rangle^2, \tag{9.13}$$

where $\langle|k_{T_\phi}|\rangle$ and $\langle|j_{T_\phi}|\rangle$ are the average values of the components of k_T and j_T perpendicular to the scattering plane.

CCOR [388] made plots of $\langle|p_{out}|\rangle^2$ versus x_E^2 (Figure 9.11a) from azimuthal distributions (as in Figure 9.3b) for four intervals of trigger p_{T_t}, 3–5, 5–7, 7–9, 9–12 GeV/c, and three values of $\sqrt{s}$, 31, 45, 62 GeV, and performed a fit to Eq. 9.13

[2] Each vector $\boldsymbol{k}_T$ has components k_{T_x} and k_{T_ϕ}; but they add randomly for the two initial state quarks giving a factor $\sqrt{2}$. See Appendix F.

[3] The same relation with all quantities as r.m.s. values is also valid.

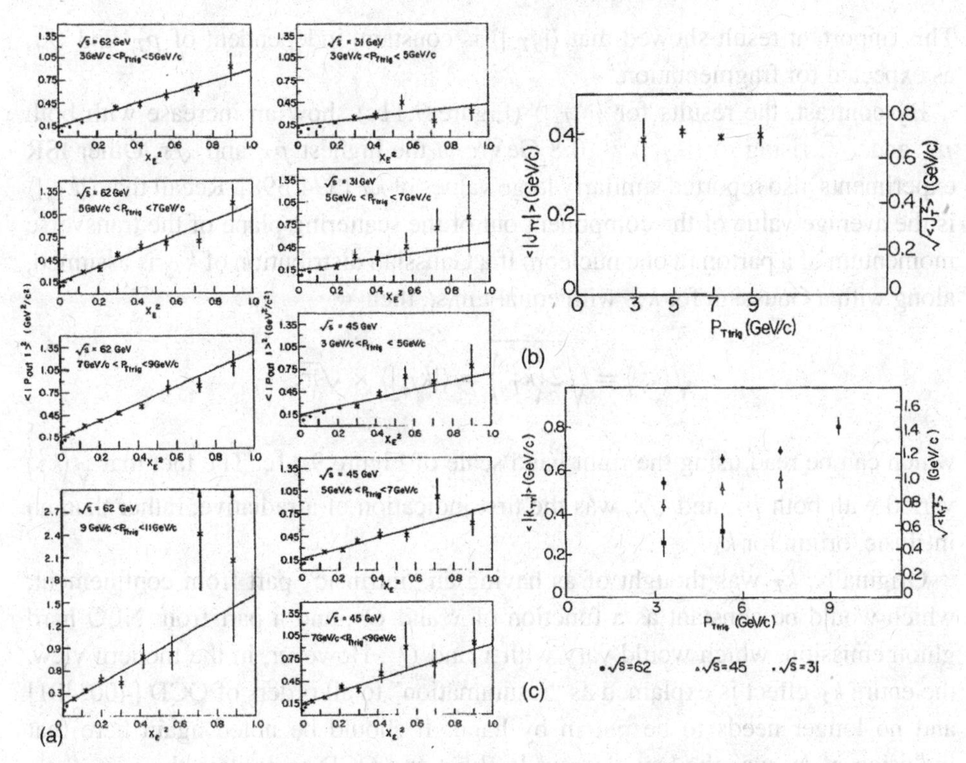

Figure 9.11 (a) CCOR [388] $\langle|p_{out}|\rangle^2$ versus x_E^2 for nine different data samples. The crosses are those points used in the fit to Eq. 9.13. The straight lines shown are obtained keeping the intercept the same for all nine data samples. (b) Fitted values of $\langle|j_{T_\phi}|\rangle$ and $\sqrt{\langle j_T^2\rangle}$ as a function of p_{T_t}. These values were all equal as a function of $\sqrt{s}$ and constrained to be equal in the fit shown. (c) $\langle|k_{T_\phi}|\rangle$ and $\sqrt{\langle k_T^2\rangle}$ as a function of p_{T_t} for the three values of $\sqrt{s}$ indicated.

for each of the nine data sets. It is clear that the data in Figure 9.11a do not satisfy Eq. 9.13 for the complete range of x_E^2. At low x_E^2 a departure from linearity is expected because j_{T_ϕ} is kinematically constrained to be small when track momenta are small. This is known as the "seagull effect" (recall Section 2.2.4). If only those points are used which correspond to $p_T > 1.4$ GeV/c, for which the kinematic constraint is small, reasonable χ^2/d.o.f. are obtained for straight line fits.

The results for $\langle|j_{T_\phi}|\rangle$ are independent of $\sqrt{s}$ and p_{T_t} (Figure 9.11b) with an average value of $\langle|j_{T_\phi}|\rangle = 0.393 \pm 0.007$ GeV/c or $\sqrt{\langle j_T^2\rangle} = 0.697 \pm 0.013$ GeV/c, assuming Gaussian distributions with the equal r.m.s. for both components, j_{T_ϕ} and j_{T_x}. These results for tracks with $p_T > 1.4$ GeV/c were originally thought to be larger than the results for jets measured in e^+e^- collisions, but actually were in agreement once the e^+e^- measurements took account of the "seagull effect" [399].

This important result showed that $\langle|j_{T_\phi}|\rangle$ is constant, independent of p_{T_t} and $\sqrt{s}$, as expected for fragmentation.

By contrast, the results for $\langle|k_{T_\phi}|\rangle$ (Figure 9.11c) show an increase with both p_{T_t} and $\sqrt{s}$, rising to $\langle|k_{T_\phi}|\rangle \sim 0.8$ GeV/c at the highest p_{T_t} and $\sqrt{s}$. Other ISR experiments also reported similarly large values of k_T [394,398]. Recall that $\langle|k_{T_\phi}|\rangle$ is the average value of the component out of the scattering plane of the transverse momentum of a parton in one nucleon. If a Gaussian distribution of k_{T_ϕ} is assumed, along with a Gaussian for k_{T_x} with equal r.m.s., then

$$\sqrt{\langle k_T^2\rangle} = \sqrt{2\langle k_{T_\phi}^2\rangle} = \langle|k_{T_\phi}|\rangle \times \sqrt{\pi},$$

which can be read using the right hand scale of Figure 9.11c. The fact that $\sqrt{\langle k_T^2\rangle}$ varied with both p_{T_t} and $\sqrt{s}$, was the first indication of a radiative, rather than an intrinsic, origin for k_T.

Originally, k_T was thought of as having an "intrinsic" part from confinement, which would be constant as a function of x and Q^2, and a part from NLO hard gluon emission, which would vary with x and Q^2. However, in the modern view, the entire k_T effect is explained as "resummation" to all orders of QCD [400,401] and no longer needs to be put in by hand. It should be noted again here that inclusion of k_T was the key element [69] beyond QCD to explain the $n \simeq 8$ x_T scaling (see Section 12.8) result of the original CCR [3] high p_T discovery and the FNAL (fixed target) experiments [273]. More recent FNAL fixed target measurements [402] and many theoretical works have used k_T as an empirical parameter to improve the comparison of measurement to NLO QCD. It is important to remember, as illustrated above, that k_T is not simply a parameter, it can be measured. It is also worthwhile to emphasize that the k_T effect is qualitatively different from NLO QCD. The Gaussian nature of k_T, which is distinctly different from the NLO power-law tail, was crisply illustrated (recall Figure 8.19) by measurements of the net transverse momentum distribution of "Drell–Yan" di-muons produced in p–p collisions [322], a process which has zero net p_T in LO and diverges in NLO but is well behaved when resummation is included [373].

9.3 Same-side and spectator region measurements

9.3.1 Same-side correlations

CCOR [387], with full azimuthal acceptance, was able to divide the detector into three regions of equal azimuthal coverage: same-side, $\Delta\phi = \pm 60°$ about the trigger; away-side, $\Delta\phi = \pm 60°$ opposite to the trigger; spectator $\Delta\phi = \pm 30°$ about

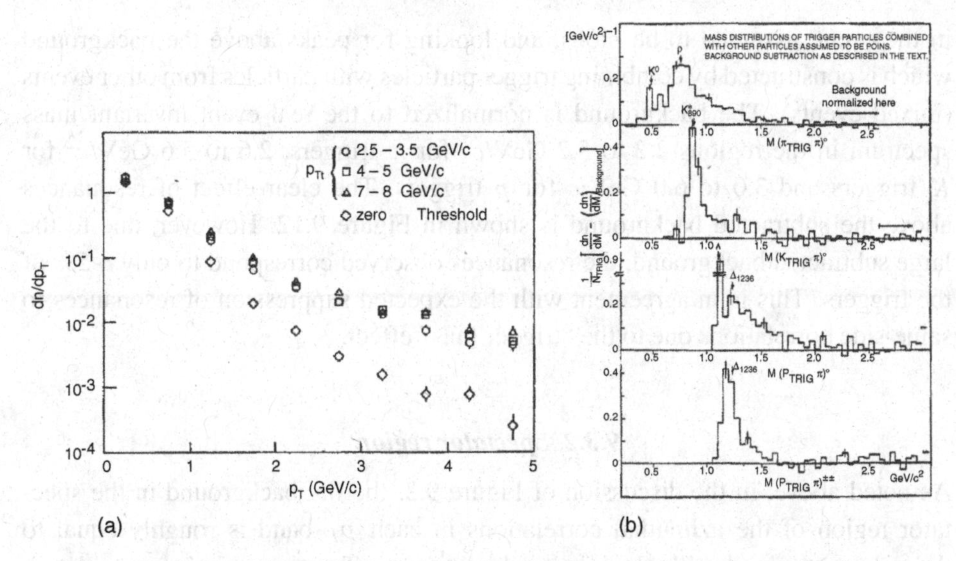

Figure 9.12 (a) CCOR [387] dn/dp_T distributions of charged particles in the same-side region for three sets of trigger p_{T_t}. The zero threshold (minimum bias) spectrum is also shown. (b) BFS [385] two particle invariant mass distributions of the systems of a trigger particle π, K, p combined with other particles observed in the SFM and assumed to be pions. The mixed-event background has been normalized to the real-event spectrum in the regions indicated and subtracted.

the two axes perpendicular to the trigger. The polar angle coverage for all these three regions was $|y| < 0.7$. The same- and away-side regions generally covered the entire jet-like peaks in the azimuthal distributions (recall Figure 9.3) while the spectator region, perpendicular to the trigger, is where the azimuthal distributions are essentially flat.

To examine the correlations of same-side charged particles with respect to the trigger, CCOR plotted the distribution of the transverse momentum p_T of all associated charged particles in the same-side region for three values of π^0 trigger p_{T_t} (Figure 9.12a).

These data indicate that the unscaled associated same-side charged particle spectra do not change as a function of trigger p_{T_t} in the range $2.5 < p_{T_t} < 8.0$ GeV/c, but that the associated spectra are quite different (much flatter) than the zero threshold (minimum bias) spectrum. This is quite different from the away-side x_E scaling and is an indication of the trigger bias.

BFS [385] took a different point of view for the same-side correlations and attempted to understand what fraction of the trigger-side correlation effect could be explained by the presence of resonances. The resonances are searched for by combining three species of trigger particles, π, K, p, with other same-side particles in the event, assumed to be pions, and looking for peaks above the background which is constructed by combining trigger particles with particles from other events (mixed events). The background is normalized to the real event invariant mass spectrum in the regions 2.2 to 5.2 GeV/c^2 for π triggers, 2.6 to 5.6 GeV/c^2 for K triggers and 3.0 to 6.0 GeV/c for p triggers. The clear effect of resonances above the subtracted background is shown in Figure 9.12. However, due to the large subtracted background, the resonancés observed correspond to only ~5% of the triggers. This is in agreement with the expected suppression of resonances to same-side correlations due to the "trigger bias" effect.

9.3.2 Spectator region

As noted above, in the discussion of Figure 9.3, the flat background in the spectator region of the azimuthal correlations in each p_T band is roughly equal to the value measured with a minimum bias trigger. To study this in more detail, CCOR [387] measured the p_T spectrum of associated particles in the spectator region as a function of p_{T_t} and compared it to the minimum bias (zero threshold) p_T distribution (Figure 9.13a). The data clearly show that the spectator conditional yield is independent of p_{T_t} and follows the inclusive (minimum bias) p_T spectrum

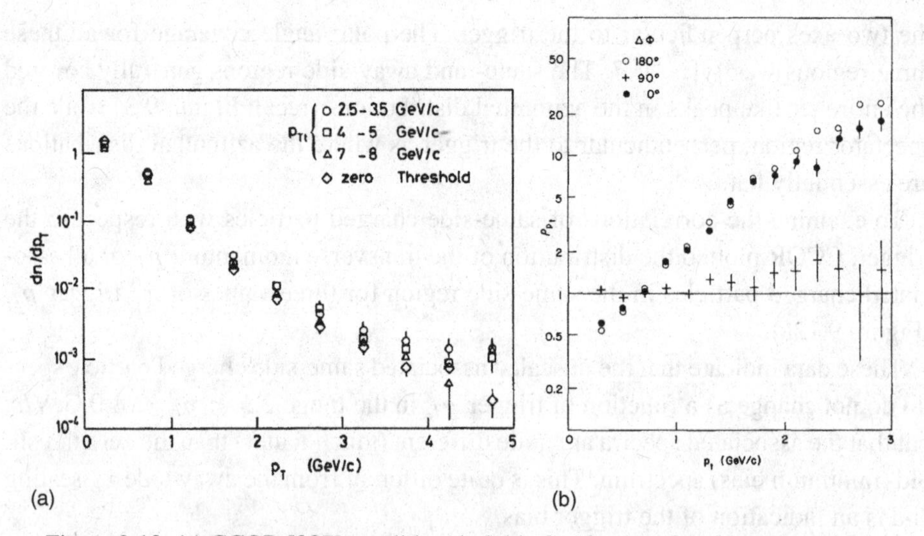

Figure 9.13 (a) CCOR [387] conditional yield, $dn/dp_T|_{p_{T_t}}$ for charged particles in the spectator region for three values of trigger p_{T_t} and minimum bias (zero threshold) collisions at $\sqrt{s} = 62.4$ GeV. (b) CERN–Saclay [394] measurement of the normalized p_T distributions in large p_{T_t} (>5 GeV/c) triggers to those in minimum bias events. Three different configurations with respect to the large p_{T_t} trigger are considered, with azimuthal separations $\Delta\phi = 0°$, 90° and 180°.

all the way out to values of p_T of 4 or 5 GeV/c which is well into the high p_T region which is dominated by hard scattering. The same effect was observed by the CERN–Saclay group [394] who plotted the ratio of the spectator yield, for a trigger with $p_{T_t} > 5$ GeV/c, to the inclusive p_T spectrum in azimuthal slices around 0°, 90° and 180° with respect to the trigger, which are smaller but similar in concept to the same-side, spectator and away-side regions of CCOR. Again the result is clear: the spectator particles which are not associated with the correlation peaks of the main high p_{T_t} collision exhibit roughly the same p_T behavior as the inclusive cross section out to regions where hard scattering dominates. One possible interpretation at the time [403] was that the spectators with $p_T \gtrsim 2$ GeV/c are caused by the scattering of the constituents remaining after the initial high p_{T_t} scattering.

It is worth noting that a later observation at larger values of $\sqrt{s} = 540$ GeV, at the CERN-SPS collider, found a "pedestal effect" [404], namely that the conditional yield in the spectator region for events with hard scattered jets was substantially higher than the minimum bias distribution. This was followed by the observation of double parton scattering at the ISR [405], and in general led to future studies of the "underlying event" [406] which continue to the present day [407].

9.4 Early direct searches for jets, isotropy of j_T

9.4.1 Isotropy of the momentum transverse to the jet axis, j_T

The variation of $\langle |p_{out}| \rangle$ with x_E is a simple and elegant way to determine the parameter j_T of jet fragmentation, but it does not prove that jets exist. However, CCOR [387] extended this analysis to show direct evidence for jets of particles with limited momentum distributed isotropically transverse to a common axis. Using the away-side region defined above, CCOR reconstructed a "jet," defined as the vector sum of all associated charged particles with transverse momenta $p_T > 0.3$ GeV/c in the away region, for events with a π^0 trigger with $p_{T_t} > 7$ GeV/c. Events were selected which have the pseudorapidity of the jet, $|\eta_{jet}| < 0.3$, and its transverse momentum $p_{T jet} > 3$ GeV/c. The first cut insures that events are centered in the apparatus. The second cut eliminates events containing only low energy spectator hadrons. The mean projected momentum of fragments transverse to the sum vector is calculated on two orthogonal axes, one of which lies in the azimuthal plane, $\langle |j_{T_\phi}| \rangle$, and one along the beam axis, $\langle |j_{T_\theta}| \rangle$. These two projections are plotted in Figure 9.14 as a function of p_T of the jet fragment for all jets with $|\sum p_{T_i}| > 3$ GeV/c. The two projected mean transverse momenta are observed to be equal to each other and limited to the value of $\langle |j_{T_\phi}| \rangle = \langle |j_{T_\theta}| \rangle = 0.35$ GeV/c, independently of the p_T of the jet fragment. This is really an indication of the jettiness of individual events, since the lower solid curve on Figure 9.14b indicates what

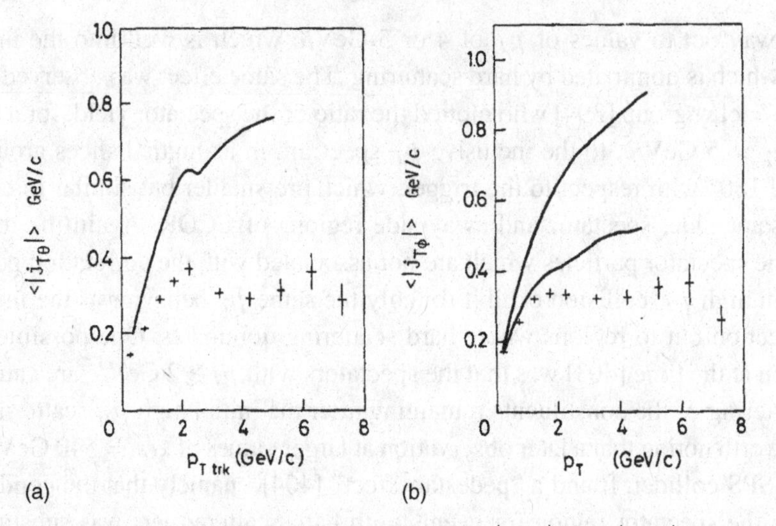

Figure 9.14 CCOR [387] measurements of (a) $\langle |j_{T\theta}| \rangle$, (b) $\langle |j_{T\phi}| \rangle$ of charged particles relative to the away-side vector sum as described in the text. The lower solid line on (b) corresponds to a Monte Carlo prediction for uncorrelated particles with $p_T \leq 5$ GeV/c distributed randomly according to the measured away-side azimuthal distribution, while the solid line on (a) and the upper solid line on (b) correspond to random flat distributions in θ and ϕ respectively.

would be expected from uncorrelated particles having the observed away azimuthal distributions shown in Figure 9.3b. The momentum radial to the jet axis can be derived from the two projections with result:

$$\begin{aligned} \langle j_T \rangle &= 0.55 \pm 0.06\,\text{GeV}/c \\ \sqrt{\langle j_T^2 \rangle} &= 0.62 \pm 0.07\,\text{GeV}/c \end{aligned} \tag{9.14}$$

in excellent agreement with the result from the $\langle |p_{out}| \rangle$ versus x_E fits (Figure 9.11a). It is interesting to notice the "seagull effect" for the low p_T tracks as well as the dip at $p_T = 3$ GeV/c corresponding to events with only one away-side track with p_T greater than the threshold.

9.4.2 Di-jets in a calorimeter – direct measurement of k_T

One of the most enlightening measurements searching for di-jet structure using hadron calorimeters in this period was made by the Fermilab–Lehigh–Pennsylvania–Wisconsin (FLPW, sometimes WPLF) collaboration, at Fermilab [408, 409], using beams of 130, 200 and 400 GeV protons and pions scattered from a liquid hydrogen target. The apparatus, shown in Figure 9.15a, consisted of a two-arm segmented hadron calorimeter array and six planes of drift chambers.

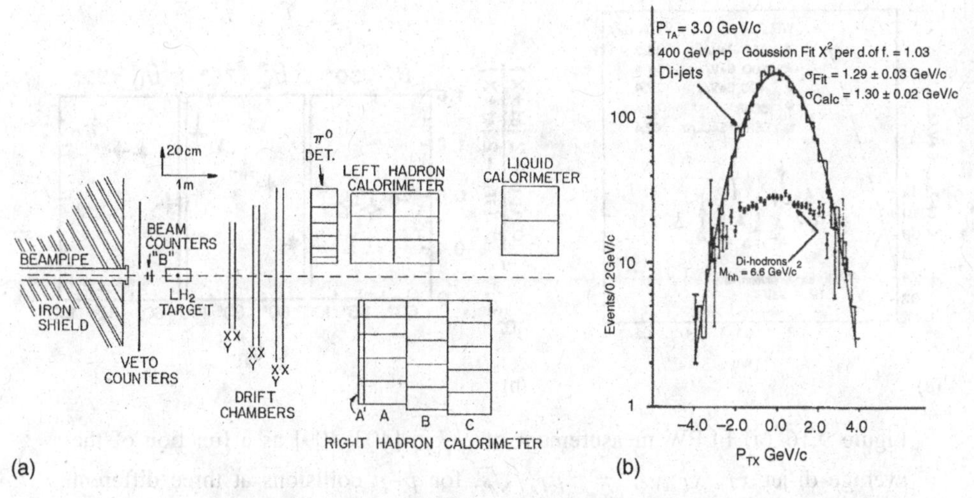

Figure 9.15 (a) Plan view of WPLF apparatus [408]. The anticounters vetoed charged particles emerging from the iron shield. The target was 45 cm of liquid hydrogen. (b) p_{T_x} distribution for di-jets with average transverse momentum $p_{T_A} = 3.0$ GeV/c centered in a band 0.8 GeV/c wide (histogram) [409]. The same distribution for $\pi^+\pi^-$ pairs with mass, $M_{hh} \equiv 2p_{T_A} = 6.6$ GeV/c^2 from the CFS experiment [410] is shown for comparison (data points).

Each calorimeter arm covered a solid angle of about 1.5 sr. A massive iron and concrete shield effectively eliminated beam halo particles and other upstream sources of background.

A jet was defined in each calorimeter as the sum of the energies measured in the individual cells weighted by the sines of their angles in the laboratory frame, i.e.

$$p_T \equiv E_T = \sum_i E_i \sin\theta_i .$$

A fiducial cut selected jets pointed at the central $\pm 10°$ in c.m.s. polar and azimuthal angles in each calorimeter. They denote the imbalance and the average of the transverse momenta of the di-jet as $p_{T_x} = p_{T\,left} - p_{T\,right}$ and $p_{T_A} = [p_{T\,left} + p_{T\,right}]/2$, respectively. The distribution of p_{T_x} for di-jets in a range of $p_{T_A} = 3.0$ GeV centered in a band 0.8 GeV wide is shown in Figure 9.15b [409] compared to a dihadron ($\pi^+\pi^-$) measurement from the Columbia–Fermilab–StonyBrook (CFS, sometimes SCF) experiment [410] with approximately the same range of p_{T_A}, which illustrates dramatically that di-jets and di-hadrons behave differently. The jets show a strong peaking about $p_{T_x} = 0$ with a Gaussian behavior. This implies that the two jets balance transverse momentum apart from the smearing, k_T, which can be obtained directly from the width of the Gaussian (Figure 9.15b) after two corrections. The corrections are from

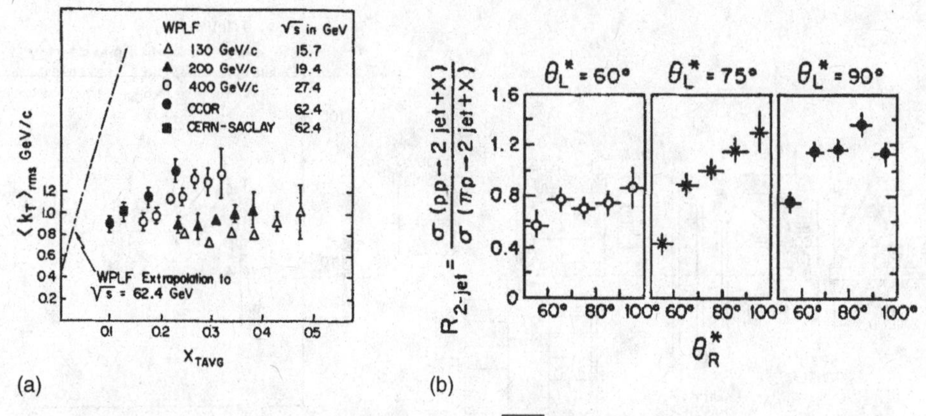

Figure 9.16 (a) FLPW measurement of $\sqrt{\langle k_T^2 \rangle}$ [403, 409] as a function of the average di-jet x_T, $x_{TAVG} = 2p_{T_A}/\sqrt{s}$, for p–p collisions at three different $\sqrt{s}$ compared to the CCOR [388] and CERN–Saclay [394] measurements at $\sqrt{s} = 62.4$ GeV. The WPLF extrapolation of their results to $\sqrt{s} = 62.4$ GeV is also indicated. (b) FLPW [408] angular correlation results for the ratio of proton induced di-jets to pion induced di-jets with $p_{T_A} \sim 2.6$ GeV/c for various combinations of c.m. scattering polar angles, θ_L^* and θ_R^*, of the individual jets.

two sources: instrumental resolution and uncollected jet fragments. The latter correction is model dependent since it depends on the assumed parameters of the jets, particularly the fragmentation function and the fragmentation transverse momentum $\langle j_T \rangle$. It should be noted as a warning that the value of $\langle j_{T_\phi} \rangle = 0.300$ GeV/c used by WPLF for the corrections [409], taken from the then available e^+e^- measurements, is 30% lower than the value measured for hadron collisions as shown above in Figure 9.11b.

The corrected results for $\sqrt{\langle k_T^2 \rangle}$ [409] are shown in Figure 9.16a as a function of $x_{TAVG} = 2p_{T_A}/\sqrt{s}$ compared to the CCOR measurements from Figure 9.11c [388]. At a given $\sqrt{s}$, the WPLF values of $\sqrt{\langle k_T^2 \rangle}$ increase with increasing p_{T_A}; but this effect is not very compelling for $\sqrt{s}$ below 27.4 GeV. The WPLF data also show that, at fixed x_{TAVG}, $\sqrt{\langle k_T^2 \rangle}$ increases with increasing $\sqrt{s}$. In the absence of quantitative theoretical predictions, the WPLF group give their own empirical extrapolation to $\sqrt{s} = 62.4$ GeV which unfortunately extrapolated wildly above the ISR measurements as shown on Figure 9.16a. Note that this discrepancy may not be due entirely to the form of the empirical relation since at fixed p_T the WPLF values of $\sqrt{\langle k_T^2 \rangle}$ show a consistent increase with increasing $\sqrt{s}$, while the CCOR results at $\sqrt{s} = 62.4$ GeV, at a given p_T, are below the values given by WPLF at $\sqrt{s} = 27.4$ GeV. Of course since any missing tracks from the jets will generally give rise to larger values of jet imbalance or p_{T_x}, it is not surprising in hindsight that the values of $\sqrt{\langle k_T^2 \rangle}$ from the early di-jet measurement are larger those from di-hadrons.

9.4.3 First evidence for constituent kinematics of the parton model

In addition to their work on transverse momentum balance of azimuthally back-to-back jets using two hadron calorimeters of 1.5 sr, the WPLF experiment also produced the first evidence for longitudinal constituent kinematics [408]. By comparing jet pairs produced in p–p collisions and π^+–p collisions at $\sqrt{s} = 15.7$ GeV, a qualitative difference could be found in the polar angle correlation of jets produced by incident protons or pions [408]. Denoting θ_L^* and θ_R^* as the c.m.s. polar angles of the jet vectors in each calorimeter relative to the direction of the incident π^+ or proton projectile, the ratio of proton induced di-jets to pion induced di-jets with $p_{T_A} \sim 2.6$ GeV/c is shown in Figure 9.16b as a function of θ_L^* and θ_R^*. As either angle moves forward, the ratio of proton to pion induced jets decreases, indicating that pion induced di-jets tend to be produced more forward. This is exactly what is expected from constituent kinematics. The pion has two quarks (actually a quark and an antiquark) but the proton has three quarks. Thus, each quark in a pion will carry a larger fraction of the total longitudinal momentum than would the quarks in a proton. Consequently the c.m. system for the constituent scattering in pion–proton collisions will have a much greater tendency to be moving forward than it would in proton–proton collisions.

9.5 Symmetric di-hadron cross sections

A major breakthrough came with the realization by both theorists and experimentalists that measurements of the cross section of symmetric di-hadrons, typically from a two particle back-to-back trigger, would be much more straightforward to understand in the parton–parton scattering model than single particle measurements. The smearing effect of k_T, which dramatically affects the single inclusive cross section due to the steeply falling spectrum, would be eliminated or strongly mitigated by a symmetric trigger, since this would select initial state configurations with overall $k_{T_x} = 0$. For theorists, this meant that the di-hadron cross sections could be calculated in the parton model or QCD, "without being influenced by the internal momentum of the constituents" [411]. In fact, such a QCD calculation [411] without k_T was in excellent agreement with the CFS [410] six-fold di-pion inclusive cross section measurement $E_1 E_2 d^6\sigma/dp_1^3 dp_2^3$, which was shown as the p_{T_x} distribution of $\pi^+\pi^-$ pairs in Figure 9.15b.

From the experimental viewpoint, experiments such as CCOR [390] and CFS [335], which were primarily studying lepton pairs using symmetric triggers in two-arm spectrometers, found it natural to reconstruct their $\pi^+\pi^-$ (CFS) or π^0–π^0 (CCOR) pairs in the same kinematic variables as they would for di-leptons, namely: invariant mass, m; the net transverse momentum of the pair, P_T; and the rapidity of the pair, Y, in the p–p c.m. system. CCOR also used $\cos\theta^*$, the polar angle of the di-pion in the parton–parton c.m. system in which the net rapidity of the pair, $Y^* = 0$. For instance, CCOR [390] accumulated "a background" of 250 000 π^0–π^0 pair events from their (electron) pair-trigger, a cluster with at least 2.5 GeV deposited in each of their two PbGl arrays (Figure 9.2b), due to the absence of electron identification at the trigger level as in CCRS (Figure 7.1). The downside of this was that $J/\Psi \rightarrow e^+e^-$ could not be measured by CCOR. However, there was an unanticipated upside from the beautiful di-hadron measurement that these large statistics enabled.

The first result presented by CCOR [390] was that the measured $\langle z_{trig}\rangle$ for pair triggers is less than for single particle inclusive triggers, both measured as described above for single triggers (compare Figure 9.4a to Figure 9.17a, especially near the trigger threshold). In simple terms, this means that symmetric triggers break the strong single particle trigger bias. This occurs because the balance

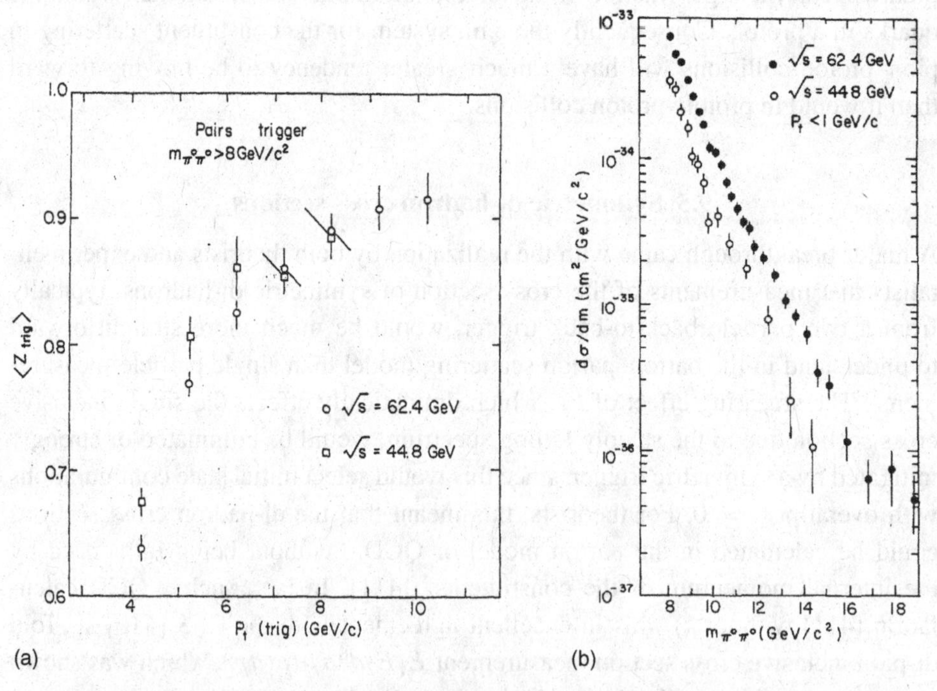

Figure 9.17 (a) CCOR [390] measurement of $\langle z_{trig}\rangle$ as a function of p_{Tt} for π^0 pairs with invariant mass, $m_{\pi^0\pi^0} > 8$ GeV/c^2. (b) $d\sigma/dm$ of π^0 pairs with $P_T < 1.0$ GeV/c as defined in Eq. 9.15.

between the steepness of the parton p_T spectrum compared to the fragmentation function is more weighted against large z in the di-hadron case, since the effect of the fragmentation function is squared for the two fragmentation functions, hence steeper. It is less probable for both fragments to go to large z than for a single fragment.

CCOR also made measurements of the invariant mass and polar angle distributions of the di-pions [390]. In order for the defined $\cos\theta^*$ of the π^0 pair to approximate more closely the c.m.s. polar scattering angle of the parton pair, a cut was made to select relatively balanced π^0 pairs with $P_T < 1.0$ GeV/c.[4] For acceptance purposes, restrictions were also placed: on the Y interval, from −0.35 to +0.35; and $|\cos\theta^*| < 0.4$ for the di-hadron cross section, $|\cos\theta^*| < 0.5$ for the polar angular distribution. The acceptance, which was independent of the invariant mass, was calculated in an array of small bins of dimension 0.05×0.05 in the Y, $\cos\theta^*$ plane, and corrected bin-by-bin by weighting the data.

The measurement of the invariant mass dependence of the inclusive di-pion cross section was presented as

$$\frac{d\sigma}{dm} = \left.\frac{d\sigma}{dmdY}\right|_{Y=0} = \frac{1}{0.7}\int_{-0.35}^{0.35} dY \int_{-0.4}^{+0.4} d\cos\theta^* \int_0^1 dP_T \frac{d^4\sigma}{dm\, dY dP_T d\cos\theta^*} \tag{9.15}$$

as shown in Figure 9.17b. It is interesting to point out that the contemporary leading order QCD calculation [411], without k_T, which was in excellent agreement with the CFS [410] $\pi^+\pi^-$ measurement at $\sqrt{s} = 27.4$ GeV, discussed above, was also in excellent agreement with the CCOR $\sqrt{s} = 62.4$ GeV measurement [403]. However, in the intervening period, Next to Leading Order (NLO) QCD calculations were found to agree with the measurements only with unreasonably small normalization and factorization scales [401]; and it was only very recently (2009), when the "all order resummation of large logarithmic corrections" was added [401], that the theory and experiment again came into agreement.

9.6 Measurement of $d\sigma^{ab\to cd}/d\cos\theta^*$ for parton–parton scattering

One of the most influential results from this series of CCOR di-hadron publications was the first measurement [390, 412] of the parton–parton scattering differential cross section, $d\sigma^{ab\to cd}/d\cos\theta^*$, as discussed originally by Bjorken [377], who thought it could only be done with di-jets. The key feature in this analysis is that

[4] The reference frame for $\cos\theta^*$ was found by transforming only along the original collision axis to the system in which the di-pion has zero net longitudinal momentum, $Y^* = 0$. If the π^0 have different transverse momenta, the two polar angles θ_1^* and θ_2^* are not back-to-back, in which case $\cos\theta^*$ is defined as the average of the two $|\cos\theta^*|$.

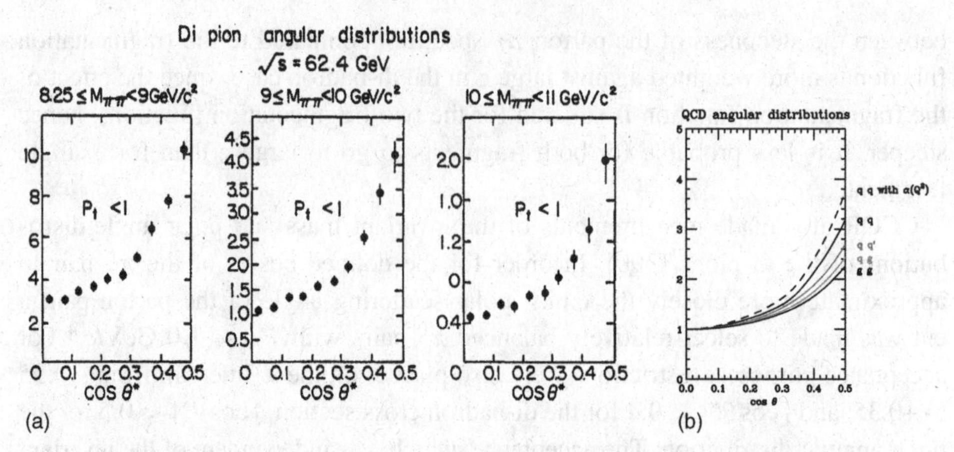

Figure 9.18 (a) CCOR measurement [390, 412] of polar angular distributions of π^0 pairs with net $P_T < 1$ GeV/c at mid-rapidity in p–p collisions with $\sqrt{s} = 62.4$ GeV for three different values of $\pi\pi$ invariant mass $M_{\pi\pi}$. Acceptance corrected number of events/1000, with $|y^* < 0.35|$ per bin of $\cos\theta^*$ in the parton–parton c.m. system. (b) QCD predictions for $d\sigma^{ab\to cd}/d\cos\theta^*$ for the elastic scattering of gg, qg, qq', qq, and qq with $\alpha_s(Q^2)$ evolution, each normalized to 1 at 90°, $\cos\theta^* = 0$.

the di-pion $\cos\theta^*$ distribution at fixed m corresponds closely to the angular distribution of the scattered partons at fixed $\hat{s}$.[5] Thus, the measured π^0 pair $\cos\theta^*$ distributions shown in Figure 9.18a could be compared directly [412], for instance, to the angular distributions of the basic subprocesses, gg, qg, qq', qq elastic scattering, predicted in QCD [413, 414] (Figure 9.18b). Notably, the measured distributions are steeper than all the QCD constituent scattering distributions [415] but are in excellent agreement with identical qq scattering provided that the increase in $\alpha_s(Q^2)$ with decreasing $\hat{Q}^2 = -\hat{t}$ at forward angles (Eq. 3.100) is taken into account (dashed curve on Figure 9.18b) as predicted by QCD [22, 23].

This measurement, presented at the 1982 ICHEP in Paris [390, 412], the same meeting at which one UA2 event was shown (Figure 11.18), which was the first true observation of high p_T di-jets in hadron collisions [416, 417], gave universal credibility to the pQCD description of high p_T hadron physics [395, 401, 418–420]. To quote the rapporteur's talk on jet production and fragmentation at this meeting [418], "Amongst the most exciting results are the direct measurement of the parton-parton scattering angular distribution and the observation of very energetic jets at the SPS collider. QCD provides a consistent description for the underlying

[5] The reason this works is that $\sqrt{\langle j_{T\phi}^2\rangle} = 0.49 \pm 0.01$ GeV/c (Figure 9.11b), so the pions are well aligned with their partons for $p_T \gtrsim 4$ GeV/c. Of course, to calculate $\hat{s}$, m^2 must be corrected for the $\langle z\rangle$ of both pions, $\sim 0.7^2$–0.9^2 from Figure 9.17.

constituent scattering processes." As if in a "phase transition," the skepticism about jets in hadron collisions,[6] evident at the famous Snowmass 1982 meeting of the US high energy physics community just three weeks earlier [421], suddenly came to an end. Since that time, QCD and jets have become the standard tools of high energy particle physics. By contrast, because of the huge background due to the large particle multiplicity in $A + A$ collisions, there were no refereed publications of jets in Au+Au collisions for the first 11 years of operation at the $\sqrt{s_{NN}} = 200$ GeV Relativistic Heavy Ion Collider (RHIC), where the techniques of single and di-hadron measurements have served as the principal probe of hard scattering.

[6] See a full discussion on this issue in Chapter 11.

10

Direct photon production

Direct single photon production at high p_T from "the inverse QCD Compton effect," i.e. the constituent reaction,

$$g + q \to \gamma + q,$$

commonly called, simply, direct-γ production, which was proposed by Fritzsch and Minkowski in 1977 [301], is one of the most beautiful hard scattering processes in p–p (Figure 10.1), or more generally in $A + B$ collisions, where A and B are distinct hadrons or nuclei.

The beauty of this reaction as a hadronic probe is that the γ ray participates directly in the hard scattering and then emerges freely and unbiased from the reaction, isolated, with no accompanying particles. Thus, no song-and-dance about jets and fragmentation is required. Furthermore, the energy of the γ ray can be measured precisely. Thus, since the scattered quark has equal and opposite transverse momentum to the direct-γ, the transverse momentum of the jet from the outgoing quark is also precisely known (modulo k_T). Also, as we have seen in Section 9.6, the direction of the jet from the away-side quark can be measured very well by a leading single particle with $p_T \gg \langle j_T \rangle \approx 0.6$ GeV/c. This means that by detection of a γ–h coincidence, where h is any hadron from the jet of the away-side quark, the full kinematics of the parton–parton scattering in a p–p collision can be determined, for the conventional assumption of zero or negligible masses of the quarks and gluons.

Let us suppose that the outgoing direct-γ is detected in the p–p c.m. system with transverse momentum p_T and rapidity y_c and an outgoing hadron (or jet) from the outgoing quark is detected at rapidity y_d, then from Section 3.4.5, Eqs. 3.116, 3.117, the solution for x_1, x_2 of the initial state quark and gluon in their respective protons is:

$$x_1 = x_T \frac{e^{y_c} + e^{y_d}}{2} \qquad x_2 = x_T \frac{e^{-y_c} + e^{-y_d}}{2}, \tag{10.1}$$

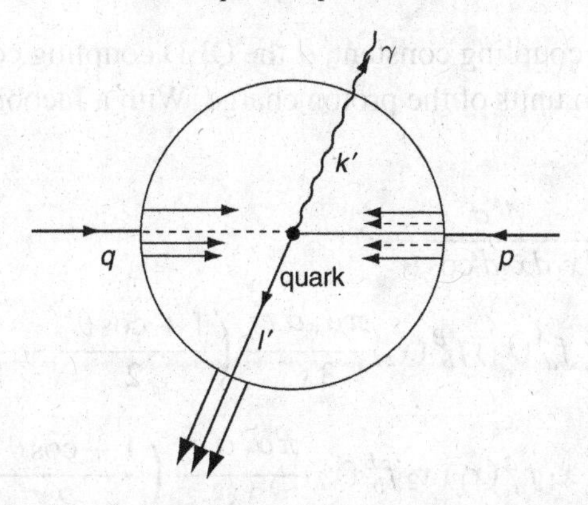

Figure 10.1 "The inverse QCD Compton effect": an incoming gluon (dashed line) from a hadron labeled q interacts with a quark from hadron p to produce a quark with four-momentum l' and a γ ray with four-momentum k' as seen in the $g+q$ c.m. system [301], where the g can be emitted from either hadron q or hadron p.

where $x_T = 2p_T/\sqrt{s}$, although, of course, we cannot tell whether x_1 or x_2 is the gluon or quark. Continuing in the same vein, the parton–parton c.m. energy $\sqrt{\hat{s}} = \sqrt{x_1 x_2 s}$, where $\sqrt{s}$ is the p–p c.m. energy; the energy of the direct-γ in the parton–parton c.m. system is $P_c^* = E_c^* = \sqrt{\hat{s}}/2$, where with help from Eqs. 3.98, 2.6:

$$p_T = p_T^* = \frac{\sqrt{\hat{s}}}{2}\sin\theta^* \qquad \sqrt{\hat{s}} = 2p_T \cosh\frac{(y_c - y_d)}{2}. \tag{10.2}$$

The c.m. scattering angle $\cos\theta^*$ is (from Eq. 3.110)

$$\cos\theta^* = \tanh\frac{(y_c - y_d)}{2}, \tag{10.3}$$

or alternatively for $\sin\theta^*$

$$p_T = p_T^* = \frac{\sqrt{\hat{s}}}{2}\sin\theta^* \qquad \sin\theta^* = \frac{x_T}{\sqrt{x_1 x_2}}. \tag{10.4}$$

Another beautiful thing about this reaction is that the cross section for the constituent scattering subprocess, $g + q \to \gamma + q$, is given by a simple analytical expression [301]

$$\frac{d\sigma^{g+q\to\gamma+q}}{d\hat{t}} = \frac{\pi\alpha_s\,\alpha\, e_q^2}{3\,\hat{s}^2}\left(\frac{\hat{s}+\hat{t}}{\hat{s}} + \frac{\hat{s}}{\hat{s}+\hat{t}}\right), \tag{10.5}$$

or using $\hat{t} = -\hat{s}(1-\cos\theta^*)/2$ (Eq. 3.100)

$$\frac{d\sigma^{g+q\to\gamma+q}}{d\cos\theta^*} = \frac{\pi\alpha_s\,\alpha\, e_q^2}{6\,\hat{s}}\left(\frac{1+\cos\theta^*}{2} + \frac{2}{1+\cos\theta^*}\right), \tag{10.6}$$

where α_s is the QCD coupling constant, α the QED coupling constant and e_q is the charge on the quark in units of the proton charge. With a Jacobian, and substitution in Eq. 9.2, we obtain:

$$\begin{aligned}\frac{d^3\sigma}{dp_T^2 dy_c dy_d} &= \frac{2}{s}\frac{d^3\sigma}{dx_1 dx_2 d\cos\theta^*} \\ &= \sum_{a,b} f_a^A(x_1) f_b^B(x_2) \frac{\pi\alpha_s \alpha e_q^2}{3\,s\,\hat{s}}\left(\frac{1+\cos\theta^*}{2}+\frac{2}{1+\cos\theta^*}\right) \\ &= \sum_{a,b} x_1 f_a^A(x_1)\, x_2 f_b^B(x_2) \frac{\pi\alpha_s \alpha e_q^2}{3\,\hat{s}^2}\left(\frac{1+\cos\theta^*}{2}+\frac{2}{1+\cos\theta^*}\right)\end{aligned} \tag{10.7}$$

which is the cross section for production of a direct-γ with p_T at y_c balanced in p_T by a jet at y_d. It is again important to emphasize that parton a with x_1 can be either the gluon or a quark, and parton b can be either a quark or the gluon, so both possibilities must be taken into account. Thus the sum over a, b includes both cases: $a = g$ with $b = u, \bar{u}, d, \bar{d}$; and $a = u, \bar{u}, d, \bar{d}$ with $b = g$, which gives for the reaction $A + B \rightarrow \gamma + q$

$$\begin{aligned}\frac{d^3\sigma}{dp_T^2\, dy_c\, dy_d} &= x_1 f_g^A(x_1)\, F_{2B}(x_2) \frac{\pi\alpha\alpha_s(Q^2)}{3\hat{s}^2}\left(\frac{1+\cos\theta^*}{2}+\frac{2}{1+\cos\theta^*}\right) \\ &+ F_{2A}(x_1)\, x_2 f_g^B(x_2) \frac{\pi\alpha\alpha_s(Q^2)}{3\hat{s}^2}\left(\frac{1-\cos\theta^*}{2}+\frac{2}{1-\cos\theta^*}\right)\end{aligned} \tag{10.8}$$

where $\cos\theta^*$ is the c.m. angle of the outgoing γ with respect to hadron A; $f_g^A(x_1, Q^2)$ (also denoted $g_A(x_1, Q^2)$) and $f_g^B(x_2, Q^2)$ are the gluon structure functions of hadron A and hadron B; and $F_{2A}(x_1, Q^2)$, $F_{2B}(x_2, Q^2)$ are exactly the $F_{2A}(x, Q^2) = x \sum_a e_a^2 f_a^A(x, Q^2)$ structure functions measured in DIS of $e + A$ (recall Eq. 9.3), where $f_a^A(x, Q^2)$ are the distributions in the number of quarks of type a, with electric charge e_a (in units of the proton charge) in hadron A. The nominal four-momentum transfer squared, Q^2, is included in the structure functions because this is now a QCD calculation which has a running coupling constant $\alpha_s(Q^2)$. Clearly for the QCD Compton effect, the reaction is s-channel and an experimentalist would think that therefore $Q^2 = \hat{s}$. However, in actual perturbative QCD calculations, there are more subtle issues, particularly when higher order effects are involved.

There is also another lowest order contributor to direct-γ production: $q + \bar{q} \rightarrow \gamma + g$. However, this is suppressed for two reasons: generally $u(x) \gg \bar{u}(x)$, $d(x) \gg \bar{d}(x)$; also, the color as well as the electric charge must balance for the annihilating quarks (as in Drell–Yan). Hence, it is a reasonable approximation to say that in p–p collisions the quark opposite the direct-γ is 8/1 u/d, since there are two u quarks, with $e_q^2 = 4/9$ and one d quark with $e_q^2 = 1/9$ in a proton. In this same approximation, measurement of the direct-γ cross section or the direct-$\gamma + h$ (or jet) cross section is a direct measurement of the gluon structure function, since the F_2 structure functions are well measured in DIS.

Of course, nothing is perfect. Direct-γ production does have one serious problem: an overwhelming background of photons from high p_T $\pi^0 \rightarrow \gamma + \gamma$ and $\eta \rightarrow \gamma + \gamma$ decays, as well as $\omega \rightarrow \pi^0 + \gamma$, etc. Also, its rate is suppressed due to the $\alpha_s\alpha$ coupling relative to the QCD purely hadronic processes which are proportional to α_s^2. However, this will be somewhat ameliorated in the observed ratio of, for example, γ/π^0, which will be enhanced relative to the ratios from the constituent subprocesses, since the π^0, as a fragment of a jet, will suffer a parent–daughter suppression (Sections 3.4.7, 9.1.1) while the γ ray will not. Extensive discussions on the calculation and elimination of these background from π^0 and η decays such as rejecting γ rays which reconstruct to the π^0 mass with other detected γ rays, or making an isolation cut, have been given previously (Sections 3.4.8, 3.5).

It is interesting to recall that in 1966, a year before the proposal of the inverse QCD Compton effect [301], direct-γ production was originally proposed by Farrar and Frautchi [298], among others [299,300], as a mechanism to explain the copious yield of prompt leptons observed in hadron collisions (Chapter 7). Internal (Dalitz) conversion of the direct photons would produce the prompt leptons. A γ/π^0 ratio of $\sim$10% would be sufficient to explain the lepton/pion ratio of $\sim 10^{-4}$ observed for $p_T > 1.0$ GeV/c at ISR energies.

Several early experiments at the ISR looked for direct photons as the source of the prompt leptons. CCRS [283], as part of their direct electron measurements, searched for low mass e^+e^- pairs from internal conversions in 1976. An upper limit for e^+e^- pairs with mass $0.35 < m_{ee} < 0.45$ GeV/c^2, and $p_T > 1.3$ GeV/c was given, which can be converted to a 95% c.l. upper limit for real photons: $\gamma/\pi^0 <$ 5% for $p_T > 1.3$ GeV/c.

Two years later, significant improvements were achieved in the measurements of low mass virtual photons in p–p collisions at the ISR as the later e^+e^- pair experiments also looked at the same-side e^+e^- pairs. The CERN–Saclay–Zurich (CSZ) Group [422], in one arm of the two-arm spectrometer used for the best ISR J/Ψ measurement (Figure 8.5c) [329], measured same-side low mass e^+e^- pairs for $p_T > 1.6$ GeV/c. The invariant mass spectrum is shown in Figure 10.2a. The background from various Dalitz decays, $\pi^0 \rightarrow \gamma e^+e^-$, $\eta \rightarrow \gamma e^+e^-$, etc., is shown and is negligible for $m_{ee} > 0.500$ GeV/c^2. In addition to the peaks corresponding to $\rho^0 + \omega \rightarrow e^+e^-$ and $\phi \rightarrow e^+e^-$, a significant e^+e^- continuum, not due to

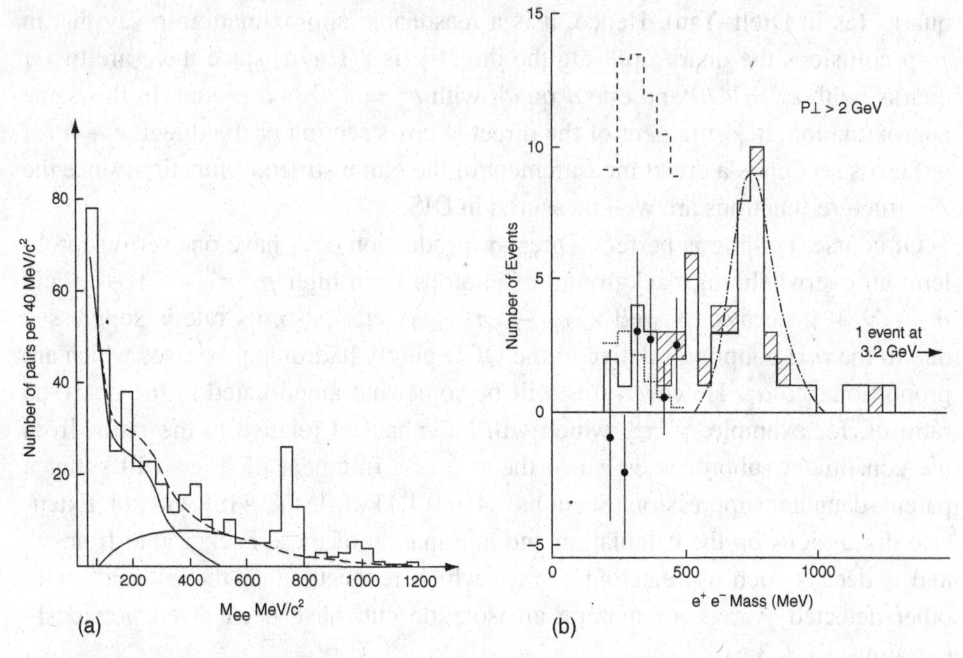

Figure 10.2 (a) CSZ [422] measurement of low mass e^+e^- pairs for $p_T > 1.6$ GeV/c. (b) AABCSY collaboration [423] invariant mass spectrum for e^+e^- pairs with $p_T > 2.0$ GeV/c. Events with $p_T > 3.0$ GeV are shaded. Dashed and dot-dash lines are explained in the text.

known Dalitz decays, is observed in the mass range $0.400 < m_{ee} < 0.600$ GeV/c^2. The authors interpreted this excess as being due to charmed particle production; however, they also quoted the upper limit of

$$\gamma/\pi^0 < 1.9\% \quad \text{at } p_T = 1.9 \text{ GeV}/c\,,$$

in case the effect were due to direct-γ production. The vector meson production observed corresponds to cross sections

$$\frac{\rho^0 + \omega}{2\pi^0} = 0.59 \pm 0.20 \quad \text{for } p_T > 2.0 \text{ GeV}/c$$

and

$$\frac{\phi}{\pi^0} = 0.12 \pm 0.05 \quad \text{for } p_T > 2.1 \text{ GeV}/c.$$

The AABCSY collaboration [423] also reported measurements on low mass e^+e^- pairs using their lithium/xenon transition radiation/liquid argon calorimeter detector, discussed previously (recall Figure 8.12) [355]. The invariant mass distribution of the observed electron pairs is shown in Figure 10.2b [423] for pairs with $2.0 < p_T < 3.0$ GeV/c. The spectrum has not been acceptance corrected, but the acceptance is largest and relatively flat from 0.30–0.50 GeV/c^2. A clear peak from the ρ^0 and $\omega \to e^+e^-$ is observed as well as a continuum for masses $0.25 \leq m_{ee} \leq 0.50$ GeV/c^2. The dotted line shows the prediction for all known Dalitz decays, and the dot-dash line the expected shape from the tails of the ρ^0 and ω. The data points represent the difference between the measured continuum and the Dalitz and ρ^0, ω contributions. The dashed line at the top left side shows the prediction for low mass pairs corresponding to $\gamma/\pi^0 = 10\%$. The conclusion is that

$$\gamma/\pi^0 \leq (0.55 \pm 0.92)\% \quad \text{for } 2.0 \leq p_T \leq 3.0 \text{ GeV}/c.$$

Thus, all the low mass virtual photon experiments were in agreement.

On the other hand, the searches for direct-γ using real photons produced incompatible results due to the difficulty of the experiments and also because they were all first generation measurements of this process and not designed expressly for the purpose. Here we concentrate on the CERN-ISR. An excellent contemporary review article [424] describes in great detail all the early (pre 1984) experiments at both Fermilab and the ISR. A session on direct photon production at the 1979 Lepton–Photon Symposium [425] also provides an interesting view of the earliest experiments.

The first search for direct single real photons was published in 1976 by a CERN group at the ISR [426], who had placed their 1 m^2 PbGl array [427] 3.6 m from the SFM collision vertex, at 90° in the c.m. system, to trigger on resolved $\pi^0 \to \gamma + \gamma$ for $2 < p_{T_t} < 4.1$ GeV/c, as previously discussed in Chapter 9 [386]. Their result [426], which gave a huge value for direct single γ production, was presented as:

$$\gamma/\pi^0 = 0.20 \pm 0.07 \pm 0.06 \quad \text{for } 2.8 \leq p_T \leq 3.8 \text{ GeV}/c.$$

The 0.07 error comes from the uncertainty in the linearity of the energy response curve of the lead glass detector, and the 0.06 from all the other sources of error. Note that the raw γ/π^0 signal observed was 0.46, which must then be corrected for single photon background from $\pi^0 \to \gamma\gamma$, $\eta \to \gamma\gamma$ and other similar decays.

Although this result is now considered to be anomalous, the experiment made some important observations which must be addressed in direct-γ measurements. Most important was the effect of non-linearity of the detector, i.e. two individual 3 GeV photon clusters must give precisely 1/2 the energy of one 6 GeV photon cluster. For instance, if the 3 GeV photons were measured correctly but the photon cluster measured as 6 GeV were really 5.7 GeV, this would give an apparent huge excess of ~60% single photons. Another check of non-linearity, as well as a test of whether the two γ from the π^0 decay background appear as a single γ

due to merging, is to plot the distribution in the energy asymmetry, α, of the two photons from reconstructed π^0 of a given momentum P, which should have a uniform distribution over the range $0 \leq \alpha \leq \beta$, where β is the velocity (divided by c) of the π^0 (recall Eqs. 3.153, 3.154). Figure 10.3a [426] shows this distribution, which is flat for the symmetric decay, $\alpha \approx 0$, thus indicating no merging problem, and drops off at large α due to the acceptance. However, the measured values in this asymmetric-pair region, $\alpha \gtrsim 0.5$, are consistently larger than predicted, which might possibly indicate a non-linearity. Another issue might be the correct subtraction of the π^0 background because this same detector gave a π^0 spectrum [427] which disagreed with all the other ISR measurements [430]. However, to be fair to the experimenters it is important to take account of their published warning:

> … the experiment was not specifically designed to detect single photons but to study the structure of events with a large transverse momentum π^0 produced at 90° [386]. In particular, the lead glass detector was primarily intended to be used as a triggering counter and no detailed study of its energy response and resolution had been made at the time of the experiment.

Results from a specifically designed real single photon search at $\sqrt{s} = 53.2$ GeV, by the Rome–Brookhaven–CERN–Adelphi (Yuan–Amaldi) group, published in 1978 [428,431] greatly clarified the situation. The detector consisted of a matrix of 9 vertical by 15 horizontal lead glass blocks, 10×10 cm^2 in area and 35 cm long,

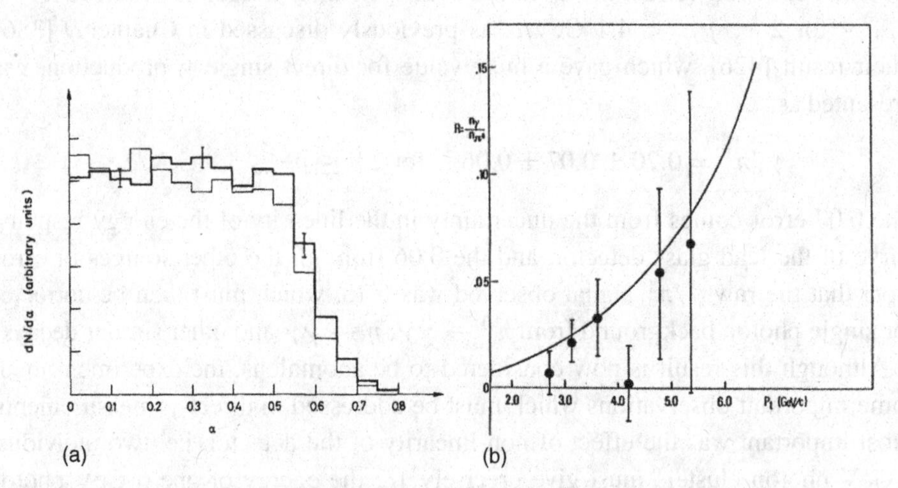

Figure 10.3 (a) The observed (heavy line) and predicted (thin line) distributions in the energy asymmetry α for photon pairs with photon energies above 400 MeV [426]. (b) Yuan–Amaldi [428] measurement of $R = n_\gamma / n_{\pi^0}$ as a function of p_T at $\sqrt{s} = 53$ GeV and $\theta^* = 90°$. The curve represents the results of a QCD prediction [429].

placed behind two matrices of scintillation counters which allowed the rejection of charged particles. Data were also collected with lead layers of various thicknesses in front of the whole detector in order to separate high energy photons from antineutrons. The whole detector, which was located at 90° with respect to the colliding beams, was remotely moveable between ~1.5 and 4.5 m from the interaction point so that systematic effects due to the geometry could be studied. As the authors most eloquently explain,

The most delicate points in the search for single gamma ray direct emission are of two different types. The first is the danger of interpreting a gamma produced in the decay of a π^0 or η as a single gamma ray either because its companion is not observed in the solid angle of the apparatus above the energy threshold of the counters or because the two gamma rays are not geometrically resolved. The second delicate problem is the background due to antineutrons annihilating in the lead glass with a large fraction of the energy emitted as π^0's.

Interestingly, the closest setup, 1.47 m from the interaction point, gave the cleanest measurement, for several reasons. The trigger was restricted to the central three vertical by five horizontal PbGl counters in which a single photon was defined as a neutral cluster in any set of 2 × 2 blocks in which no photon pair from a π^0 in the relevant energy region could be contained (i.e. for $d_{min} > 10$ cm, from Eq. 3.134, $\gamma = E/m < 2 \times 1.47/0.10 = 29.4$, so a π^0 with $E < 3.7$ GeV would not be contained). Clusters not contained in the 2 × 2 block region were called "unresolved" π^0 if compatible with π^0 decay. Also, since the trigger region was well centered in the detector, the second photon from a single photon trigger from an asymmetric π^0 decay in the measured interval 2.3–3.4 GeV would be detected with $\gtrsim$90% efficiency and the π^0 would be reconstructed.

After subtraction of the reconstructed and unresolved π^0, the calculated ~10% remaining single photons from π^0 and $\eta \rightarrow \gamma + \gamma$ decay (using the CCRS measured value $\eta/\pi^0 = 0.55 \pm 0.05$ [270]) and the measured small antineutron contamination, the results are shown in Figure 10.3b [428] as $R = n_\gamma/n_{\pi^0}$ as a function of p_T. It should be noted that the direct-γ signal extracted after background corrections (14 ± 14 events for the 1.47 m setting) corresponds to only ~10% of the observed inclusive single photon signal. The black line shown is a QCD prediction [429]. A point worthy of note is that the QCD predictions of this early era are ambiguous for reasons that will be discussed at the end of this chapter. It is not clear whether the data support the QCD prediction or are consistent with a zero value for γ/π^0. However, one definite conclusion from the data [431] is that to 95% confidence $\gamma/\pi^0 < 4\%$ for $2.3 \leq p_T \leq 3.4$ and $2.5 \leq p_T \leq 3.7$ GeV/c, which is in substantial disagreement with the first real photon result [426].

The experiment generally given credit for the discovery of direct-γ production and the first to actually claim the observation of a significant signal was the AABC experiment (R806) [432] (Figure 10.4a,b), a subset of the AABCSY collaboration at the CERN-ISR, who moved their lead/liquid argon electromagnetic calorimeter from 0.86 m to 1.65 m from the interaction point so as to increase the p_T at which they were able to resolve the two γ from π^0 decay as distinct showers up to 7 GeV/c, and then to 2.15 m for 9–12 GeV/c [433, 434]. The experiment is very similar in principle to the Yuan–Amaldi measurement [428, 431], but with much finer spatial resolution for two distinct photon showers. Their measured value for $R_\gamma = \gamma/\pi^0$ versus p_T for inclusive γ at "all energies," i.e. $30 < \sqrt{s} < 62$ GeV, is shown in Figure 10.4c, together with the calculated background; and their background subtracted direct-γ/π^0 versus p_T is shown in Figure 10.4d. These data are the first to show R_γ for direct photons "clearly rising above 0 for $p_T > 4.5$ GeV/c, thus establishing the existence of direct single photon production." In fact, the direct-γ/π^0 ratio approaches 30% for $p_T \simeq 6$ GeV/c. However, it must be emphasized that the R_γ data shown in Figure 10.4c,d are for γ and π^0

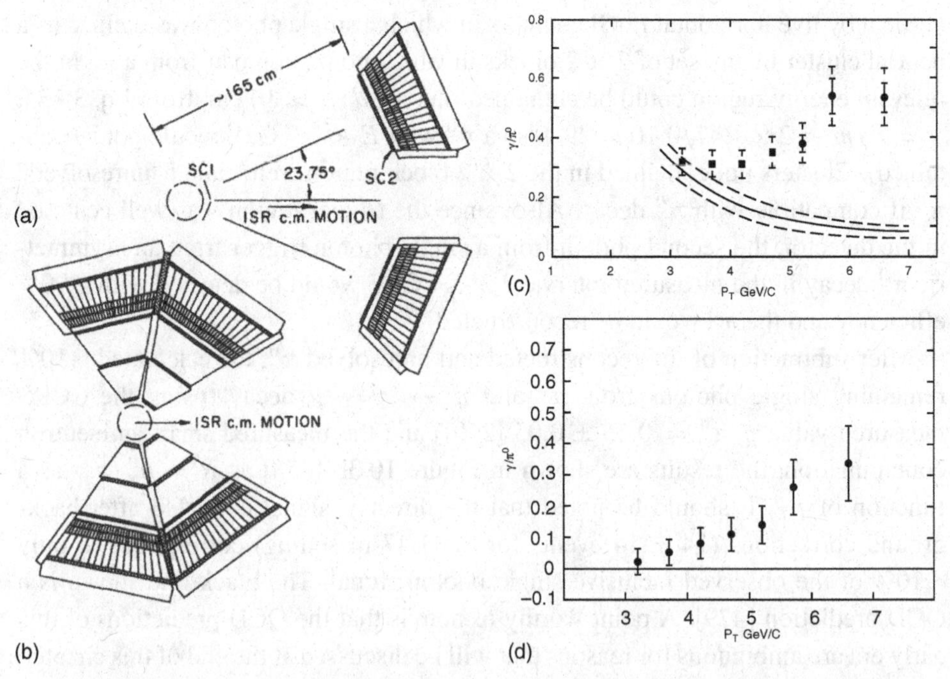

Figure 10.4 Views along the beam axis of liquid-argon–lead calorimeters: (a) arrangement for single photon experiment, "far" configuration [432]; (b) arrangement of modules for search for high mass e^+e^- pairs, "near" configuration [355]. (c) Observed [432] ratio of γ to π^0 for $30 < \sqrt{s} < 62$ GeV. (d) Final value of $R_\gamma = \gamma/\pi^0$ at $\sqrt{s} = 62$ GeV corrected for background. It is important to emphasize that the R_γ data shown in c, d are for γ and π^0 unaccompanied by other particles in a solid angle of 0.26 steradians, and therefore cannot be straightforwardly compared to inclusive direct-γ/π^0 ratios.

unaccompanied by a π^0, γ or charged particle within a solid angle of 0.26 steradian. It is also worth noting that in making their case for a discovery, the authors gave a succinct review of all the key issues in this measurement: "This signal cannot be produced by the decay of π's or η's, nor can it be explained by backgrounds, nonlinearity or merging of π^0's."

A complete opposite tack was taken by the CCOR experiment [435] using the thin-wall superconducting solenoid detector shown in Figure 9.2b (Chapter 9), which was unable to resolve single γ rays from $\pi^0 \to \gamma + \gamma$ decays in the external PbGl arrays. However, conversely, 99% of both γ rays from π^0 decays and 80% of $\eta \to \gamma + \gamma$ decays for $p_T \geq 7$ GeV/c could be captured in a cluster defined as an isolated distribution of energy in a matrix of up to 3×3 (~0.1 sr) out of a total of 168 $15 \times 15 \times 40$ cm^3 PbGl blocks arranged in 12 rows of 14 in each of two arrays located 1.4 m from the interaction point. Since the arrays were located outside the coil and cryostat of the solenoid, which amounted to $1.0X_o$ of converter, and each was preceded by a hodoscope of 12 scintillation counters (B in Figure 9.2b) mounted just outside the solenoid, a statistical determination of the average number of photons in the sample of clusters could be made by measuring the probability for the photon or group of photons in the cluster to pass through the $1.0X_o$ material without any conversion taking place.

A conversion was indicated by more than $1.5\times$ single ionization in the two B counters closest to the cluster of interest, with the condition that no charged particle tracks or other neutral cluster overlap these counters. The non-conversion probability, ν, per photon after a thickness of material t/X_o in radiation lengths is given by [140]:

$$\nu = \exp\left[-\frac{7}{9}(t/X_o)(1-\xi)\right]$$

where ξ is a small energy dependent correction. The measured non-conversion fraction is plotted in Figure 10.5 as a function of p_T, where for a single photon between 2 and 13 GeV, ν_1 varies from 0.474 to 0.462; and for two photons from the decay $\pi^0 \to \gamma + \gamma$, ν_2 varies from 0.246 to 0.221 for π^0 energies from 2 to 13 GeV, after averaging over the decay spectrum. The fraction of clusters ascribed to direct single photons can be calculated from the observed values of ν, and the values for pure single photons and for all other processes which include decays with photons from π^0, η, K_s^0, ω and η' [435]:

$$f_\gamma = \gamma/\text{all} = (\nu_{obs} - \nu_E)/(\nu_1 - \nu_E).$$

However, since the measurements in the outside array are systematically below the expected values in the lower p_T region, a calibration procedure was used. A known, but small, direct photon signal in the region $3.5 < p_T < 5.0$ GeV/c,

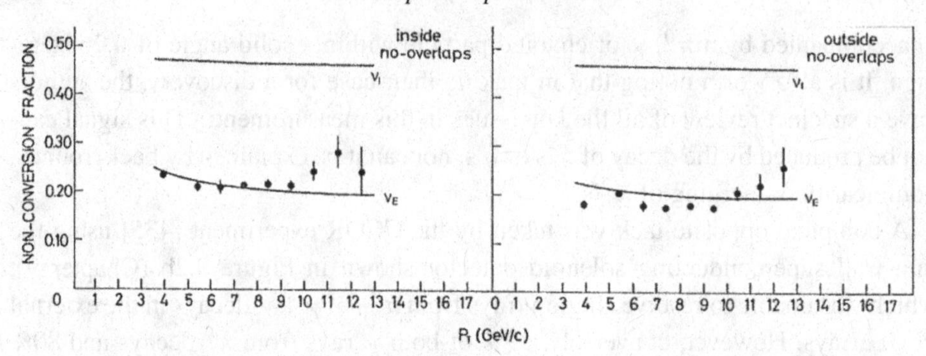

Figure 10.5 CCOR [435] measured non-conversion fraction in the inside and outside arrays for events with no overlap in the B counters. The expected values ν_1 and ν_E for single and multiple photon clusters are indicated. It should be noted that the c.m. motion was towards the outside array.

$$\langle \gamma/\pi^0 \rangle = 0.021 \pm 0.012, \tag{10.9}$$

obtained by averaging the data points in Figure 10.3b [428], was used to renormalize the expected non-conversion fractions accordingly. This procedure had the additional advantage that it eliminated two possibilities of systematic uncertainty: the absolute value of t/X_o for the coil and cryostat; and the absolute value of fractional energy lost in this material, which only applies to conversions.[1] The validity of the calibration procedure was indicated by the agreement of $f_\gamma = \gamma/\text{all}$, computed separately in the two arrays, which were then averaged to give the final result for $f_\gamma(p_T)$, shown in Figure 10.6a. Since the denominator "all" in f_γ represents the neutral clusters called "π^0" in the measurement of CCOR [269] (Figure 6.13), the direct-γ invariant cross section can be obtained without knowledge of the exact details of the cluster composition[2] simply by multiplying $f_\gamma(p_T)$ by that measurement (Figure 10.6b).

Figure 10.6b [435] was actually the first published measurement of the direct-γ cross section and the first measurement in the range $9 \le p_T \le 13$ GeV/c where the systematic error band is not too bad. In principle, the direct-γ cross section in Figure 10.6b could be corrected by a multiplicative factor of 0.8 due to the "no-overlap" cut in the B counters, if the direct-γ were purely isolated, whereas

[1] Since the lost energy is a constant fraction of the energy of the converted photon independent of p_T, the non-conversion fraction ν_1 also changes by a constant fraction due to this effect because of the power-law p_T spectrum. Thus, the absolute value of ν_1 is also fixed by the calibration procedure.

[2] The cluster composition for $p_T \ge 7$ GeV/c [435] was approximately 7% direct-γ, 62% π^0, 17% $\eta \to \gamma + \gamma$ and 14% other multi-γ decays. Thus, in principle the purely π^0 inclusive cross section could be obtained by reducing the cross sections for all the data points for $p_T \ge 7$ GeV/c in Figure 6.13 by 38%, which would not change the values of n_{eff} for $\sqrt{s} = 53.1$ and 62.4 GeV in Figure 6.14b. Also this is within the 5% systematic uncertainty of the absolute p_T scale.

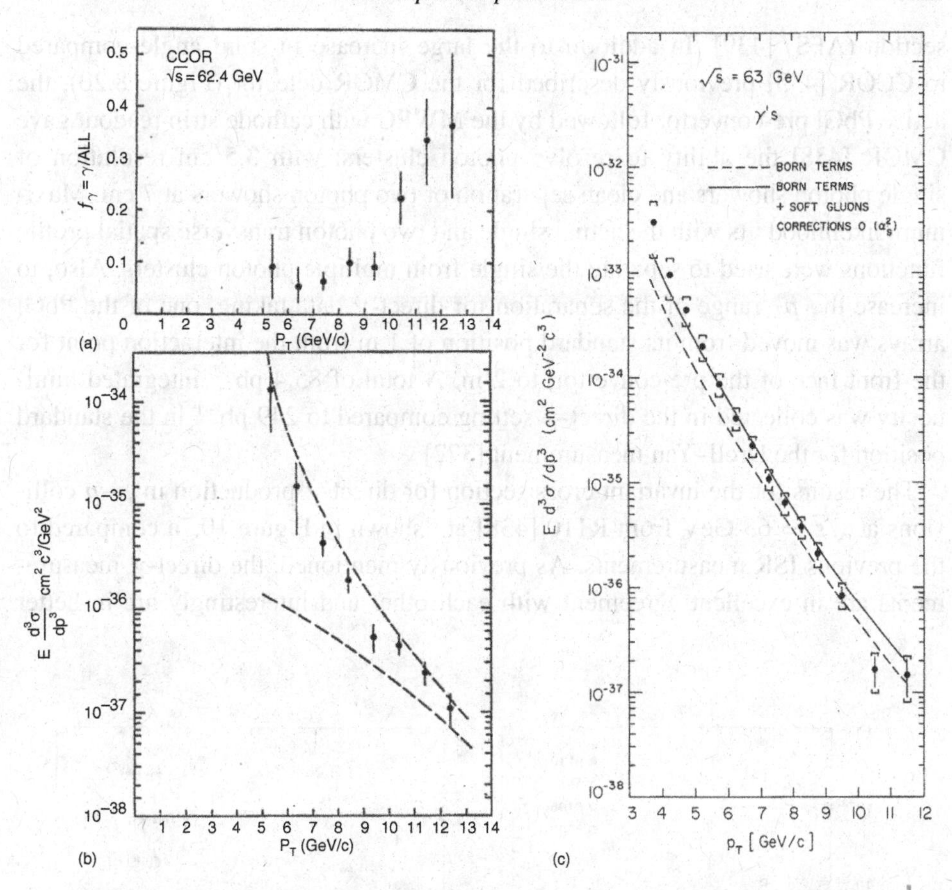

Figure 10.6 (a) CCOR [435] fraction of clusters, $f_\gamma = \gamma/\text{all}$, attributed to direct single γ production as a function of p_T. An overall systematic uncertainty of ± 0.053 by which all points may be adjusted together is not shown. (b) Inclusive direct-γ cross section. The error bars are statistical, the broken curves are the effect on the data from the ± 0.053 systematic uncertainty on f_γ. (c) Final R806 [434] cross section for direct-γ production compared to QCD calculations [436].

no correction is required if direct-γ had the same accompanying particles as π^0. No correction was made, since the relative number of associated particles was not known. In spite of these gyrations, the CCOR [435] and final R806 [434] cross sections for direct-γ production (Figure 10.6c) were in surprisingly good agreement with each other and with the QCD calculations of this early era. In fact, it was the CCOR measurement that first favored the early QCD calculations that included scale violation [436, 437]. A contemporary discussion on this issue is given in Ferbel and Molzon [424].

The final generation of ISR experiments also made measurements of the direct-γ invariant cross section (CMOR-R110) [438] as well as the direct-γ jet cross section (AFS) [439]. In addition to the large increase in solid angle compared to CCOR [435] previously described for the CMOR detector (Figure 8.20), the active PbGl pre-convertor followed by the MWPC with cathode strip readout gave CMOR [438] the ability to resolve photon clusters: with 3.5 cm resolution of single photon showers and clean separation of two photon showers at 7 cm. Maximum likelihood fits with the actual single and two photon transverse spatial profile functions were used to separate the single from multiple photon clusters. Also, to increase the p_T range of the separation for direct-γ data taking, one of the PbGl arrays was moved from its standard position of 1 m from the interaction point for the front face of the pre-convertor, to 2 m. A total of 85.4 pb^{-1} integrated luminosity was collected in the direct-γ setting compared to 249 pb^{-1} in the standard position for the Drell–Yan measurement [372].

The results for the invariant cross section for direct-γ production in p–p collisions at $\sqrt{s} = 63$ GeV from R110 [438] are shown in Figure 10.7a compared to the previous ISR measurements. As previously mentioned, the direct-γ measurements are in excellent agreement with each other and interestingly are in better

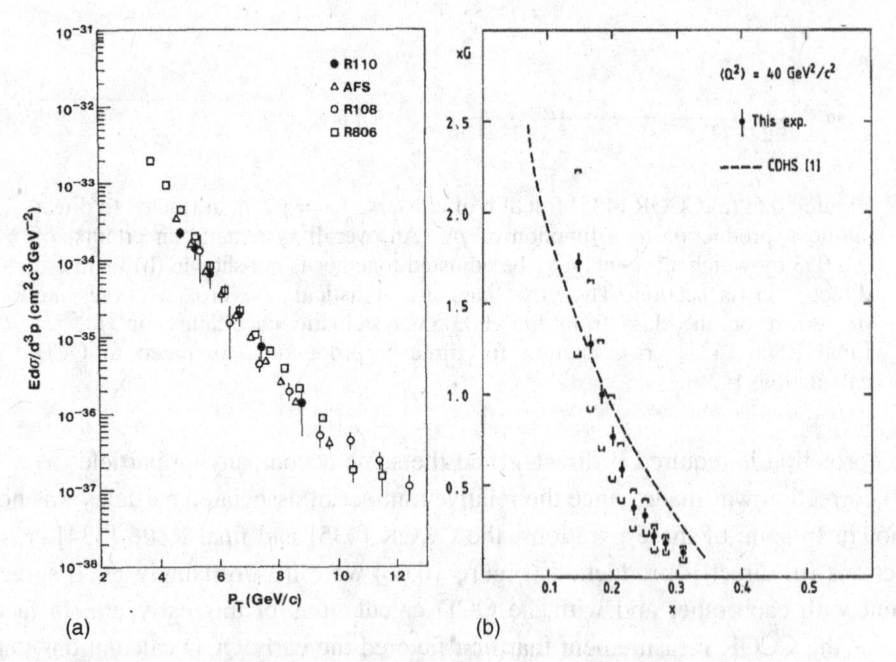

Figure 10.7 (a) CMOR-R110 [438] measurement of the direct-γ invariant cross section at $\sqrt{s} = 63$ GeV. Also shown are results from R108 [435] and R806 [434]. (b) AFS direct measurement of the gluon structure function $xg(x)$ [439] compared to a CDHS measurement [440] in DIS neutrino scattering from scaling violations of quark structure functions.

agreement with each other than the π^0 cross section measurements (Figure 6.14c). An issue worth noting about the AFS data in Figure 10.7a is that they were preliminary results that were left out of the actual publication [439] which concentrated on the direct-γ jet cross section and provided the first direct measurement of the gluon structure function $xg(x)$. This is shown in Figure 10.7b compared to an indirect measurement in DIS neutrino scattering [440] from scaling violations of quark structure functions.

Two other important results from CMOR-R110 concern the same-side and away-side direct-γ–hadron correlations (Figure 10.8) [438]. Figure 10.8a shows the azimuthal distribution of charged tracks with $p_T > 1.0$ GeV/c with respect to a trigger with $p_{T_t} > 6$ GeV/c for both direct-γ and neutral mesons. The neutral mesons show the well known same-side and away-side di-jet peaks while the direct-γ show no same-side peak, indicating that the direct-γ are isolated as predicted, with no evidence for single photons produced by Bremsstrahlung in jet fragmentation [419]. The other characteristic predicted for direct-γ production in p–p collisions, namely that the jet from the quark opposite the direct-γ is 8/1 u/d, is not so clear cut. One would think that there would be more positive charge in jets from u quarks opposite to a direct-γ than in jets opposite to a π^0; but the measured

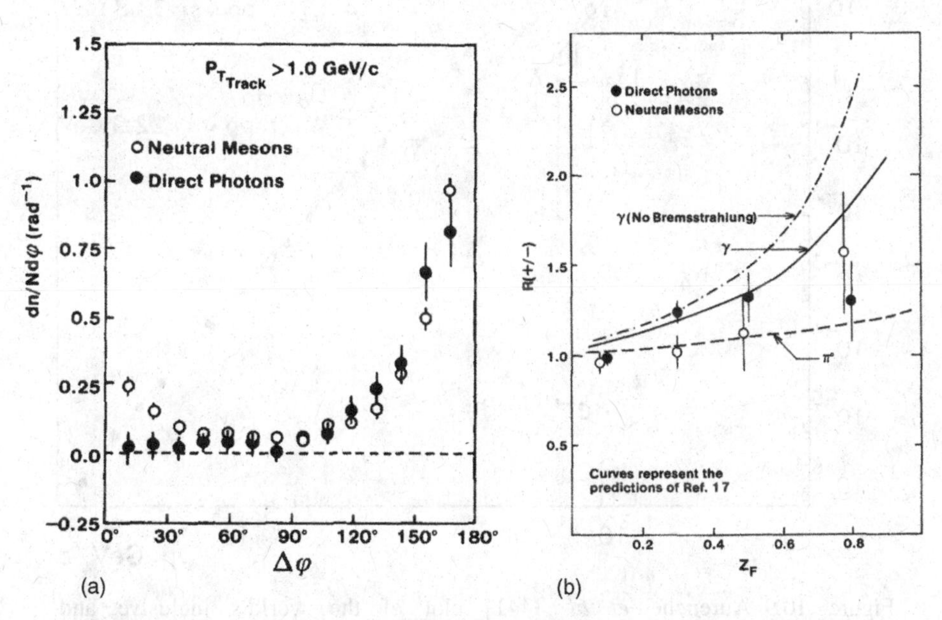

Figure 10.8 CMOR-R110 measurements [438]. (a) Azimuthal distribution of charged tracks with $p_T > 1.0$ GeV/c with respect to a trigger with $p_{T_t} > 6$ GeV/c for both direct-γ and neutral mesons. (b) Ratio of the number of positive to negative charged hadrons in the away-side peak versus x_E (called z_F) for the data in (a). Also shown are predictions from reference [419].

values of the ratio of positive to negative charged hadrons opposite to a direct-γ or a π^0 as a function of x_E (Figure 10.8b) [438] are insignificantly different. This is consistent with the AFS [439] result that there is "no difference in the fragmentation functions of recoil jets for π^0 or photon events." Evidently, distinguishing jets from u quarks using fragmentation measurements is not a simple task.

Since direct-γ production was proposed [301] as a fundamental process of QCD in which an outgoing participant in the two-to-two hard scattering, the γ ray, emerges directly and unbiased and can be measured precisely, one would think that the inclusive direct-γ production cross section would be the purest reaction

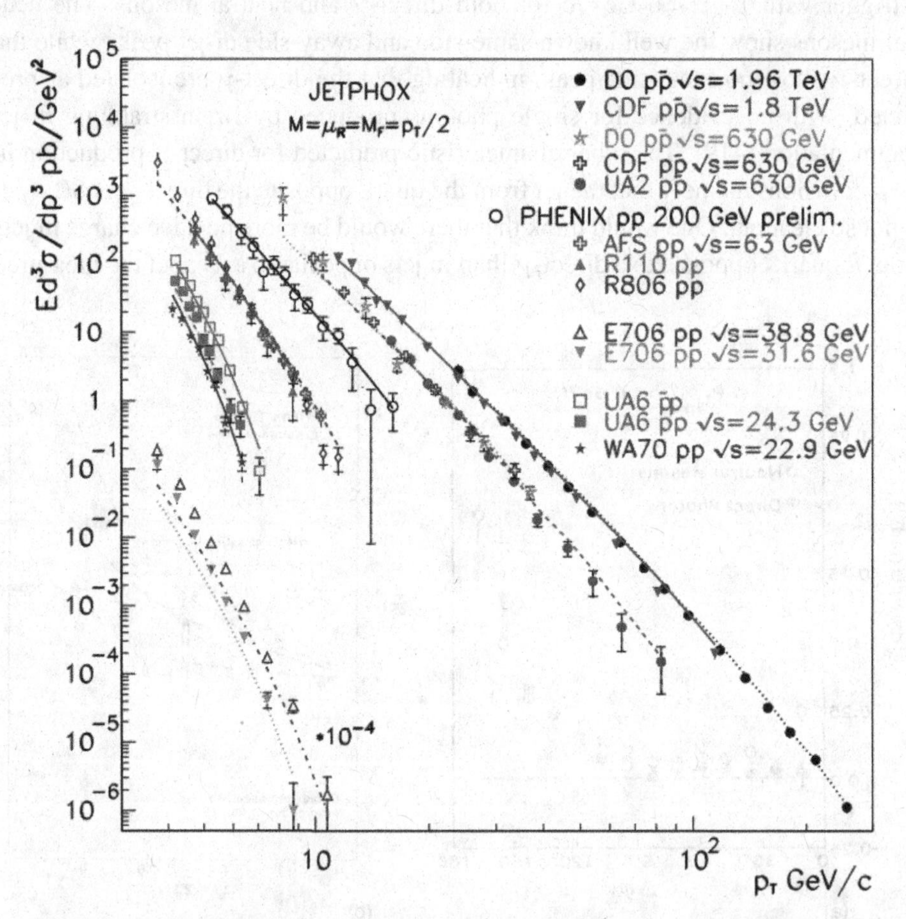

Figure 10.9 Aurenche *et al.* [441] plot of the world's inclusive and isolated direct-γ production cross sections measured in proton–proton and antiproton–proton collisions compared to their "JETPHOX" NLO predictions using BFGII (CTEQ6M) for fragmentation (structure) functions and a common scale, $p_T/2$. For clarity of the figure the E706 data are scaled by a factor of 10^{-4}. See reference [441] for details.

for testing the validity of QCD. The major uncertainty is the gluon structure function. However, the first QCD calculations [429] only used scale invariant structure functions, while subsequent calculations [436, 437] pointed out the importance of scale violations especially for the gluon structure function (e.g. Figure 10.6c).

Finally, to skip to modern times, the state of the art QCD calculation [441] for direct-γ production, which includes the latest structure functions as well as joint resummation of both threshold and recoil effects due to soft multigluon emission [442], shows excellent agreement with all available direct-γ measurements circa 2006 (Figure 10.9).[3]

[3] The only "fly in the ointment" is possibly the E706 experiment [443]. They claim that their measurement does agree with the theory, if they add empirical k_T smearing; and they derive a value of k_T from this adjustment [402] without in fact measuring k_T as was done at the CERN-ISR (Chapter 9).

11

The search for jets

11.1 Origins of E_T – the search for jets

In the decade of the 1980s, multiparticle inclusive measurements, in which many but not all of the particles from an interaction are measured, became one of the leading tools of both elementary particle physics and relativistic heavy ion physics. To quote Van Hove [124], it is indicative of the "transformation of elementary particle physics into many-body physics." The two principal multiparticle inclusive variables are the charged particle multiplicity distribution, either over all phase space or in restricted intervals of rapidity, and the transverse energy flow, or E_T distribution, where

$$E_T = \sum E_i \sin \theta_i \tag{11.1}$$

and the sum is taken over all particles emitted on an event into a fixed but large solid angle. Although these two variables are quite strongly related, their development followed rather independent paths. The charged multiplicity phenomenology was based on high cross section, "soft," multiparticle physics, while the transverse energy concept was stimulated by the desire to detect and study the jets from hard scattering with an unbiased and theoretically more efficient trigger than the single particle inclusive probe through which the constituent scattering and jet phenomena were originally discovered.

The general framework for the study of "soft" multiparticle physics was well in place by the early 1970s [445–447]. One of the important conceptual breakthroughs was the realization that the distribution of multiplicity for multiple particle production would not be Poisson unless the particles were emitted independently, without any correlation [448–451]. Instead the distribution appeared to exhibit a universal behavior when "scaled" by its average value at each energy [452]. This behavior, dubbed KNO scaling, is illustrated in Figure 11.1 for proton–proton collisions [444]. In the mid 1980s, systematic measurements by the UA5

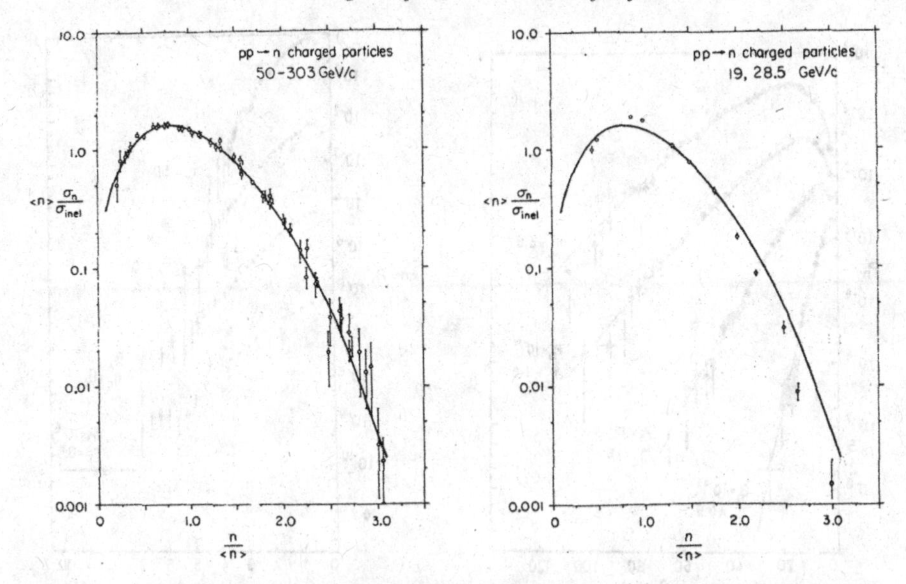

Figure 11.1 Normalized plot of the distribution in charged particle multiplicity for the reaction $pp \rightarrow n$ charged particles, measured in liquid hydrogen bubble chambers, as a function of the multiplicity n scaled by its average value $\langle n \rangle$ at each energy. The data for incident laboratory momenta of 50, 69, 102, 205, and 303 GeV/c are shown on the left, 19 and 28.5 GeV/c on the right. The curve, which is the same in both figures, is an empirical fit to the data on the left [444].

group at the CERN collider showed that KNO scaling did not hold in general [453], but that a new empirical regularity existed to describe multiplicity distributions. The multiplicity both in limited rapidity intervals and over all phase space could be described by Negative Binomial Distributions (NBD) [454–456] indicating correlations which could be described quantitatively by the NBD parameter k, which represents the first departure of a distribution from Poisson, where $\sigma^2/\mu = 1+\mu/k$ and $\mu = \langle n \rangle$ (see Appendix A). Also, the subject of multiparticle correlations was introduced [457]. The present situation for proton–(anti)proton collisions is summarized in Figure 11.2, which shows the multiplicity distributions in several pseudorapidity intervals at a c.m. energy of 540 GeV, and Figure 11.3, which shows the average multiplicity density in pseudorapidity, $dn/d\eta$, for several different c.m. energies [458].

The phenomenology of E_T measurements developed over a similar time period as that of multiplicity distributions, but was considerably more complicated and controversial. The first concept was due to Willis, in studies for ISABELLE [459], where he proposed an "impactometer" as a large solid angle (4π) non-magnetic device to measure rare events in p–p and e–p collisions (Figure 11.4).

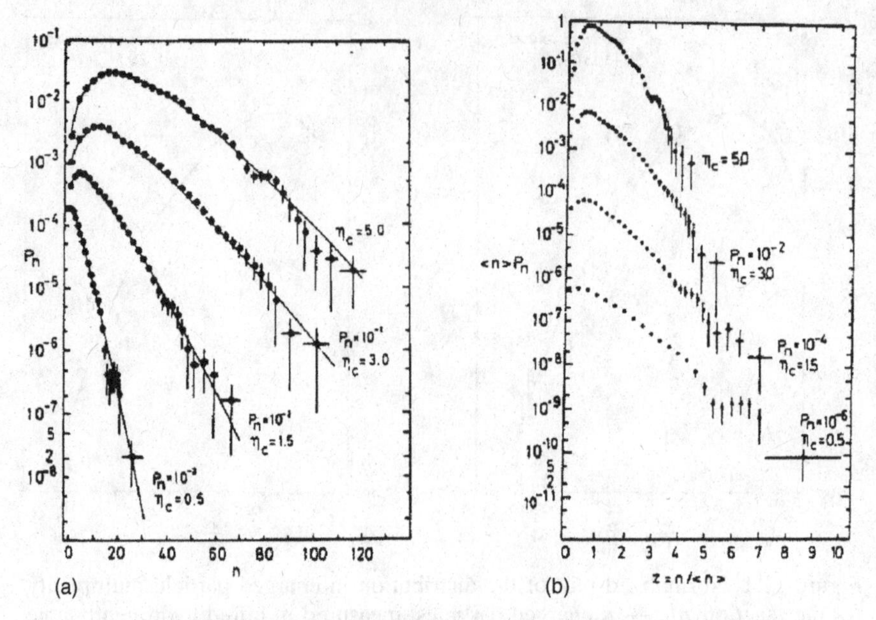

Figure 11.2 Charged multiplicity distributions in the pseudorapidity intervals $|\eta| < 0.5, 1.5, 3.0$ and 5.0 plotted versus n (a) and versus $n/\langle n\rangle$ (b). The curves in (a) illustrate negative binomial distributions with the parameters $\langle n\rangle$ and k adjusted to give the best fit [454].

The quantity "impact" was defined as the scalar sum of the transverse momenta of all particles, not just a single particle, produced in a limited azimuthal angular interval, $\Delta\psi$, at azimuthal angle ψ,

$$\Pi(\psi) = \sum_i E_i^\psi \sin\theta_i^\psi \tag{11.2}$$

where E_i is the energy of the ith particle, produced at polar angle θ_i, and the sum is taken over all polar angles. The purpose of making the detector large was to gain acceptance for the events with high p_T particles, which were produced with a low probability, but there was also the recognition that the rate for a given "impact," or total transverse momentum of many particles, would be much greater than that of a single particle of the same transverse momentum. Bjorken [460] then emphasized the use of a bank of hadron calorimeters, or other devices capable of measuring the total amount of energy emerging into small elements of solid angle, to observe the event structure of localized cores (jets) predicted for hard constituent scattering, and presented a more formal discussion of the substantial gain in cross section to be expected when measuring the entire jet at a given p_T rather than just the leading particle [379, 380, 461, 462].

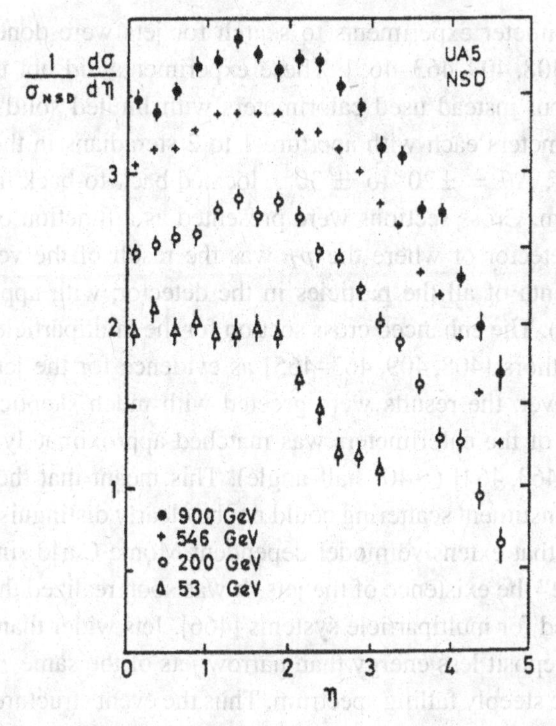

Figure 11.3 Pseudorapidity distributions for non-single diffractive events, UA5 data at four c.m. energies (pp, $p\bar{p}$) [457].

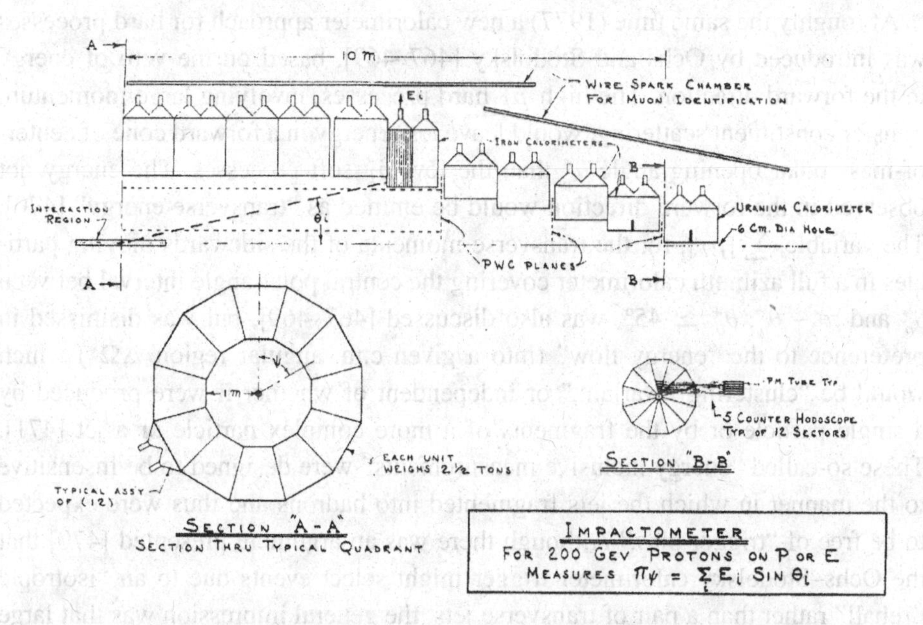

Figure 11.4 Sketch of the impactometer, from reference [459].

The first calorimeter experiments to search for jets were done at Fermilab in the late 1970s [408, 409, 463–465]. These experiments did not use full aperture (4π) detectors, but instead used calorimeters with limited solid angle, typically two small calorimeters each with aperture 1 to 2 steradians in the p–p c.m. system ($\Delta\phi = \pm 45°$, $\Delta\theta = \pm 20°$ to $\pm 30°$), located back-to-back in azimuth at 90° in the c.m. system. Cross sections were presented as a function of p_T of a single particle in the detector or where the p_T was the result of the vector sum of the transverse momenta of all the particles in the detector, with appropriate fiducial cuts (Figure 11.5). The enhanced cross section for the multiparticle state was interpreted by the authors [408, 409, 463–465] as evidence for the jets of constituent scattering. However, the results were greeted with much skepticism, principally because the size of the calorimeters was matched approximately to the expected size of the jets [463, 464] ($\sim$40° half angle). This meant that the event structure expected from constituent scattering could not be clearly distinguished from acceptance effects, so that extensive model-dependent Monte Carlo simulations had to be used to "prove" the existence of the jets. It was soon realized that "trigger bias" [462] also existed for multiparticle systems [466]. Jets wider than the calorimeter aperture would deposit less energy than narrow jets of the same p_T and would be suppressed by the steeply falling spectrum. Thus the event structure would be dominated by the calorimeter geometry. It seemed that the best way to overcome these objections was to make "geometrically unbiased" calorimeters with full azimuthal coverage.

At roughly the same time (1977) a new calorimeter approach for hard processes was introduced by Ochs and Stodolsky [467–469], based on the veto of energy in the forward direction. The high p_T hard processes, involving large momentum transfer constituent scattering, would leave less energy in a forward cone of center-of-mass polar opening angle θ^* than the low p_T soft processes. The energy not observed in the forward direction would be emitted as "transverse energy" [470]. The variable, $\sum |p_T|$, for the transverse momenta of the sidewards moving particles in a full azimuth calorimeter covering the central polar angle interval between θ_o^* and $\pi - \theta_o^*$, $\theta_o^* \simeq 45°$, was also discussed [467–469], but was dismissed in preference to the "energy flow" (into a given c.m. angular region $\Delta\Omega^*$) which would be "clustering invariant," or independent of whether it were produced by a single particle or by the fragments of a more complex particle or a jet [471]. These so-called "energy inclusive measurements" were designed to be insensitive to the manner in which the jets fragmented into hadrons and thus were expected to be free of "trigger bias." Although there was an argument presented [470] that the Ochs–Stodolsky calorimeter trigger might select events due to an "isotropic fireball" rather than a pair of transverse jets, the general impression was that large

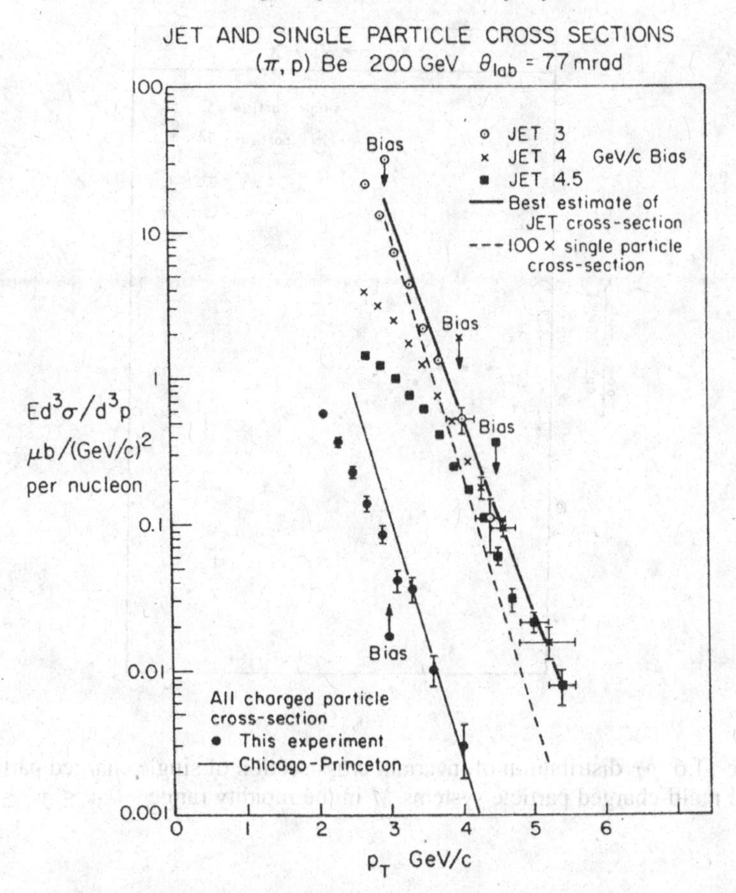

Figure 11.5 The single charged particle and "jet" cross sections averaged over the c.m. pseudorapidity region of 0.1 to 0.44. The "jet" definition is all particles into the calorimeter. The plotted "jet" data are equal amounts from proton and π^- beams while the single particle data are from a proton beam only. The calorimeter is roughly 50% efficient at the trigger threshold, and the envelope of the three "jet" biases represents the best estimate of the "jet" cross section [463,464].

aperture calorimeter triggers would be the least biased and best way to study hard scattering.

On the experimental side, the evolution to large apertures proved to be very confusing and contentious from the point of view of hard scattering. One of the first indications of a problem came from the analysis of the interactions of 110 GeV/c K^- in a hydrogen bubble chamber [472] (see Figure 11.6). The invariant cross sections of single charged particles C and multi-charged particle systems M in the c.m rapidity interval $-0.4 \leq y^* \leq 0.6$ were presented as a function

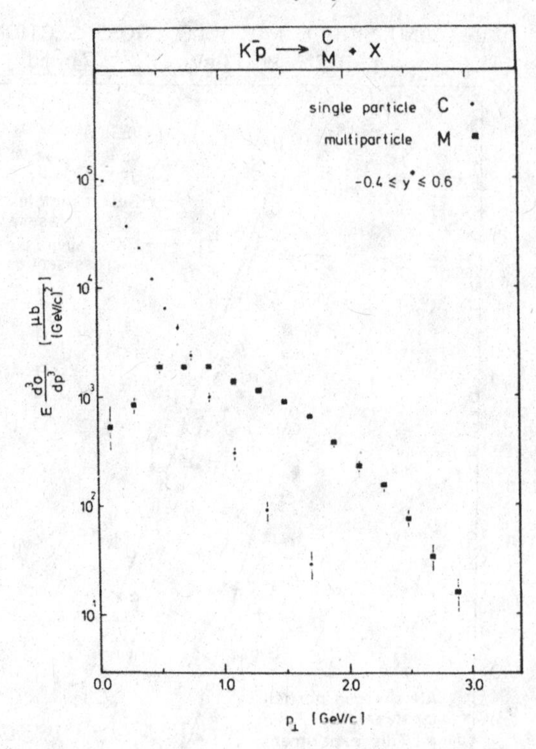

Figure 11.6 p_T distribution of invariant cross section of single charged particles C and multi-charged particle systems M in the rapidity range $-0.4 \leq y^* \leq 0.6$ [472].

of their transverse momenta [473]. Two striking features were claimed for this plot: firstly, the multiparticle cross section for $p_T > 1.5$ GeV/c was more than an order of magnitude larger than that of single particles; secondly, both distributions tended to have the same slope at large p_T. An additional striking effect was found by comparing the "minimum bias" single and multiparticle data to the charged particle data of the Fermilab calorimeter experiment [463, 464] which claimed the enhanced multiparticle cross section "proved" the existence of "jets." The data extrapolated smoothly into one another (see Figure 11.7), and had many other common features, with the important exception that the full aperture of the bubble chamber permitted the authors of reference [472] to perform an unbiased "principal axis analysis" [474, 475] of their data and determine that only a very "small fraction" of the multiparticle systems having $p_T > 2.0$ GeV/c would be considered "jet-like." Further evidence of problems came from a later Fermilab experiment [476, 477], with a somewhat larger aperture single calorimeter ($\Delta\phi^* = \pm 50°$, $\Delta\theta^* = \pm 40°$ in the p–p c.m. system), which found much larger

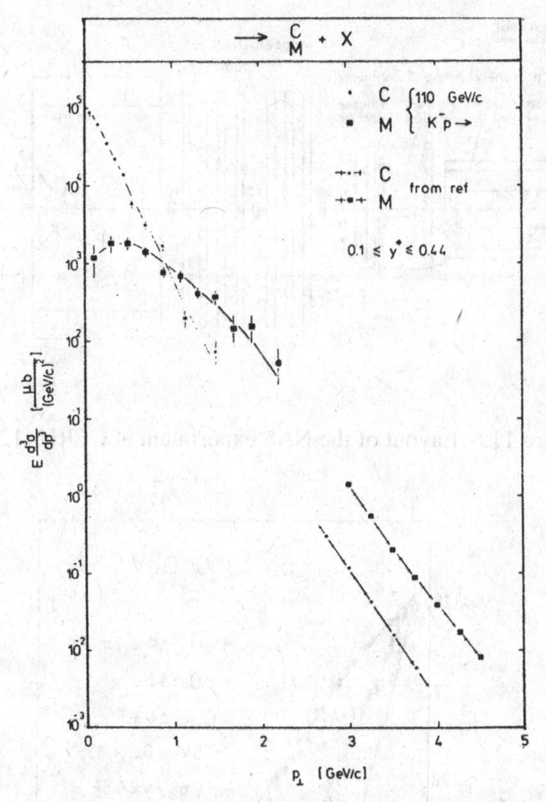

Figure 11.7 p_T distribution of invariant cross section of single charged particles C and multi-charged particle system M in the rapidity range $-0.1 \leq y^* \leq 0.44$. Also drawn are data taken from figure 23 of references [463, 464]. The dashed lines are a guide to the eye [472].

yields than the other smaller aperture experiments [408, 409, 463–465] and concluded that experiments with larger apertures tended to have "smaller slopes and larger magnitudes" of the multiparticle p_T cross section.

The coup-de-grâce to the concept of dominant hard scattering effects in large aperture calorimeters came in 1980 from the NA5 experiment at CERN [478]. This experiment also produced the first measured E_T distribution in the present day usage of the terminology [127]. The apparatus of this experiment was a streamer chamber followed by a hadron calorimeter which covered the full azimuth and c.m. polar angular interval $54° < \theta^* < 135°$ (Figure 11.8). The calorimeter was subdivided into 240 independent cells, each subtending about 9° in c.m. polar angle and 15° in azimuth. Data were taken using three triggers on the transverse energy E_T deposited in either the full azimuth, or in two smaller azimuthal regions corresponding to the earlier experiments (see Figure 11.9). As the aperture increased to the full azimuth, the cross sections at a given E_T simply

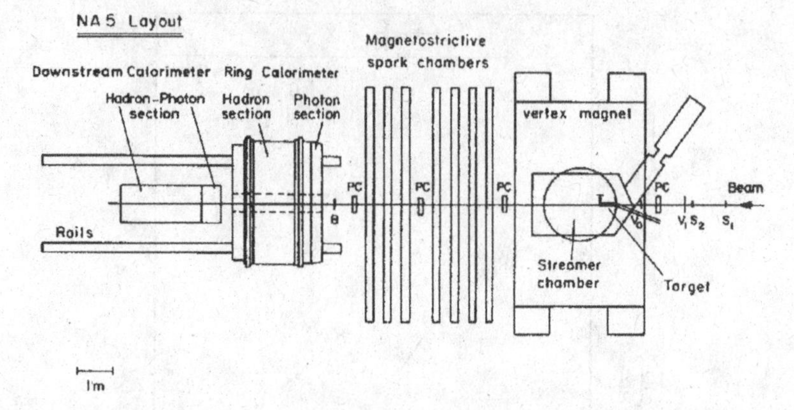

Figure 11.8 Layout of the NA5 experiment at CERN [127].

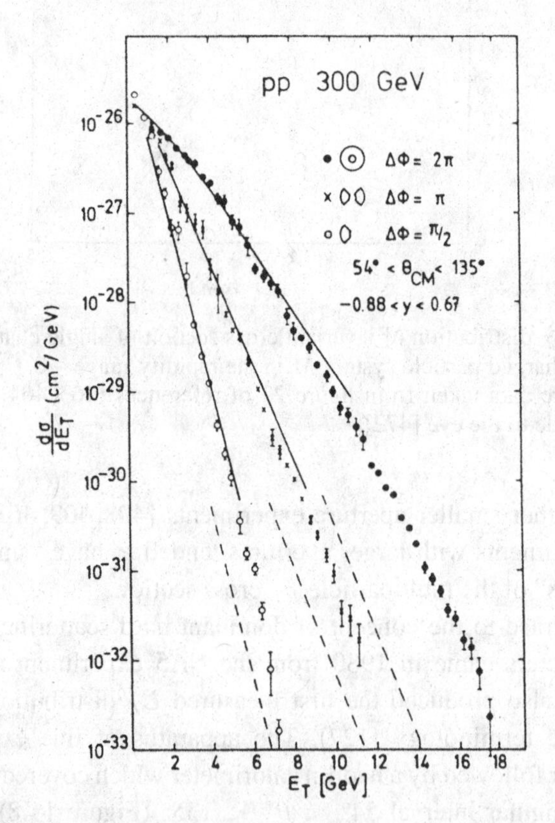

Figure 11.9 Cross sections versus transverse energy E_T measured in regions of various azimuthal acceptance $\Delta\phi$ of the calorimeter. Predictions from low p_T multiparticle production and QCD hard scattering are shown by solid and dashed curves respectively [127].

kept increasing monotonically to a value more than an order of magnitude larger than predicted for QCD jets, while the slopes decreased. Also, the full azimuth data showed no dominant jet structure. The large transverse energy observed in the calorimeter was the result of "a large number of particles with a rather small transverse momentum" [127]. This same effect and conclusion were later verified by the successor [479] to the original Fermilab "jet" experiment [463, 464] (see Figure 11.10).

In the period between 1980 and 1982, confusion reigned. There was no clear understanding of why the jets of QCD were not apparent in the full azimuth calorimeter data. Explanations involving fragments of the beam and target "spectators" [480–482] or "gluon Bremsstrahlung" [483, 484] were offered to

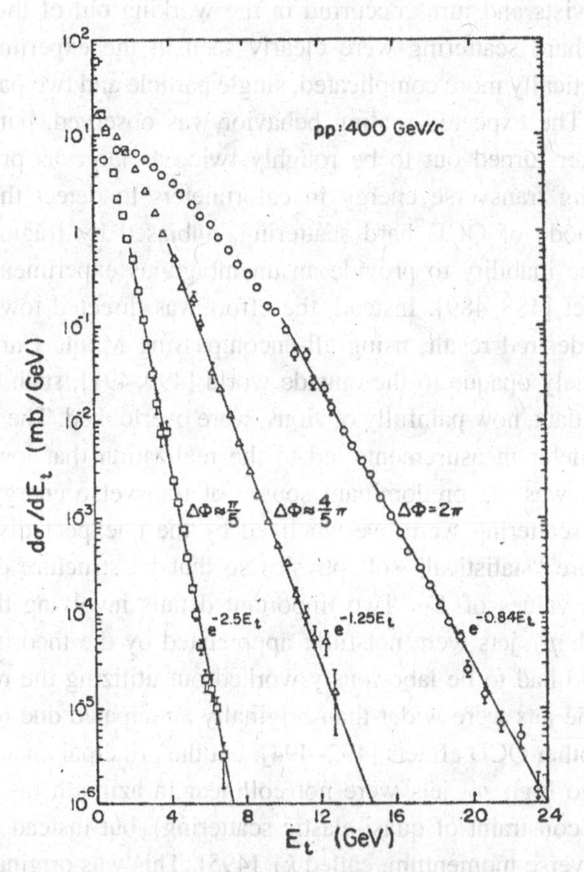

Figure 11.10 Cross sections obtained in c.m. polar angular interval $47° < \theta^* < 125°$ with large ($\Delta\phi \simeq 2\pi$, circles), medium ($\Delta\phi \simeq 2 \times 2\pi/5$, triangles), and small ($\Delta\phi \simeq \pi/5$, crosses) azimuthal acceptances as a function of transverse energy of the trigger, E_T. The error on the E_T scale is $\pm 5\%$. The lines represent exponential fits to the data [479].

explain the absence of the QCD jets; but it was not until the Paris conference of 1982 that it became generally understood that large aperture calorimeter triggers are dominated by "soft" physics [129, 130, 485]. The jets of QCD are swamped by many orders of magnitude but eventually emerge [417] to produce a dominant two jet structure in p–p or p–$\bar{p}$ collisions only at very large values of transverse energy, typically greater than 50 to 100 GeV [486, 487].

11.2 Does history provide a guide for the future?

It is tempting to draw a lesson from the historical perspective. The lesson is that the theorists were basically correct in their analysis of substructure of the proton and the prediction of hard scattering of the constituents, but that many unanticipated twists and turns occurred in the working out of the exact details. The effects of hard scattering were clearly seen in the experimentally convenient, but theoretically more complicated, single particle and two particle inclusive measurements. The expected scaling behavior was observed, but the measured scaling parameter turned out to be roughly twice as large as predicted. Initial attempts at using transverse energy in calorimeters to detect the theoretically simpler "jet" mode of QCD hard scattering, unbiased by fragmentation, were hampered by the inability to provide an unambiguous experimental demonstration of the effect [488, 489]. Instead, the effort was directed toward attempting to "prove" the desired result, using all-encompassing Monte Carlo calculations which were largely opaque to the outside world [490, 491], such that other explanations of the data, now painfully obvious, were overlooked. The advent of large aperture calorimeter measurements led to the realization that low p_T multiparticle production was the predominant source of transverse energy, and that the effects of hard scattering were overwhelmed by the unexpectedly large fluctuations of the more "statistical" soft physics so that jet structure did not emerge until very large values of E_T. Two important details involving the structure of events with high p_T jets were not fully appreciated by the theorists in their initial analysis, and had to be laboriously worked out utilizing the results of many experiments. The jets were wider than originally anticipated due to the soft fragmentation and other QCD effects [492–494]; but the principal unanticipated effect was that the two high p_T jets were not collinear in azimuth (as expected from the coplanarity constraint of quasi-elastic scattering), but instead were produced with a net transverse momentum, called k_T [495]. This was originally ascribed to intrinsic transverse momentum of the constituents of the proton [71, 391] but later turned out to be understood as due to multi-soft gluon effects [496]. The smearing of the steeply falling constituent-scattering spectrum by this extra transverse momentum, k_T, was the principal contribution [497] in changing the QCD scaling parameter from the expected value of $n = 4$ to the value $n = 8$ discussed above (Figure 6.12).

11.3 The systematics of transverse energy emission in p–p and p–$\bar{p}$ collisions

When high energy nucleons collide inelastically, the predominant mode of dissipating the energy is by multiple particle production. The produced particles are distributed relatively uniformly in rapidity, with limited transverse momentum with respect to the collision axis, leading to a description of the process as "longitudinal phase space" (recall Figure 11.3). This leads to the perverse situation that the rare events of inelastic hadron collisions, involving the scattering or production of constituents with large momentum transfer, can be treated very precisely in the framework of perturbative QCD, but that the vast majority of collisions are in the non-perturbative domain and are subject instead to a more qualitative description, based primarily on empirically observed regularities.

In the past few years, multiparticle inclusive measurements, in which many but not all of the particles from an interaction are measured, have become a leading tool in the description of the "soft" reactions which dominate the particle production process in high energy hadron collisions. The two principal multiparticle inclusive variables are the charged particle multiplicity and the transverse energy flow, taken either over all phase space or in restricted intervals of rapidity. These variables are very closely related, but this fact was not realized until quite recently (1982). The transverse energy formalism developed from the use of large aperture calorimeters to search for the rare events of "hard" scattering, while the multiplicity is one of the classical observables in the study of high energy collisions [104, 105, 498].

The formal definition of the transverse energy or E_T is

$$E_T = \sum_i E_i \sin\theta_i \quad \text{and} \quad dE_T(\eta)/d\eta = \sin\theta(\eta)\, dE(\eta)/d\eta \tag{11.3}$$

where the sum is considered to be taken over all particles emitted on an event into a fixed but large solid angle. Following the original work of NA5 [127], the traditional solid angle was typically taken as the full azimuth, $\Delta\phi = 2\pi$, and c.m. pseudorapidity interval $\Delta\eta^* \simeq \pm 0.8$. It is important to note that E_T does not have a well defined property under Lorentz transformations. Relativistically preferable and "clustering invariant" quantities have been discussed from time to time [467–469], but these are rarely used because the above definition of E_T is the most convenient for measurements in segmented calorimeters. In this case, the sum is over the energy E_i measured in the ith calorimeter cell, with average polar angle θ_i.

The relationship of the quantity "E_T" measured in calorimeters to an idealized quantity is not straightforward, particularly for collisions with nuclei, since hadron calorimeters respond to the *total* energy of produced particles but only to the *kinetic* energy of nucleons emitted from the projectile and target. Also, for nucleons, the kinetic energy $\times \sin\theta$ is typically very different from the transverse momentum (p_T) or transverse mass ($m_T = \surd(p_T^2 + m^2)$), and this difference depends strongly on the Lorentz frame. In addition, the relationship between the signal detected in a calorimeter and the true energy deposited depends strongly on the "type, energy, and shower history of the individual incident particles" [499, 500]. Thus, extensive iterative calculations must be used to sort out these effects as well as the effect of spatial and energy resolution; and different experiments use different definitions of the idealized quantity E_T to which they attempt to correct their measurement.

Three distinct varieties of "E_T" measurements have been reported in the literature. The first and "classical" method [127, 129] uses a full azimuth hadron calorimeter, typically 5 to 8 hadron absorption lengths thick, which thus measures all hadrons regardless of whether they are charged or neutral. The second method uses a track chamber device to reconstruct the momenta of all charged particles and then to construct E_T^c, the transverse energy of charged particles, usually with the assumption that all the particles are pions [485, 501]. The third method [376, 412, 502] uses an electromagnetic shower counter, typically 15 to 20 radiation lengths thick, to detect the energy of the photons from the decays of neutral mesons ($\pi^0 \to \gamma\gamma$ and $\eta^0 \to \gamma\gamma$) and has thus been called [376] "neutral transverse energy," E_T^0, or "electromagnetic" transverse energy [502], E_T^{em}. The shower counter can be a dedicated detector [376] or the electromagnetic section of a hadron calorimeter [502]. In either case, a correction may be made for the energy deposited in the shower counter by charged hadrons.

At the 1982 international conference in Paris, the systematics of the dominance of soft collisions in E_T distributions were first clearly demonstrated [129, 485]. The jets of hard scattering are inconsequential to the physics of E_T distributions, but eventually emerge to produce an important effect only at very large values of E_T [417]. The UA1 collaboration [129] at the CERN-SPS collider presented E_T distributions in their hadron calorimeter from minimum bias p–$\bar{p}$ collisions at $\sqrt{s} = 540$ GeV. Figure 11.11 shows the E_T spectrum for the full acceptance of the central calorimeter ($-3 < \eta < +3$ and $\Delta\phi = 2\pi$). The curve represents a QCD constituent level prediction (no hadronization) for hard scattering [483, 484]. Figures 11.11b to 11.11d show the effect of reducing the $\Delta\eta$, $\Delta\phi$ acceptance. All the distributions show an initial increase and then an exponential decrease with increasing E_T. The mean values and the slopes of the exponential parts of the distributions are observed to scale approximately with the $\Delta\eta$, $\Delta\phi$ acceptance.

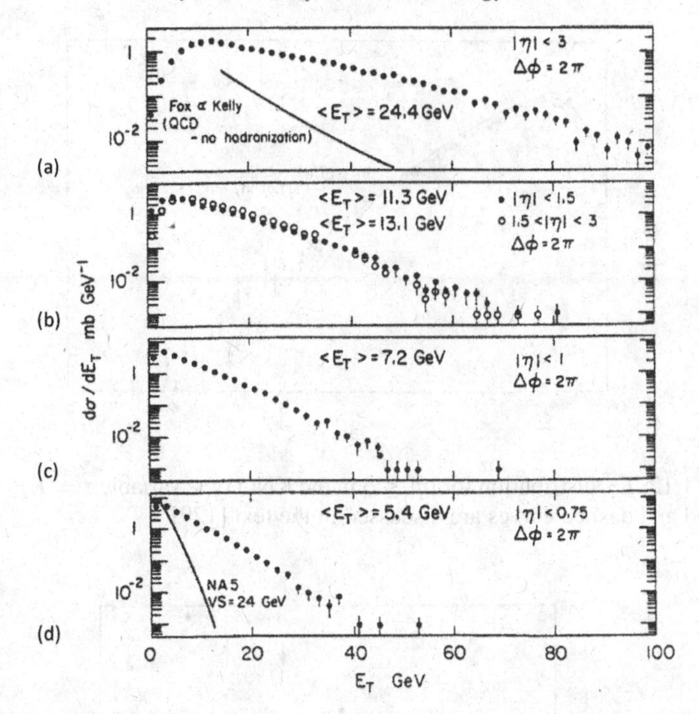

Figure 11.11 $d\sigma/dE_T$ spectra for various acceptances of the UA1 central calorimeters as shown on the figure. The average values are also indicated [129].

Such a feature is similar to the behavior of multiplicity. Figure 11.11d illustrates the original NA5 data [127] at $\sqrt{s} = 23.8$ GeV for similar acceptance. The difference between the two can largely be attributed to the higher multiplicity at the collider energy.

The relationship between the total transverse energy and the multiplicity was clearly established [129] in Figure 11.12, where the E_T distribution in the full acceptance (from Figure 11.11a) is shown in terms of a KNO type variable [452] defined as $z = E_T/\langle E_T \rangle$ and compared to the measured charged particle multiplicity distribution in the same experiment (in the KNO variable $z = n/\langle n \rangle$) shown as the solid curve. It can be seen that the two shapes are strikingly similar. The observed differences in the tail region can be accounted for by the increase of the $\langle p_T \rangle$ with charged multiplicity [503]. This effect is expected in hydrodynamical models [504] and is largely absent at c.m. energies below 63 GeV [505] (see Figure 11.13). Thus, the AFS collaboration at the CERN-ISR, using the E_T^c spectra reconstructed from charged particles observed in the c.m. rapidity range $|y| < 0.8$ in a cylindrical drift chamber covering almost 2π in azimuth from p–p collisions at $\sqrt{s} = 53$ GeV, found that the gross features of the E_T^c spectra at the ISR

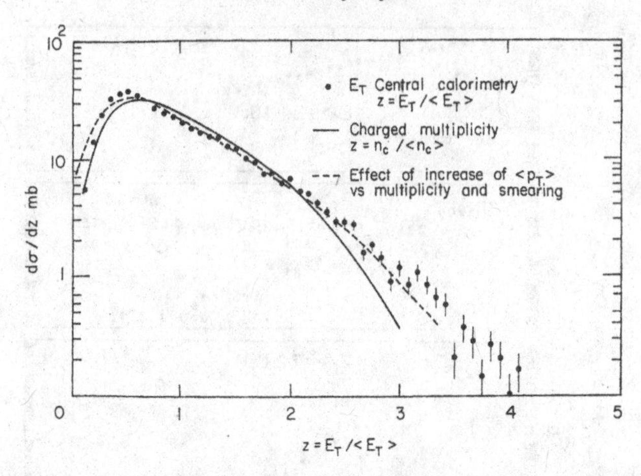

Figure 11.12 E_T distribution for $|\eta| \leq 3$ in the KNO type variable $z = E_T/\langle E_T \rangle$. The solid and dashed curves are discussed in the text [129].

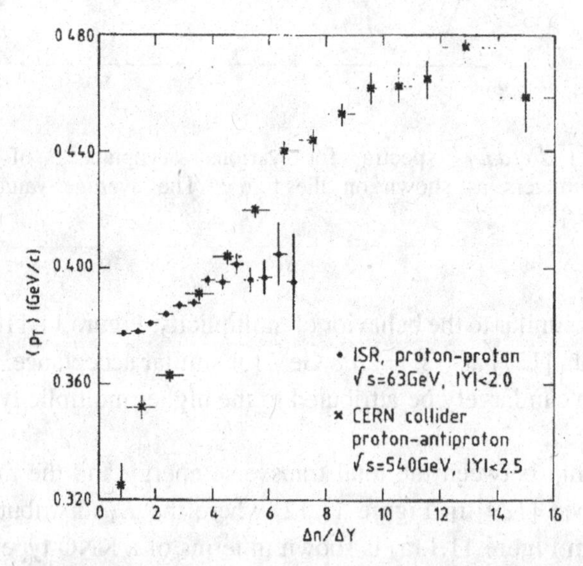

Figure 11.13 The average transverse momentum as a function of charged particle density in the central rapidity region of proton–(anti)proton collisions at ISR and CERN p–$\bar{p}$ collider energies. There is a systematic uncertainty in the $\langle p_T \rangle$ scale for ISR measurements of ~5% [505].

could be accounted for by the simple convolution of the observed inclusive p_T and charged particle multiplicity distributions, taken as uncorrelated [485] (see Figure 11.14).

The work on the E_T^c distributions was later extended by the AFS collaboration to several other c.m. energies and to p–α and α–α collisions [501]

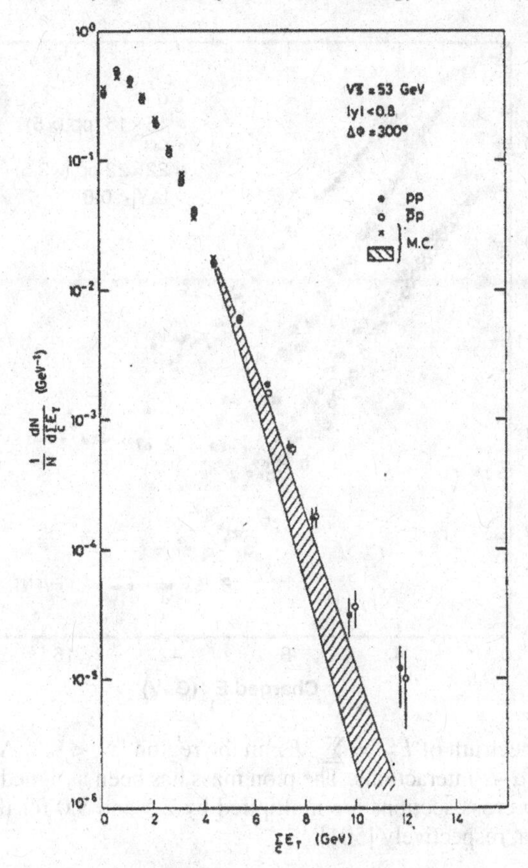

Figure 11.14 Spectrum of $E_T^c = \sum_c E_T$ in the region $|y| < 0.8$, $\Delta\phi = 300°$, for p–p and p–$\bar{p}$ collisions at $\sqrt{s} = 53$ GeV. The data are normalized to the total number of accepted events. Also shown are the results of a simple Monte Carlo calculation which folds together the inclusive multiplicity and p_T distributions, assumed to be flat in rapidity and azimuth, and independent of each other. The shaded region represents extreme limits to the high tail of the multiplicity distribution [485].

(Figure 11.15). These spectra could all be explained by the independent convolution of the observed inclusive transverse momentum and charged particle multiplicity distributions. However, a correlation is induced when the mean transverse momentum $\langle p_T \rangle$, and the mean multiplicity $\langle n \rangle$, instead of being plotted against each other, are plotted as a function of the observed E_T^c (Figure 11.16). The observed rise in $\langle p_T \rangle$ with E_T^c is not a signal for the onset of new mechanisms [506–508]. It is simply a reflection of the fact that, because of the rapid falloff of the multiplicity distribution at high multiplicities, there is a larger probability of obtaining an event of fixed E_T^c with particles of larger $\langle p_T \rangle$.

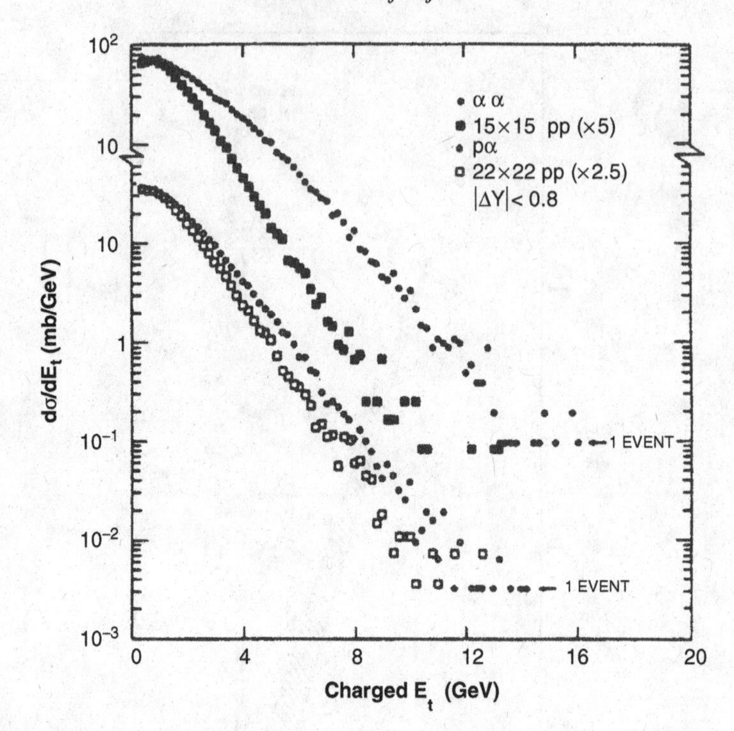

Figure 11.15 Spectrum of $E_T^c = \sum_c E_T$ in the region $|y| < 0.8$, $\Delta\phi \simeq 330°$, for p–p, p–α, and α–α interactions. The pion mass has been assigned to all charged tracks. The p–p cross sections are multiplied by 2.5 and 5.0 for the $\sqrt{s} = 31.5$ and 44 GeV data, respectively [501].

The above measurements made it clear that transverse energy distributions are built up from a large number of particles each having a rather small value of transverse momentum, typically $\simeq\langle p_T\rangle$. The strong relationship of the multiplicity and transverse energy distributions is a consequence of the fact that the transverse momentum distribution for particle production is largely independent of the rapidity and multiplicity distributions [114–116,509]. Thus, an E_T measurement is simply an analog method of counting particles: each particle produced has roughly the same value of $E_i \sin\theta_i \simeq \langle p_T\rangle$ [510].

11.4 The use of E_T distributions in p–p collisions: the study of jets

The dominance of "soft" physics in E_T distributions, once it was understood, did not prove to be a major impediment to the "geometrically unbiased" study of the jets of hard scattering in p–p and p–$\bar{p}$ collisions. The fact that the overwhelming majority of the collisions produce final states in which the transverse energy is

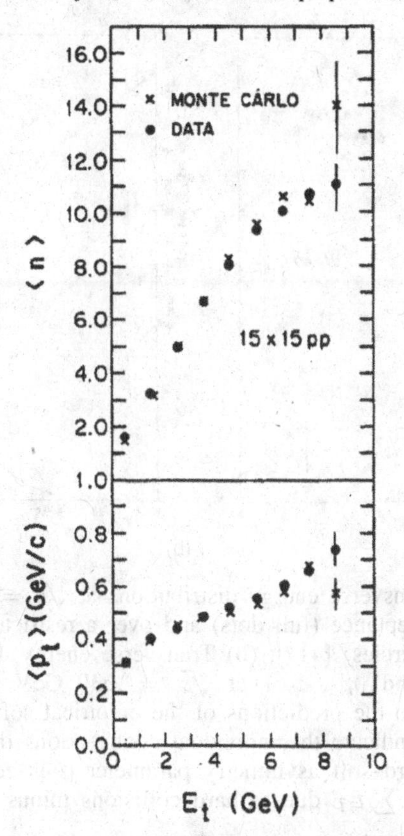

Figure 11.16 $\langle n \rangle$ and $\langle p_T \rangle$ versus E_T (actually E_T^c) for p–p $\sqrt{s} = 31.5$ GeV data. The crosses are the result of a Monte Carlo calculation choosing particles independently from the multiplicity distribution and single particle p_T distributions [501].

relatively uniformly distributed among the collision products simply means that a good hardware trigger is required to select the large values of E_T where the jets become important.

The first evidence for jets in a geometrically unbiased calorimeter experiment was presented at the 1982 Paris conference by the UA2 collaboration from p–$\bar{p}$ collisions with $\sqrt{s} = 540$ GeV at the CERN collider [416, 417]. The distribution of transverse energy E_T over the pseudorapidity interval $-1 < \eta < +1$ and an azimuthal range $\Delta\phi = 300°$ was measured using a hadron calorimeter, of total thickness 4.6 absorption lengths, segmented into 200 cells, each covering 15° in ϕ and 10° in θ. The spectrum in E_T (more properly called $\sum E_T$ in this early report) falls off exponentially for roughly 5 orders of magnitude (Figure 11.17). There were 10 events with $E_T > 60$ GeV, and these appeared to lie significantly above the exponential extrapolation. In these events, most of the transverse

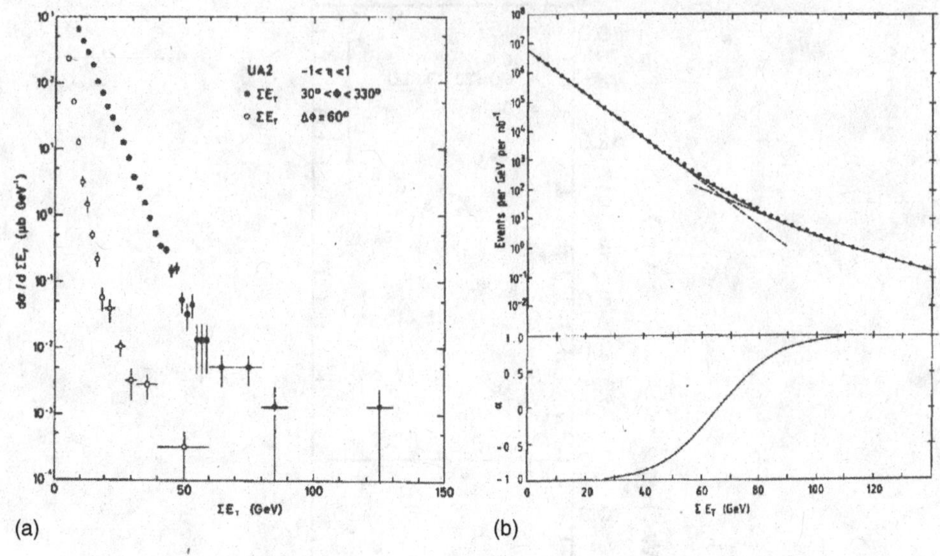

Figure 11.17 (a) Transverse energy distributions at $\sqrt{s} = 540$ GeV over the whole azimuthal acceptance (full dots) and over a restricted azimuthal region $\Delta\phi = 60°$ (open circles) [417]. (b) Transverse energy distribution over the full azimuth (2π) and $|\eta| < 1$ at $\sqrt{s} = 630$ GeV. The data (full circles) are compared to the predictions of the empirical soft–hard model (solid line). Dashed lines indicate the individual contributions of the soft and hard mechanisms. The hard–soft asymmetry parameter α is equal to the fraction of events at a given $\sum E_T$ due to hard collisions minus the fraction due to soft collisions [487].

energy was found to be contained in small angular regions, as expected for high transverse momentum hadron jets. The most spectacular example is shown in the event with the largest value of $\sum E_T$, 127 GeV (Figure 11.18). It exhibits striking features: energy is concentrated within two small regions separated in azimuth by $\Delta\phi \simeq 180°$ and towards which several collimated tracks are observed to point. In addition, the transverse energies of the two clusters are approximately equal (57 and 60 GeV). This single event removed all doubts about jets in hadron collisions. Further work [487] confirmed that there is indeed a clear break in the exponential E_T distribution (Figure 11.17b) and that the lower exponential region, spanning roughly 4 orders of magnitude in cross section, is dominated by "soft" multiparticle physics, while the larger values of $E_T > 80$ GeV are dominated by the jets of QCD "hard" collisions.

At ISR energies, the hard scattering component of E_T distributions is inconsequential except at the largest values of E_T and the highest c.m. energy, $\sqrt{s}$=63 GeV. Two experiments at the CERN-ISR made systematic studies of E_T

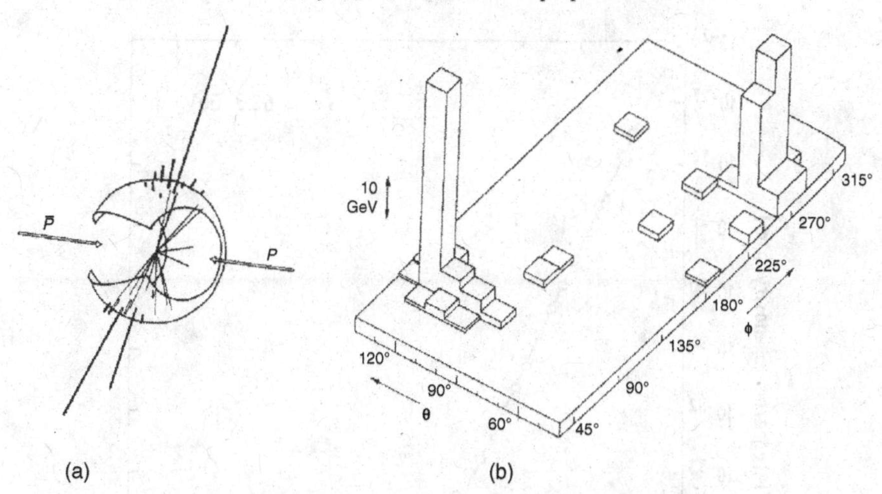

Figure 11.18 Configuration of the event with the largest value of $\sum E_T$, 127 GeV. (a) Charged tracks pointing to the inner face of the central calorimeter are shown together with the cell energies (indicated by heavy lines with lengths proportional to cell energies). (b) The cell energy distribution as a function of polar angle θ and azimuthal angle ϕ [416,417].

distributions using full azimuth geometrically unbiased calorimeters. The COR collaboration [376] used a hybrid electromagnetic calorimeter, consisting of lead glass blocks and lead-scintillator shower counters, covering 90% of 2π in azimuth with an average c.m. pseudorapidity acceptance inside this region of $\Delta\eta = \pm 0.9$, surrounding a system of tracking chambers in a thin-walled superconducting solenoid (recall Figure 8.20). The spectrum in neutral transverse energy E_T^0 at c.m. energy $\sqrt{s} = 62.4$ GeV falls exponentially and then deviates from a simple exponential above E_T^0 of 20 GeV (Figure 11.19). A principal axis analysis [474,475] of the event structure indicated a uniform distribution of E_T^0 in the lower exponential region and a dominant two jet structure for $E_T^0 > 24$ GeV. The AFS collaboration measured E_T spectra at three c.m. energies (Figure 11.20) using a full hadron calorimeter [511]. These spectra are consistent with an exponential falloff at large E_T and show no sign of a break. The measurements were made with a highly segmented calorimeter, consisting of 192 towers, a total of 3.8 hadron absorption lengths thick. The calorimeter covered the full azimuth, with average c.m. pseudorapidity acceptance roughly $\Delta\eta = \pm 0.9$, and surrounded the central tracking detector [485,501] (see Figure 11.21). The separation between "soft" and "hard" physics in these spectra, over the full ISR energy range, was determined by a study of the event shape as a function of E_T, using a principal axis analysis. A quantity, "circularity," was defined which would be unity for a totally

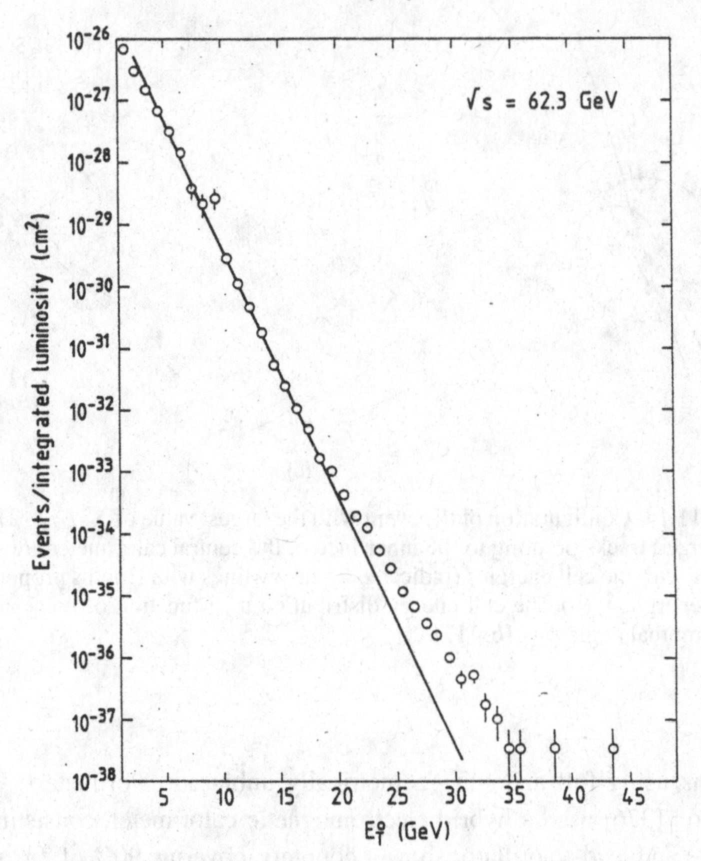

Figure 11.19 The spectrum of events/integrated luminosity versus E_T^0 at $\sqrt{s} = 62$ GeV. The line is an exponential which fits the data over the range 10 to 20 GeV in E_T^0 [376].

uniform azimuthal distribution of the components of E_T and zero if all the E_T were in two narrow jets back-to-back in azimuth. The distributions show no evidence of jets for $E_T < 25$ GeV at any c.m. energy (see Figure 11.22). For values of $E_T > 25$ GeV, there is an increase in low circularity events at $\sqrt{s} = 45$ GeV, and a similar effect at a slightly higher E_T at $\sqrt{s} = 63$ GeV, leading up to a dominance of low circularity events (corresponding to a two jet structure) for values of $E_T > 35$ GeV.

These beautiful measurements make it clear that the jets of hard scattering can indeed be observed using E_T distributions, but that hard scattering effects have negligible influence on the shape of E_T spectra in proton–(anti)proton collisions for the first four, or even six, orders of magnitude of cross section, depending on the c.m. energy. This leads to the second main application of E_T distributions:

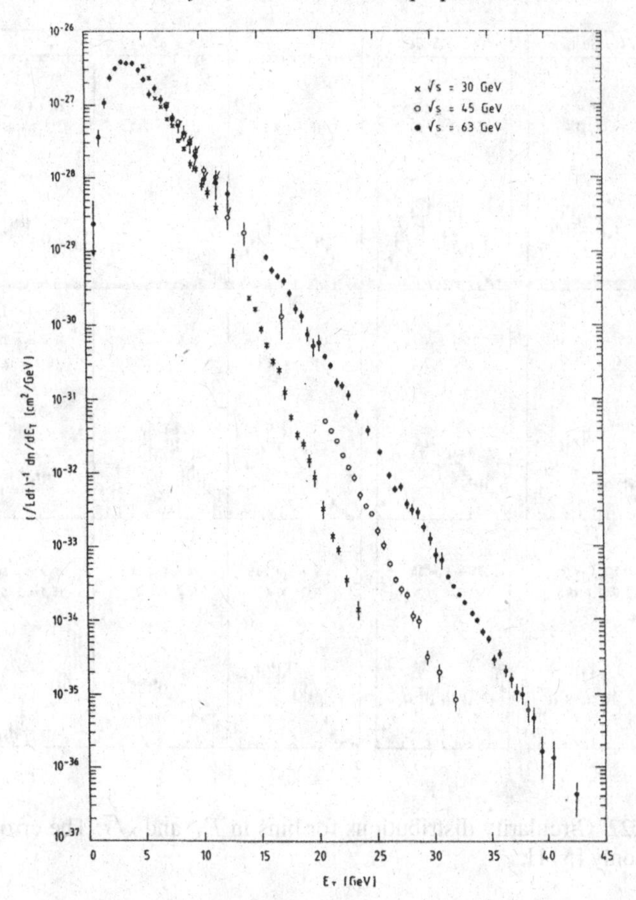

Figure 11.20 The uncorrected transverse energy spectra at $\sqrt{s} = 30$, 45 and 63 GeV. The error bars are statistical only [511].

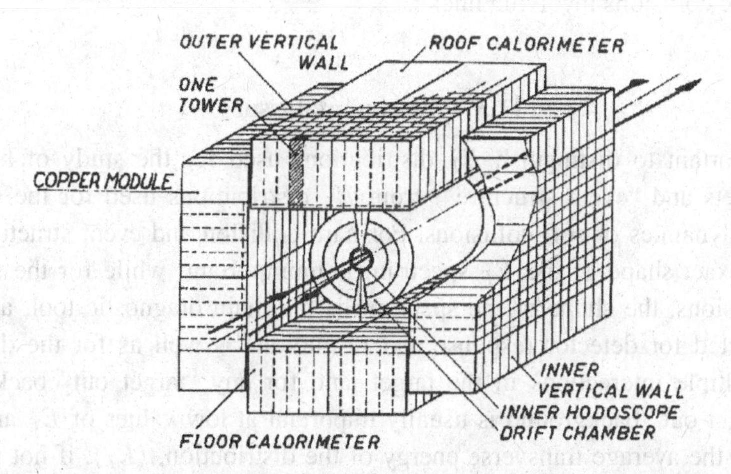

Figure 11.21 Experimental setup of the AFS collaboration [511].

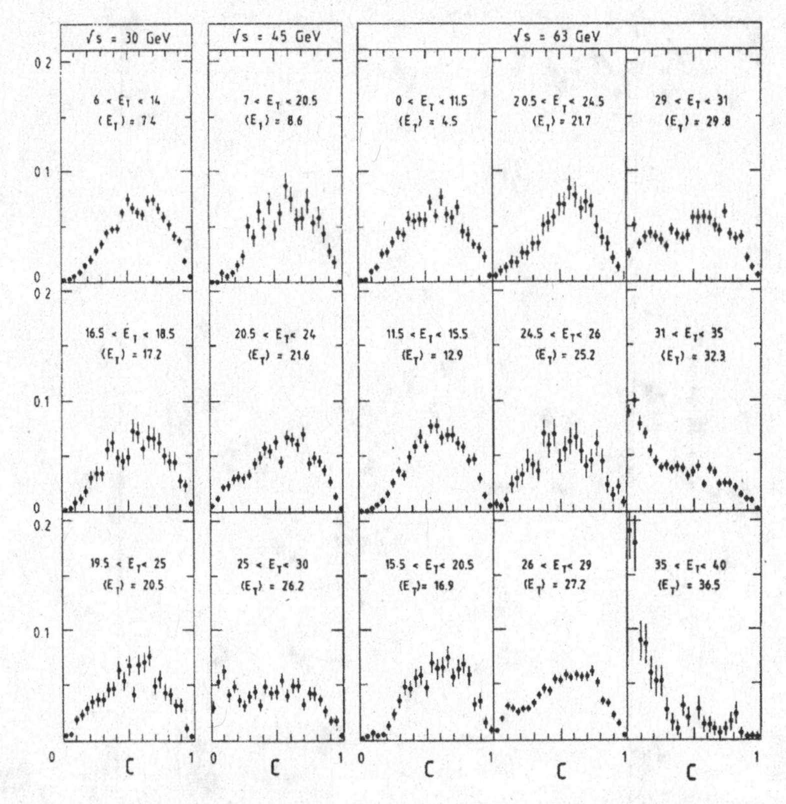

Figure 11.22 Circularity distributions for bins in E_T and $\sqrt{s}$. The error bars are statistical only [511].

the study of the reaction dynamics of multiparticle production, particularly in relativistic collisions involving nuclei.

11.5 Experimental issues

It is important to distinguish E_T distributions used for the study of hard collisions (jets and "event structure") from E_T distributions used for the study of reaction dynamics of soft collisions. For hard collision and event structure studies, the exact shape of the E_T spectrum is unimportant; while for the study of soft collisions, the shape of the spectrum is the main diagnostic tool, and must be corrected for detector response and resolution, as well as for the distortion from multiple interactions in the target, and for any "target out" background. The "target out" background is usually important at low values of E_T and tends to distort the average transverse energy of the distribution, $\langle E_T \rangle$, if not properly corrected.

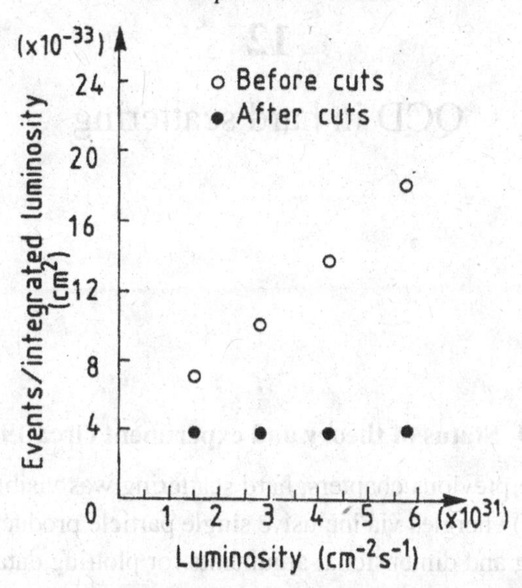

Figure 11.23 Events/integrated luminosity for a typical E_T^0 interval from Figure 11.19 versus the instantaneous luminosity. The open circles are for data without the single interaction criterion, and the closed circles are for data after the requirement of a single interaction [376].

Another important experimental issue is "pile-up" due to multiple interactions during the "gate" time of recording the energy deposited in the calorimeter. Pile-up is particularly serious in high luminosity conditions [376]. The spectrum can be severly distorted unless the pile-up is eliminated, typically by a requirement that no additional interaction take place, before or after the interaction of interest, in a time interval corresponding to plus or minus the gate width (see Figure 11.23).

12

QCD in hard scattering

12.1 Status of theory and experiment circa 1982

As discussed in the previous chapters, hard scattering was visible both at ISR and FNAL (fixed target) energies via inclusive single particle production at large $p_T \geq$ 2–3 GeV/c. Scaling and dimensional arguments for plotting data revealed the systematics and underlying physics. The theorists had the basic underlying physics correct; but many (inconvenient) details remained to be worked out, several by experiment. The transverse momentum imbalance of outgoing parton pairs, the "k_T effect," was discovered by experiment [383, 403, 512], and clarified by Feynman, Field and Fox [391].

The first modern QCD calculation and prediction for high p_T single particle inclusive cross sections, including non-scaling and initial state radiation was done in 1978 (Figure 12.1) by Jeff Owens and collaborators [69, 70] and did not include k_T smearing which, as they explained, would improve the agreement with the data for $p_T < 4$ GeV/c if included. This was closely followed and corroborated by Feynman and collaborators [71], who did include k_T smearing, under the assumption that high p_T particles are produced from states with two roughly back-to-back jets which are the result of scattering of constituents of the nucleons (partons). However, as described in Chapter 11, jets in 4π calorimeters at ISR energies or lower are invisible below $\sqrt{\hat{s}} \sim E_T \leq 25$ GeV [513]. The many false claims of jet observation in the period 1977–1982 had led to skepticism about jets and the validity of QCD in hadron collisions, particularly in the USA. This was only put to rest after the ICHEP in Paris 1982 with the single plot of the UA2 jet (Figure 11.18), and the CCOR measurement of the elementary subprocess scattering angular distribution (Figure 9.18).

12.2 Jets since 1982

Since 1982, jets have become the work-horse of high energy physics [420, 514, 515] since they do not suffer Bjorken's "parent–child" reduction in cross section

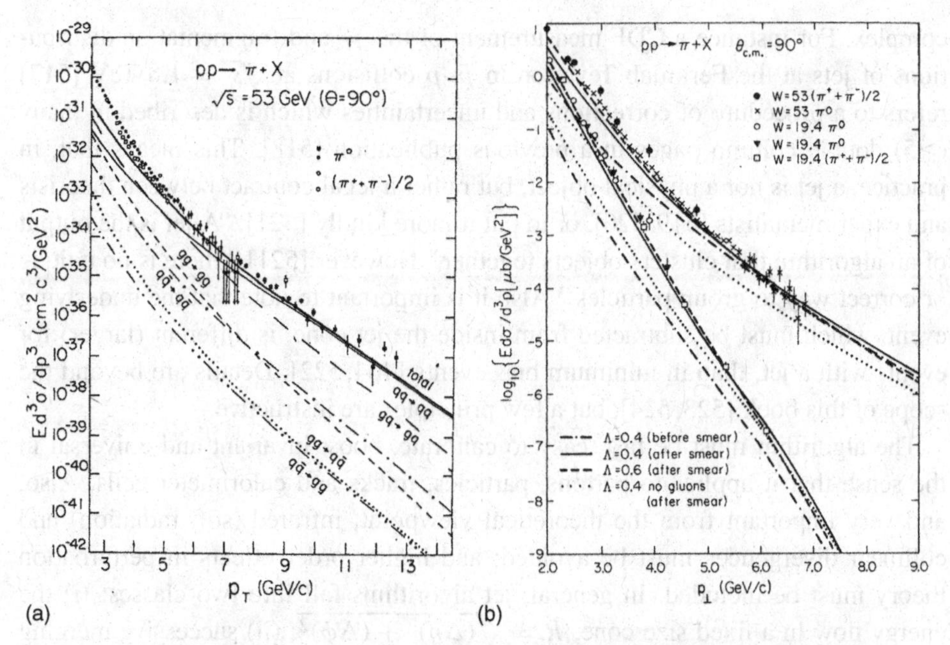

Figure 12.1 (a) Owens, Reya, Glück [69] QCD calculation of π production at $\sqrt{s} = 53$ GeV compared to BS [312], CCRS [270] and CCOR [269] measurements. Individual contributions of the various subprocesses to the total (solid curve) are indicated. (b) Feynman, Field and Fox "QCD model" [71] compared to Fermilab [273], BS [312] and CCRS [270] data, where $W = \sqrt{s}$. The solid curves are for $\Lambda = 0.4$ GeV/c and $\langle k_T \rangle = 0.848$ GeV, the results before and after smearing respectively. The broken curves show the effects of varying Λ, turning off k_T smearing and not including gluons, as indicated.

(Eq. 9.8) compared to single or two particle measurements and thus can probe much larger p_T for a given integrated luminosity. Bjorken's original idea [377] was that the transverse momentum of a scattered parton could be measured by the sum of the transverse momenta of the hadrons emerging in the direction of the struck parton which will have small transverse momenta (j_T) with respect to this axis. There was the implicit assumption that the exact details of parton fragmentation were not so important so long as one could [377] "ascertain what fraction of the parton momentum is found, on the average, in those hadrons," which he called a "core" and we now call a jet.

The details of working this out in a way that is both experimentally and theoretically acceptable has spawned a whole industry of specialists who have produced beautiful results. Nevertheless, it is important to be aware of some theoretical warnings. For example [516]: "However, for reasons of color conservation, energy-momentum conservation and quantum-mechanical interference, a jet of hadrons *cannot* be the residue of a *single* parton." Experimental warnings are even more complex. For instance a CDF measurement of the j_T and fragmentation distributions of jets at the Fermilab Tevatron in $\bar{p}$–p collisions at $\sqrt{s} = 1.8$ TeV [517] refers to a procedure of corrections and uncertainties which is described in many ($\geq$5) double column pages in a previous publication [518]. This means that, in practice, a jet is not a physical object, but rather a legal contract between theorists and experimentalists [519, 520], or to put it more kindly [521]: "A jet is the output of an algorithm that clusters objects together." However [521], "there is no unique or correct way to group particles." Also it is important to note that the underlying event, which must be subtracted from inside the jet cone, is different (larger) for events with a jet, than in minimum bias events [404, 522]. Details are beyond the scope of this book [523, 524], but a few principles are instructive.

The algorithm must be fast, easy to calibrate, boost invariant and universal in the sense that it applies to partons, particles, tracks and calorimeter cells. Also, and very important from the theoretical viewpoint, infrared (soft radiation) and collinear divergences must be avoided; and higher order effects in perturbation theory must be included. In general, jet algorithms fall into two classes: (i) the energy flow in a fixed size cone, $R = \sqrt{(\Delta\eta)^2 + (\Delta\phi)^2}$; (ii) successive merging of pairs of "proto-jets," such as the "k_T" [525] or recently popular "anti-k_T" algorithm [526]. An important experimental issue is how to correct for the different detector responses to a variety of different particles ($\pi, K, p, n, \gamma, \ldots$) with a variety of different energies and how to get the absolute p_T scale of the jet. Combined with equally complicated issues in theoretical calculations, for example for order α_s^3 [527], the results of recent jet measurements are impressively "in agreement with pQCD."

A recent measurement of the inclusive jet cross section by the D0 collaboration [528] at the Fermilab Tevatron in $\bar{p}$–p collisions at $\sqrt{s} = 1.96$ TeV is shown in Figure 12.2a as a function of p_T for six bins of jet rapidity, $|y| < 2.4$. At mid-rapidity, the data follow the characteristic hard scattering power-law out to roughly 300 GeV ($x_T \approx 0.3$) and then drop sharply. With increasing rapidity, the power-law steepens and the sharp drop occurs at lower x_T, likely due to conservation of energy. The Next to Leading Order perturbative QCD (NLO pQCD) prediction agrees incredibly well with the data, with the parameters indicated, which will be discussed below, while other caveats such as "corrected to the particle level" or "+ non-perturbative corrections" only hint at the complexity of the analysis. An even more recent measurement of the inclusive jet cross section in p–p collisions at $\sqrt{s} = 7$ TeV by the CMS experiment [529] at the CERN-Large Hadron Collider (LHC), is shown in Figure 12.2b as a function of p_T for six bins of jet rapidity, $|y| < 3$. These data show the same trends, a pure power-law out to 300 GeV/c ($x_T \approx 0.09$), but with a much weaker drop at larger p_T and much smaller changes with increasing rapidity. This is due to the increase in $\sqrt{s}$ by a factor of 3.5, which

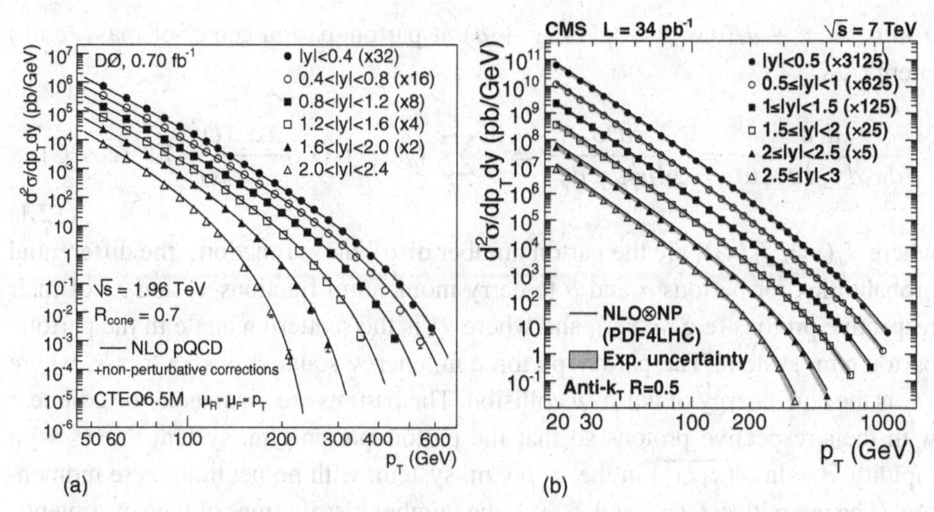

Figure 12.2 (a) D0 [528] inclusive jet cross sections at $\sqrt{s} = 1.96$ TeV as a function of jet p_T, multiplied by the factors indicated, in bins of jet rapidity, with NLO pQCD predictions. (b) CMS [529] measurements at $\sqrt{s} = 7$ TeV (data points). The NLO theoretical predictions, corrected for non-perturbative (NP) effects via multiplicative factors are superimposed. The statistical uncertainties are smaller than the symbols used to represent the data points. The gray lines are the experimental systematic uncertainty predominantly from the luminosity and jet p_T scale.

also accounts for the greater reach in jet p_T at the LHC with a factor of 20 less integrated luminosity. Interestingly, the NLO pQCD calculation which is "generally in agreement with the data" was done by the experimenters using packaged Monte Carlo programs [529] including the estimation of non-perturbative (NP) corrections for "hadronization and multiple parton interactions."

In summary, since 1982 jets and multi-jets have been the main probe of new physics and tests of the standard model in hadron collisions because they provide the largest reach in transverse momentum and hence probe the highest masses and the smallest distance scales possible. For the record, it is interesting to remember that the first measurement of three-jet production in p–p collisions was by the CCOR collaboration at the CERN-ISR [530] and the first attempts of measuring the QCD coupling constant α_s from the 3–2 jet ratio in hadron collisions were performed by UA1 [531] and UA2 [532] at the CERN-SPS collider with a later result by the AFS collaboration at the CERN-ISR [533].

12.3 The factorization theorem for pQCD

The overall p–p hard scattering cross section in lowest order perturbative QCD (pQCD) [419] is, as in the parton model (Eq. 9.2), the sum over parton reactions

$a + b \to c + d$ (e.g. $g + q \to g + q$) at parton–parton center-of-mass (c.m.) energy $\sqrt{\hat{s}}$

$$\frac{d^3\sigma}{dx_1 dx_2 d\cos\theta^*} = \frac{\hat{s}d^3\sigma}{2dp_T^2 dy_c dy_d} = \frac{1}{s}\sum_{ab} f_a(x_1) f_b(x_2) \frac{\pi\alpha_s^2(Q^2)}{2x_1x_2}\Sigma^{ab}(\cos\theta^*) \tag{12.1}$$

where $f_a(x_1)$, $f_b(x_2)$, are the parton number distribution functions, the differential probabilities for partons a and b to carry momentum fractions x_1 and x_2 of their respective protons (e.g. $u(x_2)$), and where θ^* is the scattering angle in the parton–parton c.m. system. The parton–parton c.m. energy squared is $\hat{s} = x_1x_2s$, where $\sqrt{s}$ is the c.m. energy of the p–p collision. The partons are assumed to be collinear with their respective protons so that the parton–parton c.m. system moves with rapidity $\hat{y} = \ln\sqrt{(x_1/x_2)}$ in the p–p c.m. system, with no net transverse momentum. The quantities $f_a(x_1)$ and $f_b(x_2)$, the number distributions of the constituents, are related (for the electrically charged quarks) to the structure functions measured in Deeply Inelastic lepton–hadron Scattering (DIS) (Eq. 9.3).

Equation 12.1 gives the p_T spectrum of outgoing parton c, which then fragments into a jet of hadrons, h. For measurements of inclusive single particle production of identified hadrons, e.g. π^0, one must multiply Eq. 12.1 by the fragmentation function, $D_c^h(z)$, which is the probability for a hadron h to carry a fraction $z = p^h/p^c$ of the momentum of outgoing parton c (see Appendix E for a full discussion). Then Eq. 12.1 must be summed over all subprocesses leading to a π^0 in the final state weighted by their respective fragmentation functions. In this formulation, $f_a(x_1)$, $f_b(x_2)$ and $D_c^{\pi^0}(z)$ represent the "long-distance phenomena" to be determined by experiment; while the characteristic subprocess angular distributions [413, 414], $\Sigma^{ab}(\cos\theta^*)$, and the coupling constant, $\alpha_s(Q^2) = \frac{12\pi}{25\ln(Q^2/\Lambda^2)}$ [22, 23], are fundamental predictions of QCD for the short-distance, large Q^2, phenomena.

It is important to note that Eq. 12.1 is the incoherent sum over individual parton–parton cross sections – the amplitudes are squared and then summed. This is the "impulse approximation" [54]. It follows from the assumption that the individual partons in a nucleon do not interact with each other during the short crossing time of the strongly Lorentz contracted nucleons at the large values of $\sqrt{s}$ required for large Q^2 reactions. Also the large Q^2 corresponds to a small distance so that it is assumed that only one parton in each nucleon is involved in the scattering, provided the density of partons is not too high. A further assumption is that any interactions, for example fragmentation, that occur in the final state, after the hard scattering, occur at time scales too long to interfere with it [534].

The factorization theorem [534] states that the parton distribution functions and fragmentation functions which apply in the QCD hard scattering in p–p collisions, Eq. 12.1, are the same as those measured by experiments with large momentum transfers in different systems such as DIS, Drell–Yan and e^+e^- collisions, i.e. they are universal. This theorem establishes a field theoretical basis for the methods to separate or "factorize" the long and short distance phenomena and allows a perturbation series in orders of the coupling constant $\alpha_s(Q^2)$ in analogy to Quantum Electrodynamics (QED) [535]. The price to be paid is that higher order effects come into play in perturbative QCD (pQCD), such as gluon radiation by incident partons which are assumed to be massless, singularities whenever two outgoing parton lines are parallel, etc., which must be taken into account in combination with the "DGLAP evolution" [536, 537] of the parton distribution and fragmentation functions as a function of Q^2, which is another fundamental property of pQCD. The result is that it is necessary to specify factorization scales μ for the parton distribution functions, μ_F for the fragmentation functions, in addition to a renormalization scale μ_R which governs the running of $\alpha_s(Q^2)$. The meaning of these "scales" is that pQCD is valid for $Q^2 \geq \mu, \mu_R, \mu_F$ at which the (non-perturbative) experimental values of the distribution are determined, and that all the singularities and other such problems are absorbed into the determination of the parton distribution and fragmentation functions at these scales. Then for any higher value of Q^2, the values of $\alpha_s(Q^2)$, $f_a(x, Q^2)$ and $D^h_c(z, Q^2)$ can be determined by the standard QCD renormalization group equations [534, 537]. The principal difference between pQCD and the parton model is that the parton model assumed scale invariance of $f_a(x)$ and $D^h_c(z)$ and the unknown coupling constant, while in pQCD these quantities have large non-scaling effects, i.e. dependence on Q^2.

12.4 Parton distribution and fragmentation functions

The quantities $f_a(x_1)$ and $f_b(x_2)$, the "number" distributions of the constituents, are related (for the electrically charged quarks) to the structure functions measured in Deeply Inelastic lepton–hadron Scattering (DIS), e.g.

$$F_1(x, Q^2) = \frac{1}{2} \sum_a e_a^2 f_a(x, Q^2) \quad \text{and} \quad F_2(x, Q^2) = x \sum_a e_a^2 f_a(x, Q^2) \quad (12.2)$$

where e_a is the electric charge on a quark in units of the electron charge. It is also important to note that, for example, if a is a u-quark, $f_u(x, Q^2) = u(x, Q^2) + \overline{u}(x, Q^2)$ i.e. is the sum of the u-quark and $\overline{u}$-quark parton distribution functions, and the same for $f_d(x, Q^2)$, where we consider only the light quarks u and d for this discussion. If $f_u(x, Q^2)$ and $f_d(x, Q^2)$ are determined from DIS, then $\overline{u}(x, Q^2)$, $\overline{d}(x, Q^2)$ can be determined using Drell–Yan di-lepton production in $p+p$ and $p+d$ collisions. For p–p collisions, the Drell–Yan cross section, taking into account a factor of 3 decrease due to the required color matching [322], is given by:

$$s\frac{d^2\sigma}{d\sqrt{\tau}dy} = \left(\frac{8\pi\alpha^2}{9\tau^{3/2}}\right)\sum_a e_a^2\left[f_{\bar{a}}(x_1)f'_a(x_2) + f'_a(x_1)f_{\bar{a}}(x_2)\right], \tag{12.3}$$

where $\tau \equiv m^2/s$, and for antiquarks, for example a is a $\bar{u}$-quark, $f_{\bar{u}}(x, Q^2) = \bar{u}(x, Q^2)$ and $f'_u(x, Q^2) = u(x, Q^2)$.

In principle, and sometimes in practice [538], the gluon number distribution function $g(x, Q^2)$, which does not appear in either Eq. 12.2 or 12.3 because the gluon is electrically neutral, can be determined from the direct-γ inclusive cross section by integrating Eq. 10.8, for a γ jet, over the rapidity y_d of the jet. However, the other leading order process $\bar{q}q \to \gamma + g$ must also be taken into account. In general, this is not done and $g(x)$ is usually determined from the Q^2 evolution of $F_2(x, Q^2)$ at fixed x, which is not trivial [538].

12.4.1 Parton distribution functions

In determining the Parton Distribution Functions (PDF) of protons and neutrons for the various flavor quarks, e.g. $u(x)$, $d(x)$, $s(x)$, $\bar{u}(x)$, $\bar{d}(x)$, $\bar{s}(x)$, and gluons $g(x)$, there are constraints and sum rules, dating from the quark–parton model [211,539], which still must be obeyed. For example, for the proton, the total electric charge is +1 so that

$$1 = \frac{2}{3}\int_0^1 [u(x) - \bar{u}(x)]\,dx - \frac{1}{3}\int_0^1 \left[d(x) - \bar{d}(x)\right]dx - \frac{1}{3}\int_0^1 [s(x) - \bar{s}(x)]\,dx, \tag{12.4}$$

the z component of isospin is +1/2 so

$$\frac{1}{2} = \frac{1}{2}\int_0^1 [u(x) - \bar{u}(x)]\,dx - \frac{1}{2}\int_0^1 \left[d(x) - \bar{d}(x)\right]dx, \tag{12.5}$$

and the strangeness is zero:

$$\int_0^1 [s(x) - \bar{s}(x)]\,dx = 0. \tag{12.6}$$

These equations have the solution:

$$\begin{aligned}\int_0^1 [u(x) - \bar{u}(x)]\,dx &= 2\\ \int_0^1 \left[d(x) - \bar{d}(x)\right]dx &= 1\end{aligned} \tag{12.7}$$

which gives the correct answer for the net number of u and d quarks in the proton. By isospin reflection, the PDFs of the neutron are obtained by replacing d by u and u by d, and keeping in mind that $u(x)$ is the number of up-quarks in the proton and down-quarks in the neutron. This is relevant to the momentum sum rule, which is very interesting:

$$\int_0^1 F_2^p(x)dx = \int_0^1 dx\, x \left\{\frac{4}{9}[u(x)+\bar{u}(x)] + \frac{1}{9}\left[d(x)+\bar{d}(x)\right] + \frac{1}{9}[s(x)+\bar{s}(x)]\right\}$$
$$\int_0^1 F_2^n(x)dx = \int_0^1 dx\, x \left\{\frac{1}{9}[u(x)+\bar{u}(x)] + \frac{4}{9}\left[d(x)+\bar{d}(x)\right] + \frac{1}{9}[s(x)+\bar{s}(x)]\right\}. \quad (12.8)$$

It was known [196] from the original DIS discovery [53] that $\int_0^1 F_2^p(x)dx = 0.16$. Thus, assuming zero for the strange quark content, the charge weighted momentum fraction carried by the quarks for a *uud* proton should be $8/27 + 1/27 = 1/3 = 0.33$, so that half the momentum was missing. Similarly, for the *ddu* neutron [200, 211] the value of $\int_0^1 F_2^n(x)dx = 0.12$ was again about half of the expected $2/27+4/27 = 2/9 = 0.22$. This would imply that half the momentum must be carried by neutral partons ("gluons") [211]. This point was strongly emphasized by Perkins at the 1972 ICHEP [221] who found the same deficit in neutrino scattering. Nowadays the momentum sum rule which constrains the PDFs is written as [539]:

$$1 = \int_0^1 dx\, x \left\{u(x)+\bar{u}(x)+d(x)+\bar{d}(x)+s(x)+\bar{s}(x)+g(x)\right\}. \quad (12.9)$$

For modern QCD calculations, the determination of the parton distribution functions (PDF) is a major undertaking by groups of devoted specialists since, in addition to getting the pQCD correct at LO, NLO and possibly NNLO according to the level of perturbation theory for which these PDFs will be used, one also has to cope with the many different sets of experimental data which in general do not necessarily agree with each other. Two of the most popular sources of PDFs are the CTEQ6M [540] used in Figures 10.9b and 12.2a and the new PDF4LHC collection [541] used in Figure 12.2b. The experimental data used in these analyses are typically DIS of electrons, muons and neutrinos, Drell–Yan di-lepton, $W^{\pm}$ and Z^0 production and inclusive jet production in hadron–hadron collisions. The actual data are fit to different formulas for the PDFs by the different groups. Some examples of the latest data used are shown in Figure 12.3. The scale breaking for $F_2(x)$ as a function of Q^2 (Figure 12.3a) is dramatic in the range of $Q^2 = 2p_T^2 \approx 50$–800 GeV^2 relevant to RHIC. For identified gluons and quarks (Figure 12.3b), the gluons ($xg(x)$) and sea ($xS(x)$) dominate for $x < 0.1$, and dominate even more as Q^2 is increased. For $Q^2 = 10\,000$ GeV^2 the same data show an increase in $xg(x)$ to ~30 at $x = 0.001$ from ~9 at $Q^2 = 10$ GeV^2, assuming that DGLAP evolution is valid.

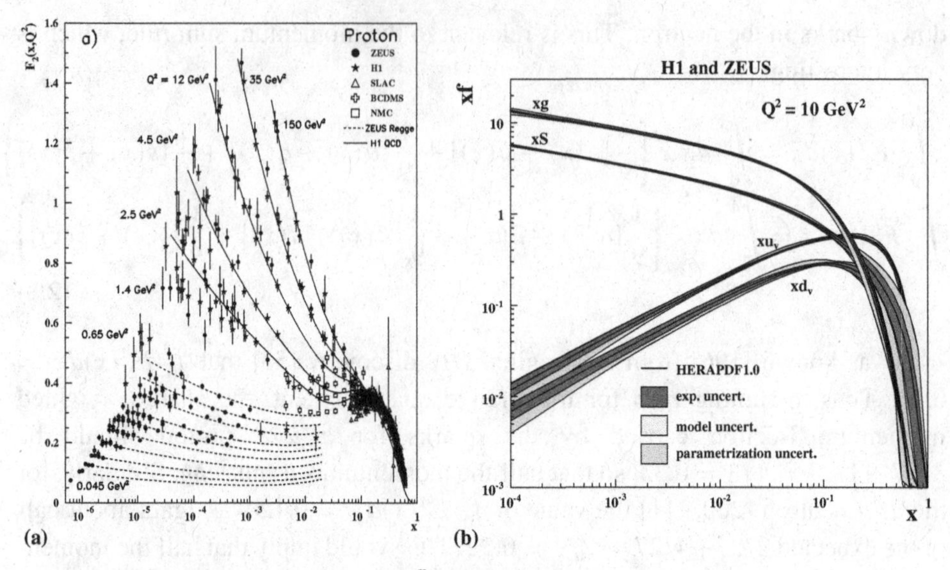

Figure 12.3 (a) Compilation of F_2^p in e–p and μ–p scattering as a function of x for various Q^2 indicated [84]. (b) PDFs for $xu_v(x)$, $xd_v(x)$, $xS(x)$, $xg(x)$ at $Q^2 = 10$ GeV^2 from combined H1 and Zeus experiments at HERA [542], where $U = u + c$, $D = d + s$, the sea $S = 2(\bar{U} + \bar{D})$, and $u_v = U - \bar{U}$, $d_v = D - \bar{D}$.

12.4.2 Fragmentation functions

The key element in measuring a fragmentation function is to know the energy of the parton that is fragmenting. For reactions such as DIS, where Q^2 and ν are measured from the scattered lepton, or $e^+ + e^- \to q + \bar{q}$ at a precisely known $\sqrt{s}$ when events with two jets are selected, the energy of the outgoing jet (which is a quark jet in both cases) is known. This is why measurements from DIS and e^+e^- provide the majority of the data for fragmentation functions. For example, in Figure 12.4a, the π^0 fragmentation function measured by the European Muon Collaboration (EMC) [543] at CERN in DIS of 200 GeV muons on a liquid hydrogen target is shown, where $Q^2 \geq 3$ GeV^2, $7.1 \leq W \leq 18.7$ GeV, and the fragmentation variable is $z = E_{\pi^0}/\nu$ (which assumes collinearity of beam, virtual photon and π^0 directions). Figure 12.4a is typical of most light parton fragmentation functions: exponential for $0.1 \leq z < 1.0$, with a steeper slope for $z \leq 0.1$ (also see Figure E.1 in Appendix E, Section E.3).

On the theoretical side, the situation in fragmentation functions is even more challenging than for parton distribution functions, so it is again a subject for specialists. To quote from a recent paper by one of the expert groups [545],

In spite of the remarkable phenomenological success of QCD as the theory of strong interactions, a detailed quantitative understanding of hadronization processes is still one of the great challenges for the theory. Hadronization is the mechanism by which a final-state quark or gluon, excited in some hard partonic interaction, dresses itself and develops into an observed final-state hadron.[1] As such, it is sensitive to physics happening at long distances and time scales, and perturbative tools that have earned QCD its present standing, simply fall short.

Nevertheless, there are several theoretical successes in describing fragmentation that must be mentioned.

In 1977, relatively soon after the advent of QCD (and two years after di-jets were generally accepted to have been seen in e^+e^- annihilation at $\sqrt{s} = 7.4$

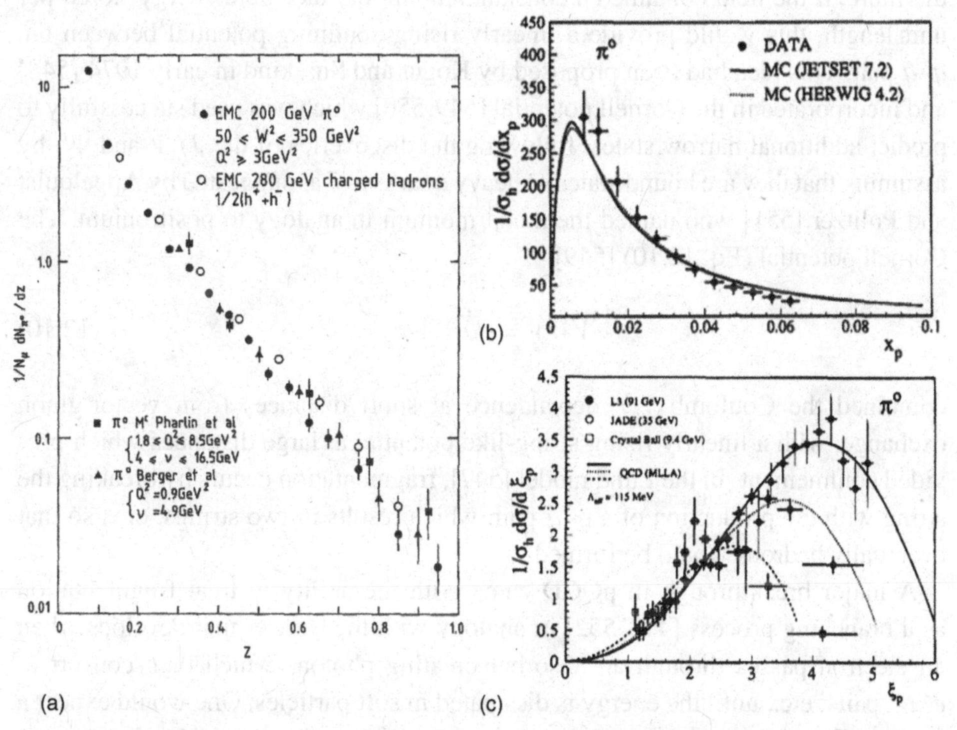

Figure 12.4 (a) Inclusive $\frac{1}{N_\mu}\frac{dN_{\pi^0}}{dz}$ measured by EMC [543] together with other data indicated. (b) Inclusive spectra of π^0 for two-jet events at the Z^0 peak in e^+e^- collisions from L3 at LEP [544] plotted in the variables $x_p = 2p/\sqrt{s}$, where p is the π^0 momentum, and $\xi_p = \ln(1/x_p)$. These are compared to previous measurements, to jet fragmentation Monte Carlos and to QCD.

[1] This is a typical example of why it is frustrating to read the literature on this subject: obviously a quark fragments to hadrons not to a hadron. Also, an "excited" quark which fragments, if different in meaning from "an outgoing massless quark" which fragments, is model dependent, since, for instance, Eq. 10.5 [301] (above) and the QCD scattering subprocesses [413,414] (Figure 12.8a–h, below) assume massless quarks and gluons. Perhaps this illustrates why it takes groups of specialists to provide fragmentation functions which match to Eq. 12.1 in a theoretically consistent framework.

GeV at SLAC [546] in agreement with the parton model but in disagreement with models of isotropic phase space), Sterman and Weinberg [494] showed that by using energy rather than particles, a final state consisting of two jets which have all but a small fraction of their energy in a pair of opposite cones of small half-angle is predicted in pQCD without use of the parton model. At roughly the same time, a clearly non-perturbative but more easily visualized Lund string model [547] was proposed, which took advantage of the idea [548] that an outgoing q–$\bar{q}$ pair from e^+e^- annihilation did not spread color flux lines over all space, as in the electromagnetic field from their electric charges, because of the three-gluon coupling which caused the color flux lines to be constrained in a thin tube-like region. Furthermore, if the field contained a constant amount of color-field energy stored per unit length, this would provide a linearly rising confining potential between the q–$\bar{q}$ pair. This idea had been proposed by Kogut and Susskind in early 1974 [548] and incorporated in the Cornell potential [549,550] which was used successfully to predict additional narrow states, following the discoveries of the J/Ψ and Ψ', by assuming that they are bound states of heavy quarks c–$\bar{c}$ as suggested by Appelquist and Politzer [551] who named them charmonium in analogy to positronium. The Cornell potential (Eq. 12.10) [549],

$$V(r) = -\frac{\alpha_s}{r} + \sigma r \tag{12.10}$$

combined the Coulomb $1/r$ dependence at short distances from vector-gluon exchange, with a linearly rising string-like potential at large distances which provided confinement. In the Lund model [547], fragmentation occurs by breaking the string with the production of a q–$\bar{q}$ pair, which results in two strings, etc., so that eventually hadrons would be formed.

A major breakthrough in pQCD came with the ability to treat fragmentation as a branching process [492, 552] in analogy with the shower that develops when an electron passes through an absorber creating photons which then convert to e^+e^- pairs, etc., until the energy is dissipated in soft particles. One would expect a shower of soft gluons and quarks in a narrow cone to be dominated by divergences in pQCD. However, it was found that due to color coherence and interference effects, the shower of soft gluons could be calculated [553] and was produced with angular ordering [554, 555] – the angular cone of each emission was restricted to be less than that of the previous emission.

This restriction in the phase space of soft gluon emission results in a suppression of emission of the softest gluons, which produces a Gaussian distribution of the momentum weighted fragmentation function, $x_p dN/dx_p = -dN/d\xi$, where $\xi = \ln 1/x_p$, with a peak away from the minimum x_p (maximum ξ) [553, 556], where x_p is the fractional particle momentum in a jet ($x_p = p/E_{jet}$). This so-called

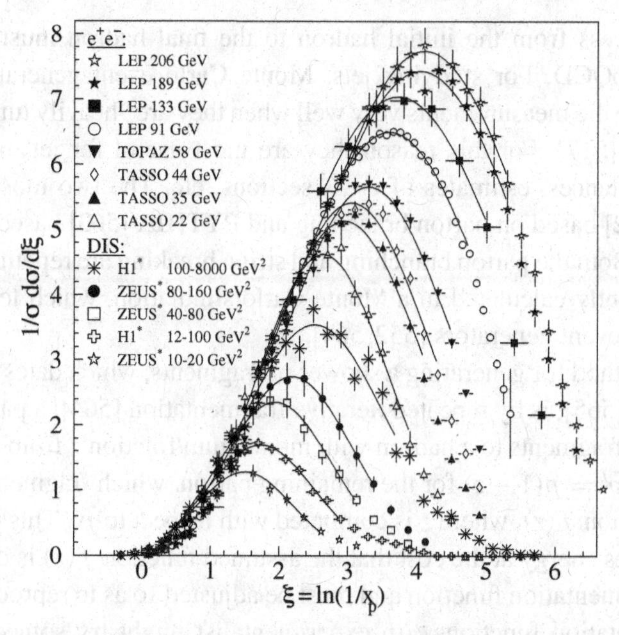

Figure 12.5 Compilation of fragmentation functions to charged particles in terms of $\xi = \ln(1/x_p)$ at several CM energies in e^+e^- and intervals of Q^2 in DIS [84]. The DIS measurements have been scaled by a factor of 2 for direct comparison to the e^+e^- results. Fits of simple Gaussians in ξ are overlaid for illustration [84].

humpback distribution of soft gluons was predicted to be directly observable experimentally by assuming local parton hadron duality [557], i.e. that the distribution of hadrons would be the same as the distribution of partons at a certain scale Q_0. Since a fragmentation function in terms of the standard fragmentation variable, z or x_p, could be easily replotted in terms of ξ (Figure 12.4b) [544], this prediction was tested and confirmed [399]. The predicted displacement of the peak to larger ξ with increasing jet energy (Figure 12.5) and cone angle [517][2] was also confirmed. Further discussion and details of this application of pQCD are best reviewed elsewhere [558].

There are several methods of providing fragmentation functions for use in pQCD calculations. For precision calculations of identified single particle inclusive cross sections, fragmentation functions are provided by expert groups [559–561] who perform a global analysis using pQCD to fit measurements in DIS, e^+e^- and hadron–hadron collisions and provide fragmentation functions at both LO and NLO. This is much more complicated than for parton distribution functions since

[2] The fact that the fragmentation function of a di-jet of a given invariant mass in hadron–hadron collisions is not unique but depends on the jet cone angle [517] is another indication that such a jet is more a legal contract than a physical object.

the entire process from the initial hadron to the final hadron must be followed according to pQCD. For studying jets, Monte Carlo event generators are used which describe the measurements very well when they are "heavily tuned" to reproduce the data [517]. For this reason they are most useful for jet measurements, detector acceptances, estimates of cross sections, etc. The two most popular are HERWIG [562] based on parton branching and PYTHIA [563] based on the Lund string model. Both the parton branching and string breaking are repetitive processes most conveniently calculated in a Monte Carlo simulation, which leads naturally to their use as event generators [552, 564].

Another method for generating a shower of fragments, which dates back to Field and Feynman [565], is by repeated iterative fragmentation [564]: a parton of initial momentum p fragments to a hadron with momentum fraction z from a distribution $f(z)$, leaving $p' = p(1-z)$ for the remaining parton, which fragments again with the same function $f(z)$, where z is computed with respect to p'. This method obviously conserves energy at the cost that the assumed function $f(z)$ is different from the actual fragmentation function and must be adjusted so as to reproduce the measured fragmentation functions. An experimentalist might try something slightly different, if writing their own Monte Carlo calculation, by using repeated independent sampling from the measured fragmentation function. This is straightforward for a single inclusive calculation, but for generating a jet would require a method to enforce conservation of energy. Another issue not mentioned in the above discussion, nor much in the literature on jets, is the momentum component j_T of the fragments transverse to the jet axis. This is an important issue in two particle correlations, which was discussed in Chapter 9.

12.5 Parton distribution functions and fragmentation functions in nuclei

As discussed in Section 4.3, measurements of DIS in nuclei were performed very shortly after the discovery in e–p scattering to test whether a nucleus with A nucleons would just be a collection of A point-like partons which would not shadow each other in DIS. The first measurements [215–218] (Figure 4.12) indeed observed a cross section which increased like $A^{1.00\pm0.01}$, which indicated that a nucleus acts like an incoherent superposition of nucleons for hard scattering so that the structure function of a nucleus of mass A is simply A times the structure function of a nucleon. This was clearly different from the total cross section for absorption of real photons ($Q^2 = 0$) on nuclei [566], with an $A^{0.91}$ dependence which indicates shadowing. The story became more complicated with further investigation.

The experiment at the BNL-AGS [218] which observed muons with an average energy 7.2 GeV scattering in various nuclear targets (Figure 4.12) did in fact notice some shadowing for $x < 0.1$, with a dependence $A^{0.963\pm0.006}$. This was followed

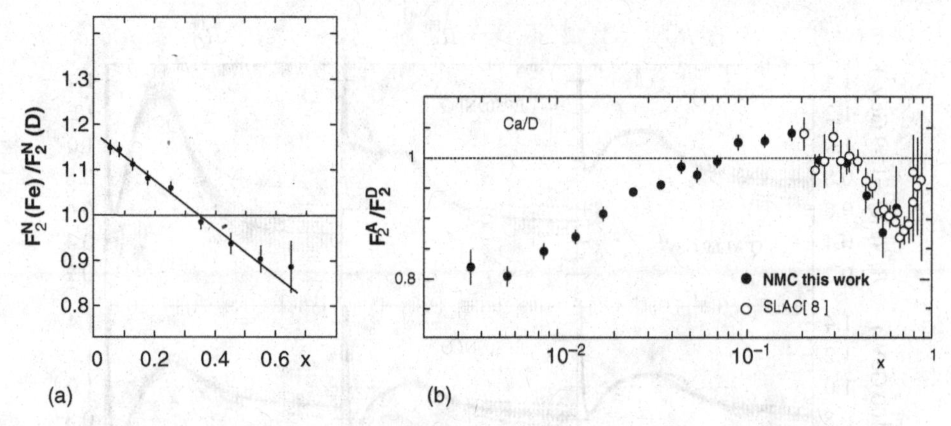

Figure 12.6 (a) EMC [567] measurement of $R_{F_2}^{Fe}(x) = (F_2^{Fe}(x)/56)/(F_2^d(x)/2)$. The Q^2 range varied as a function of x from $9 \leq Q^2 \leq 27$ GeV^2 for $x = 0.05$ to $36 \leq Q^2 \leq 170$ GeV^2 for $x = 0.65$. (b) NMC [568] measurement of $R_{F_2}^{Ca}(x)$.

nearly a decade later by the European Muon Collaboration (EMC) at CERN [567] who found such a "huge" effect (Figure 12.6a) in the scattering of 280 GeV muons in deuterium and iron that is called the "EMC effect," namely an enhancement (anti-shadowing) in the ratio of $R_{F_2}^{Fe} = (F_2^{Fe}(x)/56)/(F_2^d(x)/2) = 1.15$ at $x = 0.05$, falling linearly to a value of $\sim$0.89 at $x = 0.65$. After another decade, this was remeasured by the New Muon Collaboration (NMC) [568] at CERN, who produced a less exciting but now generally accepted measurement of $R_{F_2}^{Ca}$ (Figure 12.6b) which showed anti-shadowing of up to $R_{F_2}^{Ca} = 1.05$ for $0.08 \leq x \leq 0.2$ and shadowing for $x \leq 0.07$ and $x \geq 0.30$ reaching a value of $R_{F_2}^{Ca} = 0.8$ at $x = 0.005$ and 0.9 at $x = 0.7$. However, it should be noted that for $x < 0.01$, Q^2 varies from 1.4 down to 0.6 GeV^2, so may be outside the region of DIS.

There has been considerable progress in theoretical fits of the PDFs in nuclei, the most recent being by EPS09 [569] who follow the methods of CTEQ6M [540] and use DIS and Drell–Yan measurements as well as π^0 measurements in d+Au collisions from RHIC [570] to provide "a NLO global DGLAP analysis of nuclear parton distribution functions (nPDFs) and their uncertainties," which they expect will be used in precision analyses at RHIC and LHC. Parameterizations of the different flavor nPDFs as a function of A are used and fit to the available data. The results for R^A, the modification of the structure function of a nucleus A relative to an incoherent superposition of protons and neutrons, are shown in Figure 12.7 for Pb. They are given only for minimum bias collisions, i.e. integrated over all impact parameters. No such nPDFs as a function of centrality are available. It is also important to note that these nPDFs do not include the proposed [571] collective non-linear gluon–gluon fusion of gluons from several nucleons in a nucleus due to

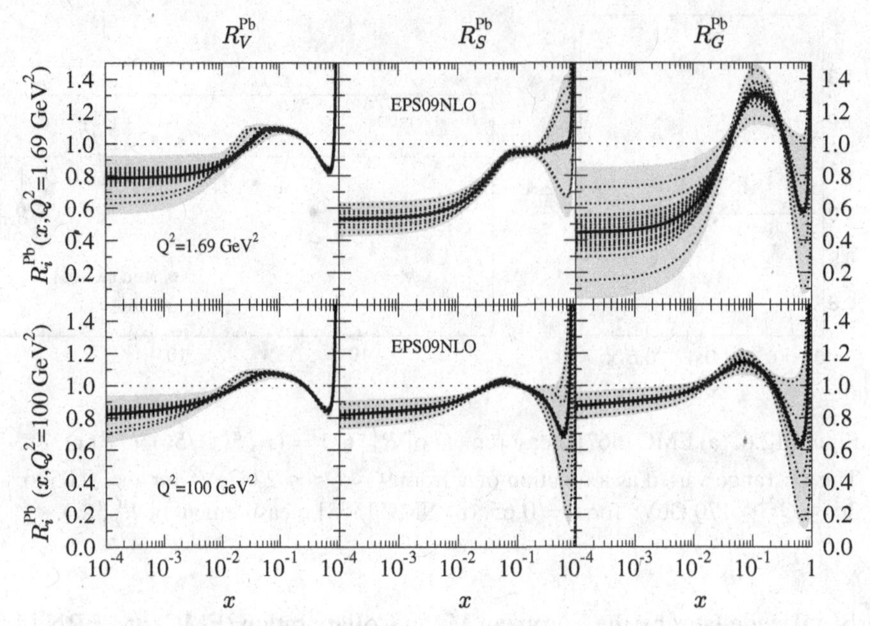

Figure 12.7 EPS09 [569] predicted nPDFs $R^{Pb}(x)$ in Pb for valence quarks V, sea quarks S, and gluons G, at the initial scale $Q_0^2 = 1.69$ GeV^2 and at $Q^2 = 100$ GeV^2. The black lines indicate the best fit and the shaded regions the uncertainties.

the high density of gluons per unit area. This effect, the so-called Color Glass Condensate (CGC), could saturate the gluon population at low x in proportion to $A^{1/3}$ in favor of an increased population and k_T at larger values of x. However the effect remains to be verified or falsified definitively by experiment.

For parton fragmentation functions, one of the first attempts at systematic measurements in nuclei was by the EMC collaboration in μ–A DIS in runs with 100, 120, 200 and 280 GeV muon beams [572]. For parton energy $\nu > 15$ GeV there was a roughly 5% suppression of hadrons with $z \geq 0.2$ in Cu, a small effect, which they attributed to interaction of the outgoing quark with the cold nuclear matter before fragmentation. Previous measurements at lower energies, notably at SLAC [573], saw increased suppression at lower $\nu \sim 10$ GeV which they also attributed to absorption of the forward going hadrons. In both cases, the issue was not about fragmentation functions but about the interaction of the outgoing quark or hadron in cold nuclear matter. Another complication in these measurements is that there is an azimuthal asymmetry of the hadrons around the virtual photon axis so that the hadrons are preferentially emitted opposite in azimuth to the scattered muon. This is due to the k_T of the quark in the nucleon [574] which makes the axis of the virtual photon–quark collision non-collinear with the axis of the virtual photon–proton collision. Thus, DIS measurements of outgoing hadrons in collisions with nuclei have not provided a clear separation of possible fragmentation function modifications from the effects of re-scattering of the outgoing quarks in the cold nuclear matter. A more recent measurement with more extensive data by the HERMES experiment at DESY with a 27.6 GeV electron beam on a gas target [575] also concludes, "A full theoretical description of hadronization in nuclei in one consistent framework, including partonic and hadronic (absorption plus rescattering) mechanisms is badly needed."

In $p + A$ collisions, the situation is no clearer because of the Cronin effect [280] which enhances the single inclusive cross section compared to the point-like scaling, $R_{pA} = \sigma_{p+A}/(A\sigma_{pp}) = 1.0$. Systematic measurements of the single and di-hadron cross sections in $p + A$ collisions at Fermilab [576–578] have shown an enhancement in tungsten relative to beryllium to $R_{pW}/R_{pBe} \sim 1.20$ for $p_T < 7$ GeV/c at $\sqrt{s} = 38.8$ GeV and then no enhancement for $8 < p_T < 12$ GeV/c for single inclusive charged particles. For symmetric charged hadron pairs, there is an enhancement of up to $\sim$1.3 for $m < 5$ GeV/c^2, a region of no enhancement for $5 < m < 8$ GeV/c^2, followed by a suppression, reducing to $R_{pW}/R_{pBe} = 0.80 \pm 0.05 \pm 0.07$ (systematic) for pair mass 12.4 GeV/c^2 which corresponds to target $x = 0.38$. This is consistent with the EMC effect of initial state shadowing within the systematic error, but a modified fragmentation at large z which would be more evident in the pair than the single cross section is "suggested" [576]. One indisputable effect observed is that R_{pW}/R_{pBe} increases with increasing out-of-plane pair momentum, p_{out}, indicative of multiple scattering of the incident nucleon or parton in the cold nuclear matter.

There have been very recent theoretical studies of fragmentation functions in nuclei [579–581] to use as a baseline to understand effects in a hot nuclear medium or the Quark Gluon Plasma (QGP). The difficulty of this problem at the end of the year 2011 is well summarized in one of the studies [581], "Most models reproduce, with different degrees of success, some features of the data, in spite of very different, even orthogonal theoretical approaches and ingredients." Clearly, on this front there is much needed to be learned in the future. Thus it is of extreme importance that measurements be made in $A + A$, p–p and $p + A$ (or $d + A$) collisions to distinguish which effects, including PDFs and fragmentation, are from vacuum, cold or hot nuclear matter.

12.6 Elements of QCD for experimentalists

In general, pQCD calculations are best left to professional theorists, but the individual elements that make up Eq. 12.1 are of great interest and importance to experimentalists, well beyond the requirement that the "long-distance non-perturbative phenomena" must be determined by experiment. It may not be too much of an exaggeration to say that over the many years since pQCD was generally accepted as the theory of the strong interactions and, in particular, of hard scattering, all experimental results so far have been found to be "in agreement with pQCD." However, on this point it is important to distinguish between what we shall call the explicit tests of QCD, the tests of the fundamental predictions such as measurement of the running coupling constant $\alpha_s(Q^2)$, the gluon self-coupling or of the fundamental subprocess cross sections

$$\left.\frac{d\sigma}{d\hat{t}}\right|_{\hat{s}} = \frac{\pi\alpha_s^2(Q^2)}{\hat{s}^2}\Sigma^{ab}(\cos\theta^*), \qquad (12.11)$$

and the implicit tests of QCD, measurements of cross section corresponding to Eq. 12.1 for example for inclusive jets, di-jets or single particles such as shown in Figure 12.2, which are predominantly tests of the factorization theorem and methods.

12.6.1 The elementary QCD subprocesses

The elementary QCD subprocesses involving quarks (q) and gluons (g) in hadron–hadron collisions and their characteristic angular distributions [413, 414], $\Sigma^{ab}(\cos\theta^*)$ are enumerated in Figure 12.8a–h. At this level, the lowest order QCD

a) $qq' \to qq'$, $\bar{q}q' \to \bar{q}q'$ $\quad \frac{4}{9}\frac{\hat{s}^2+\hat{u}^2}{\hat{t}^2}$

b) $qq \to qq$ $\quad \frac{4}{9}\left(\frac{\hat{s}^2+\hat{u}^2}{\hat{t}^2}+\frac{\hat{s}^2+\hat{t}^2}{\hat{u}^2}\right)-\frac{8}{27}\frac{\hat{s}^2}{\hat{u}\hat{t}}$

c) $\bar{q}q \to \bar{q}'q'$ $\quad \frac{4}{9}\frac{\hat{t}^2+\hat{u}^2}{\hat{s}^2}$

d) $\bar{q}q \to \bar{q}q$ $\quad \frac{4}{9}\left(\frac{\hat{s}^2+\hat{u}^2}{\hat{t}^2}+\frac{\hat{t}^2+\hat{u}^2}{\hat{s}^2}\right)-\frac{8}{27}\frac{\hat{u}^2}{\hat{s}\hat{t}}$

e) $\bar{q}q \to gg$ $\quad \frac{32}{27}\frac{\hat{u}^2+\hat{t}^2}{\hat{u}\hat{t}}-\frac{8}{3}\frac{\hat{u}^2+\hat{t}^2}{\hat{s}^2}$

f) $gg \to \bar{q}q$ $\quad \frac{1}{6}\frac{\hat{u}^2+\hat{t}^2}{\hat{u}\hat{t}}-\frac{3}{8}\frac{\hat{u}^2+\hat{t}^2}{\hat{s}^2}$

g) $gq \to gq$ $\quad -\frac{4}{9}\frac{\hat{u}^2+\hat{s}^2}{\hat{u}\hat{s}}+\frac{\hat{u}^2+\hat{s}^2}{\hat{t}^2}$

h) $gg \to gg$ $\quad \frac{9}{2}\left(3-\frac{\hat{u}\hat{t}}{\hat{s}^2}-\frac{\hat{u}\hat{s}}{\hat{t}^2}-\frac{\hat{s}\hat{t}}{\hat{u}^2}\right)$

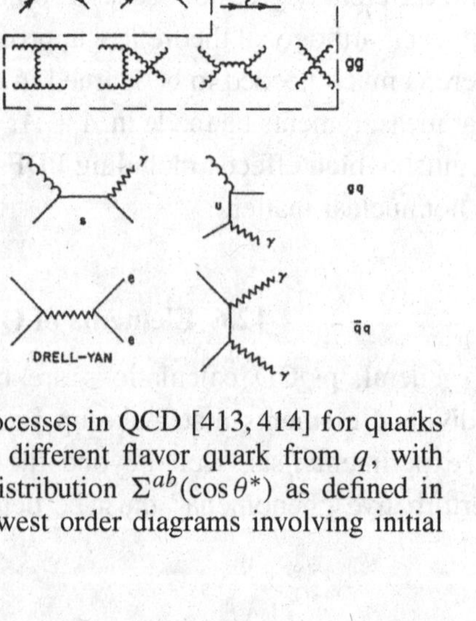

Figure 12.8 (a)–(h) Hard scattering subprocesses in QCD [413, 414] for quarks q and gluons g where the q' indicates a different flavor quark from q, with their lowest order cross section angular distribution $\Sigma^{ab}(\cos\theta^*)$ as defined in Eq. 12.11. The figures on the right are lowest order diagrams involving initial state q and g scattering.

Born diagrams, shown in Figure 12.8 (right), are analogous to the QED processes of Moller, Bhabha and Compton scattering indicated, and tests of their validity would be quite analogous to the studies of the validity of Quantum Electrodynamics (QED) which occupied the 1950s and 1960s [535]. The $\hat{t}$ and $\hat{u}$ channels and $\hat{s}$ and $\hat{t}$ channels are indicated for the qq and $\bar{q}q$ diagrams respectively. It is important to note that the diagrams in the dashed box are unique to QCD since the gluons carry color charge and interact with each other while the photons in QED do not carry electric charge and so do not self-interact.

Although the different combinations of $\hat{s}$, $\hat{t}$ and $\hat{u}$ for the cross sections in Figure 12.8a–h may at first seem to be formidable, it can be seen by recalling that $\hat{t} = -\hat{s}(1 - \cos\theta^*)/2$, $\hat{u} = -\hat{s}(1 + \cos\theta^*)/2$, that indeed the $\Sigma^{ab}(\cos\theta^*)$ are nothing other than angular distributions. For example, for $qq' \rightarrow qq'$, Figure 12.8a:

$$\Sigma^{qq'}(\cos\theta^*) = \frac{4}{9}\frac{\hat{s}^2 + \hat{u}^2}{\hat{t}^2} = \frac{4}{9}\left[\left(\frac{2}{1 - \cos\theta^*}\right)^2 + \left(\frac{1 - \cos\theta^*}{1 + \cos\theta^*}\right)^2\right]. \quad (12.12)$$

12.7 Explicit tests of QCD: $\alpha_s(Q^2)$, $\Sigma^{ab}(\cos^*)$

The two primary examples of explicit tests of QCD are the measurement of $\alpha_s(Q^2)$ and of the fundamental QCD subprocess angular distributions $\Sigma^{ab}(\cos\theta^*)$.

The asymptotic freedom of QCD is based on the negative value for the "β" function of the renormalization group for non-Abelian (with gluon self coupling) Yang–Mills gauge theories [22, 23] which gives the leading order result [391, 413, 582, 583]:

$$\alpha_s(Q^2) = \frac{12\pi}{(11N_c - 2n_f)\ln(Q^2/\Lambda^2)} = \frac{12\pi}{25\ln(Q^2/\Lambda^2)} \quad (12.13)$$

for $N_c = 3$ colors and $n_f = 4$ flavors of quarks (u, d, s, c). Thus QCD predicts the absolute value of the coupling constant and its Q^2 evolution to small values at large Q^2 (asymptotic freedom) as well as a logarithmic divergence at low Q^2 indicating confinement, with a cutoff parameter, Λ, which must be fixed by experiment. Thus an explicit test of QCD is to observe the Q^2 evolution of Eq. 12.13 and to determine a unique value of Λ for all Q^2 and in all reactions. This again is a job for specialists and the methods will not be covered in detail here except to say that hadron–hadron collisions have not made a major contribution to the precise determination of $\alpha_s(Q^2)$ because [584], "Sizeable uncertainties in the absolute normalization of measured jet cross sections and in the theoretical predictions limit the accuracy with which the measurements can be compared with QCD."

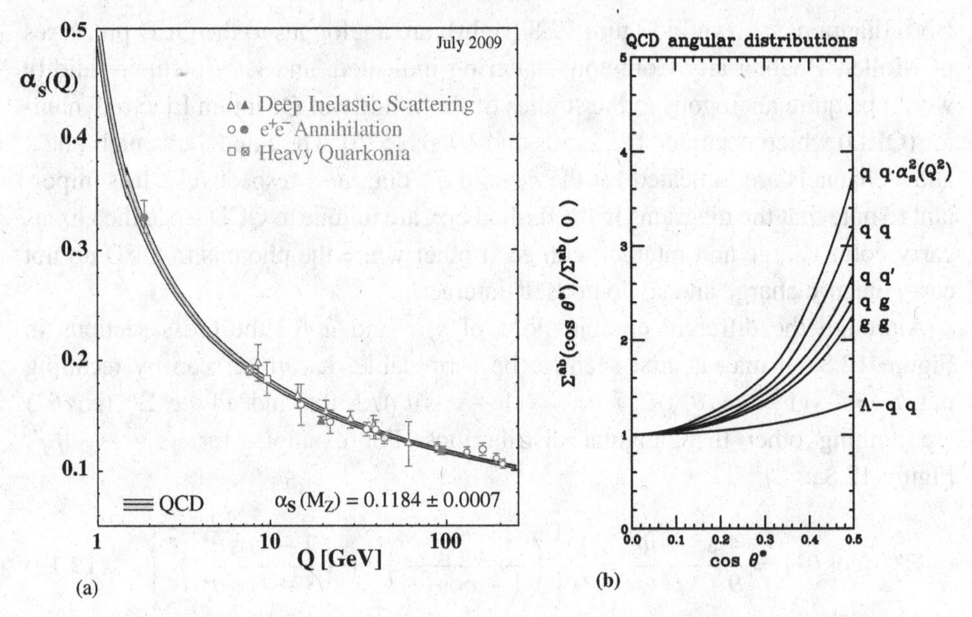

Figure 12.9 (a) Measurements of $\alpha_s(Q^2)$ for the reactions indicated [586]. The band represents the range of the QCD predictions for the combined world average of $\alpha_s(M_{Z^0})$. Full symbols are results based on N3LO QCD, open circles are based on NNLO, open triangles and squares on NLO QCD. The cross-filled square is based on lattice QCD. (b) Elementary QCD subprocess angular distributions $\Sigma^{ab}(|\cos\theta|^*)/\Sigma^{ab}(|\cos\theta|^* = 0)$ from Figure 12.8 for gg, qg, qq', qq, qq with $\alpha_s(Q^2 = |\hat{t}|)$ evolution for $\hat{s} = 12.5\ (\text{GeV})^2$ and $\Lambda_{\text{QCD}} = 0.1$ GeV [412]. Also shown is the interference term $(\Lambda - qq)$ between a hypothetical qq contact interaction with scale Λ_c and QCD [587–589] which should be added to the QCD qq distribution.

As described in a series of thorough review articles by Bethke [584–586] the best measurements of $\alpha_s(Q^2)$ come from τ lepton and Υ decays, jet production and structure function evolution in DIS, hadronic event shapes and jet production in e^+e^- annihilation, and measurements of the ratio of partial widths $\Gamma(Z^0 \to \text{hadrons})/\Gamma(Z^0 \to e^+ + e^-)$ as shown in Figure 12.9a [586]. The data are in excellent agreement with QCD evolution with the parameter $\Lambda = 213 \pm 9$ MeV in the $\overline{MS}$ factorization scheme [582] and a precision world-average value of $\alpha_s(M_{Z^0}) = 0.1184 \pm 0.0007$. This is a fantastic explicit precision test of the validity of QCD. It also tests the factorization theorem since the different reactions give consistent values of $\alpha_s(Q^2)$.

Hadron–hadron collisions provide a different explicit test of QCD, a direct measurement of the fundamental subprocess angular distributions. This can be accomplished independently of the absolute normalization and the details of parton distribution or fragmentation functions by measuring the angular distribution of nearly back-to-back di-pions or di-jets with large invariant mass and small pair p_T as a function of the c.m. scattering angle θ^*, for fixed invariant mass ($m \propto \hat{s} = x_1 x_2 s$) and rapidity, normalized to the value at $\theta^* = 90°$. Figure 12.9b is a plot of the elementary QCD constituent scattering angular distributions $\Sigma^{ab}(\cos\theta^*)$ from Figure 12.8 for gg, qg, qq', qq, normalized by $\Sigma^{ab}(\cos\theta^* = 0)$. By convention, for the case of either identical initial (e.g. p–p) or final (e.g. π^0–π^0 or jet–jet) states, the distribution for $\cos\theta^* \leq 0$ is added to that of $\cos\theta^* \geq 0$ (which does not affect the angular distribution of identical partons) and measurements are presented as a function of $|\cos\theta^*|$. To maintain the correct integral of the cross section over $\cos\theta^*$, the integral is taken over the region $0 \leq |\cos\theta^*| \leq 1$.

As discussed in Section 9.6, this measurement was first performed by the CCOR collaboration [390] using π^0 pairs with invariant mass in the range $8.5 \leq m_{\pi\pi} \leq 18$ GeV/c^2 and pair $p_T < 1$ to 3 GeV/c depending on $m_{\pi\pi}$ (Figure 9.18), and was presented at the ICHEP in Paris 1982 [412] at the same meeting at which the iconic plot of the UA2 di-jet event (Figure 11.18) [416] was first presented. Figure 12.10a shows a plot from the rapporteur's proceedings [418] of one CCOR $\cos\theta^*$ distribution [412], $11 \leq m_{\pi\pi} \leq 12$ GeV/c^2, on which he superimposed the prediction for $qq \to qq$ from Figure 12.8b, which lies reasonably on top of the data. For the lower di-pion invariant masses, shown in Figure 9.18, e.g.

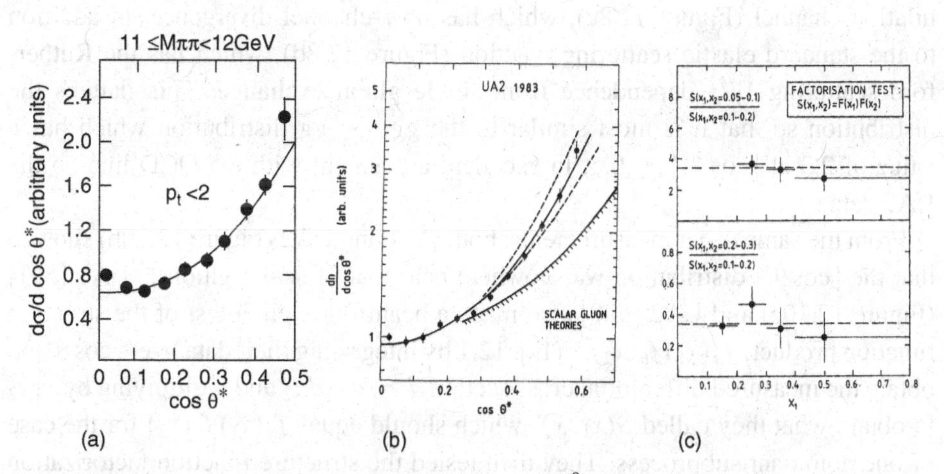

Figure 12.10 (a) CCOR $\cos\theta^*$ distribution for π^0 pairs with $11 \leq m_{\pi\pi} \leq 12$ GeV/c^2 [412] with the prediction for $qq \to qq$ added by the rapporteur, Günter Wolf, in the proceedings [418]. (b) UA2 $\cos\theta^*$ distribution for di-jets [590]. All QCD subprocesses (except the pure $\hat{s}$-channel $\bar{q}q \to \bar{q}'q'$, Figure 12.8c) separately normalized to the data, lie in the area between the two dashed curves. The solid line is the overall QCD prediction and the region of scalar gluon theories is indicated. (c) UA1 [591] ratio $S(x_1, \Delta x_2)/S(x_1, \Delta' x_2)$ for di-jets with invariant $m_{jj} \gtrsim 70$ GeV for the values of Δx_2 and $\Delta' x_2$ indicated. In each case the data are consistent with a constant independent of x_1 (dashed line).

$9 \leq m_{\pi\pi} \leq 10$ GeV/c, the experimenters found that by including the QCD evolution of $\alpha_s(Q^2)$ in the $qq \to qq$ distribution, with $Q^2 = |\hat{t}|$ for elastic scattering, the agreement with the steepness of the data at more forward angles was substantially improved. This resolved the issue of the disagreement of the simple qq curve with the data which had initially puzzled them. The reason for a steeper distribution is that as the scattering angle θ^* for qq elastic scattering decreases, $Q^2 = |\hat{t}|$ also decreases, which makes $\alpha_s(Q^2 = |\hat{t}|)$ increase, as shown in Figure 12.9a. This steepens the qq curve for more forward angles, as shown in Figure 12.9b. Thus, measurements of $\Sigma^{ab}(\cos\theta^*)$ can be used both to test the validity of the elementary QCD subprocess distributions and to study the evolution of $\alpha_s(Q^2)$, where $Q^2 = |\hat{t}| = \hat{s}(1 - \cos\theta^*)/2$ for qq elastic scattering at a fixed value of $\hat{s}$.

The first measurements of $\Sigma^{ab}(\cos\theta^*)$ with di-jets were published two years later by UA1 [591] and UA2 [590] (Figure 12.10b) in p–$\bar{p}$ collisions at $\sqrt{s} = 540$ GeV at the CERN-Sp$\bar{\text{p}}$S collider. These were "in agreement with QCD" but somewhat different from CCOR (Figure 12.10a). The careful reader will note that the value at $\cos\theta^* = 0.5$ for UA2 is 2.3 for QCD and a maximum of 2.7 for the steepest subprocess, while for CCOR the qq, or better fitting $qq \cdot \alpha_s^2(Q^2)$, values (Figure 12.9b, Figure 9.18) are 3.2 and 3.73, respectively, much larger than the UA2 values. However, this is actually a success of QCD because the UA2 measurement is in p–$\bar{p}$ collisions where the dominant $\bar{q}q$ reaction has an annihilation channel (Figure 12.8c), which has no $\hat{t}$-channel divergence, in addition to the standard elastic scattering reaction (Figure 12.8d), which has the Rutherford scattering $1/\hat{t}^2$ dependence from single gluon exchange. This flattens the distribution so that it is most similar to the $gg \to gg$ distribution which has a value of 2.3 at $|\cos\theta^*| \leq 0.5$, in excellent agreement with the QCD line on the UA2 data.

From the same di-jet measurements, both UA1 and UA2 (Figure 12.10b) showed that the $|\cos\theta^*|$ distribution was nowhere near that of scalar gluons.[3] UA1 [591] (Figure 12.10c) and UA2 [590] also made a beautiful explicit test of the structure function product, $f_a(x_1)f_b(x_2)$, in Eq. 12.1 by integrating their data over $|\cos\theta^*|$ to obtain the measured differential cross section $d^2\sigma/dx_1dx_2$ and multiplying by x_1x_2 to obtain what they called $S(x_1, x_2)$ which should equal $f_a(x_1)f_b(x_2)$ for the case of one dominant subprocess. They then tested the structure function factorization by measuring the ratio $S(x_1, \Delta x_2)/S(x_1, \Delta' x_2)$, where Δx_2 and $\Delta' x_2$ represent two different intervals of x_2, which should have no x_1 dependence since $f_a(x_1)$ should

[3] This had actually been shown a year earlier by the CDHW collaboration at the CERN-ISR [592] using the rapidity difference $\Delta y_{cd} = y_c - y_d$ between a trigger π^+ and an away-side charged particle at the split field magnet. Although the effect had huge statistical significance it did not have the visual impact of Figure 12.10b, which they could have made with the simple conversion of Δy_{cd} to $\cos\theta^* = \tanh(\Delta y_{cd}/2)$ (Eq. 10.3) since this equation was mentioned in their paper.

drop out in the ratio. This was actually observed by both UA1 (Figure 12.10c) [591] and UA2 [590].

With regard to the beautiful QCD "inverse Compton effect," the reaction $g+q \to \gamma+q$ which was discussed in detail in Chapter 10, it is difficult, if not impossible, to find in the literature a measurement of the $\hat{s}$- and $\hat{u}$-channel dominated "Compton" angular distribution (Eq. 10.8) which is much less divergent than the predominant $\hat{t}$-channel processes (Figure 12.8a–b,d–h) which lead to the Rutherford scattering like angular distributions. UA2 [593] did publish one measurement of the ratio of the $|\cos\theta^*|$ distributions of direct-γ-jet and π-jet systems in $\bar{p}$–p collisions at $\sqrt{s} = 630$ GeV, where the "π" represents multi-photons which come predominantly from π^0 and η decay and were identified by the conversion probability in a preshower converter (similar to the method used by CCOR [435] discussed in Chapter 10). Figure 12.11a [593] shows the ratio of direct-γ-jet to π-jet angular distributions as a function of $|\cos\theta^*|$ for pair invariant masses >30 GeV/c^2. The result agrees with the expected weaker forward peaking of direct-γ-jet to π-jet as indicated by the curve. However, it is important to realize that this is in $\bar{p}$–p collisions, where the dominant production process is not the Compton subprocess but the $\bar{q}q \to \gamma + g$ subprocess which cannot be neglected because of the large valence $\bar{q}$ distribution in the antiproton, where $f_{\bar{q}}^{\bar{p}}(x) \equiv f_q^p(x)$ by C (charge conjugation) invariance. In fact, the ratio of the direct-γ cross section in $\bar{p}$–p to p–p collisions at $\sqrt{s} = 24.3$ GeV was measured [595] to be a factor of 2 to 3 over

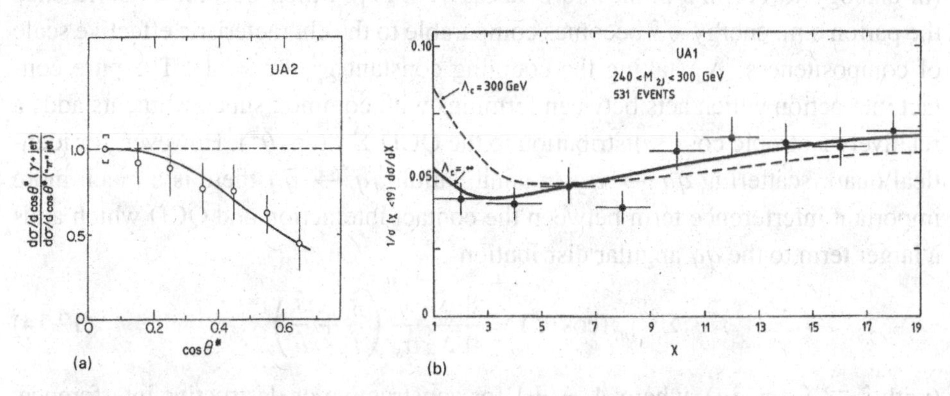

Figure 12.11 (a) UA2 [593] measurement of the ratio of direct-γ-jet to "π"-jet angular distributions as a function of $|\cos\theta^*|$ in $\bar{p}$–p collisions at $\sqrt{s} = 630$ GeV. The curve is the ratio of the weighted sums of the lowest order QCD subprocesses leading to γ+jet and jet+jet final states [593]. (b) UA1 [594] measurement of the angular distribution $(1/\sigma)(d\sigma/d\chi)$ for di-jets with invariant mass $240 < m_{jj} < 300$ GeV, where $\chi = \exp(y_c - y_d)$ and y_c, y_d are the rapidities of the two jets. The solid curve represents the QCD prediction with $Q^2 = p_T^2$ and the dashed curve corresponds to a contact interaction with $\Lambda_c = 300$ GeV (Eq. 12.14).

the range $4.1 \leq p_T \leq 7$ GeV/c, while the ratio for π^0 production was unity. Also the measurement of the difference in direct-γ cross sections between $\bar{p}p$ and pp is directly sensitive to the $\bar{q}q \to \gamma + g$ subprocess. This beautiful measurement, which illustrates the importance of the annihilation component for direct-γ production in $\bar{p}$–p collisions, was performed by the UA6 experiment [595] using p and $\bar{p}$ collisions in a hydrogen gas jet in the Sp$\bar{p}$S ring. While relevant to tests of QCD, it is not directly relevant to heavy ion collisions.

The examples given above illustrate how the richness of the nucleon substructure and the different QCD constituent reactions is revealed in the explicit tests of QCD, which is hidden by the simple phrase "agreement with QCD" of the implicit tests.

12.7.1 Does pQCD break down? – the search for new physics

Although uniquely suited to measure the elementary QCD subprocess angular distributions, $\Sigma^{ab}(\cos\theta^*)$, which come at leading order in hadron–hadron collisions, hadron colliders provide a better way to test QCD by using the large cross section of di-jets to search explicitly for violations of pQCD at large Q^2. An example of such a violation of pQCD which is especially suited to angular distribution measurements is the possibility that quarks are composite, i.e. bound states of more fundamental subconstituents [587, 588]. The force that binds these subconstituents will also mediate a new interaction between bound states of these subconstituents leading to a point-like four-fermion contact interaction between quarks [587, 588] (in analogy to Fermi's point-like β decay [18, 19]) which becomes visible once the parton c.m. energy $\sqrt{\hat{s}}$ becomes comparable to the characteristic effective scale of compositeness, Λ_c (taking the coupling constant $g_c^2/4\pi = 1$). The pure contact interaction which acts between fermions with common subconstituents adds a relatively isotropic $\cos\theta^*$ distribution to the QCD $\Sigma^{ab}(\cos\theta^*)$. However, for identical quark scattering $qq \to qq$ or annihilation $\bar{q}q \to \bar{q}q$ there is a much more important interference term between the contact interaction and QCD which adds a larger term to the qq angular distribution

$$\Sigma^{qq}_{c-int}(\cos\theta^*) = \frac{8}{9}\frac{\hat{s}}{\Lambda_c^2}\frac{A}{\alpha_s}\left(\frac{\hat{s}}{\hat{t}} + \frac{\hat{s}}{\hat{u}}\right) \tag{12.14}$$

(with $\hat{s} \rightleftarrows \hat{u}$ for $\bar{q}q$), where $A = \pm 1$ for constructive or destructive interference. This term only diverges like $1/\hat{t}$ instead of $1/\hat{t}^2$ so is dramatically flatter than QCD, thus most visible at small values of $|\cos\theta^*|$ (see $\Lambda - qq$ on Figure 12.9b).[4]

[4] Another intriguing aspect of composite models is that they generally violate parity because the scale Λ_c is much greater than the weak interaction scale M_W. The model discussed here [587, 588] is in fact an explicitly parity violating left–left interaction. Using parity violation can increase the sensitivity to the contact interaction because the QCD background vanishes, leaving only the QCD-weak interference as background. As this is evidently not relevant to heavy ion collisions, we refer the reader elsewhere for details [589].

This aspect was emphasized in the second measurement of $\Sigma^{ab}(\cos\theta^*)$ with di-jets done by UA1 two years later [594]. For this analysis they first showed a $\cos\theta^*$ plot for di-jets of invariant mass $180 < m_{jj} < 240$ GeV which agreed with the QCD prediction for $Q^2 = p_T^2 = (0.5 - 1)|\hat{t}|$ but excluded the case $Q^2 = \hat{s}$ which does not evolve with scattering angle. They then switched to a transformed plot in the variable χ (Eq. 12.15), which had been introduced in 1984 by Combridge and Maxwell [596],

$$\chi \equiv \frac{\hat{u}}{\hat{t}} = \frac{1+\cos\theta^*}{1-\cos\theta^*} = \cot^2\frac{\theta^*}{2} = e^{(y_c - y_d)}, \tag{12.15}$$

who pointed out several nice properties of this new variable: (i) it is simply related to the difference in rapidity of the two outgoing partons; (ii) the angular distribution in the variable χ, $d\sigma/d\chi = 0.5(1 - \cos\theta^*)^2 d\sigma/d\cos\theta^*$, is a constant for Rutherford scattering (Eq. 4.4); (iii) the dominant subprocesses $qq \to qq$, $qg \to qg$, $gg \to gg$ all have approximately the same χ (ergo $\cos\theta^*$) distribution, characteristic of "the gluon self-coupling which gives rise to $\hat{t}$-channel gluon exchange in every subprocess," which becomes exact at large χ. The UA1 measurement (Figure 12.11b) [594] shows that the exact QCD prediction in terms of χ, with $Q^2 = p_T^2$, is nearly constant and in excellent agreement. A deviation from QCD due to a contact interaction would be visible by extra counts at small values of $\cos\theta^*$ ($\chi \approx 1 - 2$) due to the addition of the interference term (Eq. 12.14), as shown for $\Lambda_c = 300$ GeV by the dashed line on Figure 12.11b, which is far from the data. This gives a 95% confidence limit that $\Lambda_c > 415$ GeV. All subsequent measurements of the constituent scattering angular distribution at hadron colliders have used the variable χ, with the most recent limit from di-jets with $520 < m_{jj} < 2000$ GeV by the ATLAS experiment in $\sqrt{s} = 7$ TeV p–p collisions at the LHC [597] being $\Lambda_c > 9.5$ TeV at 95% confidence. This means that there is no visible deviation from QCD in hard scattering for values of $Q^2 = |\hat{t}|$ up to 2 million (GeV/c)^2 (e.g. compare Figures 4.8, 4.9) corresponding to point-like quarks down to distance scales $\lesssim 1.4 \times 10^{-4}$ fm or to 2.1×10^{-5} fm, with 95% confidence, from the limit on Λ_c.

12.8 x_T scaling

Another explicit test of QCD without the need to know the details of the structure functions, fragmentation functions or coupling constant is provided by x_T scaling. This is primarily sensitive to the quantum exchange governing the reaction (Eq. 6.12) and also has sensitivity to scale breaking (Eq. 6.13). Although precision direct tests of pQCD in hadron–hadron collisions are best done with the constituent scattering angular distributions, they require measurements with two nearly back-to-back large p_T hadrons or large p_T jets. For measurements of single particle or single jet inclusive p_T distributions, x_T scaling [266, 268] provides a totally data driven test of whether pQCD or some other underlying subprocess is at work, as well as providing a compact quantitative way to describe the data using the effective index, $n_{eff}(x_T, \sqrt{s})$, from Eq. 6.13, which we repeat here for convenience:

$$E\frac{d^3\sigma}{dp^3} = \frac{d^3\sigma}{p_T dp_T dy d\phi} = \frac{1}{p_T^{n_{eff}(x_T,\sqrt{s})}} F\left(\frac{p_T}{\sqrt{s}}\right) = \frac{1}{\sqrt{s}^{n_{eff}(x_T,\sqrt{s})}} G(x_T), \quad (12.16)$$

where $Ed^3\sigma/dp^3 = \sigma^{inv}(p_T, \sqrt{s})$ is the invariant cross section for inclusive particle production with transverse momentum p_T at c.m. energy $\sqrt{s}$ and $x_T \equiv 2p_T/\sqrt{s}$. It is important to emphasize that the effective power, $n_{eff}(x_T, \sqrt{s})$, is different from the power n of the invariant cross section, which varies with $\sqrt{s}$ (which it must if x_T scaling is to hold).

For pure vector gluon exchange, or without the evolution of α_s and the structure and fragmentation functions in QCD, $n_{eff} = 4$ as in Rutherford scattering (which can be seen from Eq. 12.1). However, as discussed in Chapter 6, due to the non-scaling in QCD, the measured value of n_{eff} depends on the x_T value and the range of $\sqrt{s}$ used in the computation:

$$n_{eff}(x_T, \sqrt{s_1}, \sqrt{s_2}) = \frac{\ln\left[\sigma^{inv}(x_T, \sqrt{s_1})/\sigma^{inv}(x_T, \sqrt{s_2})\right]}{\ln\left[\sqrt{s_2}/\sqrt{s_1}\right]}. \quad (12.17)$$

For instance in Figure 6.14a,b it was shown that $n_{eff}(x_T = 0.3, \sqrt{s_1} = 53.1, \sqrt{s_2} = 62.4) = 4.9 \pm 0.4$ while $n_{eff}(x_T = 0.3, \sqrt{s_1} = 30.7, \sqrt{s_2} = 53.1) = 6.8 \pm 0.3$ (including the systematic error). As can be seen in Figure 12.1b, this difference is due to the k_T effect which raises the p_T spectrum at the lower $\sqrt{s}$ more than at the larger $\sqrt{s}$ because of the flattening of the p_T spectrum with increasing $\sqrt{s}$. Hence for a precision value of $n_{eff}(x_T)$ at some value of $\sqrt{s}$ it would be important to use values of $\sqrt{s_1}$ and $\sqrt{s_2}$ close together as in 53.1 to 62.4 GeV, although the better agreement with pQCD for the 53.1/62.4 comparison is due to the smaller k_T effect as well as the smaller $\sqrt{s}$ interval.

The status of the invariant cross section as a function of p_T for non-identified charged hadrons, $h = (h^+ + h^-)/2$, for the range of c.m. energies, $\sqrt{s}$, available in 1988 was shown in Figure 2.4 repeated here for convenience (Figure 12.12a). This was very important at the startup of RHIC, in the year 2000, with Au+Au collisions at nucleon–nucleon c.m. energy $\sqrt{s_{NN}} = 130$ GeV, where there were no existing p–p data. The p–p comparison data were provided by using x_T scaling [598] (Figure 12.12b [599]) to interpolate between the available data, with a value of $n_{eff} = 6.3$ in the range of interest $0.03 \leq x_T \leq 0.06$.

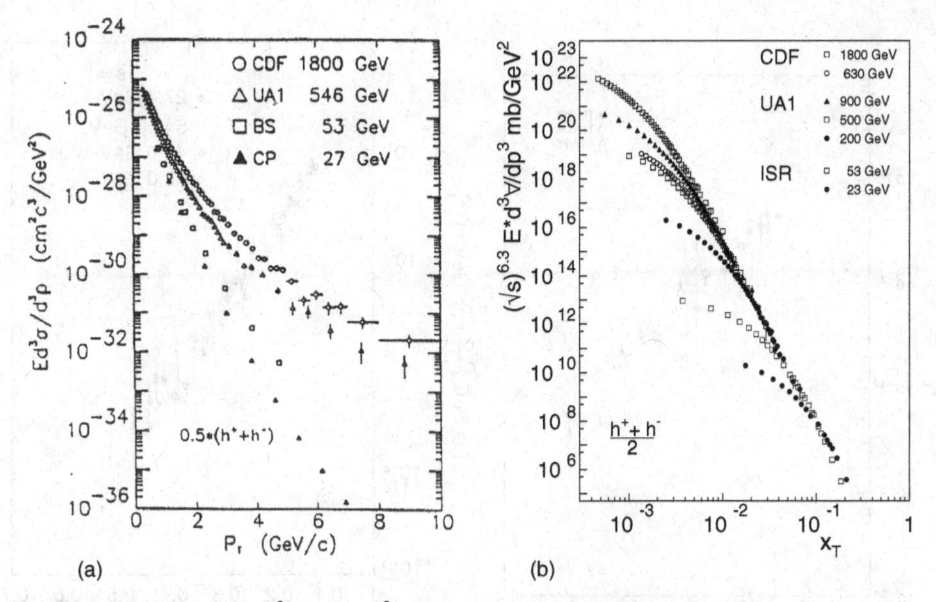

Figure 12.12 (a) $Ed^3\sigma(p_T)/d^3p$ at mid-rapidity as a function of $\sqrt{s}$ in p–p and $\bar{p}$–p collisions from CDF compilation in 1988 [113]. (b) Log–log plot of $\sqrt{s}(\mathrm{GeV})^{6.3} \times Ed^3\sigma/d^3p$ versus $x_T = 2p_T/\sqrt{s}$ with the data of PHENIX [599].

For the other reactions measured in this period, namely direct-γ [438] and jet production [600], the x_T scaling for values of $x_T \approx 0.3$ gave values of $n_{eff} \approx 5.0$ for both direct-γ and jets (Figure 12.13), not obviously different from each other nor from π^0 production at $x_T = 0.3$ (Figure 6.14a,b). More modern studies also show no difference for direct-γ [605] and jets [606], but with a smaller $n_{eff} \approx 4.5$, and also involve an interesting story (Figure 12.14). The interesting story is that the CDF measurement for jets at 1.8 TeV [607] initially was somewhat higher than the pQCD prediction and it was found useful [608] to plot the x_T scaled jet cross section $E_T^3 d^2\sigma/dE_T d\eta$ for the theoretical predictions as a function of $x_T = 2E_T/\sqrt{s}$ [606] to try to avoid as much as possible uncertainties from the various parton distribution functions used. Note that the theory curves do, in fact, show little difference for the different PDFs used. The n_{eff} values are 4.45 for the experimental value of 1.6 and 4.56 for the theoretical value of 1.8 for the scaled cross sections. The solution to the disagreement turned out to be a new, improved functional form for the gluon PDF by the CTEQ group [609] followed by a later overall improvement in their methods [610].

For the single particle inclusive data, the latest measurements of x_T scaling from the PHENIX experiment at RHIC [611] nicely confirm the early results, but with identified particles. Figure 12.15 shows the x_T scaling results from p–p collisions at $\sqrt{s} = 200$ and 62.4 GeV from PHENIX [611], plus other data as indicated, for

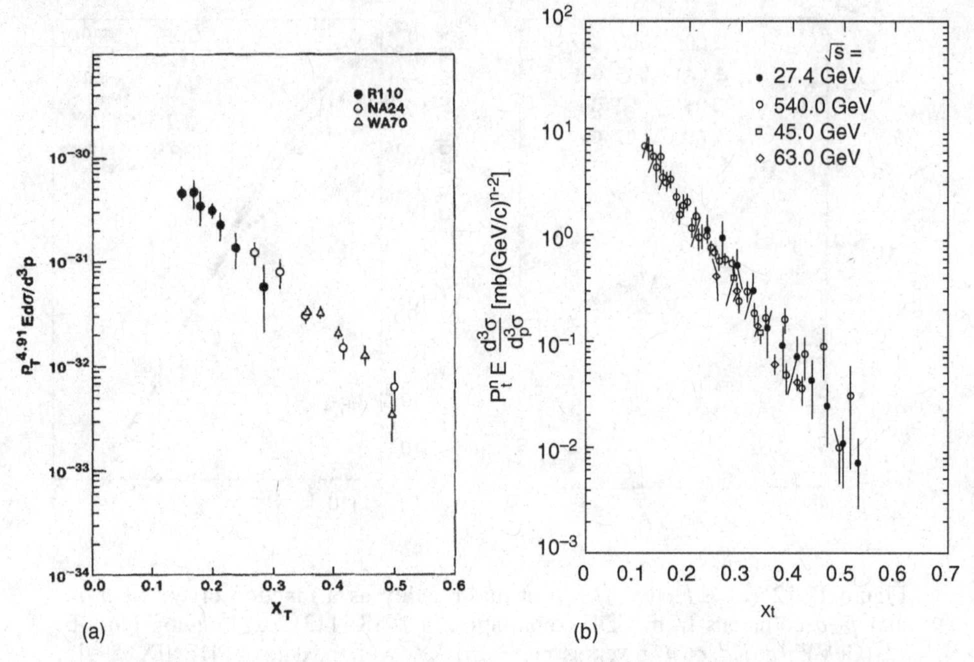

Figure 12.13 (a) CMOR-R110 [438] plot of $p_T^{4.91} Ed^3\sigma/dp^3$ for direct-γ production at $\sqrt{s} = 63$ GeV together with lower energy data from NA24 [601] and WA70 [602] at $\sqrt{s} = 23.8$ and 23.0 GeV respectively. (b) Plot of $p_T^{5.1} Ed^3\sigma/dp^3$ versus x_T for inclusive single jet production by the E609 (originally WPLF) collaboration [600] at $\sqrt{s} = 27.4$ GeV from Fermilab, together with higher energy data from AFS [603] and UA2 [604] at CERN.

(a) protons, with $n_{eff} = 6.52 \pm 0.59$, (b) antiprotons, with $n_{eff} = 6.15 \pm 0.05$, (c) $\pi^\pm, \pi^0$ with $n_{eff} = 6.35 \pm 0.23$, where the power n_{eff} was determined in each case by a fit of the data shown to Eq. 12.16 with $G(x_T) = x_T^{-m}$. Thus n_{eff} is within the range 6.3–6.5 for pions, protons, and antiprotons, in agreement with the previous estimate shown in Figure 12.12b. The data points deviate from the x_T scaling in the transverse momentum region of $p_T < 2$ GeV/c as indicated on Figure 12.15a,b,c. This deviation indicates the transition from hard scattering to soft multiparticle production for $p_T < 2$ GeV/c. Figure 12.15d [612] shows $n_{eff}(x_T, 62.4, 200)$ determined point-by-point from only the PHENIX data [611] using Eq. 12.17 and including both the statistical and systematic errors. This clearly indicates the need for identified $\pi^\pm$, p, $\bar{p}$ and kaon data at higher values of p_T to complement the π^0 data which are shown in full in Figure 12.16a,b. An interesting question for the π^0 data is whether $n_{eff}(x_T, 62.4, 200)$ will drop to $\sim$5 at $x_T = 0.3$ following the trend of $n_{eff}(x_T, 53.1, 62.4)$ in Figure 6.14a,b.

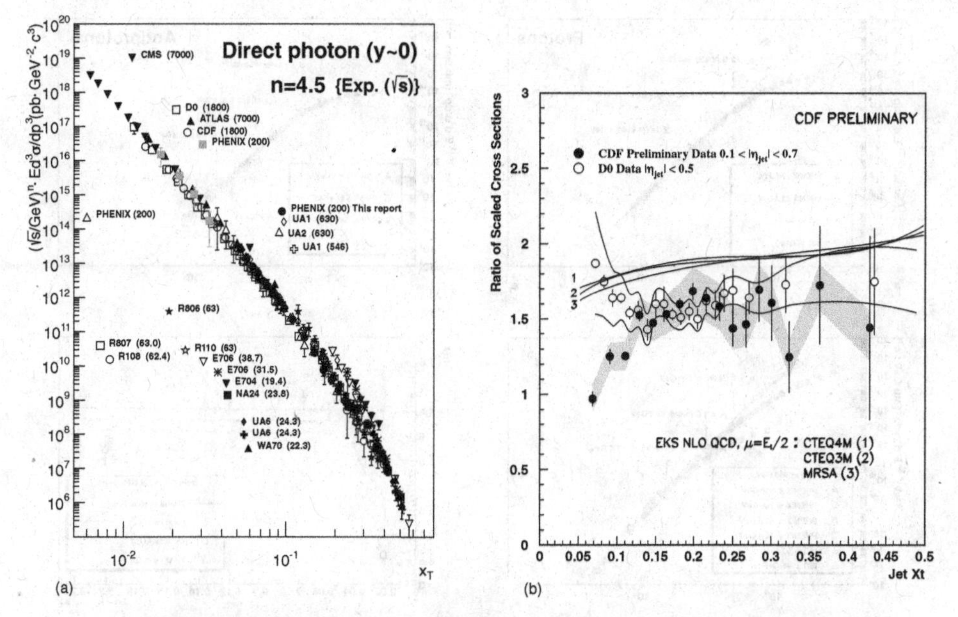

Figure 12.14 (a) PHENIX compilation [605] of x_T scaling of direct photon data in p–p and p–$\bar{p}$ collisions. The quantity plotted is $(\sqrt{s})^n \times Ed^3\sigma/dp^3(x_T)$ with $n = 4.5$. The experiment and $\sqrt{s}$ (GeV) are indicated in the legend. (b) x_T scaling of jet cross sections measured in p–$\bar{p}$ collisions by CDF and D0 [606]. The quantity plotted is the ratio of p_T^4 times the invariant cross section (actually $E_T^3 d^2\sigma/dE_T d\eta$) as a function of $x_T = 2E_T/\sqrt{s}$ for $\sqrt{s} = 630$ and 1800 GeV both for the experiment and the theory (with the indicated PDFs). The shaded region indicates the CDF systematic uncertainty.

One interesting contemporary thread which dates back to the famous $n_{eff} = 8$ issue in the original high p_T discovery (Chapter 6) is whether, in addition to the predominant evidence for the validity of pQCD, there are also other reactions at work which have a steeper p_T dependence, the so-called "higher-twist" contributions, such as predicted in the Constituent Interchange Model (CIM) (Section 6.3) [266, 613]. These would diminish relative to pQCD with increasing p_T but could be seen in x_T scaling [614], i.e. measurements at a fixed x_T but different values of p_T for two different values of $\sqrt{s}$. This is illustrated in a simple model of an additional p_T^{-6} component added to the QCD p_T^{-4} component in a two-component mixture where for $p_T = 1$ GeV/c the p_T^{-6} component is 50 times larger than the p_T^{-4} component (Figure 12.16c dotted line). Also shown is the NLO pQCD prediction (solid line) as well as values of $n_{eff}(x_T = 0.2)$ extracted from pion data at the two values of $\sqrt{s}$ shown and plotted at the geometric mean of the two values of p_T. One nice thing about this plot is that one can compare

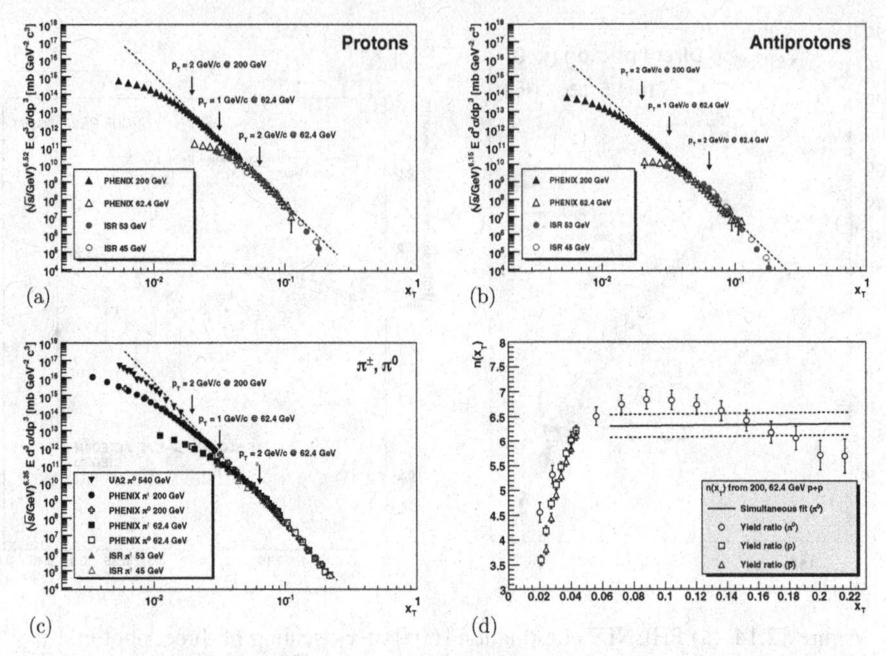

Figure 12.15 x_T scaled spectra [611] at mid-rapidity for (a) protons, (b) antiprotons, (c) $\pi^\pm, \pi^0$ from $\sqrt{s} = 200$ and 62.4 GeV. Only statistical uncertainties of the data are shown. Data from lower and higher $\sqrt{s}$ are indicated. The dashed curves are the fitting results. (d) x_T scaling power n_{eff} as determined from the ratios of yields as a function of x_T, for (open circle) neutral pions (Figure 12.16a,b) [612], (open square) protons, and (open triangle) antiprotons using p–p data at $\sqrt{s} = 200$ and 62.4 GeV. The error of each data point is from the systematic and statistical errors of p_T spectra.

any previous, present or future measurements to pQCD at $x_T = 0.2$ with a linear vertical scale over a huge range of $\sqrt{s}$ from 20 to 2000 GeV on a logarithmic horizontal scale to look for a higher twist effect with a characteristic dependence on p_T compared to pQCD. Another nice thing is that it is easy to see issues which need to be resolved. For instance, the difference in $n_{eff}(x_T = 0.2, 62.4, 200) = 6.0$ compared to the NLO value $n_{eff} = 5.4$ corresponds to a cross section ratio $\sigma^{inv}(x_T = 0.2, 62.4)/\sigma^{inv}(x_T = 0.2, 200) \approx 2.0$, i.e. a factor of 2.0 larger than predicted in NLO pQCD, which in the model corresponds to $\sigma^{inv}(x_T = 0.2, 62.4)$ a factor of 2.28 larger than NLO pQCD which should be easy to see directly from the data, and $\sigma^{inv}(x_T = 0.2, 200)$ a factor of 1.14 larger than NLO pQCD which is probably within the experimental error. In fact, a direct comparison of the PHENIX measurement at $\sqrt{s} = 62.4$ with NLO pQCD [612] does show that the data are higher by a factor of 2 than the theory at $x_T = 0.2$ for factorization scale $\mu = p_T$ but also that the NLO prediction varies by a factor of 2 larger and smaller with variation of μ from $p_T/2$ to $2p_T$. Hence the apparent deviation from NLO pQCD

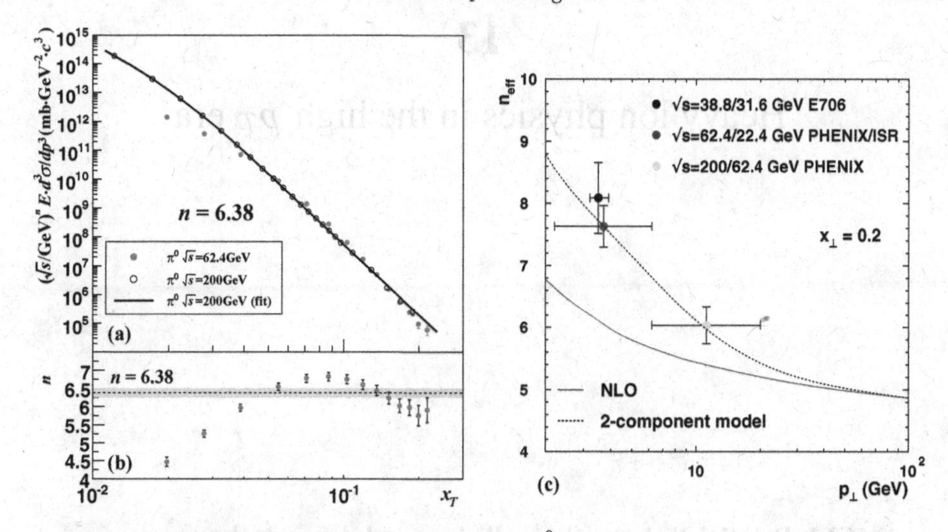

Figure 12.16 (a) Invariant cross section of π^0 production in p–p collisions at $\sqrt{s} = 62.4$ and 200 GeV as a function of x_T scaled by $(\sqrt{s}/\mathrm{GeV})^n$ with $n = 6.38$ [612]; the solid line is a parameterization of the $\sqrt{s} = 200$ GeV data. (b) $n_{eff}(x_T = 0.2, 62.4, 200)$ obtained from Eq. 12.17 [612]. The error bars show the statistical and systematic uncertainties of the data points, with the exception of the combined overall normalization uncertainty of the measurements at 62.4 and 200 GeV/c which is shown as the shaded band. (c) p_T dependence of n_{eff} of pions at $x_T = 0.2$ calculated [614] from measurements at the indicated $\sqrt{s}$, from NLO pQCD (solid line) and in the two-component model discussed in the text (dotted line).

at $p_T = 11$ GeV/c apparently shown on Figure 12.16c is not significant once the systematic uncertainties in the pQCD calculation are taken into account.

These discussions illustrate the strength of direct tests of pQCD such as x_T scaling to see deviations which may then be sorted out by the more detailed professional calculations. It also illustrates the important point that at present NLO pQCD still comes with significant errors of orders of $\pm$ a factor of 2 in calculations of the single particle invariant cross section and that the excellent Monte Carlo event generators which are now the repository of accumulated knowledge still need to be "heavily tuned" to reproduce the data [517].

13

Heavy ion physics in the high p_T era

13.1 Relativistic heavy ion collisions and dense nuclear matter

As discussed in Section 1.7, the collisions of relativistic heavy ions were predicted to provide the means of obtaining superdense nuclear matter in the laboratory [80–83], leading to a phase transition from a state composed of nucleons containing bound quarks and gluons to a state of deconfined quarks and gluons in chemical and thermal equilibrium, covering the entire volume of the colliding nuclei, dubbed the Quark Gluon Plasma (QGP) [79] because its properties were thought to be most similar to that of an ionized gas. In the terminology of high energy physics, this is called a "soft" process, related to the QCD confinement scale

$$\Lambda_{QCD}^{-1} \simeq (0.2\,\text{GeV})^{-1} \simeq 1\,\text{fm}. \tag{13.1}$$

In a dense medium of many free (color) charges, as discussed originally by Collins and Perry [78], "long range interactions are screened by many-body effects," i.e. the many charges in the medium are attracted to a test charge and screen its potential. This is called Debye screening and can be included straightforwardly [615] in the Cornell potential (Eq. 12.10) [549] for bound states of heavy quarks c–$\bar{c}$,

$$V(r) = -\frac{4}{3}\frac{\alpha_s}{r} + \sigma r \rightarrow -\frac{4}{3}\frac{\alpha_s}{r} e^{-\mu_D r} + \sigma \frac{(1 - e^{-\mu_D r})}{\mu_D} \tag{13.2}$$

where $\mu_D = \mu_D(T) = 1/r_D$ is the Debye screening mass [615]. For $r \ll 1/\mu_D$ a quark feels the full color charge and a Coulomb $1/r$ potential, but for $r \gg 1/\mu_D$, the quark is free of the potential as well as the string tension, so is effectively deconfined.[1] Such considerations [616] led Matsui and Satz in 1986 [617] to propose.

[1] An important point to note about the screened Coulomb term is that it represents the Yukawa potential [222] for the exchange of a quantum of mass μ_D.

J/Ψ suppression as the "gold plated" signature of deconfinement. In the QGP, the $c, \bar{c}$ interaction is screened so that the $c, \bar{c}$ do not bind to make a J/Ψ but go their separate ways and eventually pick up other quarks at the periphery to become *open charm*. The prediction of this dramatic effect drove the design of the detectors both at RHIC and the LHC. More about this later.

In relativistic statistical physics "an increase in temperature also provides an increase in density" [616], so that a large density or large energy density corresponds to high temperature. The theoretical treatment of gauge theories like QCD at high temperature is done using thermodynamic variables which represent macroscopic properties of a hot QCD medium. The partition function and free energy F are calculated (starting with Feynman [618]) analogously to the classical method, $F = -kT \ln Z$ where Z is the partition function, or sum over states, which is of the form $Z \propto \left\langle e^{-(E-\sum_i \mu_i Q_i)/kT} \right\rangle$, where k is Boltzmann's constant and μ_i are chemical potentials associated with conserved charges Q_i, except that the Hamiltonian operator is used for the energy E and special handling is required [619]. The thermodynamic properties of the medium, including possible phase transitions, are calculated "in the usual fashion" [615] by derivatives of F.[2] Thus Eq. 13.2 can be rewritten in terms of temperature, and with increasing temperature, T, in analogy to increasing Q^2, the strong coupling constant α_s becomes smaller. However, at the relevant scale [620], $\alpha_s(2\pi T) \approx \alpha_s(1\,\text{GeV}) \approx 0.5$ (Figure 12.9a), turns out to be too large for pQCD calculations so that QCD calculations for QGP production are done using lattice gauge theory, numerical solution of the QCD Lagrangian on a lattice [621].

Lattice gauge calculations support the screening of the Coulomb-like QCD potential and deconfinement for heavy quarks, see Figure 13.1a [622] – when the potential becomes constant with increasing radius, the binding force vanishes [624]. Presumably, the most accurate predictions of the QGP phase transition are also given by lattice QCD, see Figure 13.1b [623]. The sharp increase of energy density ϵ/T^4 above the transition temperature T_c indicates the phase transition from nucleon degrees of freedom to the much larger quark, gluon and color degrees of freedom. However, the critical energy density at the transition temperature, $T_c \sim 150$–170 MeV, from this calculation is stated to be known only with large errors [624], $\epsilon_c = (0.3$–$1.3)$ GeV fm^{-3}. In general, calculations of T_c at mid-rapidity, where μ_b, the baryon chemical potential, is zero ($\bar{p}/p \approx 1$), are in the range $165 \leq T_c \leq 185$ MeV [625] with $2.5 \lesssim \mu_D/T \lesssim 5$ for $4 \leq T/T_c \leq 1$ [626].

[2] The mathematics is closely related to cumulants in mathematical statistics since F and the cumulant generating function of mathematical statistics are both the logarithm of the expectation value of an exponential function of a probabilistic variable (see Eq. A.37 in Appendix A).

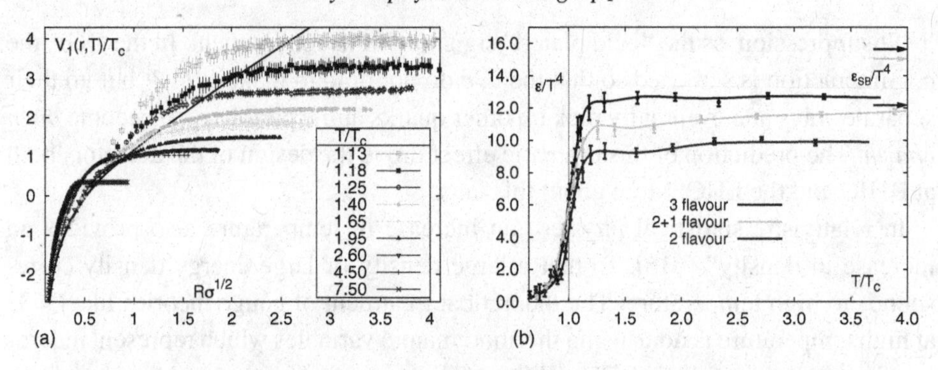

Figure 13.1 (a) Lattice gauge calculations [622] of heavy quark potential $V(r)$ as a function of T/T_c, where T_c is the transition temperature. The solid line is the normal ($T = 0$) potential. (b) Lattice calculation [623] of energy density, ϵ/T^4 as a function of the number of active flavors: two flavor (u, d), three flavor (u, d, s).

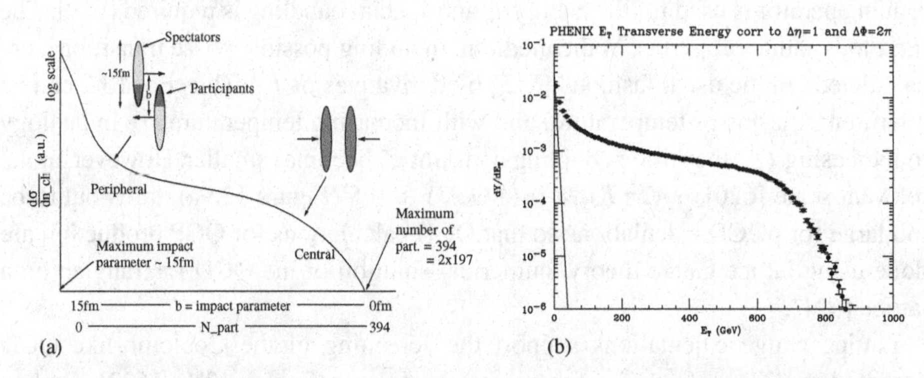

Figure 13.2 (a) Schematic of a collision of two nuclei with radius R and impact parameter b. The curve with the ordinate labeled $d\sigma/dn_{ch}$ represents the relative probability of charged particle multiplicity n_{ch} which is directly proportional to the number of participating nucleons, N_{part}. (b) Transverse energy (E_T) distribution in Au+Au (data points) and p–p collisions (line) at $\sqrt{s_{NN}} = 200$ GeV from PHENIX [627].

13.2 Experimental issues in $A + A$ compared to p–p collisions

As noted in Section 2.1, and shown in actual events from the STAR and PHENIX detectors at RHIC in Figure 2.1, the main challenge of experimental design of RHI collisions is the huge multiplicity in $A+A$ central collisions compared to p–p collisions. A schematic drawing of a collision of two relativistic Au nuclei is shown in Figure 13.2a. In the center-of-mass system of the nucleus–nucleus collision, the two Lorentz contracted nuclei of radius R approach each other with impact parameter b. In the region of overlap, the "participating" nucleons interact with each other, while in the non-overlap region, the "spectator" nucleons simply continue on their original trajectories and can be measured in Zero Degree Calorimeters (ZDC), so that the number of participants can be determined.

13.2.1 Centrality

The degree of overlap is called the centrality of the collision, with $b = b_{min} = 0$ being the most central, and $b = b_{max} \approx 2R$ the most peripheral. The maximum time of overlap is $\tau_0 = 2R/\gamma c$ where γ is the Lorentz factor and c is the speed of light. The energy of the inelastic collision is predominantly dissipated by multiple soft (low p_T) particle production, where n_{ch}, the number of charged particles produced, is directly (but not necessarily linearly) proportional to the number of participating nucleons (N_{part}) as sketched in Figure 13.2a: the smaller the impact parameter, the more the colliding nuclei overlap, so that more nucleons participate in the collision, which increases both the produced multiplicity n_{ch} and E_T. Thus, n_{ch} or the total transverse energy E_T in central Au+Au collisions is roughly A times larger than in a p–p collision, as shown in the measured transverse energy spectrum in the PHENIX detector for Au+Au compared to p–p collisions (Figure 13.2b).

The centrality of a collision can be determined from **any** measured variable that is monotonic in impact parameter [628], for example the E_T distribution in Figure 13.2b. The Lorentz contracted relativistic nuclei are roughly black disks of radii $R = 1.2A^{1/3}$ fm [173], so that the total inelastic interaction cross section for an $A + A$ collision, where A is the number of nucleons, is just geometrical:

$$\sigma_{int}^{AA} = \pi b_{max}^2 = 4\pi R^2. \tag{13.3}$$

Since the impact parameter cannot be measured, the centrality of a collision is defined by the upper percentile of a measured distribution, with the most central collisions producing the largest n_{ch} and E_T, i.e. top 10–20%-ile, top 5%-ile of n_{ch} or E_T (where unfortunately the "-ile" is not usually explicitly stated). For instance, in Figure 13.2b, the top 5%-ile is $E_T \geq 560$ GeV and the top 1%-ile is $E_T \geq 680$ GeV, the knee of the distribution. The upper percentile of the distribution is, to a good approximation, simply related to the probability that the impact parameter $b \leq b_0$ in a collision:

$$\mathcal{P}(b \leq b_0) = \int_0^{b_0} 2\pi b db/4\pi R^2 = \left(\frac{b_0}{2R}\right)^2 \tag{13.4}$$

so the top 5%-ile means $b_0^2 \leq 0.05 \times (2R)^2$, etc. The centrality of a collision is also expressed in terms of the number of participating nucleons, N_{part}. Typically the distribution in the number of participating nucleons in an $A + A$ collision as well as other such "nuclear geometrical" quantities can be computed to high precision in the "impulse approximation" [629] using the Glauber [630] method in a Monte Carlo calculation (see Appendix C).

In addition to the large multiplicity and centrality determination, there are two other issues in RHI physics which are different from p–p physics:

(i) space-time issues, both in momentum space and in coordinate space – for instance what is the spatial extent of fragmentation? is there a formation time before a parton fragments?
(ii) huge azimuthal anisotropies of particle production (colloquially collective flow) most evident in non-central collisions which are interesting in their own right but which cause a serious complication for studies of hard scattering, particularly two particle correlations.

13.2.2 Space-time and quantum mechanical issues

As just discussed, the interaction of two relativistic heavy ions can be viewed initially as the superposition of successive independent collisions of the participating nucleons in the overlap region [630]. Collective effects may subsequently develop due to rescattering of the participants with each other or with produced particles. Conceivably, a cascade of interacting particles could develop. However, the actual situation is more fascinating.

Space-time and quantum mechanical issues play an important role in the physics of RHI collisions and the considerations are somewhat different in the longitudinal and transverse directions. In both QED and QCD, large transverse momenta correspond to small impact parameters and are thus rare. Because of their rarity, particles with large transverse momenta were not expected to play much of a role in RHI physics. In the longitudinal direction, because of the large γ factors involved, small excitations/de-excitations of the colliding nucleons take place over long distances. When a nucleon with momentum, p_L, mass, M, makes a collision, the only thing it can do consistent with relativity and quantum mechanics is to get excited to a state with invariant mass $M^* \geq M$, with roughly the same energy and reduced $p'_L = p_L - \Delta p_L$, where $\Delta p_L^2 = -\Delta m_T^2 \simeq -\Delta m^2$ from Eq. 2.3. By the uncertainty principle, a distance $\Delta z = \hbar/\Delta p_L = \gamma\beta/\Delta m$ (equal to 14 fm for $\gamma = 10$ and $m = m_\pi$) is required to resolve Δm. This is also true in the transverse direction with similar consideration for a π from a fragmenting parton.[3]

The large γ factor in relativistic collisions ensures that the excited nucleons pass through the entire target nucleus before de-exciting into for example a nucleon plus a pion. Thus, nuclei are transparent to relativistic nucleons; and pions are produced outside the target nucleus, avoiding a nuclear cascade. For instance, in the collision of a relativistic proton with a 15 interaction length thick lead (Pb) brick

[3] It is interesting to note that the same arguments for the formation time of a J/Ψ in order to resolve its very narrow width, $\Delta E = \Gamma = 93$ keV, give $\Delta t = \hbar/\Delta E = 2122$ fm/c, much longer than the lifetime of a QGP.

or hadron calorimeter, a cascade develops and all particles are absorbed. Nothing comes out the back. By contrast, in the collision of a relativistic proton with a Pb nucleus, which is roughly 15 interaction mean-free-paths thick through the center, the (excited) proton comes out the back! This is relativity and quantum mechanics in action [125, 126].

Similar arguments explain why the number of participants N_{part} is relevant to the bulk of particle production in $A + A$ collisions. When a high energy nucleon passes through a nucleus, it can make several successive collisions and does not have time to de-excite between collisions. Further assumptions – that the excited nucleon interacts with the same cross section as an unexcited nucleon and that the successive collisions of the excited nucleon do not greatly affect the excited state or its eventual fragmentation products – lead to the conclusion that the multiplicity in nuclear interactions should be proportional to the total number of projectile and target participants, N_{part}, rather than to the total number of collisions. This is called the Wounded Nucleon Model (WNM) [631], which works well at mid-rapidity at c.m. energy $\sqrt{s_{NN}} \approx 20$ GeV, where it was discovered [632], but otherwise serves as a reasonable guide. Details are beyond the scope of this book [125, 126].

The same arguments do not apply to hard scattering, the production of particles (e.g. π) with large p_T. Since hard scattering is point-like, with distance scale $1/p_T < 0.1$ fm, much less than the size of a nucleon or nucleus, the scattering cross section in $p + A$ collisions is simply A times larger than in p–p collisions, and in $A + A$ collisions A^2 times larger than the p–p cross section, where A represents the number of nucleons in the nucleus. For $A + A$ or $B + A$ collisions in a limited range of impact parameters, i.e. centrality interval f, the factor is $\langle T_{AB}\rangle_f$, the purely geometric overlap integral of the nuclear thickness functions (see Appendix C). The precise relation is that the ratio of the measured number of hard scattering events in the centrality class f, $d^2N^{hard}_{AB}(p_T)|_f/p_T dp_T dy$, to the total number $N^{evt}_{AB}|_f$ of AB inelastic events in centrality class f, equals $\langle T_{AB}\rangle_f$ times the cross section for the hard process in p–p collisions, $d^2\sigma^{hard}_{pp}/p_T dp_T dy$,

$$(1/N^{evt}_{AB}|_f)\, d^2N^{hard}_{AB}(p_T)|_f/p_T dp_T dy = \langle T_{AB}\rangle|_f \times d^2\sigma^{hard}_{pp}/p_T dp_T dy. \quad (13.5)$$

It is important to realize that Eq. 13.5 relates the fractional yield of high p_T particles per $B + A$ interaction, with centrality f, to a geometrical factor, $\langle T_{AB}\rangle_f$, times the *cross section* for the high p_T particles in p–p collisions. This relation is also called binary collision or N_{coll} scaling because the number of binary nucleon–nucleon (N–N) collisions (in the Glauber sense) in a $B + A$ interaction of centrality class f is simply

$$N^f_{coll} = \langle T_{AB}\rangle_f \times \sigma_{NN} \quad (13.6)$$

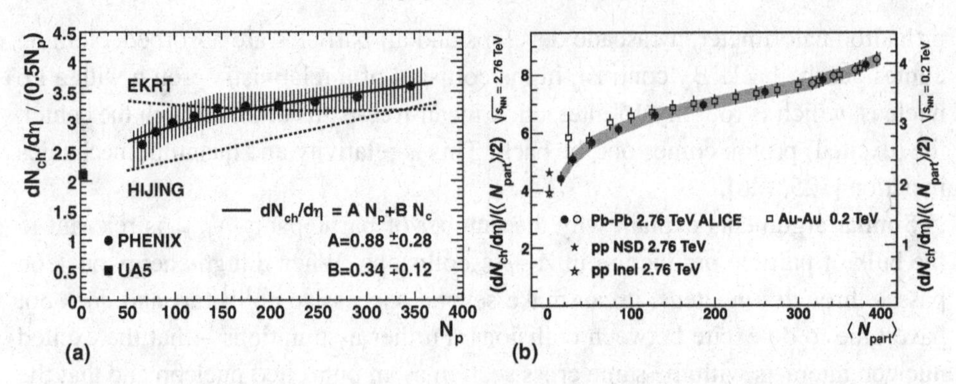

Figure 13.3 (a) PHENIX [633] measurement of the charged particle pseudorapidity density per participant pair at mid-rapidity in Au+Au collisions at $\sqrt{s_{NN}} = 130$ GeV as a function of the number of participants (filled circles). The shaded area represents the correlated systematic errors. The solid line is the empirical fit discussed in the text. (b) ALICE [634] measurement of the dependence of $dN_{ch}/d\eta/(0.5N_{part})$ on the number of participants for Pb+Pb collisions at $\sqrt{s_{NN}} = 2.76$ TeV (circles). Also shown is the RHIC average [627] for the same measurement in Au+Au collisions at $\sqrt{s_{NN}} = 200$ GeV (squares) for which the right hand scale should be used which differs from the left hand scale for the LHC data by a factor of 2.1.

where σ_{NN} is the N–N inelastic cross section at $\sqrt{s_{NN}}$, the nucleon–nucleon c.m. energy of the $B + A$ collision.

Returning to the bulk of particle production, Figure 13.3 shows measurements of the multiplicity density at mid-rapidity, $dN_{ch}/d\eta$ (which, as discussed in Chapter 11, is made up predominantly of soft, low p_T, particles), in p–p and $A + A$ collisions as a function of centrality from both RHIC [633] in Au+Au collisions and the LHC [634] in Pb+Pb collisions, at vastly different c.m. energies. The results are presented as $dN_{ch}/d\eta$, per participant pair (so as to be able to compare directly with p–p measurements), as a function of centrality expressed as N_{part}. If the WNM were the whole story, then $dN_{ch}/d\eta/(0.5N_{part})$ would be constant and equal to $dN_{ch}^{pp}/d\eta$. The apparently identical relative increase of $dN_{ch}/d\eta/(0.5N_{part})$ with increasing centrality found at RHIC ($\sqrt{s_{NN}} = 200$ GeV) and LHC ($\sqrt{s_{NN}} = 2.76$ TeV) is striking and indicates that nuclear geometry is the predominant effect. The dependence has been parameterized [633], in a fit shown in Figure 13.3, as $dN_{ch}^{AA}/d\eta = (dN_{ch}^{pp}/d\eta)\,[xN_{part}/2+(1-x)N_{coll}]$ with the implication that a certain fraction x of the production is by soft processes (pure WNM), and $(1 - x)$ by hard scattering.[4] A representation of the data more in keeping

[4] The split into soft and hard components with the implication that hard scattering contributes to $N_{\rm ch}$ or E_T distributions is not reasonable (e.g. Figure 11.17). See references [635, 636] for further discussions.

with the evident nuclear geometric dependence is given by the constituent quark model [635, 637, 638], similar to the previous additive quark model [639, 640] and WNM [631] except that the fundamental element of particle production is a wounded quark or quark-participant. In this geometrical framework, the data in Figure 13.3 represent the increase in the number of quark participants (N_{q-part}) per wounded nucleon (N_{part}) with increasing centrality, which evidently does not change much with the variation of $30 \lesssim \sigma_{NN} \lesssim 60$ mb over the entire range of c.m. energies $4.8 \leq \sqrt{s_{NN}}(\text{GeV}) \leq 2760$ so far measured in RHI collisions.

13.2.3 Collective flow

A distinguishing feature of $A + A$ collisions compared to either p–p or $p + A$ collisions is the collective flow observed. This effect is seen over the full range of energies studied in heavy ion collisions, from incident kinetic energy of 100A MeV to c.m. energy of $\sqrt{s_{NN}} = 2.76$ TeV [641, 642]. Collective flow, or simply flow, is a collective effect which cannot be obtained from a superposition of independent N–N collisions.

Immediately after an $A + A$ collision, the overlap region defined by the nuclear geometry is almond shaped (see Figure 13.4a) with the shortest axis along the impact parameter vector. Due to the reaction plane breaking the ϕ symmetry of the problem, the semi-inclusive single particle spectrum is modified by an expansion in

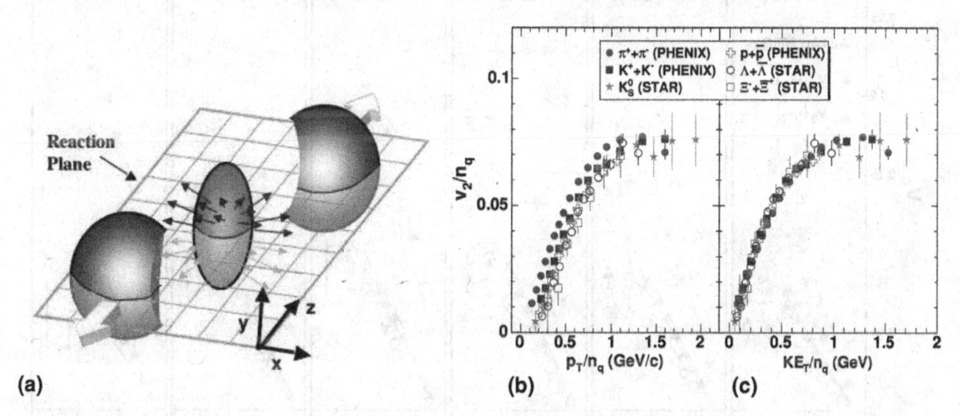

Figure 13.4 (a) Almond shaped overlap zone generated just after an $A + A$ collision where the incident nuclei are moving along the $\pm z$ axis. The reaction plane by definition contains the impact parameter vector (along the x axis) [645]. Measurements of elliptic flow (v_2) for identified hadrons plotted as v_2 divided by the number of constituent quarks n_q in the hadron as a function of (b) p_T/n_q, (c) KE_T/n_q [646].

harmonics [643] of the azimuthal angle of the particle with respect to the reaction plane, $\phi - \Phi_R$ [644], where the angle of the reaction plane Φ_R is defined to be along the impact parameter vector, the x axis in Figure 13.4a:

$$\frac{Ed^3N}{dp^3} = \frac{d^3N}{p_T dp_T dy d\phi} = \frac{d^3N}{2\pi\ p_T dp_T dy}\left[1 + \sum_n 2v_n \cos n(\phi - \Phi_R)\right]. \quad (13.7)$$

The expansion parameter v_2, called elliptic flow, is predominant at mid-rapidity. In general, the fact that flow is observed in final state hadrons shows that thermalization is rapid so that hydrodynamics comes into play before the spatial anisotropy of the overlap almond dissipates. At this early stage hadrons have not formed and it has been proposed that the constituent quarks flow [647], so that the flow of a hadron should be proportional to the number of its constituent quarks n_q, in which case v_2/n_q as a function of p_T/n_q would represent the constituent quark flow as a function of constituent quark transverse momentum and would be universal. However, in relativistic hydrodynamics, at mid-rapidity, the transverse kinetic energy, $m_T - m_0 = (\gamma_T - 1)m_0 \equiv KE_T$, rather than p_T is the relevant variable, and in fact v_2/n_q as a function of KE_T/n_q seems to exhibit nearly perfect scaling [646] (Figure 13.4c).

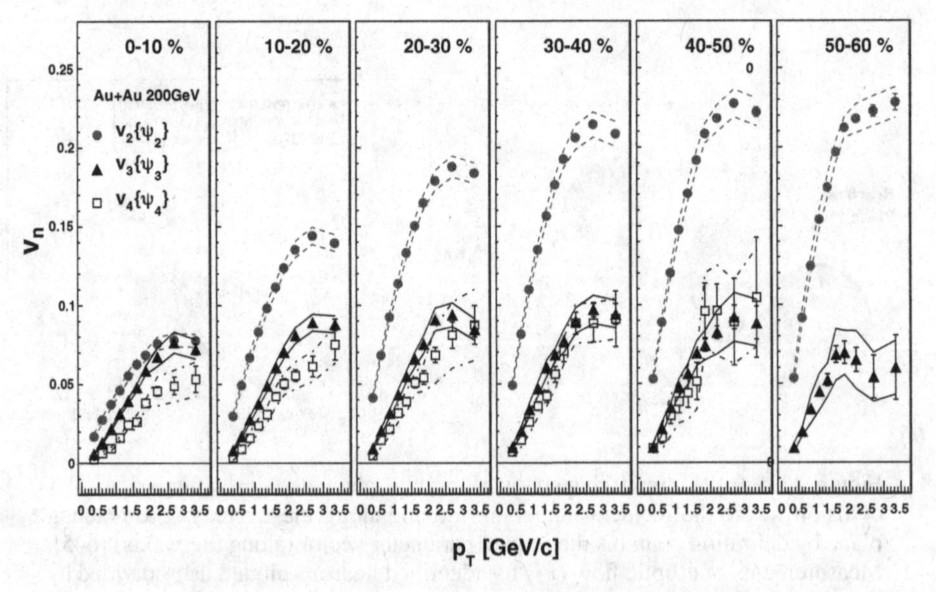

Figure 13.5 PHENIX [656] measurements of the v_n parameters using Eq. 13.7 (with the appropriate reaction plane) as a function of p_T for different centrality slices in $\sqrt{s_{NN}} = 200$ GeV Au+Au collisions.

The fact that the flow persists for $p_T > 1$ GeV/c implies that the viscosity is small [648], perhaps as small as a quantum viscosity bound from string theory [649], $\eta/s = 1/(4\pi)$ where η is the shear viscosity and s the entropy density per unit volume. This has led to the description of the "sQGP" produced at RHIC as "the perfect fluid" [650].

13.2.3.1 Triangular flow, odd harmonics

For the first 10 years of operation of RHIC, and dating back to the Bevalac, all the experts thought that the odd harmonics in Eq. 13.7 would vanish by the symmetry $\phi \rightarrow \phi + \pi$ of the almond shaped overlap region [651] (Figure 13.4a). However, in 2010, an MIT graduate student and his Professor in experimental physics, seeking (at least since 2006) how to measure the fluctuations of v_2 in the PHOBOS experiment at RHIC, realized that fluctuations in the collision geometry on an event-by-event basis, i.e. the distribution of participants from event-to-event, did not respect the average symmetry. This resulted in what they called "participant triangularity" and "triangular flow," or v_3 in Eq. 13.7, which they measured [652]. A Brazilian group had shown in 2009 that v_3 does appear in an event-by-event hydrodynamics calculation without jets [653,654], but the MIT group [652] was the first to show it with real data.

Many experiments at both RHIC and the LHC presented their first measurements of v_3 at Quark Matter 2011 [655], e.g. Figure 13.5 [656], and these were among the most exciting results at the conference. There are two striking observations from Figure 13.5 which indicate that fluctuations of the initial collision geometry are driving the observed v_3:

(i) the centrality dependence of $v_3(p_T)$ is weak as one would expect from fluctuations, but $v_2(p_T)$ which is most sensitive to the geometry of the "almond" shaped overlap region tracks the change in eccentricity with centrality;
(ii) for the most central collisions (0–10%), where the overlap region is nearly circular so that all the v_n are driven by fluctuations, $v_2(p_T)$, $v_3(p_T)$, $v_4(p_T)$ are comparable.

The fact that the observed collective flow of final state particles follows the fluctuations in the initial state geometry again points to real hydrodynamic flow of a nearly perfect fluid.

14
RHIC and LHC

14.1 The road to RHIC

The roads to RHIC and the LHC are highly intertwined. The great experimental discoveries of the late 1960s and early 1970s – DIS in e–p, hard scattering in p–p collisions, J/Ψ – inspired a surge of proposals for new accelerators to study the new phenomena. Of key importance in this development were the two major theoretical discoveries: QCD in 1973 as the theory of the strong interactions; and the unification of electromagnetic and weak interactions by the Glashow–Weinberg–Salam model [95–97] a few years earlier. The Glashow–Weinberg–Salam model [100] added two new neutral particles to the unified "electroweak" interaction:

(i) a vector boson, Z^0, as the carrier of a neutral current weak interaction, which predicted such previously unobserved reactions as $\nu_\mu + N \rightarrow \nu_\mu +$ hadrons, with no final state μ, via the exchange of a Z^0;
(ii) a neutral scalar "Higgs" boson which "spontaneously" broke the electroweak symmetry of the gauge theory Lagrangian and gave mass to the Z^0 and $W^\pm$ bosons while keeping the photon massless.[1]

Brookhaven (BNL) was first in the accelerator competition [657, 658]. In 1971, convinced by the success of the proton–proton collider concept at the CERN-ISR, BNL proposed a 200 × 200 GeV (later 400 × 400 GeV) p–p collider with a high luminosity of $\mathcal{L} = 10^{33}$ cm^{-2} s^{-1} using superconducting magnets, the Intersecting Storage Accelerator, or ISABELLE. The energy and associated luminosity were specified under Leon Lederman's influence [659] using Bjorken scaling, as discussed above (Figure 5.8), so as to discover either the W boson or to detect other fundamental modifications of high energy weak interactions implied by unitarity –

[1] The Higgs mechanism was preferred for electroweak symmetry breaking because it also gave mass to the fermions, such as the $\mu^\pm$ meson and b quark; however there are other methods of electroweak symmetry breaking in the boson sector only [100], which do not give mass to the fermions.

to quote Lederman [659], "at this luminosity ISABELLE is guaranteed to make at least one fundamental discovery!" As later noted by Rubbia [660], "ISABELLE was the first accelerator in which the production of W–Z was a primary goal." BNL's proposal was followed by Fermilab's proposal in 1972 to build a 1 TeV ring of superconducting magnets, "the Doubler," to double the energy of the fixed target machine [661–663], as well as by many summer studies about possible experiments at these machines.

The discovery of weak neutral currents in neutrino experiments at CERN in 1973–1974 [664, 665] dramatically changed the ground rules and the shape of the playing field because the measured ratio of neutral/charged currents gave a clear mass range where the predicted W and Z bosons should exist ($M_{W^\pm} \simeq 65$ GeV/c^2, $M_{Z^0} \simeq 80$ GeV/c^2 [666]). This inspired a new series of accelerator studies and proposals to find the $W^\pm$ and now the Z^0. These were the Large Electron–Positron collider (LEP) at CERN [666, 667], with c.m. energy $\sqrt{s} \approx 200$ GeV, a conventional circular machine with a huge circumference of 27 km; and a novel 50×50 GeV linear collider at SLAC, the SLC, proposed in 1980 [668], in which the SLAC linac would accelerate both electrons and positrons and bring them into collision in a collider ring composed of small aperture magnets with very strong alternating gradient focussing which fit nicely on the SLAC site. The strong magnets and small bending radius could be tolerated because the beams only made a single pass and were discarded after the collision.

However, in 1976, Carlo Rubbia and collaborators proposed "a colliding beam experiment" [247], at both Fermilab and the CERN-SPS, to convert the pulsed accelerator ring into a proton–antiproton storage ring/collider by injecting and storing antiprotons. This had been done previously for electron–positron collisions at the Cambridge Electron Accelerator [669], but making and storing antiprotons is a lot more difficult than making positrons because the produced antiprotons must be "cooled" to reduce their large initial momentum spread [660]. CERN quickly took up this idea and was able to capitalize on its great accelerator and hadron collider experience to solve the technical problems (the crucial technical development being "stochastic cooling" pioneered at the ISR) and build the SPS collider, which went into operation in 1981 and produced the major discoveries of jets in hadron collisions in 1982 (Figure 11.18), as well as the W (Figure 5.3) and Z bosons [670, 671] in 1983.

Construction of the SPS collider did not stop the LEP project, which was approved in 1981 with the huge 27 km circumference tunnel, completed in 1988, with the first collisions in 1989 [672]. The director general, Herwig Schopper, had pushed for as large a tunnel as possible [667], "in view of the long-term future of CERN and of a later project, a proton collider [a Large Hadron Collider (LHC)] in the LEP tunnel, a possibility which had been considered already at that time [673]."

In distinction to previous CERN projects, LEP was to be constructed within the basic CERN budget, with no additional funds. This necessitated closing and dismantling the ISR in 1983, "a particularly painful decision" [667], and curtailing other activities. Ironically, in spite of these restrictions, a new program of accelerating heavy ions in the SPS was approved by Schopper in 1983 [674], in part because of contributions from LBL and GSI of the ion source, pre-injector and plastic-ball detector [675] but partly due to some "exciting results" from α–α collisions in the CERN-ISR [676]. LEP was very successful in measuring the properties of the $W^{\pm}$ and Z^0 bosons in the period from first collisions in 1989 to its closure at the end of 2000 to make room for construction of the LHC [677].

The situation in the USA was not that smooth. ISABELLE was approved for construction in 1978 but ran into serious magnet problems in 1979. Meanwhile, at Fermilab, a sector (1/6) of the Doubler was approved as an R&D project. Although R&D magnets for the Doubler had previously run into problems [662, 663], these had been solved by 1979, leading to a successful test of the sector. Also, in 1980, the proposal for the SLC was submitted. Although ISABELLE's magnet problems were solved in 1981 with the development of the Palmer magnet [678] (which has been used in every subsequent superconducting accelerator), the tightening of budgets coupled with the magnet problems led the US high energy physics community to smell "blood in the water" and organize a meeting at Snowmass, Colorado, from June 28 to July 16, 1982, to "assess the future of elementary particle physics, to explore the limits of our technological capabilities, and to consider the nature of future major facilities for particle physics in the U.S." [421].

Without going into details, following Snowmass 1982 and the discovery of the W and Z bosons at CERN in January 1983, the USA chose to proceed with the Fermilab Doubler/Tevatron-Collider and the SLC. ISABELLE was canceled in favor of pursuing the "Desertron," proposed by Bob Wilson (R. R. Wilson, builder of Fermilab [663]) at Snowmass 1982 [679], a 20×20 TeV p–p collider, later known as the Superconducting Super Collider (SSC). The SSC was approved in 1987 and canceled in 1993 [680] – a sad story for High Energy Physics (HEP) in the USA. However, on the world scene, the US decision to maintain vitality in HEP by approving a proposal [680] for a major commitment to the LHC, which amounted to $531 million of accelerator and detector components [681], allowed the LHC construction to proceed to its full energy 14 TeV in a single stage [677].

In a story worthy of Cinderella, the rejection of ISABELLE (and BNL) by the US HEP community became the good fortune of the US nuclear physics community because on the very day that ISABELLE was officially canceled, July 11, 1983, the Nuclear Science Advisory Committee (NSAC) was having a workshop [682] "to recommend the next major construction project to follow the just approved 4 GeV electron accelerator, CEBAF at the future Jefferson Lab." The committee seized the opportunity and proposed [682] "to build a colliding beam heavy ion accelerator in the CBA [new name for ISABELLE with the Palmer Magnet] tunnel." This recommendation appeared in the final report as [683] "We identify a relativistic heavy ion collider as the highest priority for the next major facility to be constructed, with the potential of addressing a new scientific frontier of fundamental importance." BNL immediately went into high gear while all the assets (e.g. the tunnel, the cryogenic plant), the resources and the experience of ISABELLE/CBA were still intact. This was led by the Director, Nick Samios [684], who appointed a "Task Force on Relativistic Heavy Ion Physics" on July 14 to meet the following day, followed by Task Forces, Workshops, moving the third Quark Matter meeting from Helsinki to BNL in September 1983 [685], etc. The rest is indeed history [684] that need not be repeated here except to note that RHIC was formally proposed in 1984, approved for construction in 1990–1991, with the first collisions in 2000.

As noted above, CERN had not been idle during this period and had approved a program of light and heavy ion experiments, originally proposed for the PS [686], but moved to the SPS, first for technical reasons [675], but after 1983 [687] because, in preparation for RHIC, BNL started a fixed target heavy program at the AGS, using the existing Tandem Van de Graaff as the injector via a new transfer line [688], which began running in the fall of 1986. These two programs were the first to provide heavy ion collisions at sufficiently high energies that particle production rather than nuclear breakup was the dominant inelastic process. Both these programs provided some interesting results [125, 126, 689] but perhaps more importantly provided the training of outstanding young physicists.

Before moving on to discuss the application of hard scattering to $A+A$ collisions at RHIC and the LHC, it is worthwhile to review the "exciting results" from α–α collisions in the CERN-ISR, mentioned above, which helped convince CERN, in 1983, to approve the SPS heavy ion program.In 1979, Martin Faessler [690, 691] and collaborators proposed to measure p–α and α–α collisions in the CERN-ISR using the SFM. This resulted in runs with α–α at $\sqrt{s_{NN}} = 31$ GeV and p–α at $\sqrt{s_{NN}} = 44$ GeV in 1980, with a subsequent run in 1983 with α–α, p–α, d–d and p–p interactions all at the same $\sqrt{s_{NN}} = 31$ GeV. The high energy physicists at the ISR had the option of turning off their detectors and resting a few weeks or continuing to operate their detectors for the nuclear collisions, with the possibility of exciting new results. They all opted to continue. Exactly the same option and, predictably, exactly the same outcome occurred at the LHC 30 years later.

Leaving aside the many interesting α–α soft physics results [125, 126], the "exciting results" mentioned above provide an important lesson for RHIC and the LHC, which merits a discussion. They concerned the measurements of the Cronin effect at large p_T (Figure 14.1a) [692] for which the ratio, R, of the p–α/p–p cross

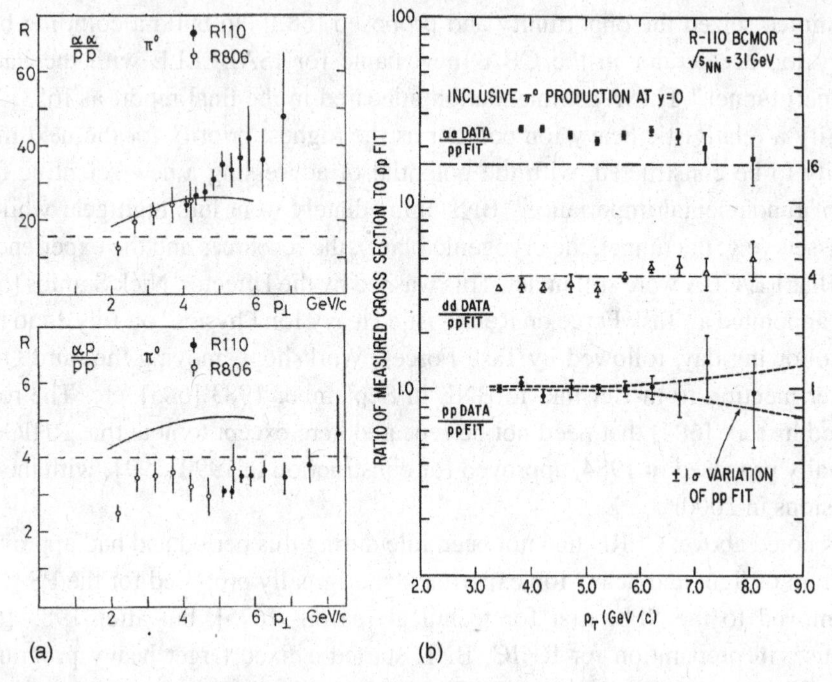

Figure 14.1 (a) Ratio of cross sections in α–p and α–α interactions to the cross sections in p–p interactions as a function of p_T: $R\,[(\alpha p \to \pi^0 + X)/(pp \to \pi^0 + X)]$ at $\sqrt{s_{NN}} = 44$ GeV and $R\,[(\alpha\alpha \to \pi^0 + X)/(pp \to \pi^0 + X)]$ at $\sqrt{s_{NN}} = 31$ GeV, compiled by Faessler [692]. (b) BCMOR measurements [693] of the inclusive π^0 cross sections in α–α, d–d and p–p collisions at $\sqrt{s_{NN}} = 31$ GeV divided by a fit to the p–p data. For explanation of solid lines and dashes, see text.

sections as a function of p_T were below the naive extrapolation $A \times B$ (dashes) and the extrapolation $(AB)^\alpha$ of the Cronin effect (solid line) using the $\alpha(p_T)$ measured by Cronin and collaborators at Fermilab (Figure 6.17) [257]. By contrast, for α–α collisions the R110 data for $p_T > 4$ GeV/c [694] were clearly well above the extrapolation and theoretical predictions [695]. To quote Faessler [695], "If this trend is confirmed, it could eventually signify that something very interesting is going on in nucleus nucleus collisions."

Well, it was not confirmed; it was simply an error in the extrapolated p–p cross section because there were no p–p comparison data at the same $\sqrt{s_{NN}}$. This indicates the importance of contemporaneous measurements of $A + A$ and p–p comparison data. The situation was rectified in the 1983 run. The R110 (now BCMOR) $\alpha\alpha/pp$ ratio [693] (Figure 14.1b) shows no marked p_T dependence over the measured range, $3.7 \leq p_T \leq 9.0$ GeV/c, but clearly exhibits the Cronin "anomalous enhancement" with $\alpha\alpha/pp$ cross section ratio, $\langle R \rangle = 23.8 \pm 0.4$, a factor of 1.5 greater than $A^2 = 16$, with a possible decrease for $p_T > 5$ GeV/c.

This is in stark contrast to the previous measurement (Figure 14.1a) [694], where the erroneous continuous rise of $\langle R \rangle$ to values of approaching 40 had caused great excitement.

To close this section with an anecdote: the error in reference [694], which should be evident to any reader of this book (Chapter 6, Eqs. 6.12, 6.13), was noted at the time by a co-author of this book (MJT) who was also a co-author of that paper. It was not followed up or corrected because he was too busy making magnets at ISABELLE at the time – in retrospect, perhaps a lucky break. This shows that sometimes wrong results have a bigger impact than correct results because they are exciting; but this does not excuse making mistakes.

14.2 Proposals for experiments

As discussed above, the construction of RHIC was approved to start in the financial year 1991 (October 1, 1990). In preparation there had been several workshops, detector R&D projects and a call for Letters of Intent (LOI) for experiments, in April 1990, leading to a major workshop held at BNL in July 1990 [696], "with the main goal of forming collaborations to prepare and submit letters of intent." The LOIs were to be submitted by September 28, 1990 for review by the Program Advisory Committee (PAC). The PAC responded two months later to all the nine original LOIs with individual letters containing comments, criticisms, suggestions for improvement and an independent cost estimate. Revised LOIs were due on July 15, 1991, taking into account suggestions from the PAC review as well as a constraint of two large (≈$30M each) and two small (≈$5M each) experiments imposed by the new Associate Director of High Energy and Nuclear Physics, Mel Schwartz (Nobel Laureate 1988, Figure 5.2), who had returned to BNL and Columbia for the exciting start of the new accelerator.

Decision day was on September 3, 1991, when the LOI for the Solenoid Tracker At RHIC (STAR) (originally, "An Experiment on Particle and Jet Production at Midrapidity" [697]) was approved [698] to proceed with a proposal to build "a large TPC detector" ($30M) which would concentrate on hadron physics, while the three competing lepton–photon LOIs were all rejected [698]: "The Committee decided to place the emphasis on a detector designed to study electrons and photons emerging from the QGP. . . The Laboratory appointed Sam Aronson as Spokesman and Project Director with the charge of developing a new collaboration to design and build such a detector" ($30M[2]). This became the PHENIX experiment [699].

[2] This money was from RHIC project funds. It was hoped/expected that resources outside BNL could be found to augment this sum. This type of planning led to a series of incremental upgrades for the RHIC detectors due to funding constraints.

The two small experiments approved were the BRAHMS experiment, a moveable spectrometer over the range $0 \lesssim y \lesssim 4$ with $\pi^{\pm}$, $K^{\pm}$, $p^{\pm}$ separation in the range p_T of 2–4 GeV/c, depending on rapidity; and PHOBOS, an experiment designed to study particle production down to very low p_T and over the full phase space. This was accomplished with large arrays of silicon detectors and a two-arm spectrometer near mid-rapidity with particle identification for $p_T < 1$ GeV/c, "where the majority of particle production is expected to occur." Full details of the RHIC detectors are given in reference [700].

The need for "resources outside BNL" to augment the RHIC project funds was met in many ways. However, one in particular stands out as an incredible success, worthy of mention. While RHIC was being designed, the issue of whether to include p–p collisions was raised. Fermilab was afraid that this would become competition to their collider and some influential nuclear physicists thought this was just an excuse to "resurrect the CBA" [682]. The issue was solved, in part, by a proposal from a group of experimental, theoretical and accelerator physicists with a common interest in spin, "The RHIC Spin Collaboration," which started in 1990 [701, 702] (but which dates back to Snowmass 1982 [703]) to make RHIC a polarized proton collider. To make a long story short, the proposal was strongly supported by both PHENIX and STAR and was approved by the RHIC PAC in October 1993 following a positive technical review that deserves to be quoted. "The proposal has the flavor of the application of an ingenious technological invention (siberian snakes) to make possible exciting physics research (polarization physics) reminiscent of the application of stochastic cooling to obtain $\bar{p}p$ beams for W and Z at the CERN SPS" [704]. The best part is that in 1995, Rikagaku Kenkyusho (RIKEN, The Institute of Physical and Chemical Research) of Japan agreed to fund the spin hardware in RHIC, including the siberian snakes and spin rotators, and a second muon arm for spin for PHENIX.[3] In April 1987, RIKEN also established and funded the RIKEN BNL Research Center, located at BNL, "dedicated to the study of strong interactions, including spin physics, lattice QCD and RHIC physics through the nurturing of a new generation of young physicists," with founding Director T. D. Lee.

The situation in Europe for planning experiments at the LHC was again more orderly than the process at RHIC. It started with a Large Hadron Collider workshop sponsored by the European Committee for Future Accelerators (ECFA) in Aachen, Germany, from October 4–9, 1990, and proceeded with further workshops, detector

[3] Here it should be noted that the Nuclear Physics Division of the US Department of Energy did rise to the occasion, once construction of the detectors had started, and provided "Additional Experimental Equipment (AEE)" money (~$35M over 5 years, starting in the financial year 1996) for a muon arm in PHENIX and an EM calorimeter and Silicon Vertex Tracker (SVT) in STAR which had been de-scoped from the original proposals.

R&D, LOI and proposals. This procedure is nicely described by ALICE and the other LHC experiments in a recent publication [705], so need not be repeated here. The result, especially with the US commitment in lieu of the SSC [681], was perhaps the greatest collection of detectors ever assembled, which included ALICE which was designed to study nucleus–nucleus collisions, as well as the ATLAS and CMS particle physics detectors which were meant to discover the Higgs boson and hopefully new and unexpected physics, such as supersymmetry, at the highest c.m. energy likely to be available for the foreseeable future; this turned out also to play an important role in RHI physics, particularly in the realm of hard scattering (Figure 14.2).

The ALICE detector (Figure 14.2a), which was the only one specifically designed for heavy ion collisions, is based on a large $\sim$7 m radius, $B = 0.5$ T, solenoid magnet from the L3 experiment at LEP for a central detector, containing a Time Projection Chamber (TPC) for full azimuthal coverage over a pseudorapidity range $|\eta| \leq 0.88$, with particle identification using Time Of Flight (TOF) and dE/dx, transition radiation detectors and fine grain (PHOS) and coarse grain (EMCAL) electromagnetic calorimeters. A Si vertex detector (ITS) complements the charged particle tracking in the TPC, with a minimum of material in the sensitive tracking volume $\approx 10\% X_o$. A very elegant muon arm in the forward direction $2.5 < \eta < 4$ consists of a dipole magnet and a complex arrangement of absorbers. The PHOS, which covers $|\eta| < 0.12$ and $\Delta\phi = 100°$ is made of $PbWO_4$ scintillating crystals located 4.6 m from the vertex with segmentation $\Delta\eta \times \Delta\phi = 0.005 \times 0.005$ capable of resolving $\pi^0 \rightarrow \gamma\gamma$ up to 27–54 GeV (Eq. 3.134) with resolution $\sigma_E/E \approx 1\% \oplus 3.6\%/\sqrt{E(\mathrm{GeV})}$ over a broad dynamic range 0.5–150 GeV [707]. The Pb scintillator EMCAL which covers $|\eta| < 0.7$, $\Delta\phi = 107°$ has respectable segmentation about a factor of 3 larger $\Delta\eta \times \Delta\phi = 0.014 \times 0.014$, and "moderate" energy resolution $\sigma_E/E \sim 10\%/\sqrt{E(\mathrm{GeV})}$ primarily for the study of jets.

The ATLAS and CMS detectors are "conventional" but beautifully executed HEP collider detectors based on superconducting solenoids ($r = 1.2$ m, $B = 2$ T, ATLAS; $r = 3$ m, $B = 3.8$ T, CMS) with silicon trackers, and hermetic EM and hadron calorimeters covering $|\eta| < 4.9$, with higher resolution better segmented EM calorimeters in the mid-rapidity region $|\eta| < 3$. The calorimeters in the two detectors differ in technique, with ATLAS using liquid argon EM and forward hadronic calorimeters and CMS using $PbWO_4$ EM calorimeters. The segmentation $\Delta\eta \times \Delta\phi$ is 0.017×0.017 for CMS and $\approx 0.025 \times 0.025$ for ATLAS but varies with the depth in the calorimeter. Both use scintillator-Fe or brass sampling hadron calorimeters. The notable difference between the detectors is the muon system, which consists of the return yoke in the CMS detector (Figure 14.2c) and a huge external superconducting toroid for ATLAS (hence the name Compact Muon Solenoid for CMS with 15 m outside diameter detector compared to 20 m for

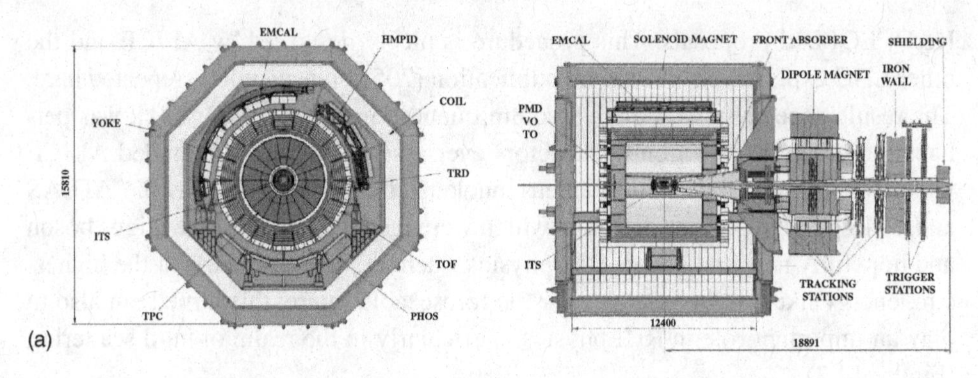

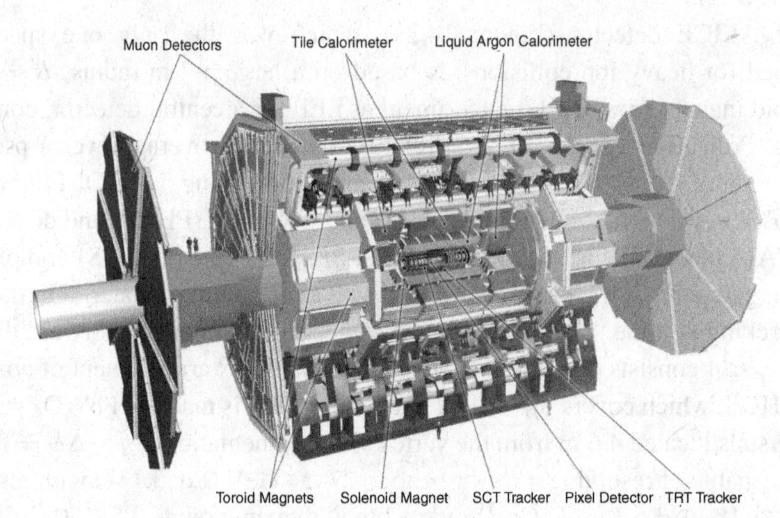

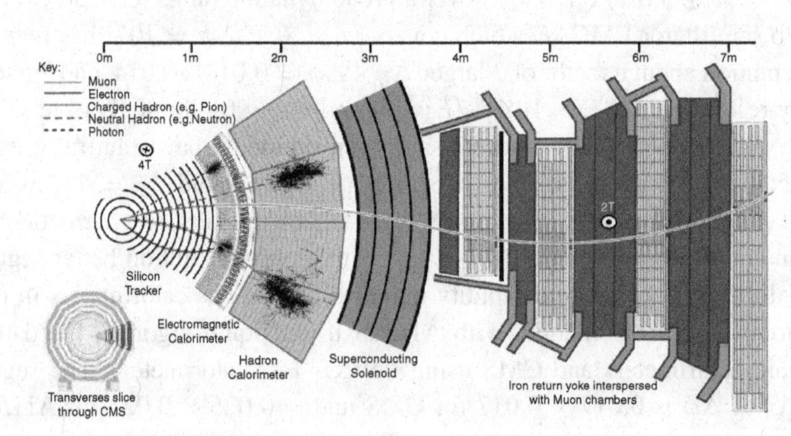

Figure 14.2 LHC detectors: (a) ALICE [705]; (b) ATLAS [705]; (c) CMS (slice view) [706]. For details see references.

ATLAS although the solenoid for ATLAS is actually smaller). For more details consult the references [705].

14.3 Hard scattering at RHIC

14.3.1 The RHIC machine and experiments

In the year 2000, RHIC [700], which is composed of two independent rings of superconducting magnets 3.8 km in circumference, began operation. RHIC is a versatile accelerator–collider which has collided beams of Au+Au, d+Au, Cu+Cu and polarized *p*–*p*, in runs from 2000–2011, at 12 different c.m. energies ranging from $\sqrt{s_{NN}} = 7 - 200$ GeV in $A + A$ collisions and up to $\sqrt{s} = 500$ GeV in polarized *p*–*p* collisions; U+U and Cu+Au collisions are scheduled for 2012. As noted previously, in addition to being the first heavy ion collider, RHIC is also the first polarized *p*–*p* collider supplying both longitudinally and transversely polarized protons by virtue of spin rotators (for both PHENIX and STAR), in addition to the "siberian snakes" which preserve the polarization (Figure 14.3).

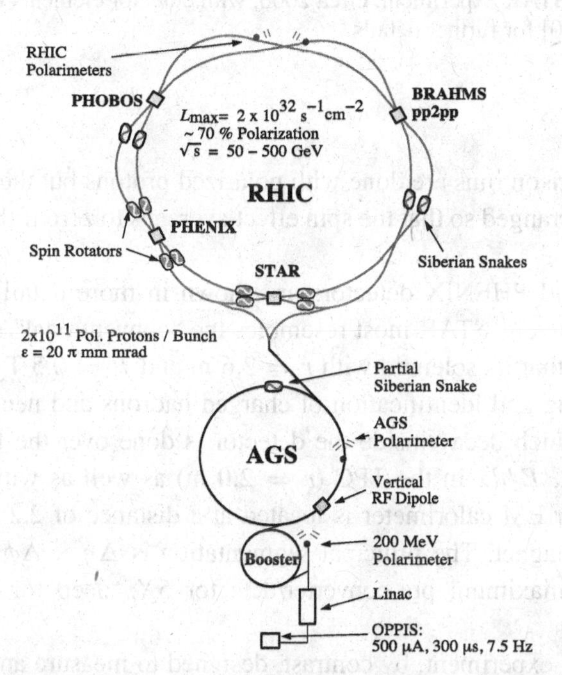

Figure 14.3 Schematic diagram of the RHIC machine [700] emphasizing the equipment for polarized *p*–*p* collisions. For RHI collisions the injection for the first 11 years has been from the Tandem Van de Graaff via the transfer line to the AGS. Starting with 2012, ions will be supplied by a new Electron Beam Ion Source (EBIS) located right next to the booster.

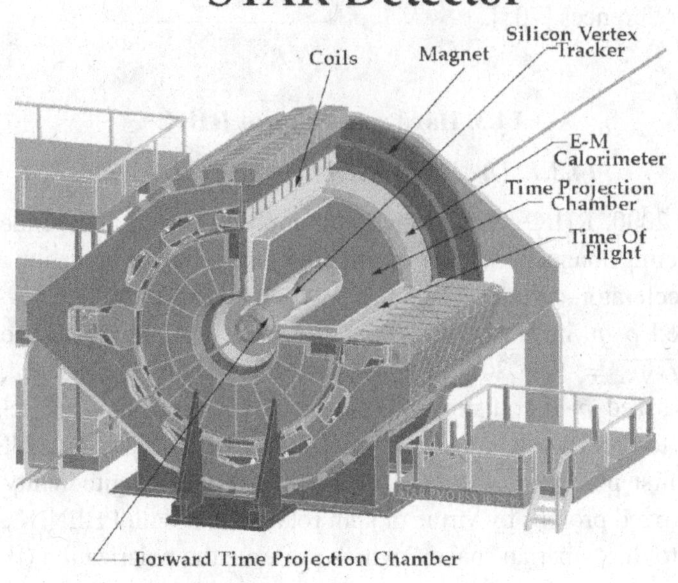

Figure 14.4 STAR experiment, circa 2000, with detector elements indicated. See reference [700] for further details.

All p–p comparison runs are done with polarized protons but the bunch-to-bunch spin pattern is arranged so that the spin effect averages to zero if the polarization is ignored.

The STAR and PHENIX detectors are shown in more detail in Figures 14.4 and 14.5 respectively. STAR most resembles the "conventional" solenoid collider detector except that its solenoid with $r = 2.6$ m and $B = 0.5$ T is not superconducting. Tracking and identification of charged hadrons and neutral Λ and other such baryons which decay inside the detector is done over the full azimuth and $|\eta| < 1.0$ with dE/dx in the TPC ($r = 2.0$ m) as well as with TOF counters. A Pb scintillator EM calorimeter is located at a distance of 2.2 m from the vertex inside the magnet. The principal segmentation is $\Delta\eta \times \Delta\phi = 0.05 \times 0.05$ with a shower-maximum pre-converter/detector $5X_o$ deep for improved π^0/γ separation.

The PHENIX experiment, by contrast, designed to measure and trigger on rare processes involving leptons, photons and identified hadrons at the highest luminosities, is quite non-conventional, with a very high granularity, highly selective, special purpose detector covering a smaller solid angle at mid-rapidity $|\Delta\eta| < 0.35$, $\Delta\phi = 2\times\pi/2$ rad together with two muon detectors at forward and backward

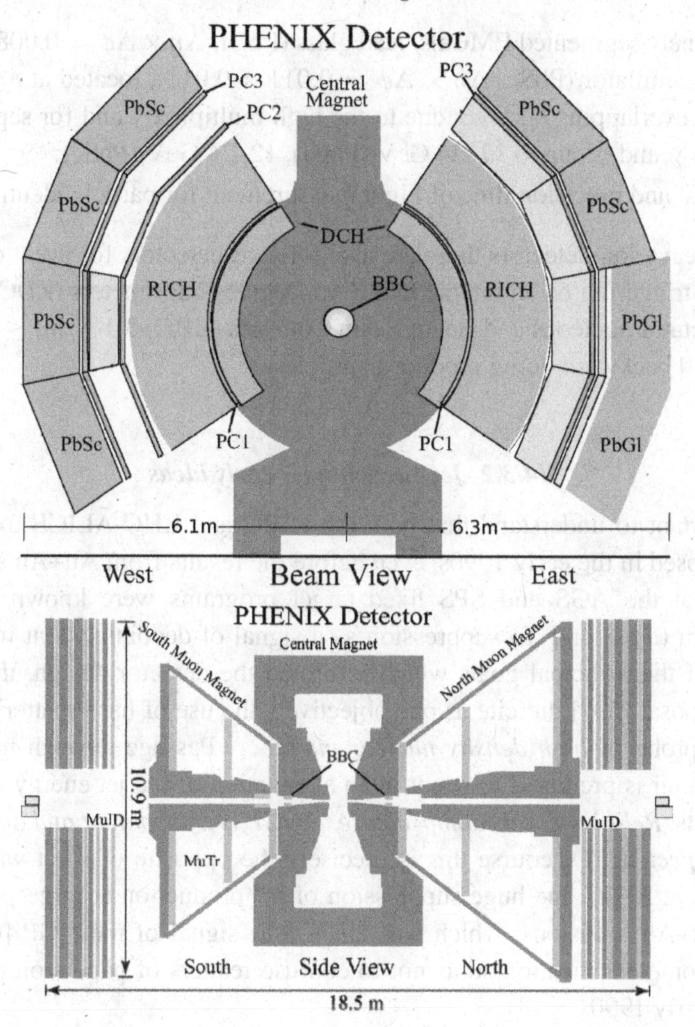

Figure 14.5 PHENIX experiment, circa 2006, with detector elements mentioned in the text. See reference [700] for further details.

rapidity $1.1 < |\Delta\eta| < 2.2$–2.4 and $\Delta\phi = 2\pi$ rad. Special features of the central detector include:

(i) a minimum of material (0.4% X_o) in the aperture to avoid photon conversions;

(ii) possibility of zero magnetic field on axis to prevent de-correlation of e^+e^- pairs from photon conversions;

(iii) ElectroMagnetic Calorimeter (EMCal) and Ring Imaging Cherenkov Counter (RICH) for $e^{\pm}$ identification and level-1 $e^{\pm}$ trigger;

(iv) two finely segmented EMCals, lead glass (PbGl, $\Delta\eta \times \Delta\phi = 0.008 \times 0.008$), lead scintillator (PbSc, $\Delta\eta \times \Delta\phi = 0.011 \times 0.011$), located at $r = 5$ m, to avoid overlapping showers due to the high multiplicity and for separation of single γ and π^0 up to 17–34 GeV (PbGl), 12–24 GeV (PbSc);
(v) EMCal and precision time of flight measurement for particle identification.

As in all heavy ion detectors there are also forward detectors for event characterization and triggering on an interaction: Zero Degree Calorimeters (ZDC) $|\eta| \gtrsim 6$ to detect spectator nucleons and Beam Beam Counters (BBC) $3.1 \leq |\eta| \leq 3.9$ for far forward and backward going produced particles.

14.3.2 *Jet quenching – early ideas*

It is important to understand that both the RHIC and LHC-ALICE experiments were proposed in the early 1990s, even before the results from Au+Au and Pb+Pb collisions at the AGS and SPS fixed target programs were known. Although quarkonium (J/Ψ and Υ) suppression as a signal of deconfinement in the QGP was one of the principal goals which informed the detector design, the original STAR proposal [697] did cite as one objective: "the use of hard scattering of partons as a probe of *high density nuclear matter*... Passage through hadronic or nuclear matter is predicted to result in an attenuation of the jet energy and broadening of jets. *Relative to this damped case, a QGP is transparent and an enhanced yield is expected.*" Of course this is precisely the opposite of what was actually discovered at RHIC, the huge suppression of π^0 production at large p_T [598] in central Au+Au collisions, which was taken as a signal of the QGP [650]. This deserves some clarification so as not to confuse readers of papers on the subject from the early 1990s.

It was Bjorken, in 1982 [708], who first suggested that high energy quarks and gluons propagating through a QGP suffer "differential energy loss via elastic scattering from quanta in the plasma." This is similar to ionization loss in QED (Eqs. 3.4, 3.5). Miklos Gyulassy and collaborators [709, 710], in 1990, then calculated the collisional energy loss, dE/dx, of a quark in an ideal QGP of temperature, T, and found it to be quite small, $\approx$0.1 GeV fm^{-1}. Also they found that radiative energy loss was strongly suppressed by the LPM effect [709, 711]. For dense nuclear matter, based on the μ–A DIS measurements (for $\nu \lesssim 10$ GeV) discussed in Section 12.5, and theoretical arguments, they estimated $dE/dx \approx 1$ GeV fm^{-1}, which would thus reduce by a factor of 10 if the matter became deconfined. This was the source of the statement in the STAR proposal.

However, this type of "jet quenching" was greeted with disbelief by the old-time ISR physicists who had gathered a few months later in Strasbourg to discuss

"Quark-Gluon Plasma Signatures" [712], who were familiar with the Cronin effect, more like jet enhancement in nuclear matter (e.g. recall Figure 14.1). This was reinforced by the general understanding of quark and hadron interactions in nuclei (as discussed in Section 13.2.2) that a struck nucleon or parton is relatively unaffected by being struck again except for possibly picking up a little momentum transverse to its trajectory [713]. Also there was a general consensus that leading particles were the only way to detect jets at RHIC because in central Au+Au collisions there would be an estimated $\pi \Delta R^2 \times \frac{1}{2\pi}\frac{dE_T}{d\eta} \sim 375$ GeV in one unit of a nominal jet cone $\Delta R = \sqrt{(\Delta\eta)^2 + (\Delta\phi)^2}$, nearly four times the maximum possible jet energy.

Well, we now know (and shall show below) that the consensus was correct for RHIC, where there has been only one published jet cross section measurement [714] in the first 11 years of operation (and it was in p–p not $A + A$ collisions); while all the hard scattering measurements were done using single particle inclusive and two particle correlation measurements, as pioneered at the ISR where hard scattering was discovered. On the theoretical side, great progress was made throughout the 1990s so that the propagation of energetic partons in hot and cold QCD matter was given a firm basis in pQCD [715, 716]. The pairs of high p_T outgoing partons produced by hard scattering, with their color charge fully exposed, were predicted to lose energy by coherent (LPM) gluon Bremsstrahlung if the medium being traversed also has a large density of similarly exposed color charges (i.e. a QGP). To quote the authors [716], "Numerical estimates of the loss suggest that it may be significantly greater in hot matter than in cold. This makes the magnitude of the radiative energy loss a remarkable signal for quark-gluon plasma formation."

LPM radiation in QED [135] and QCD are quite different because a radiated gluon can again radiate since it is also color charged. Also the gluon radiation is coherent, with the many scattering centers inside the formation length acting as a single center [716]. One striking prediction resulting from the coherence is that the energy loss, dE/dx, should be proportional to the formation length L so that the total energy loss should be proportional to L^2. Another prediction in this vein is the reduced radiative energy loss by heavy quarks (as in QED), the "Dead Cone Effect" [717], which results [717] "in much smaller heavy-quark quenching."

This new idea of jet quenching, just as RHIC was beginning to operate, set the stage for hard scattering being the principal probe of the QGP. Fortunately both PHENIX and STAR had been designed with an eye towards making high p_T measurements which were an important part of the RHIC spin program. The principal hard scattering reactions measured in p–p collisions at RHIC are shown in Figure 14.6, namely π^0 production [718], jet production [714], direct single $e^{\pm}$ production from the decay of heavy quarks [719] and direct-γ production [153],

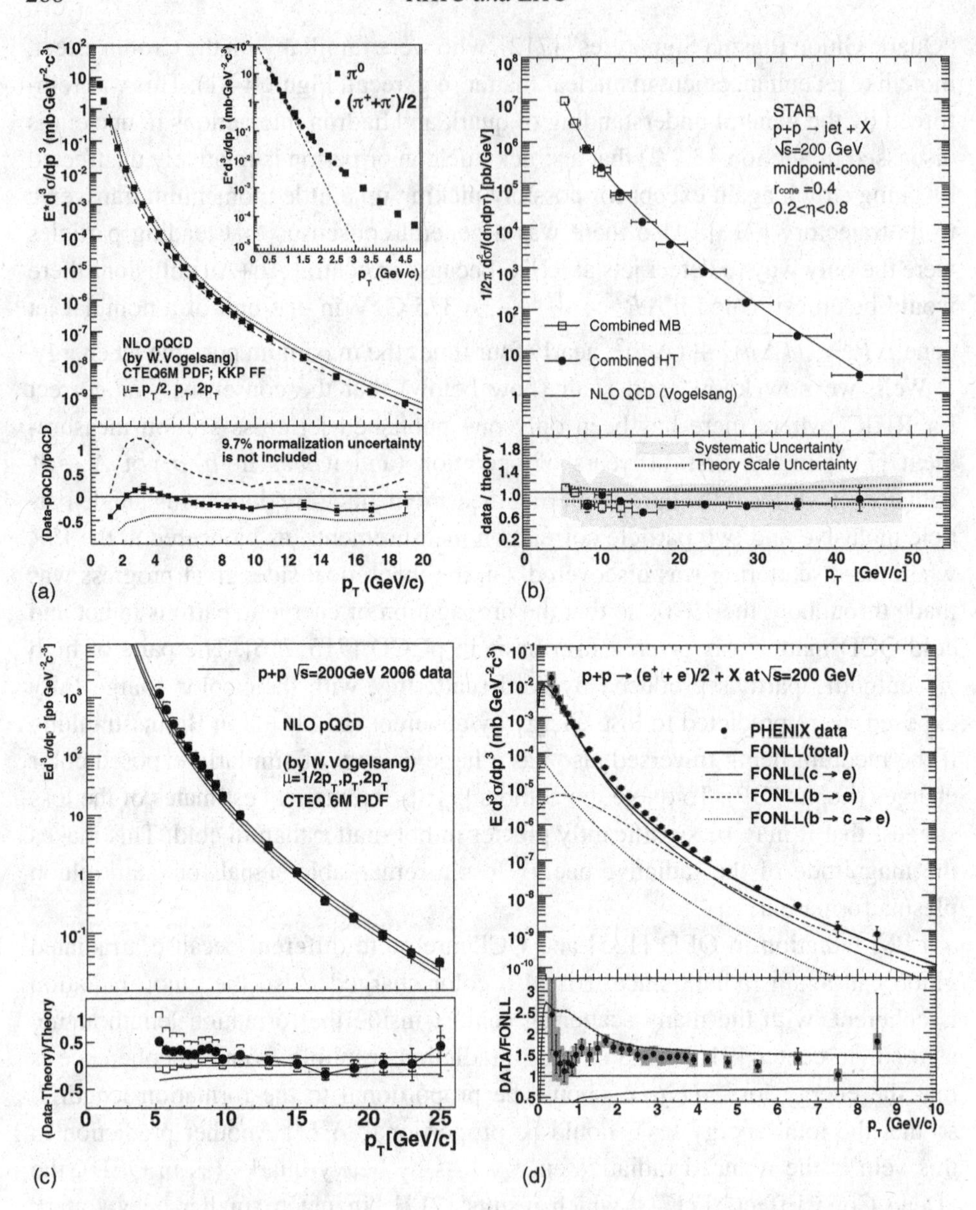

Figure 14.6 Measurements of the single inclusive invariant cross section $Ed^3\sigma/dp^3$ as a function of p_T in p–p collisions at $\sqrt{s} = 200$ GeV for production of (a) π^0 [718], (b) jet [714], (c) direct γ [605], (d) direct single $e^{\pm}$ [720]. On each plot pQCD predictions are shown, while a bottom panel for each reaction shows the ratio of the data to the theory (or 1 minus the ratio).

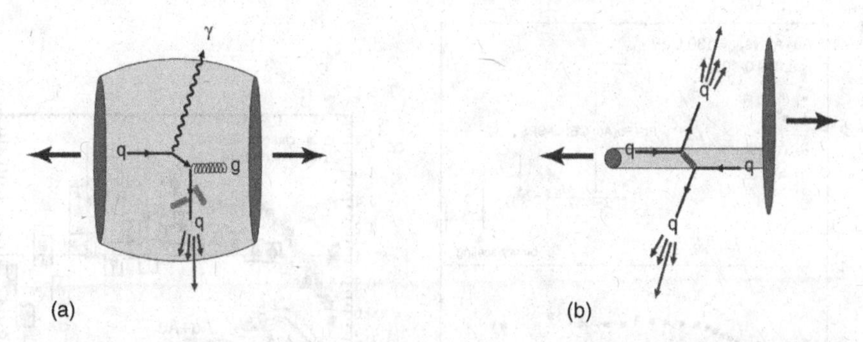

Figure 14.7 (a) Two outgoing nuclei indicated by the dark gray disks after a collision which produces a medium (light gray) in which an outgoing parton pair from an initial hard scattering may interact. (b) A $p+A$ or $d+A$ collision in which the medium is limited (one nucleon wide) or non-existent, so that any interaction of outgoing partons is minimal.

which are all in agreement with pQCD as shown. As of the end of the year 2011, all of these except for the jet measurement had been used as p–p comparison data in publications of the same measurements performed in $A + A$ collisions.

14.3.3 Hard scattered partons as probes in RHI collisions

At RHIC, in heavy ion collisions, hard scattering of partons from the initial collision turned out to be the principal in situ probe of the medium produced. As sketched in Figure 14.7a, in a hard collision, e.g. $g + q \rightarrow \gamma + q$, the outgoing γ which does not interact with the medium since it is both color and electrically neutral emerges directly, without energy loss, while the color charged quark, which initially balanced transverse momentum with the γ, may lose energy in the medium. This can be measured by comparing the p_T of the jet from the quark fragmentation to that of the γ ray. (Detailed discussions will be given later.) To make sure that any effect measured in an $A + A$ collision is from the medium and not an effect of cold nuclear matter such as shadowing of the parton distribution function (recall Figure 12.7), the baseline for any cold nuclear matter effect is provided by $p + A$ or $d + A$ collisions (Fig. 14.7b) in which no (or a very limited) medium is produced.

14.3.4 The major discovery at RHIC – jet quenching

The discovery at RHIC [598] that π^0 produced at large $p_T > 3$ GeV/c are suppressed by roughly a factor of 5 compared to point-like scaling from p–p collisions is arguably the major discovery in relativistic heavy ion physics. The suppression

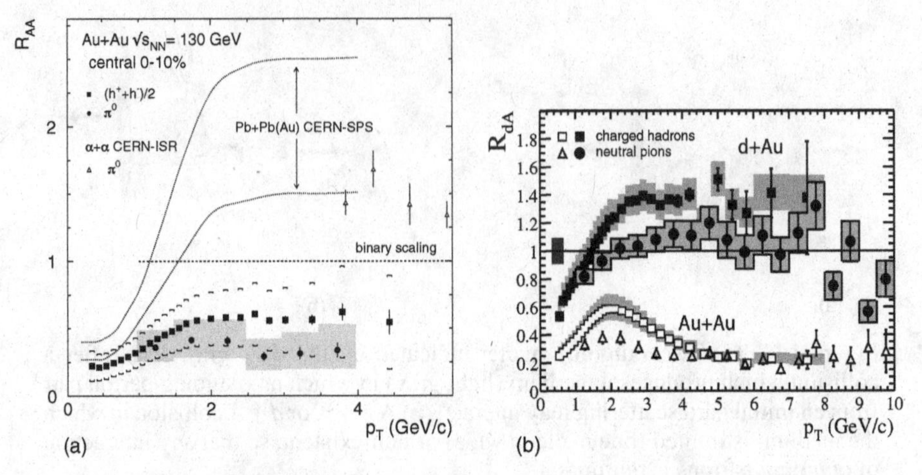

Figure 14.8 (a) PHENIX [598] measurement R_{AA} for charged hadrons $h^{\pm}$ (squares) and π^0 (dots) in Au+Au central (0–10%) collisions at $\sqrt{s_{NN}} = 130$ GeV. Systematic uncertainties are shown by bands. Also shown are R_{AA} of the minimum bias inclusive cross section for α–α at $\sqrt{s_{NN}} = 31$ GeV from Figure 14.1b [693] (triangles) as well as an estimate from central Pb+Pb and Pb+Au collisions at $\sqrt{s_{NN}} = 17$ GeV [721] shown as a band of uncertainty. (b) R_{dAu} for π^0 and $h^{\pm}$ in d+Au minimum bias collisions at $\sqrt{s_{NN}} = 200$ GeV and R_{AA} for central (upper 10%-ile) Au+Au collisions at $\sqrt{s_{NN}} = 200$ GeV [722].

is represented (Figure 14.8a) by the nuclear modification factor, R_{AA}, which is the ratio of the measured semi-inclusive yield of for example π^0 in a given centrality class f in $A + A$ collisions (recall Eq. 13.5) divided by $\langle T_{AA}\rangle_f$ times the *cross section* for the high p_T particles in p–p collisions:

$$R_{AA} = \frac{d^2N^{\pi}_{AA}(p_T)|_f / p_T dp_T dy\, N^{evt}_{AA}|_f}{\langle T_{AB}\rangle\,|_f \times d^2\sigma^{\pi}_{pp}/p_T dp_T dy}. \tag{14.1}$$

For pure point-like hard scattering, labeled binary scaling on Figure 14.8a, $R_{AA} = 1$. For the Cronin effect, as observed in α–α collisions at the CERN-ISR (Figure 14.1) [693], and in $p + A$ collisions [257], $R_{AA} > 1$. This is why suppression in central Au+Au collisions is so exciting, it is exactly the opposite of the effect seen in cold nuclear matter at lower $\sqrt{s_{NN}}$.

In order to verify that the suppression at RHIC was due to the medium produced in Au+Au collisions and not an effect in the cold matter of an individual nucleus, a measurement in d+Au collisions was performed where no suppression was observed as verified by all four experiments. PHENIX results [722] are shown in Figure 14.8b: in d+Au, there is a Cronin effect, $R_{dAu} > 1$, for charged hadrons, while $R_{dAu} \approx 1$ for the π^0. It is important to note that for Au+Au collisions

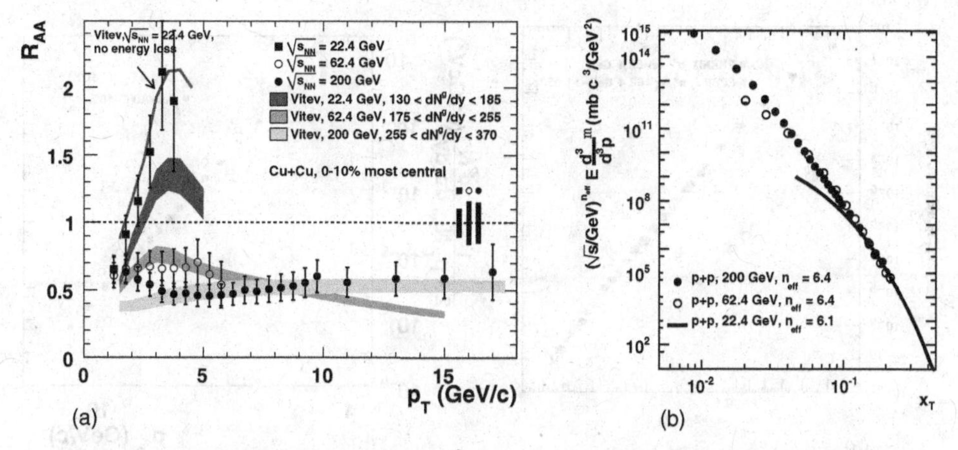

Figure 14.9 (a) PHENIX R_{AA} for π^0 in central (0–10%) Cu+Cu collisions at $\sqrt{s_{NN}} = 200$, 62.4 and 22.4 GeV [723]. For details on the model curves, consult reference [723]. (b) x_T scaling with $n_{eff} = 6.1$–6.4 for the measured π^0 cross sections in p–p collisions at $\sqrt{s} = 62.4$ and 200 GeV, together with the $\sqrt{s} = 22.4$ GeV comparison spectrum obtained with a QCD based fit [723, 724].

(Figure 14.8) the R_{AA} of non-identified charged hadrons $h^{\pm}$ and π^0 are also considerably different for $p_T < 5$ GeV/c. This shows that one cannot treat unidentified $h^{\pm}$ as if they were $\pi^{\pm}$ (see further discussion below).

In order to determine the c.m. energy where suppression overtakes the Cronin effect, measurements were made in Cu+Cu collisions at several values of $\sqrt{s_{NN}}$. Figure 14.9a shows that R_{AA} for central (0–10%) Cu+Cu collisions is comparable at $\sqrt{s_{NN}} = 62.4$ and 200 GeV, but that there is no suppression, actually a Cronin enhancement [257], at $\sqrt{s_{NN}} = 22.4$ GeV. This indicates that the medium which suppresses jets is produced somewhere between $\sqrt{s_{NN}} = 22.4$ GeV, the SPS fixed target highest c.m. energy, and 62.4 GeV. It is important to note that PHENIX did not measure the reference p–p spectrum at $\sqrt{s} = 22.4$ GeV but used a QCD based fit [724] to the world's data on charged and neutral pions which was checked against PHENIX p–p measurements at 62.4 [725] and 200 GeV using x_T scaling (Figure 14.9b) [723].

A key issue in this fit is that the data at $\sqrt{s} = 22.4$ GeV were consistent with each other and with pQCD [724] except for one outlier which was excluded based on the experience from a previous global fit [726] to the world's data at $\sqrt{s} = 62.4$ GeV where there are large disagreements (recall Figure 6.14) and for which there was no basis for eliminating outliers or adjusting the absolute p_T scales. When PHENIX measured the p–p reference spectrum at 62.4 GeV [725], it agreed with the CCOR [269] measurements shown in Figure 6.14a to within the systematic error of the absolute p_T scales (see Figure 14.10a), but disagreed significantly with

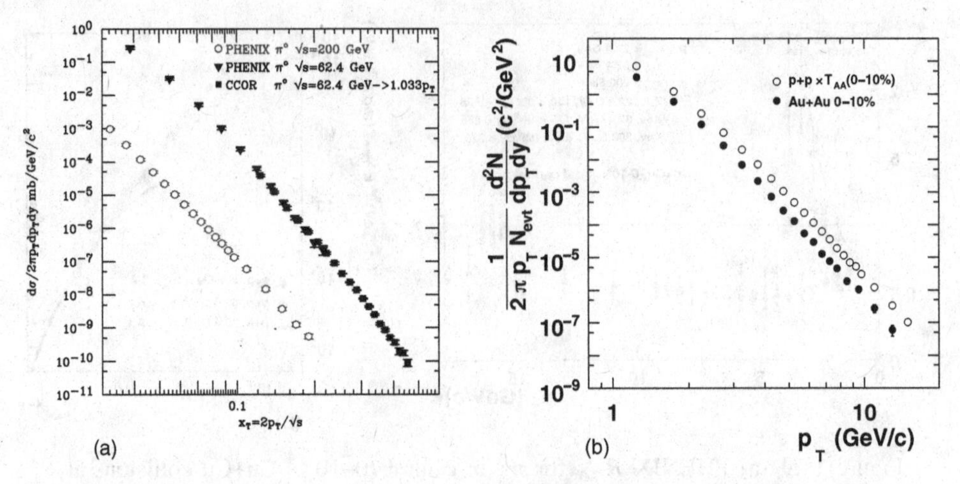

Figure 14.10 (a) $Ed^3\sigma/dp^3$ versus x_T for PHENIX mid-rapidity π^0 at $\sqrt{s} =$ 200 GeV in p–p collisions [718] plus PHENIX [725] and CCOR-ISR [269] measurements at $\sqrt{s} = 62.4$ GeV, where the absolute p_T scale of the ISR measurement has been corrected upwards by 3% to agree with the PHENIX data. (b) π^0 p–p data versus p_T at $\sqrt{s} = 200$ GeV from (a) multiplied by $\langle T_{AA}\rangle$ for Au+Au central (0–10%) collisions compared to semi-inclusive π^0 invariant yield in Au+Au central (0–10%) collisions at $\sqrt{s_{NN}} = 200$ GeV [727].

the global fit at 62.4 GeV [726]. This, once again, indicates the importance of measurements at the same value of $\sqrt{s_{NN}}$ of p–p and $p+A$ (or $d+A$) comparison spectra for $A+A$ measurements, all in the same experiment.

14.3.5 Parton absorption or energy loss?

From the measurement of R_{AA} alone, it is impossible to know whether partons have lost energy, which would shift the spectrum to lower p_T, thus causing suppression because of the steeply falling spectrum, or whether a certain fraction $1 - R_{AA}$ of partons were absorbed in the medium while the others exited relatively unscathed, i.e. "punched through" the medium with minimal effect. In order to understand what is actually happening in the medium, it is important to look at the individual p_T distributions which go into the calculation of R_{AA}. The invariant cross sections for π^0 production in p–p collisions at $\sqrt{s} = 200$ [718] and 62.4 GeV [725] at RHIC are shown in Figure 14.10a in comparison to ISR measurements [269] at $\sqrt{s} = 62.4$ GeV. The data are shown on a log–log plot to emphasize the almost pure power-law dependence $Ed^3\sigma/dp^3 \propto 1/p_T^n$ of the invariant cross section over large regions of x_T where, as shown above (Figure 12.2), the power n increases at larger x_T. The difference between the predominant $n = 8.1$ at $\sqrt{s} = 200$ GeV and $n \approx 10$ at $\sqrt{s} = 62.4$ GeV is apparent on the plot.

In Figure 14.10b, the 200 GeV p–p data, multiplied by the point-like scaling factor $\langle T_{AA} \rangle$ for (0–10%) central Au+Au collisions are compared to the semi-inclusive invariant π^0 yield in central (0–10%) Au+Au collisions at $\sqrt{s_{NN}} = 200$ GeV and, amazingly, the Au+Au data follow the same power-law as the p–p data but are suppressed from the point-like scaled p–p data by a factor of ~5, independent of p_T. This explains the striking feature shown in Figure 14.8b that for π^0 at $\sqrt{s_{NN}} = 200$ GeV the nuclear modification factor $R_{AA} \approx 0.2$ is remarkably constant for $4 \leq p_T \leq 10$ GeV/c. However, there is another way to understand these two curves: the remarkably parallel (on a log–log plot) T_{AA} scaled p–p and Au+Au invariant p_T spectra at $\sqrt{s_{NN}} = 200$ GeV can also be interpreted to show that the Au+Au spectrum is shifted down in p_T from the point-like scaled p–p spectrum by a fraction, $\Delta p_T / p_T \equiv S_{loss}$ (as defined by PHENIX [727]), which appears to be constant, independent of p_T. In other words, a π^0 that would have been produced with a transverse momentum p'_T in the reference p–p spectrum is detected with $p_T = p'_T(1 - S_{loss})$ due to having lost a fraction S_{loss} of its transverse momentum in the medium, which is independent of the initial p'_T. This would not be surprising if the radiative energy loss in QCD were similar to that in QED in that the fractional energy loss $\Delta E/E$ of a parton with energy E just depended on the effective "radiation length" of the medium and was independent of the parton energy E.

It is important to realize that the effective fractional energy loss, S_{loss}, estimated from the shift in the p_T spectrum is actually less than the real average energy loss at a given p_T, i.e. the observed particles have p_T closer to the original value than to the average. This effect is similar to that of "trigger bias" (Section 9.1.1) where, due to the steeply falling p_T spectrum, the $\langle z \rangle$ of detected single inclusive particles is much larger than the $\langle z \rangle$ of unbiased jet fragmentation. Similarly, for a given observed p_T, the events at larger p'_T with larger energy loss are lost under the events with smaller p'_T (with a much larger cross section) with smaller energy loss. The fractional shift of the spectra, S_{loss}, is easily related to the ratio R_{AA} at the same p_T [727]:

$$R_{AA}(p_T) = \frac{d\sigma/dp_T}{d\sigma/dp'_T\big|_{p'_T=p_T}} = \frac{(1 - S_{loss})^{n-1}}{(1 - d(p_T S_{loss})/dp_T)}, \tag{14.2}$$

where the power is $n-1$ because the shift is in the $d\sigma/dp_T$ spectrum. For a constant fractional energy loss, Eq. 14.2 reduces to:

$$R_{AA}(p_T) = (1 - S_{loss})^{n-2}, \tag{14.3}$$

i.e. a constant fractional energy shift S_{loss} gives a constant R_{AA} as apparently observed.

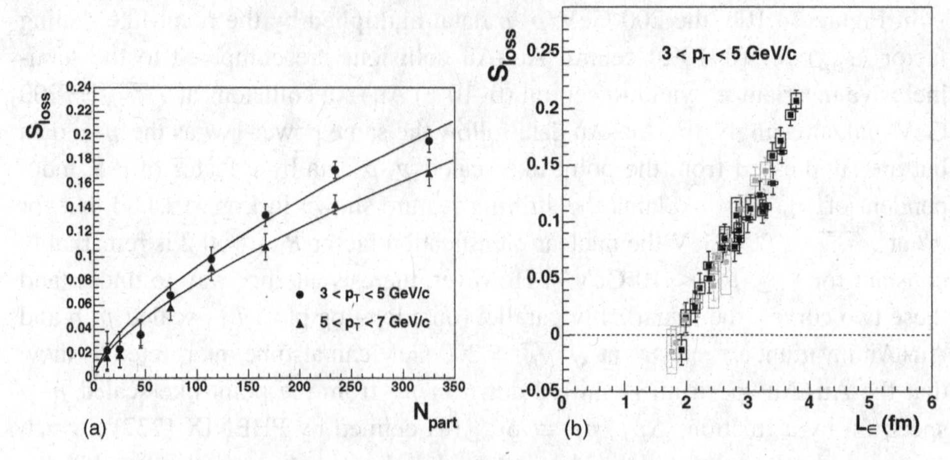

Figure 14.11 (a) Effective fractional energy loss S_{loss} versus centrality (N_{part}) [727]. Lines are fits of the form $N_{part}^{2/3}$ for each p_T range. (b) S_{loss} plotted versus L_ε (see text), where each point represents a centrality–$\Delta\phi$ combination from all six centrality classes, with six $\Delta\phi$ bins in each class [727].

The measured fractional energy shift S_{loss} for π^0 in Au+Au collisions at $\sqrt{s} =$ 200 GeV [727] as a function of centrality (N_{part}) is shown in Figure 14.11a for two different p_T ranges, $3 < p_T < 5$ GeV/c and $5 < p_T < 7$ GeV/c. There appears to be a small decrease in S_{loss} with increasing p_T, but the main observation from Figure 14.11a is that S_{loss} increases approximately like $N_{part}^{2/3}$, roughly like the average radius of the overlap almond squared. Is this evidence of coherence [716]?

To attempt a better measurement of the effective fractional energy loss as a function of the distance traversed and to separate the effect of the density of the medium and path length traversed, PHENIX studied the dependence of the π^0 R_{AA} on the angle $\Delta\phi$ with respect to the reaction plane (recall Figure 13.4a) and centrality. For a given centrality, variation of $\Delta\phi$ gives a variation of the path length traversed for fixed initial conditions, while varying the centrality allows the density of the medium to vary. Six angles $\Delta\phi$ and six centralities (10–20%, 20–30%,…, 60–70%) were measured for two ranges of p_T, 3–5 GeV/c and 5–8 GeV/c. Many ways were tried to characterize the path length weighted by the density, but one of the most informative plots (Figure 14.11b) was made with the most naive variable, L_ε, the distance from the edge to the center of the elliptical overlap zone of the Au+Au collision, to represent the average path length of a parton in the medium. This assumes a static medium and ignores the change in density with centrality. Figure 14.11b appears to show that: (i) S_{loss} is universal and is a linear function of L_ε for all centrality classes and both p_T ranges; (ii) the fractional energy loss S_{loss} appears to go to zero for $L_\varepsilon \leq 2$ fm for both p_T ranges, suggesting a formation time effect, which has generally not been taken into account in parton energy loss models (see, however, reference [728]).

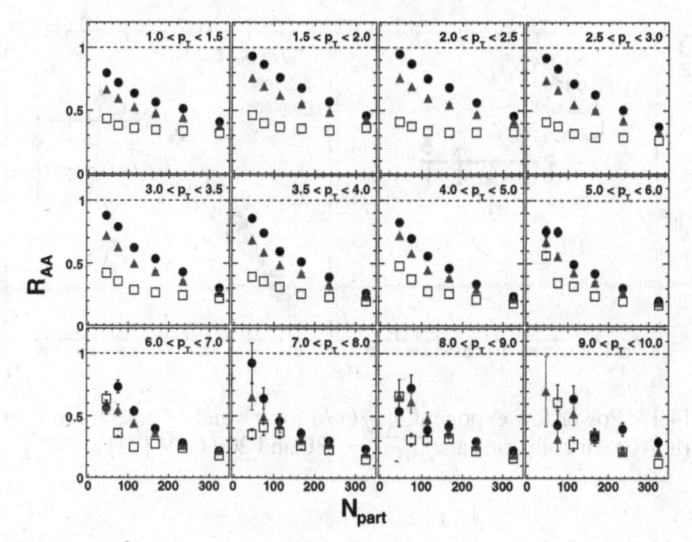

Figure 14.12 $R_{AA}^{\pi^0}$ in reaction plane bins as a function of p_T and centrality (N_{part}) in Au+Au collisions at $\sqrt{s_{NN}} = 200$ GeV [729]. Filled circles represent $R_{AA}(0 < \Delta\phi < 15°)$ (in-plane), open squares $R_{AA}(75 < \Delta\phi < 90°)$ (out-of-plane) and filled triangles $R_{AA}(30 < \Delta\phi < 45°)$.

In a later measurement which further demonstrated the sensitivity of R_{AA} to the path length traversed in the medium without resort to an ad hoc path length measure, PHENIX [729] observed a striking difference in the behavior of the dependence of the in-plane $R_{AA}(\Delta\phi \sim 0, p_T)$ for π^0 as a function of centrality compared to the dependence of the $R_{AA}(\Delta\phi \sim \pi/2 \text{ rad}, p_T)$ in the direction perpendicular to the reaction plane (Figure 14.12). The R_{AA} perpendicular to the reaction plane is relatively constant with centrality, while the variation of R_{AA} in the direction parallel to the reaction plane shows a strong centrality dependence. This is a clear demonstration of the sensitivity of R_{AA} to the length traversed in the medium, which is relatively constant as a function of centrality perpendicular to the reaction plane but depends strongly on the centrality parallel to the reaction plane. In principle, this method could be used to extract the true length and density dependence of energy loss in the medium [730] but this has not yet been accomplished.

14.3.6 The importance of particle identification

14.3.6.1 The baryon anomaly and x_T scaling

Many high energy and RHI physicists tend to treat non-identified charged hadrons $h^{\pm}$ as if they were $\pi^{\pm}$. While this may be a reasonable assumption in p–p collisions, it is clear from Figure 14.8 that the suppression of non-identified charged hadrons and π^0 is very different for $p_T < 5$ GeV/c.

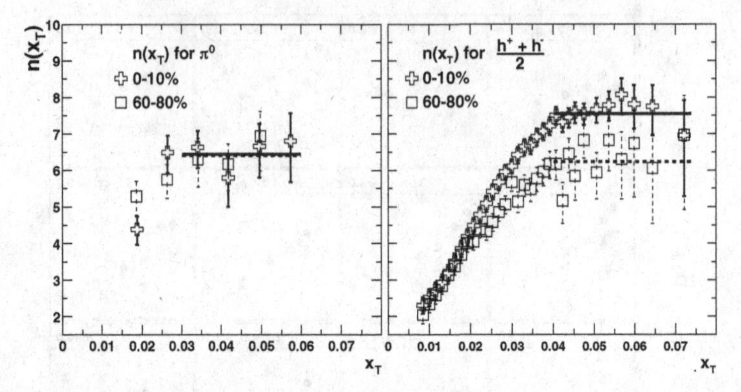

Figure 14.13 Power-law exponent $n_{eff}(x_T)$ for π^0 and $h^{\pm}$ spectra in central and peripheral Au+Au collisions at $\sqrt{s_{NN}} = 130$ and 200 GeV [731].

If the production of high p_T particles in Au+Au collisions were the result of hard scattering according to pQCD, then x_T scaling should work just as well in Au+Au collisions as in p–p collisions and should yield the same value of the exponent $n_{eff}(x_T, \sqrt{s})$. The only assumption required is that the structure and fragmentation functions in Au+Au collisions should scale, in which case Eq. 6.13 still applies, albeit with a $G(x_T)$ appropriate for Au+Au. In Figure 14.13, $n_{eff}(x_T, \sqrt{s_{NN}})$ in Au+Au is shown for π^0 and $h^{\pm}$ in peripheral and central collisions, derived by taking the ratio of $Ed^3\sigma/dp^3$ at a given x_T for $\sqrt{s_{NN}} = 130$ and 200 GeV, in each case [731].

The π^0 exhibit x_T scaling, with the same value of $n_{eff} = 6.3$ as in p–p collisions, for both Au+Au peripheral and central collisions. The x_T scaling establishes that high p_T π^0 production in peripheral and central Au+Au collisions follows pQCD as in p–p collisions, with parton distributions and fragmentation functions that scale with x_T, at least within the experimental sensitivity of the data. The fact that the fragmentation functions scale for π^0 in Au+Au central collisions indicates that the effective energy loss must scale, i.e. $\Delta E(p_T)/p_T$ is a constant, which is consistent with the constant value of $R_{AA}(p_T)$ for $p_T > 4$ GeV/c (Figure 14.8b) as discussed above (Eq. 14.3).

The deviation of $h^{\pm}$ in Figure 14.13b from x_T scaling in central Au+Au collisions is indicative of and consistent with the strong non-scaling modification of particle composition of identified hadrons observed in Au+Au collisions as a function of centrality in the range $2.0 \leq p_T \leq 4.5$ GeV/c, where particle production in p–p collisions is the result of jet fragmentation. This is called the baryon anomaly. As shown in Fig. 14.14a the p/π^+ and $\bar{p}/\pi^-$ ratios as a function of p_T increased dramatically to values ~1 as a function of centrality in Au+Au collisions at RHIC [732]. This is nearly an order of magnitude larger than had ever been

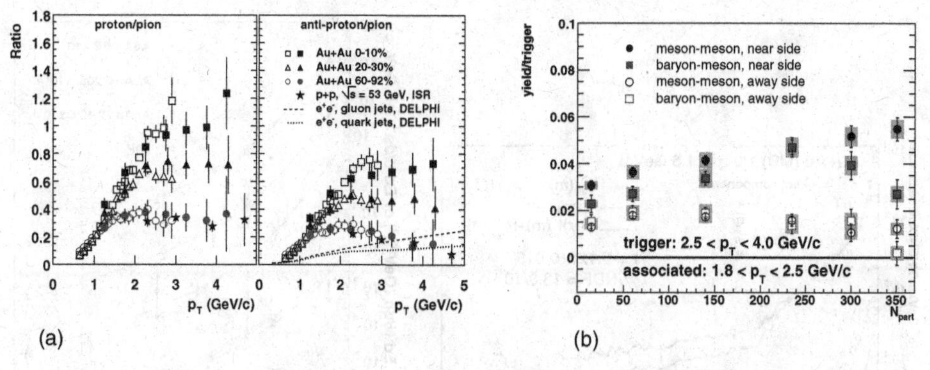

Figure 14.14 (a) p/π^+ and $\bar{p}/\pi^-$ as a function of p_T and centrality from Au+Au collisions at $\sqrt{s_{NN}} = 200$ GeV [732] compared to other data indicated. (b) Conditional yields, per trigger meson (circles), baryon (squares) with $2.5 < p_T < 4$ GeV/c, of associated mesons with $1.7 < p_T < 2.5$ GeV/c integrated within $\Delta\phi = \pm 0.94$ radian of the trigger (near side, full) or opposite azimuthal angle (open), for Au+Au collisions at $\sqrt{s_{NN}} = 200$ GeV [734].

seen previously either in fragmentation of jets in e^+e^- collisions or in the average particle composition of the bulk matter in Au+Au central collisions [733].

This "baryon anomaly" was beautifully explained as due to the coalescence of an exponential (thermal) distribution of constituent quarks (also known as the QGP) [735, 736]. Unfortunately, measurements of correlations of $h^\pm$ in the range $1.7 \le p_{T_a} \le 2.5$ GeV/c associated with identified meson or baryon triggers with $2.5 \le p_{T_t} \le 4.0$ GeV/c showed the same near-side and away-side peaks and yields (Figure 14.14b) characteristic of di-jet production from hard scattering [734, 737], which would be absent in soft coalescence, apparently ruling out this beautiful model.

14.3.6.2 The first measurement anywhere of direct-γ at low p_T

Internal conversion of a photon from π^0 and η decay is well known and is called Dalitz decay [287–289]. Perhaps less well known in the RHI community is the fact that for any reaction (e.g. $q + g \to \gamma + q$) in which a real photon can be emitted, a virtual photon (e.g. e^+e^- pair of mass $m_{ee} \ge 2m_e$) can also be emitted. This is called internal conversion and is generally given by the Kroll–Wada formula [738, 739]:

$$\frac{1}{N_\gamma}\frac{dN_{ee}}{dm_{ee}} = \frac{2\alpha}{3\pi}\frac{1}{m_{ee}}\left(1 - \frac{m_{ee}^2}{M^2}\right)^3 \times$$

$$|F(m_{ee}^2)|^2\sqrt{1 - \frac{4m_e^2}{m_{ee}^2}}\left(1 + \frac{2m_e^2}{m_{ee}^2}\right), \qquad (14.4)$$

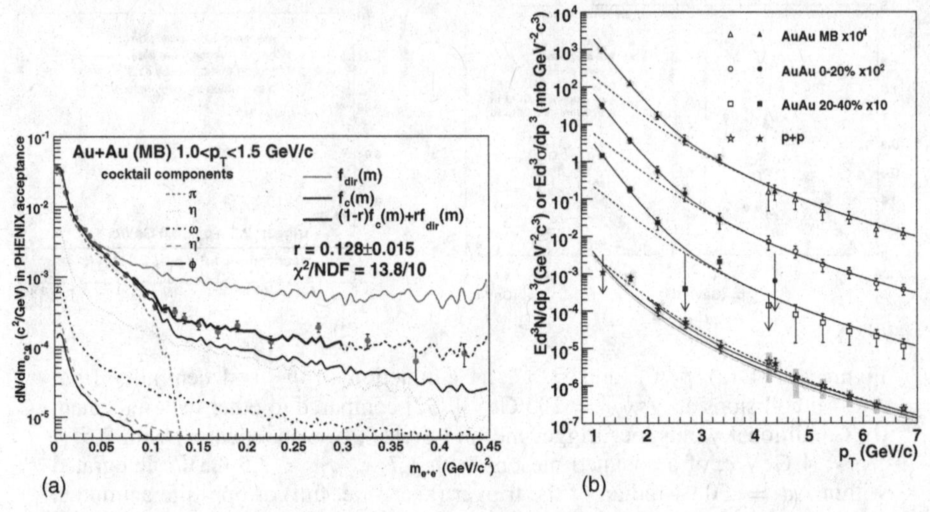

Figure 14.15 (a) Invariant mass ($m_{e^+e^-}$) distribution of e^+e^- pairs from Au+Au minimum bias events for $1.0 < p_T < 1.5$ GeV/c [739]. Dashed lines are Eq. 14.4 for the mesons indicated. The lowest solid line is $f_c(m)$, the total di-electron yield from the sum of contributions or "cocktail" of meson Dalitz decays; the highest solid line is $f_{dir}(m)$ the internal conversion $m_{e^+e^-}$ spectrum from a direct photon ($M \gg m_{e^+e^-}$). The middle solid line is a fit of the data to the sum of cocktail plus direct contributions in the range $80 < m_{e^+e^-} < 300$ MeV/c^2. (b) Invariant cross section (p–p) or invariant yield (Au+Au) of direct photons as a function of p_T [739]. Filled points are from virtual photons, open points from real photons. Lines on p–p data are described in the text. The dashed line on the Au+Au data is the fit to the p–p data (dashed line) multiplied by $\langle T_{AA} \rangle$. The solid line on the Au+Au data is the sum of a fitted exponential added to the $\langle T_{AA} \rangle$ scaled p–p fit.

where M is the mass of the decaying meson or the effective mass of the emitting system. The dominant terms are on the first line of Eq. 14.4: the characteristic $1/m_{ee}$ dependence, and the cutoff of the spectrum for $m_{ee} \geq M$ (Figure 14.15a) [739]. Since the main background for direct single γ production is a photon from $\pi^0 \to \gamma + \gamma$, selecting $m_{ee} \gtrsim 100$ MeV/c^2 effectively reduces the background by an order of magnitude by eliminating the background from π^0 Dalitz decay, $\pi^0 \to \gamma + e^+ + e^-$, at the expense of a factor $\sim$1000 in rate. This allows the direct photon measurements to be extended (for the first time in both p–p and Au+Au collisions) below the value of $p_T \sim 4$ GeV/c, possible with real photons, down to $p_T = 1$ GeV/c (Figure 14.15b) [739], which is a real achievement. The solid lines on the p–p data are QCD calculations which work down to $p_T = 2$ GeV/c. The dashed line is a fit of the p–p data to the modified power law $B(1 + p_T^2/b)^{-n}$, used in the related Drell–Yan [740] reaction, which flattens as $p_T \to 0$.

The relatively flat, non-exponential, spectra for the direct-γ and Drell–Yan reactions as $p_T \to 0$ is due to the fact that there is no soft physics production process

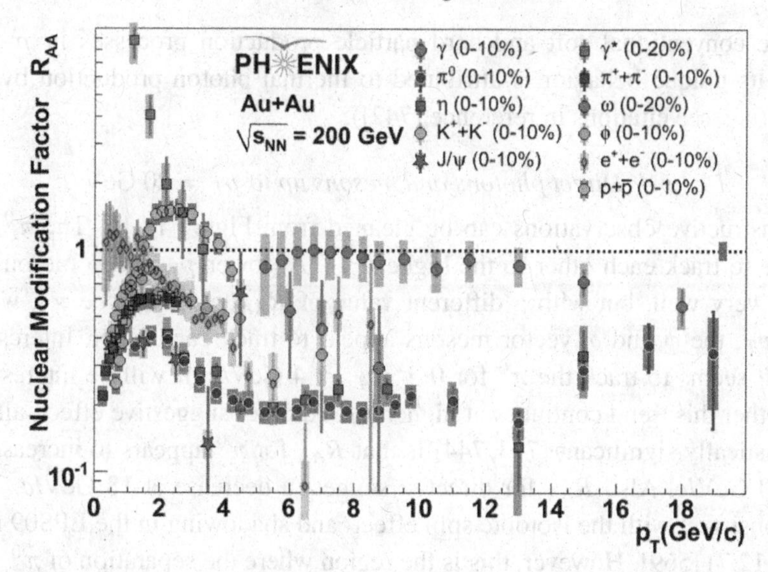

Figure 14.16 Nuclear modification factor, $R_{AA}(p_T)$, for all identified particles so far measured by PHENIX in central Au+Au collisions at $\sqrt{s_{NN}}$ = 200 GeV [741].

for them, only production via the partonic subprocesses, $g + q \rightarrow \gamma + q$ and $\bar{q} + q \rightarrow e^+ + e^-$, respectively. This is quite distinct from the case for hadron production, e.g. π^0, where the spectra are exponential as $p_T \rightarrow 0$ in p–p (Figure 14.6a) and A+A collisions due to soft production processes. Thus, for direct-γ in Au+Au collisions, the exponential spectrum of excess photons above the $\langle T_{AA}\rangle$ extrapolated p–p fit is unique and therefore suggestive of a thermal source.

14.3.6.3 Low p_T versus high p_T direct-γ and other identified particles

The unique behavior of direct-γ at low p_T in Au+Au relative to $p + p$ compared to any other particle is illustrated more dramatically by examining the R_{AA} of all particles measured by PHENIX in central Au+Au collisions at $\sqrt{s_{NN}} = 200$ GeV (Figure 14.16) [741]. For the entire region $p_T \leq 20$ GeV/c so far measured at RHIC, apart from the $p + \bar{p}$ which are enhanced in the region $2 \leq p_T \lesssim 4$ GeV/c (the baryon anomaly), the production of *no other particle* is enhanced over point-like scaling. The behavior of R_{AA} of the low $p_T \leq 2$ GeV/c direct-γ is totally and dramatically different from all the other particles, exhibiting an order of magnitude exponential enhancement as $p_T \rightarrow 0$. This exponential enhancement is certainly suggestive of a new production mechanism in central Au+Au collisions different from the conventional soft and hard particle production processes in p–p collisions. Its unique behavior is attributed to thermal photon production by many authors (e.g. see citations in reference [742]).

14.3.6.4 *Direct photons and mesons up to $p_T = 20$ GeV/c*

Other instructive observations can be gleaned from Figure 14.16. The π^0 and η continue to track each other to the highest p_T. At lower p_T, the ϕ meson tracks the $K^{\pm}$ very well, but with a different value of $R_{AA}(p_T)$ than the π^0, while at higher p_T, the ϕ and ω vector mesons appear to track each other. Interestingly, the J/Ψ seems to track the π^0 for $0 \leq p_T \leq 4$ GeV/c; it will be interesting to see whether this trend continues at higher p_T. Another suggestive effect, although not statistically significant [743, 744] is that R_{AA} for π^0 appears to increase from 10 to 20 GeV/c. Also R_{AA} for direct-γ suggest a decrease at 18 GeV/c. This is not inconsistent with the isotopic spin effect and shadowing in the EPS09 nPDFs (Figure 12.7) [569]. However, this is the region where the separation of $\pi^0 \to \gamma\gamma$ and γ in the PHENIX EMCal is difficult. Improved measurements of both direct-γ and π^0 in the range $10 < p_T < 20$ GeV/c at RHIC are possible and are of the utmost importance.

14.3.6.5 *J/Ψ suppression, still golden?*

"Anomalous suppression" of J/Ψ was found in $\sqrt{s_{NN}} = 17.2$ GeV Pb+Pb collisions at the CERN-SPS [745] (Figure 14.17a). This is the CERN fixed target heavy ion program's main claim to fame: but the situation has always been complicated because the J/Ψ is suppressed in $p + A$ collisions as shown in Figure 8.16, above. For example, in $\sqrt{s_{NN}} = 38.8$ GeV $p + A$ collisions [322] the Drell–Yan cross section in $p + A$ collisions exhibits point-like scaling, $A^{1.0}$, while the J/Ψ and Υ cross sections per nucleon are suppressed by an amount A^{α} with $\alpha = 0.920 \pm 0.008$ for both J/Ψ and Ψ' [362] and $\alpha = 0.96 \pm 0.01$ for both the Υ_{1s} and Υ_{2s+3s} [363]. This is called a Cold Nuclear Matter (CNM) effect and is shown as the line with $\alpha = 0.92$ on Figure 14.17a. The "anomalous suppression" is the difference between the data point at $AB = 208^2$ and the line, provided that the CNM effect is the same at $\sqrt{s_{NN}} = 17.2$ and 38.8 GeV.

At RHIC, PHENIX was designed to measure J/Ψ production down to $p_T = 0$, and does so via e^-e^+ at mid-rapidity in the range $|y| < 0.35$, and via $\mu^+\mu^-$ at forward rapidity, $1.2 < |y| < 2.2$, which allows a measurement of the total J/Ψ cross section integrated over rapidity as well as $d\sigma/dy$. An example of a nice J/Ψ and Ψ' peak at mid-rapidity is given in Figure 14.17b [746].

The dramatic difference in π^0 suppression from SPS to RHIC c.m. energy (Figure 14.9a) is not reflected in J/Ψ suppression, which is nearly identical at mid-rapidity at RHIC compared to the NA50 measurements at the SPS

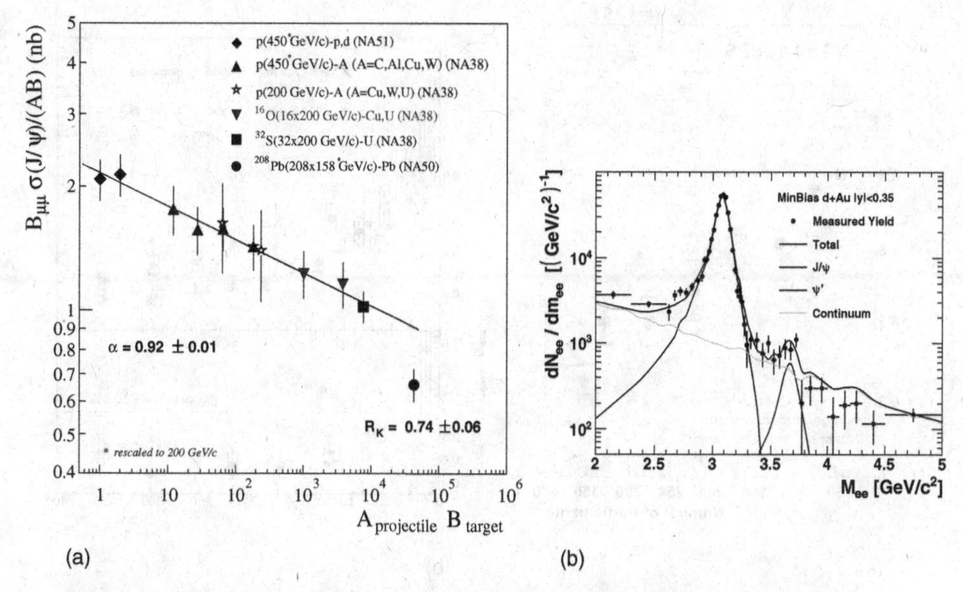

Figure 14.17 (a) Total cross section for J/Ψ production divided by AB in $A+B$ collisions at 158–200A GeV [745]. (b) PHENIX [746] invariant mass distribution of e^+e^- pairs at mid-rapidity in $\sqrt{s} = 200$ GeV d–Au collisions.

(Figure 14.18a) [747]. Also, at RHIC [749], the J/Ψ are more suppressed at forward rapidity than at mid-rapidity, which is opposite to the expectation that the suppression should be less at forward rapidity, where the energy density is lower. This casts new doubt on the value of J/Ψ suppression as a probe of deconfinement, in addition to the previous complication that J/Ψ are already suppressed (compared to point-like scaling) in $p + A$ and $B + A$ collisions (Figure 14.17a).

One possible explanation is that c and $\bar{c}$ quarks in the QGP coalesce (like the idea in references [735, 736]) to regenerate J/Ψ, miraculously making the observed R_{AA} equal at SPS and RHIC c.m. energies [750]. The good news is that such models predict the vanishing of J/Ψ suppression or even an enhancement ($R_{AA} > 1$) at LHC energies [751]. This would be spectacular, if observed, but more related to the large number of c and $\bar{c}$ quarks produced, rather than a proof of deconfinement.

14.3.6.6 A major surprise/discovery from heavy quark production

PHENIX was specifically designed to be able to detect charm particles via direct single $e^{\pm}$ from their semi-leptonic decay since this went along naturally with $J/\Psi \rightarrow e^+ + e^-$ detection and because the single particle reaction avoided the huge combinatoric background in Au+Au collisions [752]. The direct single $e^{\pm}$ measurement in p–p collisions at $\sqrt{s} = 200$ GeV [720] was shown in Figure 14.6d

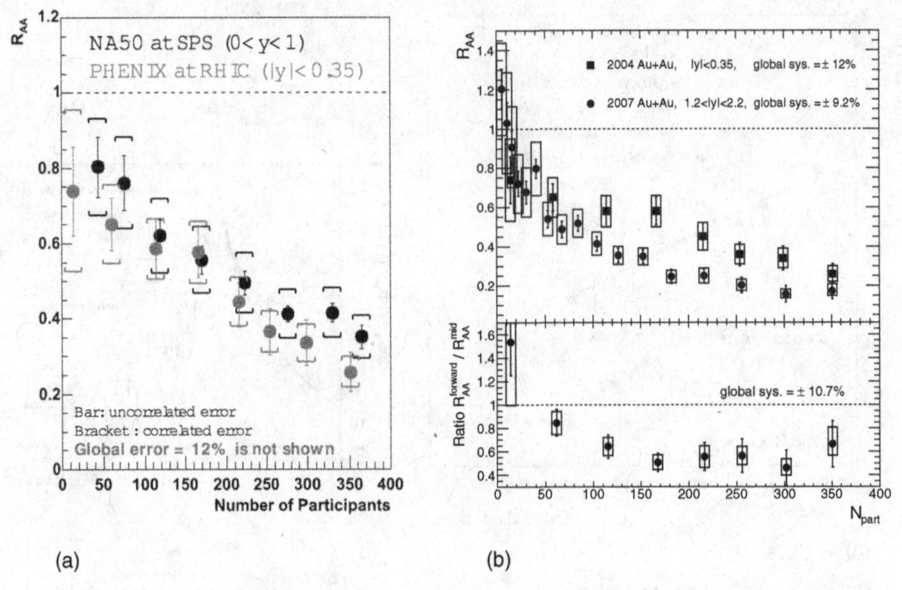

Figure 14.18 (a) J/Ψ suppression relative to p–p collisions (R_{AA}) as a function of centrality (N_{part}) at RHIC [747, 748] and at the CERN-SPS [745]. (b) R_{AA} of J/Ψ versus N_{part} at RHIC ($\sqrt{s_{NN}} = 200$ GeV) measured by PHENIX [749]. The lower panel shows the the ratio of forward rapidity (circles) to mid-rapidity (squares) measurements of the upper panel. Open boxes represent point-to-point correlated systematic error. The global normalization uncertainties are indicated.

in agreement with a QCD calculation [753] of c and b quarks as the source of the direct single $e^{\pm}$ (heavy flavor $e^{\pm}$).

In Au+Au collisions, a totally unexpected and dramatic result was observed (Figure 14.19a) [754]. The direct single $e^{\pm}$ from heavy quarks are suppressed the same as the π^0 from light quarks (and gluons) in the range $4 \leq p_T \leq 9$ GeV/c where b and c contributions are roughly equal [755] (see Figure 14.19b). Also the decay $e^{\pm}$ from the heavy quarks exhibit significant collective flow (v_2). Both these observations indicate a very strong interaction and possible thermalization of the heavy c and b quarks with the medium, which is difficult to understand if the free quarks and gluons in the medium have much smaller masses. The apparent equal suppression of light and heavy quarks also strongly disfavors the pQCD radiative energy loss explanation of jet quenching because, naively and theoretically [717], heavy quarks should radiate much less than light quarks and gluons in the medium. However, it opens up a whole range of new possibilities including string theory [756]. One of the far-out ideas, proposed by Nino Zichichi [757], was that the Higgs boson does not give mass to the fermions (see footnote 1 in Section 14.1) in which case [757], "since the origin of the quark masses is still not

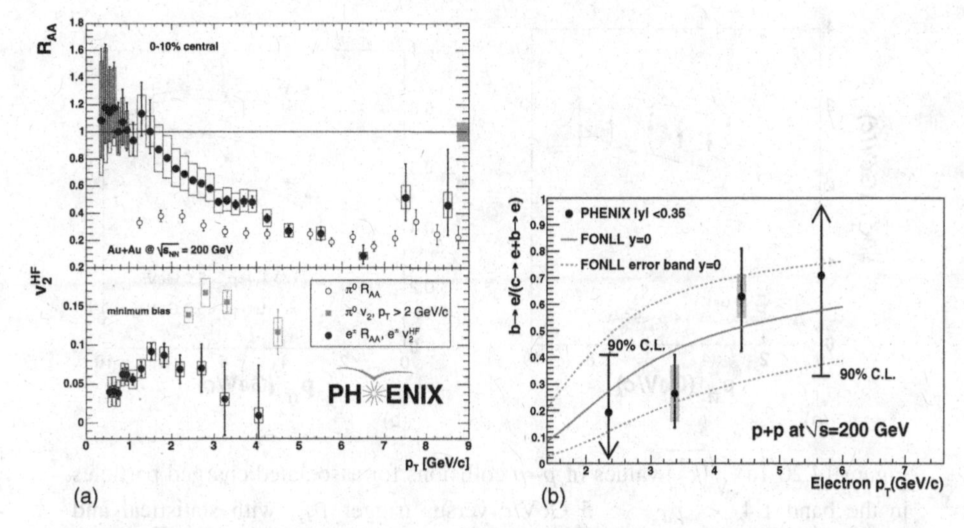

Figure 14.19 (a) Upper panel shows R_{AA} as a function of p_T for direct single $e^{\pm}$ [754] and π^0 in Au+Au central (0–10%) collisions at $\sqrt{s_{NN}} = 200$ GeV. Lower panel shows v_2 of heavy flavor $e^{\pm}$ compared to π^0. (b) Fraction of heavy flavor $e^{\pm}$ from b quark relative to total $e^{\pm}$ from $b + c$ quarks as a function of electron p_T from $e^{\pm}$–$K^{\mp}$ correlations [755] compared to the QCD calculation [753].

known, it cannot be excluded that in a QCD coloured world [QGP], the six quarks are nearly massless ..." Obviously, this would be the simplest explanation of the apparent equal suppression of "heavy" and "light" quarks in the QGP and to some readers may not be as far-out as string theory. The main advantage of this explanation is that it can be directly falsified in p–p collisions if a Higgs is found with the fermion Yukawa couplings [100] (should be known by the end of 2012) or in $A + A$ collisions where the energy loss of light and heavy quarks is in principle measurable (see discussion below).

14.3.7 *Two particle correlations, p–p collisions*

Results from two particle correlation measurements at RHIC in $\sqrt{s} = 200$ GeV p–p collisions were consistent with those at $\sqrt{s} = 62.4$ GeV, from the CERN-ISR, with some quantitative differences. The azimuthal correlation functions at RHIC looked very much like those in Figure 9.3. To recall the terminology, the azimuthal correlations are measured from associated particles, with transverse momenta p_{T_a}, which are fragments of jets, with transverse momenta $\hat{p}_{T_a}$, opposite in azimuth to trigger particles with transverse momenta p_{T_t} from jets with $\hat{p}_{T_t}$. The $\sqrt{\langle k_T^2 \rangle}$ was larger at $\sqrt{s} = 200$ GeV compared to 62.4 GeV (Figure 14.20a), but most interestingly, the $\sqrt{\langle j_T^2 \rangle}$ was the same as a function of the trigger p_{T_t} at both

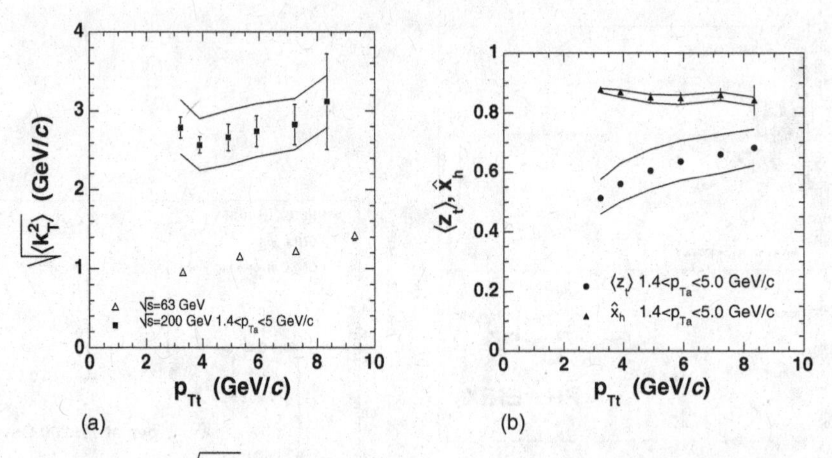

Figure 14.20 (a) $\sqrt{\langle k_T^2\rangle}$ values in p–p collisions for associated charged particles in the band $1.4 < p_{T_a} < 5$ GeV/c versus trigger p_{T_t}, with statistical and systematic errors, measured by PHENIX [149] in p–p collisions at $\sqrt{s} = 200$ GeV compared to CCOR [269] measurements at $\sqrt{s} = 62.4$ GeV. (b) The values of $\langle z_t\rangle$ and $\hat{x}_h$ used in the PHENIX measurement, shown with statistical and systematic errors.

c.m. energies (and also the same at the LHC at $\sqrt{s} = 7$ TeV, see Figure 14.31 below).

There was, however, one significant difference that led to a major discovery which overturned a firm belief from ISR days (as discussed in Section 9.2.1) that the x_E distribution represented the fragmentation function of the away jet from a trigger particle with p_{T_t} once a correction was made for $\langle z_{trig}\rangle$ (also denoted $\langle z_t\rangle$), the mean z of the trigger (recall Figure 9.6). The difference was that, because of the larger c.m. energy at RHIC, the inclusive single particle p_T distribution was flatter, $n = 8.1$ at $\sqrt{s} = 200$ GeV compared to $n \approx 10$ at $\sqrt{s} = 62.4$ GeV, so that $\langle z_t\rangle$ was smaller (i.e. < 1) at $\sqrt{s} = 200$ GeV. Thus, the approximate formula used at the ISR (Eq. 9.13) in which $\langle z_t\rangle$ had been neglected had to be replaced by the more exact formula also given by Feynman, Field and Fox [391]:

$$\frac{\langle z_t(k_T, x_h)\rangle}{\langle \hat{x}_h(k_T, x_h)\rangle}\sqrt{\langle k_T^2\rangle} = \frac{1}{x_h}\sqrt{\langle p_{out}^2\rangle - \langle j_{T_\phi}^2\rangle (1 + x_h^2)}, \tag{14.5}$$

where x_h ($\hat{x}_h$) is the ratio of the associated particle (parton) transverse momentum to the trigger particle (parton) transverse momentum:

$$x_h \equiv \frac{p_{T_a}}{p_{T_t}} \qquad \hat{x}_h = \hat{x}_h(k_T, x_h) \equiv \frac{\hat{p}_{T_a}}{\hat{p}_{T_t}}, \tag{14.6}$$

and in the collinear limit, $\Delta\phi \to \pi$ rad (Eq. 9.11), $x_E \to x_h$. Note that the hadronic variable x_h is measured on every event and the other quantities on the right hand side of Eq. 14.5, $\left\langle j_{T_\phi}^2 \right\rangle$ ($\langle p_{out}^2 \rangle$), can be extracted directly from the measured same-side (away-side) azimuthal correlation functions. On the left hand side of Eq. 14.5, the partonic variable $\hat{x}_h$ is a function of both k_T and x_h, as is the "trigger bias" $\langle z_t \rangle$. Thus, the solution of Eq. 14.5 for $\sqrt{\langle k_T^2 \rangle}$ is an iterative process. The results are shown in Figure 14.20b [149].

The discovery occurred because PHENIX tried to use the measured x_E distribution, corrected by $\langle z_t \rangle$, as the fragmentation function, in order to compute the $\langle z_t \rangle$ in Eq. 14.5 by iteration. There were serious difficulties with convergence which led to the discovery that the x_E distribution is not sensitive to the shape of the fragmentation function (see Appendix E). Instead it was found that in the collinear limit ($p_{T_a} \to x_E p_{T_t}$, $x_E \to x_h$), the shape of the x_E distribution is given by the power n of the partonic and inclusive single particle p_T spectrum ($Ed^3\sigma/dp^3 \propto p_T^{-n}$):

$$\left. \frac{dP_\pi}{dx_E} \right|_{p_{T_t}} \approx \langle m \rangle_h (n-1) \frac{1}{\hat{x}_h} \frac{1}{(1+\frac{x_E}{\hat{x}_h})^n}, \tag{14.7}$$

where the only dependence on the fragmentation function is in $\langle m \rangle_h$, the mean multiplicity of the species of measured particles in the jet.

The true meaning of the x_E distribution turned out to be enormously useful. Equation 14.7 relates the ratio of the transverse momenta of the away and trigger particles, $p_{T_a}/p_{T_t} = x_h \approx x_E$, which is measured, to the ratio of the transverse momenta of the away jet to the trigger jet, $\hat{p}_{T_a}/\hat{p}_{T_t}$, which can thus be deduced. Although derived for p–p collisions, Eq. 14.7 should work just as well for $A+A$ collisions since the only assumptions are independent fragmentation of the trigger and away jets with the same exponential fragmentation function and a power-law parton $\hat{p}_{T_t}$ distribution. The only other (and weakest) assumption is that $\hat{x}_h$ is constant for fixed p_{T_t} as a function of x_E. Thus in $A+A$ collisions, Eq. 14.7 for the x_E distribution provides a method of measuring the ratio $\hat{x}_h = \hat{p}_{T_a}/\hat{p}_{T_t}$ and hence the relative energy loss of the away-side to the same-side jet assuming that both jets fragment outside the medium with the same fragmentation function as in p–p collisions (which was found out to be true at the LHC [758, 759], see below).

14.3.7.1 Two particle direct-γ–hadron correlations

The key to measure the fragmentation function of the jet of particles from an outgoing hard scattered parton is to know the energy of the original parton which fragments. This is straightforward for di-jet events in $e^+e^- \to q+\bar{q}$ collisions as pioneered at the LEP (recall Figure 12.4b) [544]. In p–p or $A+A$ collisions the analogous reaction is direct-γ–hadron correlations from the constituent reaction

$g + q \to \gamma + q$. The γ is not the fragment of a parton but a direct participant in the two-to-two subprocess (recall Figure 14.7), which emerges freely and unbiased from the reaction, isolated, with no accompanying particles (even in a QGP because it is both electrically and color neutral), and whose energy can be measured to high precision. Since the scattered quark in the reaction has equal and opposite p_T to the direct-γ, the transverse momentum of the jet from the outgoing quark at the point of interaction is also precisely known (modulo k_T). Thus in p–p collisions, γ–h correlations can be used to measure the fragmentation function of the jet from the outgoing quark, which is 8/1 u quark with a small gluon contamination from the reaction $\bar{q} + q \to \gamma + g$ in p–p and $A + A$ collisions (but which is obviously a much larger effect in $\bar{p}$–p collisions).

Beyond leading order, single γ may be produced in the fragmentation process [419]. However, measurements at both the CERN-ISR [438] (recall Figure 10.8) and RHIC show that this is a small effect. In Figure 14.21 the effect of an isolation cut on the direct photon cross section of Figure 14.6c [605] was measured to be negligible, $<10\%$, in agreement with the NLO calculations shown. The main utility of the isolation cut is to reduce the background of photons from hadronic decays by a factor of order 2.5.

The isolation of direct-γ is shown more dramatically in the measurement of $\gamma-h$ correlations compared to π^0-h correlations in p–p collisions at $\sqrt{s} = 200$ GeV (Figure 14.21b) [760]. The strong near-side correlation for π^0 triggers is largely absent for direct-γ triggers as expected from a sample dominated by photons produced directly in the hard scattering. Above 7 GeV/c this measurement [760]

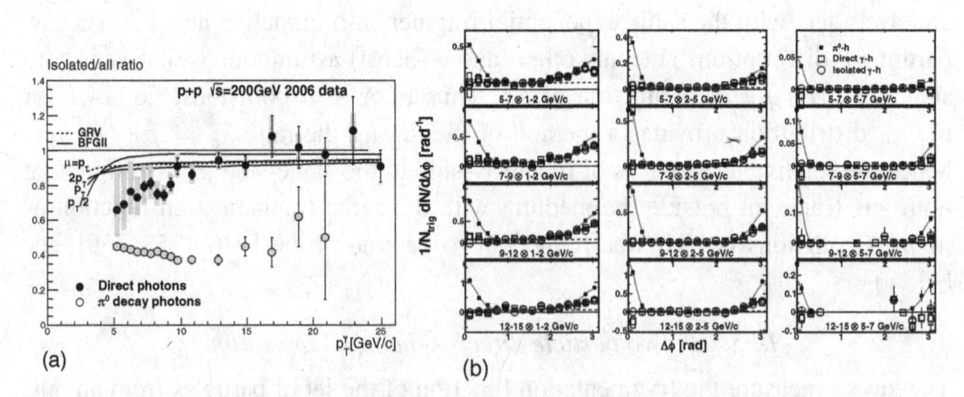

Figure 14.21 (a) Ratio of isolated direct-γ to all direct-γ (solid points) from cross section measurement of Figure 14.6c [605]. Also shown is the ratio of isolated photons from π^0 decays to all π^0 decay photons. (b) Direct-γ–h and π^0–h azimuthal correlations in p–p collisions at $\sqrt{s} = 200$ GeV, with p_{Tt} from 5–15 GeV/c, and p_{Ta} from 1–7 GeV/c as indicated [760].

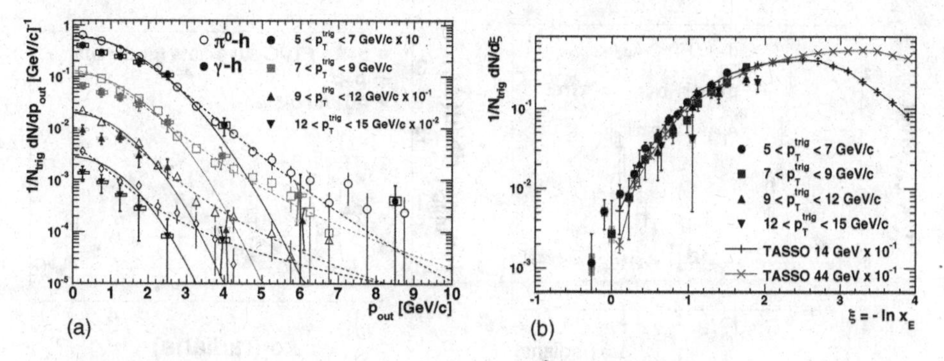

Figure 14.22 (a) p_{out} distributions from $\pi^0 - h$ and isolated direct-γ–h correlations in p–p collisions at $\sqrt{s} = 200$ GeV [760]. (b) $\xi = \ln 1/x_E$ distributions from the isolated direct-γ–h data for all p_{Tt} ranges combined, compared to e^+e^- collisions at $\sqrt{s} = 14$ and 44 GeV [761].

constrains the near-side yield of particles with $p_T \geq 1$ GeV/c associated with direct-γ to be smaller than 15% of the yield associated to π^0.

The p_{out} and x_E distributions from these γ–h correlation functions [760] (Figure 14.22) are also quite interesting. The p_{out} distributions for π^0–h and isolated direct-γ–h correlations appear to be nearly identical and both show the features of a Gaussian distribution for $p_{out} < 3$ GeV/c, which is the k_T effect, now thought to be due to resummation of soft gluons [441], and a power-law tail from NLO hard gluon emission. This is the first time that the k_T effect has been measured with direct-γ, and the $\sqrt{\langle k_T^2 \rangle}$ is essentially identical in π^0 and direct-γ production.

From the discussion above, the x_E distribution of the $h^{\pm}$ in direct-γ–h correlations should be a measurement of the u quark fragmentation function. To make this more apparent, the x_E distributions of the away-hadrons calculated from Figure 14.21b and converted to $\xi = -\ln x_E$ distributions are shown in Figure 14.22b. They are in quite excellent agreement with the dominant u quark fragmentation functions measured in e^+e^- collisions at $\sqrt{s}/2 = 7$ and 22 GeV [761], which cover a comparable range,in jet energy, when normalized empirically to account for the limited PHENIX acceptance for the away jet.

14.3.8 Two particle correlations, Au + Au collisions

One of the first and still most striking measurements of two particle correlations in Au+Au collisions was presented by STAR at the Quark Matter 2002 conference [762]. In a later published measurement [764] which includes a d+Au measurement [763] (Figure 14.23a), the conditional probability – given a trigger particle with p_{T_t} between 4 and 6 GeV/c – of detecting an associated particle with

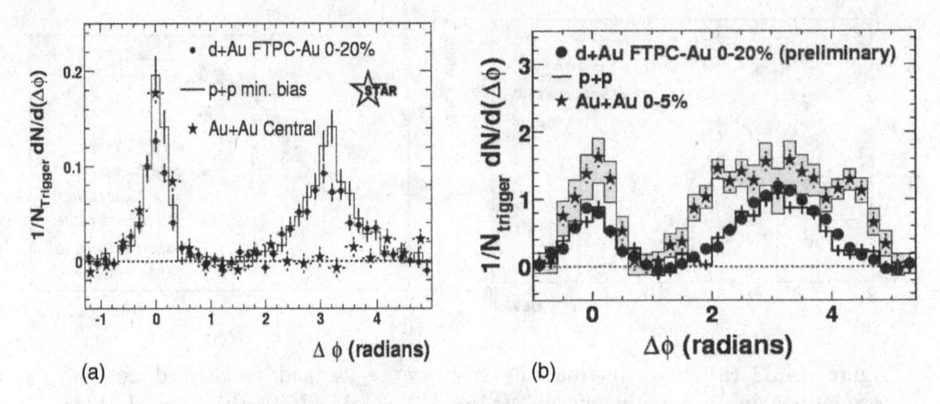

Figure 14.23 STAR conditional probability two particle correlation functions with flow modulated background subtracted: (a) measurements in d+Au [763], p–p and Au+Au central [764] collisions at $\sqrt{s_{NN}} = 200$ GeV with $4 < p_{T_t} < 6$ GeV/c and $2\,\text{GeV}/c < p_{T_a} < p_{T_t}$; (b) STAR data with the same trigger p_{T_t} but with $0.15 < p_{T_a} < 4$ GeV/c [765].

p_{T_a} in the range $2\,\text{GeV}/c < p_{T_a} < p_{T_t}$ is shown for central Au+Au collisions at $\sqrt{s_{NN}} = 200$ GeV as a function of the azimuthal angle difference, $\Delta\phi$, of the two particles. The usual two particle azimuthal correlation function (expressed as the conditional probability) becomes more complicated for $A + A$ collisions: the sum of the background of particles randomly associated to the trigger, which is modulated by the common hydrodynamic flow (represented only by $v_2(p_T)$), plus the jet correlation function for $A + A$ collisions:

$$C_2^{AuAu}(\Delta\phi) = C_2^{jet}(\Delta\phi) + B\left[1 + 2v_2(p_{T_t})v_2(p_{T_a})\cos(2\Delta\phi)\right]. \tag{14.8}$$

The background is subtracted on Figure 14.23 which shows only the jet correlation function, $C_2^{jet}(\Delta\phi)$. The trigger-side correlation peak in central Au+Au collisions and d+Au central collisions appears to be the same as that measured in p–p collisions[4] but the away-side jet correlation in Au+Au appears to have vanished. This observation appears consistent with a large energy loss in the medium, or a medium that is opaque to the propagation of high momentum partons, as originally indicated by the suppression observed [598] in single particle inclusive measurements for $p_{T_t} > 3$ GeV/c (recall Figure 14.8).

Although the apparent vanishing of the away jet in central Au+Au collisions was fantastic from a public relations perspective, it is misleading from a scientific viewpoint as it suggests that the away-jet was totally absorbed by the opaque medium.

[4] The fact that the number of associated particles on the trigger side is the same in both Au+Au and p–p collisions is not trivial. It shows that absorption of hadrons is not the cause of jet-quenching since then all the hadrons would have been absorbed roughly equally. This is consistent with the quantum mechanical argument on resolving a pion given above (Section 13.2.2).

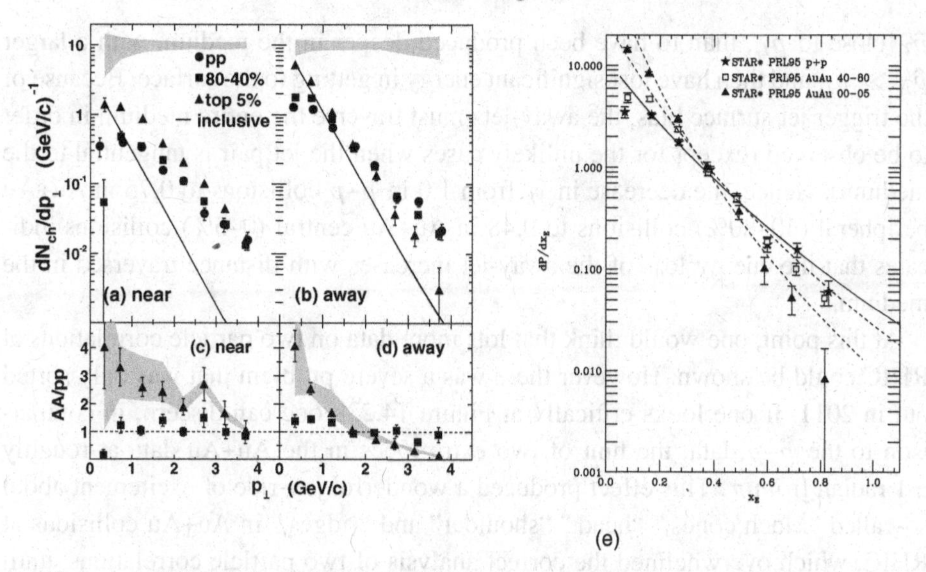

Figure 14.24 STAR measurement [766] of transverse momentum ($p_\perp$) distribution of associated charged hadrons for a trigger charged hadron with $4 < p_{T_t} < 6$ GeV/c for p–p, Au+Au peripheral (80–40%), Au+Au central (top 5%) collisions at $\sqrt{s_{NN}} = 200$ GeV: (a) near-side, (b) away-side; (c,d) I_{AA} ratio of A–A to p–p $p_\perp$ distributions for near side (c), and away side (d). (e) Data from (b) plotted as dP/dx_E compared to Eq. E.31 with $\hat{x}_h = 1$ for p–p, $\hat{x}_h = 0.75$ for Au+Au peripheral, and $\hat{x}_h = 0.48$ for Au+Au central [767].

Later work presented by STAR at Quark Matter 2004 [765,766] with $4 < p_{T_t} < 6$ GeV/c and $0.15 < p_{T_a} < 4$ GeV/c showed that the away jet did not disappear, it just lost energy and the away-side correlation peak became much wider than in p–p collisions (Figure 14.23b).

This first measurement of two particle correlations in Au+Au collisions provided an opportunity to test the validity/utility of Eq. 14.7. Figure 14.24b shows the STAR measurement [766] of the away-side p_{T_a} distribution, given p_{T_t}, from the data shown in Figure 14.23b. In Figure 14.24e [767], these measurements are plotted as an $x_E \approx p_{T_a}/p_{T_t}$ distribution and shown together with Eq. 14.7, with $n = 8.1$, scaled to match the data, which are beautifully consistent with no relative energy loss of the two jets in p–p collisions, $\hat{x}_h = 1$. By contrast, in Au+Au collisions, agreement with the data is obtained with a ratio of away/trigger jet momenta of 0.75 in peripheral (40–80%) and 0.48 in central (0–5%) collisions. This indicates a clear relative energy loss of the away jet compared to the trigger jet, which increases with increasing centrality. However, the trigger jets in Au+Au collisions are surface biased by the falling power-law p_T spectrum, an effect analogous to "trigger bias" (Section 9.1.1). The jets which give trigger particles with observed p_{T_t} are more likely to be produced near the surface and lose little energy, with

$\hat{p}_{T_t}$ close to p_{T_t}, than to have been produced deeper in the medium with a larger $\hat{p}_{T_t} \gg p_{T_t}$ and then have lost significant energy in getting to the surface. Because of the trigger jet surface bias, the away-jets must traverse the entire medium in order to be observed (except for the unlikely cases when the jet pair is tangential to the medium). Hence, the decrease in $\hat{x}_h$ from 1.0 in p–p collisions to 0.75 in Au+Au peripheral (40–80%) collisions to 0.48 in Au+Au central (0–5%) collisions indicates that the energy loss of the away-jet increases with distance traversed in the medium.

At this point, one would think that lots more data on two particle correlations at RHIC could be shown. However there was a severe problem that was only sorted out in 2011. If one looks critically at Figure 14.23b one can discern, in comparison to the p–p data, the hint of two extra lobes in the Au+Au data at roughly ± 1 radian from π. This effect produced a wonderful joy-ride of excitement about so-called "Mach cones," "head," "shoulder" and "ridges," in Au+Au collisions at RHIC, which overwhelmed the correct analysis of two particle correlations, until the discovery in 2011 that v_3 should be added to the background modulation in Eq. 14.8, which seemed to eliminate all the artifacts [652, 768, 769]. Fortunately, the background effect is reduced at higher p_{T_t} and p_{T_a}, so that there are a few interesting and instructive results.

PHENIX [770, 771] has made measurements of π^0–$h^{\pm}$ correlations, where $h^{\pm}$ denotes non-identified charged particles, in p–p and Au+Au collisions at $\sqrt{s_{NN}} = 200$ GeV for which the x_E distributions[5] are shown in Figure 14.25a. The energy loss is measured by the fact that the Au+Au spectrum is steeper than the p–p spectrum as more typically shown by the ratio of the spectra in A–A/p–p (Figure 14.25b) [770], called I_{AA}.

All I_{AA} distributions, if they go down to low enough associated p_{T_a}, exhibit the same shape as shown in Figure 14.25b: (i) a constant for larger values of p_{T_a}, which indicates away-partons which punch through (or are emitted tangentially to) the medium, without energy loss ($\hat{x}_h \approx 1$), and then fragment the same as p–p jets (vacuum fragmentation); (ii) a steep curve for values of $p_{T_a} \leq 2$–3 GeV/c, which indicates vacuum fragmentation of jets from away-partons which have lost energy in the medium ($\hat{x}_h < 1$). Also in Figure 14.25b, the p_{T_a} distributions are calculated using the full away-side azimuth $|\Delta\phi - \pi| < \pi/2$ rad opposite the trigger (circles) as well as a restricted "head" region $|\Delta\phi - \pi| < \pi/6$ rad (squares). The fact that the results are identical shows that the p_{T_t} is large enough that the artifacts have vanished [770].

The solid lines on the p–p and Au+Au data points in Figure 14.25a are fits to Eq. 14.7 with values of the fitted parameters [771] shown on the figure, where N

[5] x_E distributions are also denoted $z_T \equiv p_{Ta}/p_{Tt} = x_h$ in some articles [772, 773].

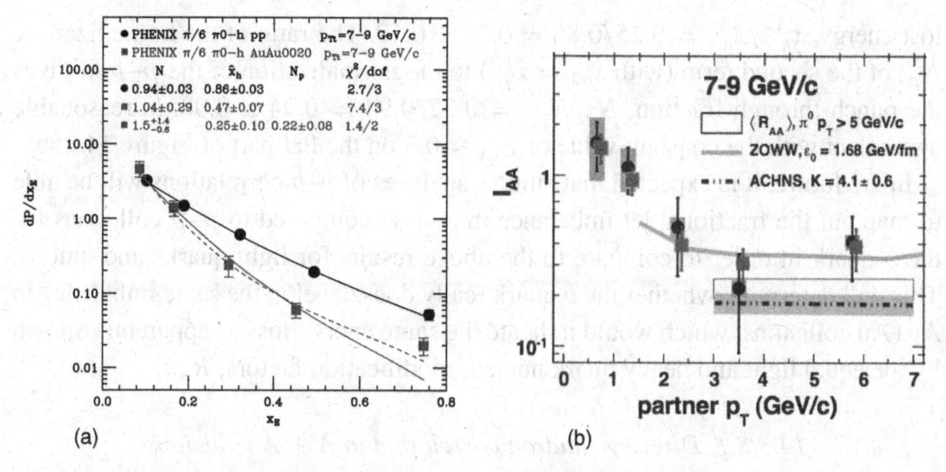

Figure 14.25 (a) PHENIX [771] x_E distributions in narrow azimuthal cut $|\Delta\phi - \pi| < \pi/6$ rad from p–p (circles) and central (0–20%) Au+Au collisions (squares) for $7 < p_{T_t} < 9$ GeV/c together with fits with parameters indicated. (b) $I_{AA}(p_{T_a})$ = ratio of Au+Au/p–p data from (a) versus p_{T_a} (squares) [770]. Circles represent full away-side azimuth $|\Delta\phi - \pi| < \pi/2$ rad. The long box indicates that R_{AA} is smaller than the flat part of I_{AA} for the same p_{T_t}. The various lines on the figure are discussed in reference [770].

represents $\langle m \rangle_h$ in Eq. 14.7. In general, the values of $\hat{x}_h^{pp}$ in p–p collisions do not equal 1 but range between $0.8 < \hat{x}_h^{pp} < 1.0$ due to k_T smearing and the range of x_E covered. For the fixed range of associated p_{T_a}, 0.7–5.8 GeV/c, used in reference [770], the lowest $p_{T_t} = 4$–5 GeV/c trigger provides the most balanced same-side and away-side jets, with $\hat{x}_h \approx 1.0$, while as p_{T_t} increases up to 9–12 GeV/c, for the fixed range of p_{T_a}, the jets become unbalanced towards the trigger side in p–p collisions due to k_T smearing. Thus, the p_{T_t} and x_E ranges are kept identical for the p–p and Au+Au comparison. Furthermore, in order to take account of the imbalance ($\hat{x}_h^{pp} < 1$) observed in the p–p data, the ratio $\hat{x}_h^{AA}/\hat{x}_h^{pp}$ is taken as the measure of the energy of the away jet relative to the trigger jet, the relative jet balance in $A + A$ compared to p–p collisions.

It is important to note that in Figure 14.25a, the away jet energy balance in Au+Au relative to p–p, $\hat{x}_h^{AA}/\hat{x}_h^{pp} = 0.47/0.86 = 0.54 \pm 0.08$, is significantly less than 1, indicating energy loss of the away jet in the medium. Also since the away jet may suffer a distribution of energy losses for a given trigger jet $\hat{p}_{T_t}$, due to variations in the path length through the medium, $\hat{x}_h^{AA}$ should be understood as $\langle \hat{x}_h^{AA} \rangle$. In order to take account of the flat part of the I_{AA} distribution in Figure 14.25b (the presumed punch-through partons), a fit was performed by adding a second term of the form of Eq. 14.7 to the fit function with the value of $\hat{x}_{h2}$ constrained to be equal to the p–p value $\hat{x}_h^{pp}$. This is represented by the dashed line on Figure 14.25a. The value of $\hat{x}_h^{AA} = 0.25$ for this fit indicates a larger energy loss for the parton that has lost energy, $\hat{x}_h^{AA}/\hat{x}_h^{pp} = 0.25/0.86 = 0.29 \pm 0.12$. The ratio of the normalization, N_p, of the second term (with $\hat{x}_{h2} = \hat{x}_h^{pp}$) to the normalization of the p–p fit gives the punch-through fraction, $N_p/N_{pp} = 0.22/0.94 = 0.24 \pm 0.08$, in reasonable agreement with the constant value of $I_{AA} \approx 0.3$ on the flat part of Figure 14.25b.

In the future, it is expected that similar analyses of b–$\bar{b}$ correlations will be able to map out the fractional jet imbalance in $A + A$ compared to p–p collisions for the b quark in order to compare to the above results for light quarks and gluons. This will determine whether the b quark really does develop the same imbalance in Au+Au collisions, which would indicate the same energy loss as apparently shown by the equal light and heavy quark nuclear modification factors, R_{AA}.

14.3.8.1 Direct-γ–hadron correlation in $A + A$ collisions

Both STAR and PHENIX have inconclusive measurements of γ–h correlations at RHIC in Au+Au central collisions at $\sqrt{s_{NN}} = 200$ GeV. The results are presented in different formats.

STAR [773] (Figure 14.26a) presents I_{AA} for direct-γ–h and π^0–h correlations as a function of $z_T = p_{T_a}/p_{T_t}$ (equivalent to x_E or x_h) for $8 \leq p_{T_t} < 16$ GeV/c. In both cases, only the flat part of the I_{AA} distribution is measured, with equal values of $I_{AA} \approx 0.3$, in agreement with the PHENIX I_{AA} measurement for $\pi^0 - h$ in roughly the same p_{T_t} range [770] (Figure 14.25b). As in π^0–h correlations, a flat I_{AA} distribution in direct-γ–h correlations would be indicative of the fragmentation function of a fraction of away partons which have not lost appreciable energy in the medium and fragmented in vacuum.

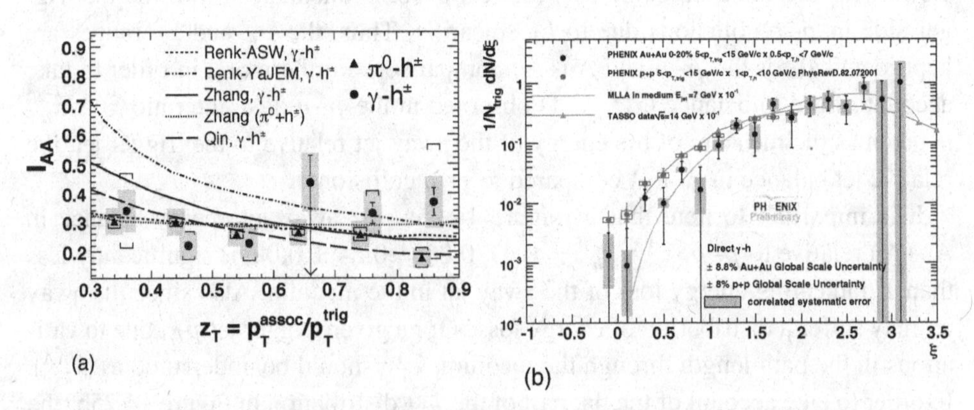

Figure 14.26 (a) STAR [773] measurement of I_{AA} versus $z_T = p_{T_a}/p_{T_t}$ for direct-γ–h and π^0–h correlations in central (0–10%) Au+Au collisions. (b) PHENIX preliminary [774] $\xi = -\ln x_E$ distribution from direct-γ–h measurements with $5 < p_{T_t} < 15$ GeV/c in central (0–20%) Au+Au collisions in comparison to p–p and TASSO data from Figure 14.22b.

PHENIX has presented a preliminary version (Figure 14.26b) of direct-γ–h correlations [774] in the form of a fragmentation function $\xi = -\ln x_E$ of the away jet in central (0–20%) Au+Au collisions in comparison to the p–p measurement from Figure 14.22b. The Au+Au measurement agrees nicely with the TASSO data for $\xi > 2$ ($x_E < 0.14$) (where the p–p data have run out), which would seem to indicate $I_{AA} \approx 1$ in this region, while for $\xi \leq 1$ ($x_E > 0.37$) the Au+Au data appear to fall below the p–p data. However, the errors are too large to draw a conclusion. Improved measurements of direct-γ–h correlations at RHIC are greatly desirable and would likely be very informative.

14.4 Hard scattering at the LHC

The first circulating beam in the LHC was achieved on September 10, 2008. Unfortunately, nine days later, a faulty splice in a magnet interconnect caused a huge quench which seriously damaged the machine. After more than a year of repairs and consolidations to ensure that such an incident could never recur, the machine once again circulated beam on November 20, 2009 with first collisions on November 23, 2009, and a run of several weeks at the end of 2009 at $\sqrt{s} = 0.9$ TeV. The LHC then began full operation in p–p collisions at $\sqrt{s} = 7$ TeV on March 30, 2010 [775], with Pb+Pb runs at $\sqrt{s_{NN}} = 2.76$ TeV in November 2010 and 2011, with a short p–p comparison run at $\sqrt{s} = 2.76$ TeV in March 2011. A p+Pb comparison run is planned for November 2012. The LHC machine has run well, with ever increasing luminosity, and will begin the 2012 p–p run at $\sqrt{s} = 8$ TeV, followed by a shut down at the end of 2012 for further repairs to allow the full design energy of $\sqrt{s} = 14$ TeV and design luminosity $\mathcal{L} = 10^{34}$ cm^{-2} s^{-1} to be achieved. Startup is expected for a full energy run in 2015.

14.4.1 Quenching of fully reconstructed jets

One positive outcome of the one year delay was that when the first collisions took place, the detectors were in excellent shape and produced outstanding physics results in an incredibly short time. The most spectacular example of this was the ATLAS discovery of jet quenching, with fully reconstructed jets, which was obvious on the on-line event display, almost as soon as the LHC switched over to Pb+Pb running in 2010 [776]. In p–p collisions,[6] the jets in the "Lego" plot were roughly balanced in transverse energy E_T while for Pb+Pb collisions at $\sqrt{s_{NN}} = 2.76$ GeV a large number of events were observed in which the jets were obviously highly

[6] Measurements of hard scattering using jets as tests of pQCD and in searches for new physics are evidently part of the LHC p–p program and some results have been mentioned above (Figure 12.2 [529]). The present section concentrates only on those measurements in the LHC-Ions program or other relevant measurements.

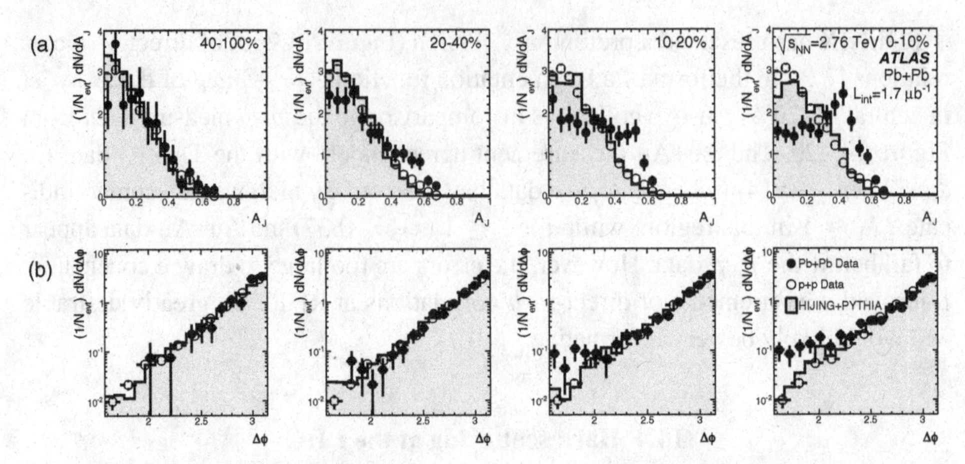

Figure 14.27 (a) ATLAS [776] plots of the A_J distribution of di-jets with $E_{T_1} > 100$ GeV and $E_{T_2} > 25$ GeV, for the highest energy jet in the opposite hemisphere, as a function of centrality in Pb+Pb collisions at $\sqrt{s_{NN}} = 2.76$ TeV (black points). Open points are measurements from p–p collisions at $\sqrt{s} = 7$ TeV and filled histograms are simulated p–p di-jets embedded in heavy ion events and reconstructed. (b) Distribution of $\Delta\phi$, the azimuthal angle between the two jets.

unbalanced [776]. Another feature of these unbalanced jets was that the leading and secondary jets were primarily back-to-back in azimuth.

ATLAS characterized the di-jet asymmetry with a new quantity,

$$A_J \equiv (E_{T_1} - E_{T_2})/(E_{T_1} + E_{T_2}) \approx (1 - \hat{x}_h^{AA})/(1 + \hat{x}_h^{AA}).$$

Figure 14.27 [776] shows plots of the A_J distribution, and the distribution of $\Delta\phi$ the azimuthal angle between the two jets, in both Au+Au and p–p collisions where the jets have been reconstructed with "anti-k_T" algorithm with radius parameter $R = 0.4$ [776]. The increasing asymmetry of di-jets as a function of centrality is evident. For the azimuthal distributions, apart from the most central events where there is a small tail at large $\Delta\phi$, the distributions for $\pi - \Delta\phi < 1$ radian appear to be identical, which implies that there is little or no angular scattering of the partons while they traverse the medium and lose energy. This might be difficult to reconcile with the predicted pQCD radiative energy loss [716] in which the strong medium effect of coherent LPM gluon radiation also causes the partons to multiple scatter.

CMS [758] also measured the di-jet imbalance in Pb+Pb collisions and found out something else very interesting about the associated jets which had lost energy in traversing the medium: they had exactly the same fragmentation function [759] (Figure 14.28) (and j_T) as the same energy jets in p–p collisions. This means that the partons lose energy in the medium but then fragment normally to jets outside the medium. The fragmentation function is not modified in the medium.

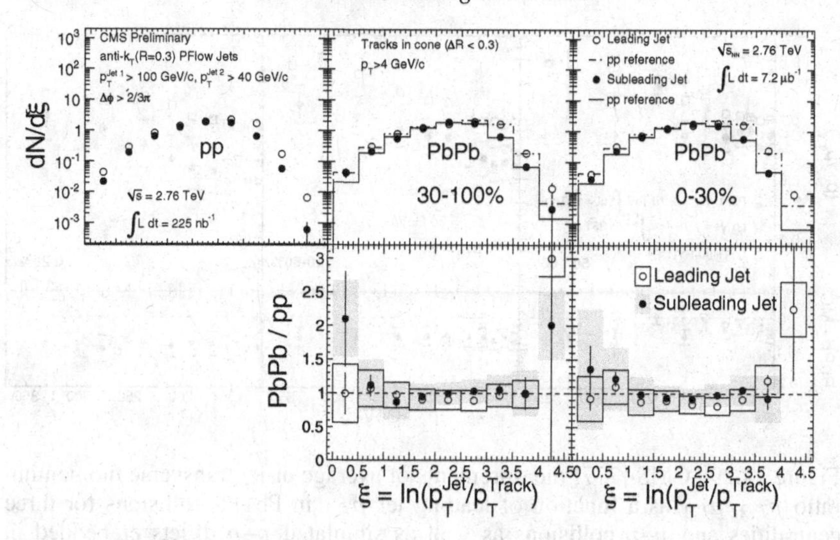

Figure 14.28 CMS [759] charged particle fragmentation functions for jets produced in p–p and in both peripheral (30–100%) and central (0–30%) Pb+Pb collisions at $\sqrt{s_{NN}} = 2.76$ TeV. Bottom panels are the ratio of Pb–Pb to p–p fragmentation functions.

This is consistent with the quantum mechanical arguments discussed above (Section 13.2.2), but came as a surprise to many people. Another interesting observation in the same papers [758,759] was that a large fraction of the energy lost by the outgoing parton is carried by soft particles ($p_T < 2$ GeV/c) and radiated at large angles $\Delta R > 0.8$ with respect to the associated subleading jet, much larger than the $R = 0.5$ cone used in the jet definition.

In a more recent publication, CMS [777] made a systematic study of the di-jet imbalance as a function of centrality and leading jet $p_{T,1}$ in the range $120 \leq p_{T,1} \leq 350$ GeV/c, for subleading jets with $p_{T,2} > 30$ GeV/c. Results were presented in the form of the ratio of jet transverse momenta, $\hat{x}_h = \hat{p}_{T,2}/\hat{p}_{T,1}$, as function of $\hat{p}_{T_1}$ measured in Pb+Pb and p–p collisions at $\sqrt{s_{NN}} = 2.76$ TeV as well as from simulated p–p di-jets embedded in heavy ion events and reconstructed (Figure 14.29). The results indicate that $\hat{x}_h^{PbPb} - \hat{x}_h^{ppMC} \approx 0.10 \pm 0.05$ in central (0–20%) Pb+Pb collisions, which is amazingly constant over the full range of $\hat{p}_{T_1}$ studied, and is also constant at a lower level, $\hat{x}_h^{PbPb} - \hat{x}_h^{ppMC} \approx 0.06 \pm 0.02$, for less central (20–50%) collisions (the errors are systematic and point-to-point correlated). This means that the directly measured fractional jet imbalance in $A + A$ relative to p–p collisions $(1 - \hat{x}_h^{AA}/\hat{x}_h^{pp})$ in the jet $\hat{p}_T$ range from 120–350 GeV/c at $\sqrt{s_{NN}} = 2.76$ TeV is roughly constant at 0.14 ± 0.07 in central collisions, which is much smaller than the PHENIX results of $1 - \hat{x}_h^{AA}/\hat{x}_h^{pp} = 0.46 \pm 0.08$, for jet $\hat{p}_T$ ($\approx p_{T_t}/\langle z_t \rangle$) range 10–20 GeV/$c$ in central (0–20%) Au+Au collisions at $\sqrt{s_{NN}} = 200$ GeV,

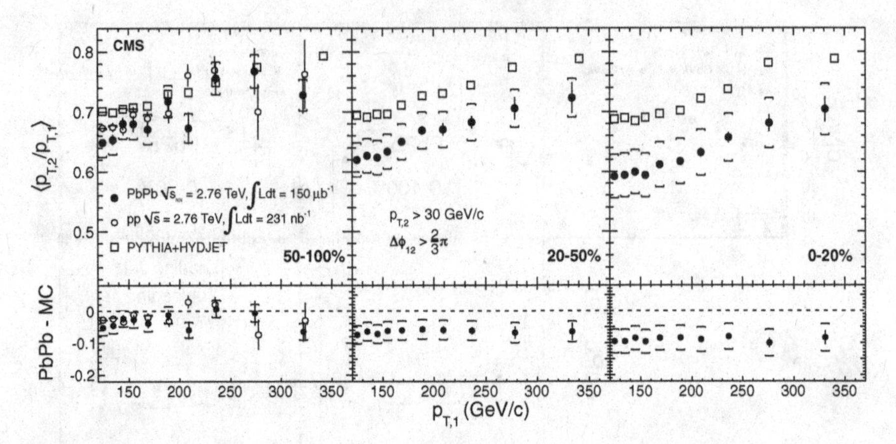

Figure 14.29 CMS [777] measurements of average di-jet transverse momentum ratio $p_{T,2}/p_{T,1}$ as a function of leading jet $p_{T,1}$ in Pb+Pb collisions for three centralities and p–p collisions as well as simulated p–p di-jets embedded in heavy ion events.

deduced from two particle correlation measurements (recall Figure 14.25). Based on single particle inclusive R_{AA} measurements, to be discussed next, this is clearly an indication of a strong p_T dependence of the QCD energy loss (fractional jet energy imbalance) of a parton in a QGP, which is again different from the analogy to radiative energy loss in QED where $\Delta E/E$ is independent of the energy E of electrically charged particles.

14.4.2 Suppression of high p_T single particles at the LHC

The measurements of suppression of high p_T single particles at the LHC are also eye-opening. The ALICE measurements [778] of R_{AA} for unidentified charged hadrons $h^{\pm}$ in the range $5 \leq p_T \leq 20$ GeV/c for central (0–5%) Pb+Pb collisions at $\sqrt{s_{NN}} = 2.76$ TeV (Figure 14.30a) [779] are, incredibly, nearly exactly on top of the PHENIX preliminary π^0 measurements in the same p_T range [743, 744]. However, the ALICE data have much smaller errors so that the increase from $R_{AA} = 0.2$ at $p_T = 10$ to $R_{AA} \sim 0.35$ at 20 GeV/c is statistically significant. However, what is really impressive is to realize that the equality of the R_{AA} at the LHC and RHIC in the range $5 \leq p_T \leq 20$ GeV/c corresponds to a fractional shift in the p_T spectrum, S_{loss}, a factor of ~1.5 larger at the LHC than at RHIC, in this p_T range, with presumably larger fractional jet imbalance, because the $1/p_T^n$ power-law spectrum is flatter at $\sqrt{s} = 2.76$ TeV, with $n \approx 6$, compared to $n = 8.1$ at $\sqrt{s} = 200$ GeV (recall Eq. 14.3).

The reduction of suppression (rise in R_{AA}) with increasing p_T continues at the LHC, approaching $R_{AA} \approx 0.6$ for $p_T = 50$–100 GeV/c, as shown by CMS [780]

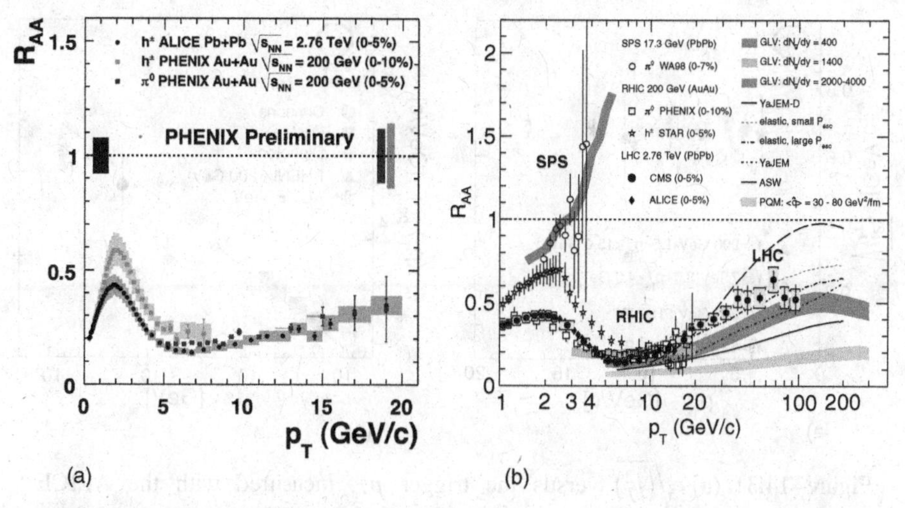

Figure 14.30 (a) ALICE measurements [778] of R_{AA} versus p_T for non-identified charged particles in central (0–5%) Pb+Pb collisions at $\sqrt{s_{NN}} = 2.76$ TeV compared to PHENIX measurements [779] of π^0 and $h^{\pm}$ in Au+Au collisions at $\sqrt{s_{NN}} = 200$ GeV. (b) CMS measurements [780] of R_{AA} of non-identified charged particles in central (0–5%) Pb+Pb collisions compared to measurements at lower c.m. energies and theoretical curves as indicated. See reference [780] for details.

(Figure 14.30b). The reduction of suppression at $p_T \sim 100$ GeV/c is not surprising, considering that the fractional di-jet imbalance for $p_T > 120$ GeV/c observed at the LHC is smaller than for $10 \leq p_T \leq 20$ GeV/c at RHIC, and likely even smaller compared to this p_T range at the LHC because of the larger fractional shift in the spectrum.

14.4.3 Two particle correlations and direct-γ production

ALICE has performed two particle correlation measurements in p–p collisions [781] by the methods used at the ISR and RHIC and has obtained results for the jet fragmentation transverse momentum $\sqrt{\langle j_T^2 \rangle}$ and k_T, expressed as $\langle p_T \rangle_{pair} = \sqrt{2}\,\langle k_T \rangle$, in agreement with the trends from lower $\sqrt{s}$ (Figure 14.31), namely, $\langle p_T \rangle_{pair}$ keeps increasing with $\sqrt{s}$ but the jet fragmentation transverse momentum $\sqrt{\langle j_T^2 \rangle}$ as a function of p_{T_t} does not change with $\sqrt{s}$.

In Pb+Pb collisions, ALICE [782] measured I_{AA} from two particle correlations of non-identified trigger and associated charged particles with $8 < p_{T_t} < 15$ GeV/c and 3 GeV/$c < p_{T_a} < p_{T_t}$ (Figure 14.32a). This range is nearly the same as for the PHENIX [770] and STAR [773] measurements at RHIC (Figures 14.25b, 14.26a). For the LHC measurement, only the flat part of the

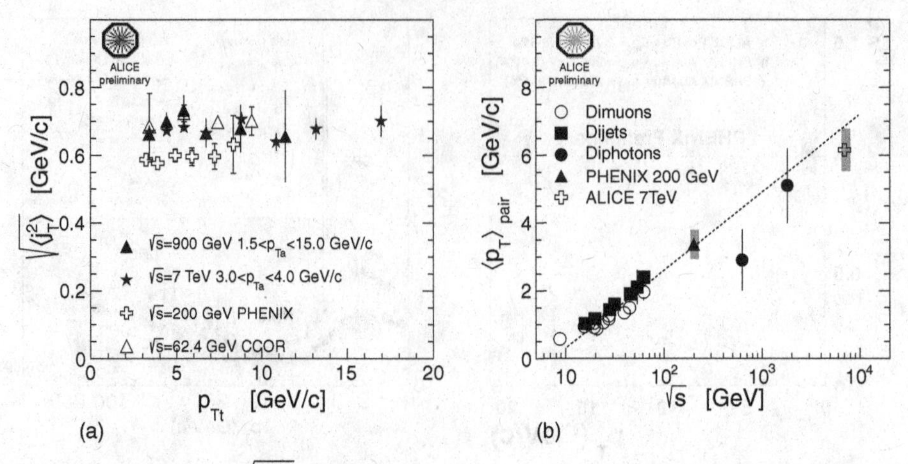

Figure 14.31 (a) $\sqrt{\langle j_T^2 \rangle}$ versus the trigger p_{T_t} measured with the ALICE experiment [781] and compared to the values from CCOR and PHENIX [149, 388]. (b) $\langle p_T \rangle_{pair}$ measured by ALICE at 7 TeV [781], compared to measurements at other $\sqrt{s}$ [149, 402].

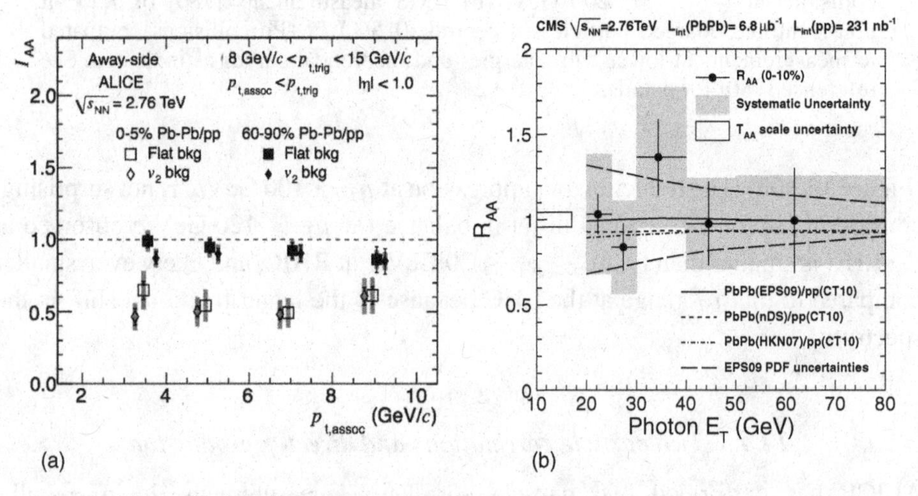

Figure 14.32 (a) ALICE [782] measurement of away-side I_{AA} from two particle correlations of non-identified trigger and associated charged particles with $8 < p_{T_t} < 15$ GeV/c and 3 GeV/$c < p_{T_a} < p_{T_t}$ in central (0–5%) and peripheral (60–90%) Pb+Pb collisions at $\sqrt{s_{NN}} = 2.76$ TeV. Background subtraction with and without v_2 modulation is shown and makes no difference. (b) CMS [783] measurement of R_{AA} for isolated direct photons in central (0–10%) Pb+Pb collisions as a function of the transverse momentum (E_T) of the photon.

distribution is shown, with constant $I_{AA} \approx 0.6$. Thus, one cannot see whether there is an increase in I_{AA} for $p_{T_a} < 3$ GeV/c. What is most interesting compared to the RHIC measurements is that, at the LHC, $I_{AA} \approx 0.6$ is much larger than $R_{AA} \approx 0.2$ in the same p_{T_t} range, while, at RHIC, this difference is considerably smaller

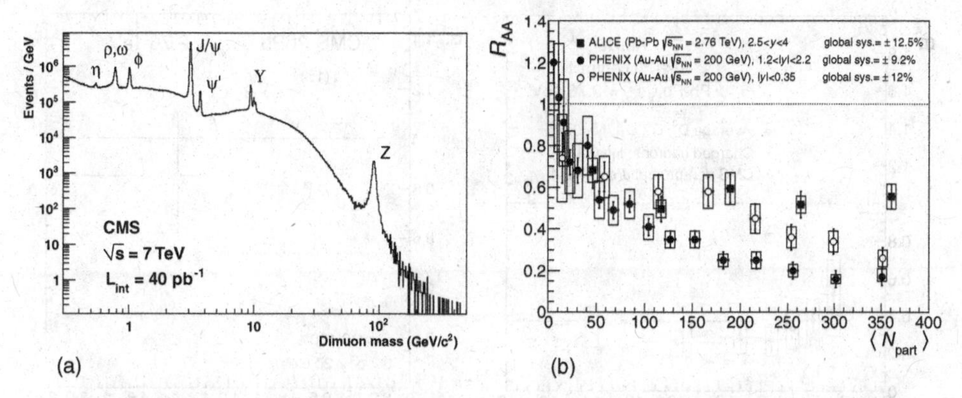

Figure 14.33 (a) CMS [784] di-muon invariant mass spectrum in $\sqrt{s} = 7$ TeV p–p collisions. (b) ALICE measurement of R_{AA} of J/Ψ production down to $p_T = 0$ in the rapidity range $2.5 \leq y \leq 4$ as a function of centrality for Pb+Pb collisions at $\sqrt{s_{NN}} = 2.76$ TeV compared to PHENIX data at $\sqrt{s_{NN}} = 200$ GeV nearer and at mid-rapidity [749].

for π^0–h correlations: $I_{AA} \approx 0.3$, for the same $R_{AA} \approx 0.2$. The challenge to explain this difference should provide a greater understanding of energy loss in the medium as well as surface and tangential biases and punch-through effects.

One LHC measurement with a straightforward explanation is the CMS measurement of R_{AA} for "isolated prompt photons" in central (0–10%) Pb+Pb collisions [783] (Figure 14.32b). The measurement shows that $R_{AA} \approx 1$ (within large statistical and correlated systematic errors) over the range $20 \leq p_T \leq 80$ GeV/c, as expected.

14.4.4 Heavy quark and quarkonium suppression

The beautiful CMS [784] di-muon spectrum in p–p collisions (Figure 14.33a), with impressive resolution over more than three decades in invariant mass, is an example of the di-lepton capabilities of the LHC detectors. Measurements of J/Ψ suppression from the ALICE forward ($2.5 \leq y \leq 4$) di-muon spectrometer [751] (Figure 14.33b) show little centrality dependence but, notably, show less suppression in central collisions than the mid-rapidity measurements at RHIC and considerably less suppression than the forward RHIC measurements ($1.2 \leq |y| \leq 2.2$) [749]. The authors [751] suggest that this is evidence for the recombination of the c and $\bar{c}$ quarks in the deconfined medium. Measurements at mid-rapidity where presumably more $c - \bar{c}$ quarks are produced could provide confirming evidence.

The actual measurement of charm D meson production at mid-rapidity by ALICE at the LHC [785] (Figure 14.34a) confirms the suppression, by a factor

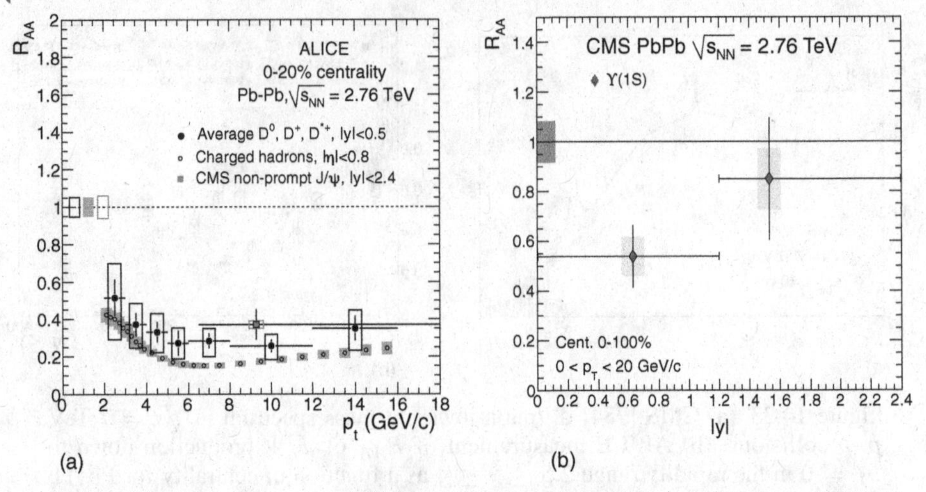

Figure 14.34 (a) R_{AA} of ALICE [785] D mesons, $h^{\pm}$, and CMS [786] non-prompt J/Ψ in central (0–20%) Pb+Pb collisions. (b) CMS [786] measurement of $\Upsilon(1S)$ suppression versus rapidity in minimum bias Pb+Pb collisions.

3 to 4 for $p_T > 5$ GeV/c in central (0–20%) Pb+Pb collisions and equality with the $h^{\pm}$ suppression, as discovered at RHIC. Also shown on Figure 14.34a is the CMS measurement [786] of the suppression of non-prompt J/Ψ from hadrons with b quarks, with comparable suppression to hadrons from c and light quarks, a very exciting result, which keeps the mystery of heavy quark suppression very much alive.

A new quarkonium effect was also opened up by CMS at the LHC with the first measurement of $\Upsilon(1S)$ suppression [786], as shown for minimum bias Pb+Pb collisions as a function of rapidity in Figure 14.34b. $\Upsilon(1S)$ suppression is not supposed to occur in a deconfined QGP because it is the most tightly bound state of all quarkonia and should resist dissociation up to a temperature as high as twice the transition temperature, $2T_c$ [620]. Thus the observed suppression of $\Upsilon(1S)$ might rule out the entire model of quarkonium suppression via Debye screening [617]. However, the observed effect is consistent with the suppression of $\Upsilon(1S)$ from the decay of the $\Upsilon(2S)$ and $\Upsilon(3S)$, whose suppression was observed previously by CMS [787]. This is the beginning of another long chapter in the saga of quarkonium suppression.

14.5 Conclusion

Hard scattering has been one of the most fruitful probes in high energy p–p collisions and has become perhaps even more important as a probe in the considerably more complicated environment of high energy heavy ion collisions. Much has been learned in the study of the dense nuclear matter produced and the search for the Quark Gluon Plasma (QGP), but there are still many unanswered questions. For instance, in Enrico Fermi's classic textbook on nuclear physics dating from 1949 [788], chapter 1 of 10 chapters is "Properties of nuclei" and chapter 2 is "Interaction of radiation with matter." If the present book were a textbook on "Dense nuclear matter and the QGP" one still could not answer most of the questions of chapter 1 raised by Gordon Baym in his review of the scientific motivations for RHIC at Quark Matter 2001 [682]. Regarding chapter 2, "Interaction of QCD radiation with QCD matter," while it is clear that color charged partons lose energy in a medium "of similarly exposed color charges" [716], the exact fundamental mechanisms with experimentally verified formulas, as in Section 3.2.1 for QED, simply do not exist at this time, not to mention all the collective effects that may occur in a such a medium. Clearly much still remains to be learned in the very exciting times that lie ahead.

Appendix A
Probability and statistics

A.1 Probability

The theory of probability starts with a known population or "probability density function" (or probability distribution) which can then be used to predict the properties of samples drawn from this population or distribution. In experimental science, we are usually confronted with the results of an experiment or experiments (i.e. the sample) from which we attempt to draw conclusions about the unknown population or probability density function, usually in terms of parameters. The a priori probability of any outcome is simply defined as the fraction of that particular outcome to all possible outcomes. For instance to find the probability of an outcome in games of chance, such as rolling a pair of dice, the first thing to do is to enumerate all possible outcomes.

A random variable or variate is a numerical valued variable defined on a sample space, i.e. it is a number assigned to the outcome of a chance occurrence. The sample space is defined as the set of all possible outcomes of the chance occurrence.

The frequency function or probability density function for a continuous random variable x is a function $f(x)$ that posesses the following properties:

(i)

$$f(x) \geq 0,$$

(ii)

$$\int_{x_{min}}^{x_{max}} f(x)dx = 1$$

where x_{min} and x_{max} are the minimum and maximum possible values of x,

(iii)

$$\int_{a}^{b} f(x)dx = P\{a \leq X \leq b\}$$

where a and b are any two values of x with $x_{min} \leq x < b \leq x_{max}$.

Then

$$F(x) = \int_{x_{min}}^{x} f(x)dx = P\{x_{min} \leq X \leq x\} \tag{A.1}$$

and

$$f(x) = \frac{dF(x)}{dx}. \tag{A.2}$$

A.1.1 Permutations and combinations

Permutations A permutation is defined as an arrangement of objects in a particular order. For instance suppose there is a jar containing n balls labeled $1, \ldots, n$. The number of different outcomes from drawing all n objects out of the jar, where the order counts is n for the first ball, $\times(n-1)$ for the second ball, $\ldots \times 1$ for the last ball. Thus the number of permutations of n objects is $n(n-1)(n-2)\ldots 1 \equiv n!$, where $n!$ is denoted n-factorial.

Suppose we only want to draw m balls out of the jar with n balls, like they do in the lottery, the number of ways to do this in order is denoted ${}_nP_m$, where

$${}_nP_m = n(n-1)\ldots(n-m+1). \tag{A.3}$$

By multiplying and dividing Eq. A.3 by $(n-m)!$ one can see that:

$${}_nP_m = \frac{n!}{(n-m)!}. \tag{A.4}$$

Combinations Now suppose that we do not care about the order of the m objects, this is called a combination. Since the number of permutations of the m objects is $m!$, we can write the number of different combinations (without regard to order) of m objects from a sample of n objects, denoted ${}_nC_m$, as:

$${}_nC_m = \frac{{}_nP_m}{m!} = \frac{n!}{(n-m)!\,m!}. \tag{A.5}$$

The combinatoric factor is often written in another way:

$${}_nC_m = \begin{pmatrix} n \\ m \end{pmatrix} = \frac{n!}{(n-m)!\,m!}. \tag{A.6}$$

The most common use of combinatoric factors is in products of sums and power series. For instance

$$(a+b)^n = \sum_{m=0}^{n} \begin{pmatrix} n \\ m \end{pmatrix} a^{n-m} b^m = \sum_{m=0}^{n} \frac{n!}{(n-m)!\,m!} a^{n-m} b^m. \tag{A.7}$$

From this simple equation, we get the sum rule for combinatoric factors by setting $a = b = 1$

$$2^n = \sum_{m=0}^{n} \binom{n}{m} = \sum_{m=0}^{n} \frac{n!}{(n-m)!\,m!}, \tag{A.8}$$

where $0! \equiv 1$.

Permutations when some objects are alike Suppose the jar contains m_1 balls labeled 1, m_2 balls labeled 2, ... m_k balls labeled k, how many different outcomes can I get when taking all n objects out of the jar? The total number of permutations is $n!$; but for each of the numbers $1, \ldots, k$, the permutations of the same numbered balls $m_1!, m_2!, \ldots, m_k!$ are not distinguishable. Thus the total number of permutations of n objects of which there are $m_1, m_2, \ldots m_k$ alike is:

$$\frac{n!}{m_1!m_2!\cdots m_k!} \equiv \binom{n}{m_1, m_2, \ldots, m_k}. \tag{A.9}$$

Equation A.9 is known as the multinomial coefficient because it is the combinatoric factor used in expanding the multinomial function:

$$(a_1 + a_2 + \cdots + a_k)^n = \sum_{m_1+m_2+\cdots+m_k=n} \binom{n}{m_1, m_2, \ldots, m_k} a_1^{m_1} a_2^{m_2} \cdots a_k^{m_k}. \tag{A.10}$$

We can also get a sum rule for the multinomial coefficients by setting $a_1 = a_2 = \cdots = a_k = 1$:

$$k^n = \sum_{m_1+m_2+\cdots+m_k=n} \frac{n!}{m_1!m_2!\cdots m_k!}. \tag{A.11}$$

A.1.2 Bayes' rule and conditional probability

Let A and B be two possible outcomes with probabilities $P(A)$ and $P(B)$. Bayes' rule defines the conditional probabilities, where $P(A.and.B)$ is the probability for both outcomes to occur:

$$P(A.and.B) = P(A) \times P(B)|_A = P(B) \times P(A)|_B. \tag{A.12}$$

The a priori or prior probabilities $P(A)$ and $P(B)$ are very different than the conditional probabilities $P(A)|_B$, the conditional probability of A given that B has occurred, and $P(B)|_A$, the conditional probability of B given that A has occurred. However the conditional probabilities are simply related to each other:

$$P(A)|_B = \frac{P(A) \times P(B)|_A}{P(B)} = P(B)|_A \times \frac{P(A)}{P(B)}. \tag{A.13}$$

A.1.2.1 Example of Bayes' rule

Suppose an event D could be caused by n possible causes, $C_1, \ldots, C_n$, which are mutually exclusive and exhaustive (they exhaust the sample space, i.e. they represent all possible causes), with known probabilities, $P(D)|_{C_1}, \ldots, P(D)|_{C_n}$. For example, suppose D is death, C_1 is cancer, C_2 is auto-accident, C_3 is heart attack... C_n is all others. Given that D has happened, what is the probability that it was caused by C_k:

$$P(C_k)|_D = P(D)|_{C_k} \times \frac{P(C_k)}{P(D)} = \frac{P(C_k) \times P(D)|_{C_k}}{P(D)}.$$

Since all the outcomes are mutually exclusive:

$$P(D) = P(D.and.C_1) + \cdots + P(D.and.C_n)$$

$$P(D) = P(C_1) \times P(D)|_{C_1} + \cdots + P(C_n) \times P(D)|_{C_n}$$

$$P(C_k)|_D = \frac{P(C_k) \times P(D)|_{C_k}}{P(C_1) \times P(D)|_{C_1} + \cdots + P(C_n) \times P(D)|_{C_n}}.$$

Bayes' rule is usually mis-applied by taking all the prior probabilities equal.

A.1.3 Expectation values and moments

(i) **Expectation value** The expectation value for a variate x from a probability distribution $f(x)$ is defined as:

$$\langle x \rangle \equiv \int_{-\infty}^{\infty} x f(x) dx \rightarrow \sum_{i=1}^{n} x_i f_i \tag{A.14}$$

where the integral is for continuous variables, nominally taking values from $-\infty$ to ∞; and the sum is for discrete valued random variables with n possible outcomes $x_1, \ldots, x_n$ with probability f_i such that $\sum f_i = 1$.

Similarly for any function $g(x)$,

$$\langle g(x) \rangle \equiv \int_{-\infty}^{\infty} g(x) f(x) dx \rightarrow \sum_{i=1}^{n} g(x_i) f_i. \tag{A.15}$$

Note that for a constant, c, $\langle c \rangle = c$. Also $\langle \langle x \rangle \rangle = \langle x \rangle$ since $\langle x \rangle$ is a constant, i.e. does not depend on x, in the integrand of Eq. A.15.

(ii) **Moments of a distribution** The expectation value, also known as the mean, μ, or mean value, is the first moment of the distribution, more correctly the first moment of the distribution about the origin. The kth moment of a probability density function $f(x)$ about the origin is defined as:

$$\mu'_k \equiv \langle x^k \rangle \equiv \int_{-\infty}^{\infty} x^k f(x) dx \rightarrow \sum_{i=1}^{n} x_i^k f_i. \quad \text{(A.16)}$$

Clearly, $\mu'_0 = 1$, $\mu'_1 \equiv \mu = \langle x \rangle$.

Also $\langle \mu \rangle \equiv \mu$, since when used in the integral μ or $\langle x \rangle$ does not depend on x.

(iii) **Central moments, or moments about the mean** Apart from the 0th and first moments, it is usually much more convenient to take moments about the mean. The kth moment of $f(x)$ about its mean is defined (n.b. without the $'$) as:

$$\mu_k \equiv \langle (x-\mu)^k \rangle \equiv \int_{-\infty}^{\infty} (x-\mu)^k f(x) dx \rightarrow \sum_{i=1}^{n} (x_i - \mu)^k f_i$$

$$\mu_k = \int_{-\infty}^{\infty} (x - \langle x \rangle)^k f(x) dx \rightarrow \sum_{i=1}^{n} (x_i - \langle x \rangle)^k f_i. \quad \text{(A.17)}$$

Thus, for any distribution:

$$\mu_1 = \langle x - \langle x \rangle \rangle = \langle x - \mu \rangle = \langle x \rangle - \langle \mu \rangle = \mu - \mu = 0 \quad \text{(A.18)}$$

$$\mu_2 = \langle (x - \langle x \rangle)^2 \rangle = \langle x^2 - 2x \langle x \rangle + \langle x \rangle^2 \rangle = \langle x^2 \rangle - 2 \langle x \rangle \langle x \rangle + \langle x \rangle^2$$
$$\mu_2 = \langle (x - \langle x \rangle)^2 \rangle = \langle x^2 \rangle - \langle x \rangle^2 \equiv \sigma^2. \quad \text{(A.19)}$$

The kth moment about the mean (Eq. A.17) can also be expanded in the same way with the result:

$$\mu_k = \langle (x - \langle x \rangle)^k \rangle = \sum_{m=2}^{k} \frac{k!}{(k-m)!m!} (-\langle x \rangle)^{k-m} (\langle x^m \rangle - \langle x \rangle^m). \quad \text{(A.20)}$$

(iv) **Those special moments** In addition to the second moment about the mean, $\mu_2 \doteq \sigma^2$ (Eq. A.19) which is called the variance. The central moments μ_3 and μ_4 have special names, when properly normalized.

Skewness

$$\gamma_1 \equiv \frac{\mu_3}{\mu_2^{3/2}} = \frac{\mu_3}{\sigma^3} \quad \text{(A.21)}$$

is called the coefficient of skewness, or simply the skewness.

Kurtosis

$$\gamma_2 \equiv \frac{\mu_4}{\mu_2^2} - 3 = \frac{\mu_4}{\sigma^4} - 3 \tag{A.22}$$

is called the coefficient of kurtosis, or simply the kurtosis. Both the skewness and kurtosis are defined so that they are zero for a Gaussian. The skewness and kurtosis represent what are also called normalized cumulants, which will be discussed further below.

A.2 The moment generating functions

Let x be a random variate and t a parameter and consider

$$M'_x(t) \equiv \langle e^{xt} \rangle = \int_{R_x} e^{xt} f(x) dx \tag{A.23}$$

where R_x, e.g. $-\infty$ to ∞, represents the sample space of $f(x)$.

Now expand e^{xt}:

$$M'_x(t) = \langle e^{xt} \rangle = \left\langle \sum_{k=0}^{\infty} \frac{(xt)^k}{k!} \right\rangle \tag{A.24}$$

$$M'_x(t) = \langle e^{xt} \rangle = \sum_{k=0}^{\infty} \frac{t^k}{k!} \langle x^k \rangle. \tag{A.25}$$

The point of this definition is that all the moments of $f(x)$ can be derived from $M'_x(t)$ by differentiation:

$$\frac{d^p M'_x(t)}{dt^p} = \sum_{k=p}^{\infty} \frac{k(k-1)\cdots(k-p+1)}{k!} t^{k-p} \langle x^k \rangle \tag{A.26}$$

$$\frac{d^p M'_x(t)}{dt^p} = \sum_{k=p}^{\infty} \frac{t^{k-p}}{(k-p)!} \langle x^k \rangle, \tag{A.27}$$

since if $k \geq p$, $k! = k(k-1)\cdots(k-p+1)\left[(k-p)\cdots 1\right]$, where the term in square brackets is $(k-p)!$. Hence to find the term $\langle x^p \rangle$, we set $t = 0$. In other words:

$$\left. \frac{d^k M'_x(t)}{dt^k} \right|_{t=0} = \langle x^k \rangle. \tag{A.28}$$

We can also substitute Eq. A.28 back into Eq. A.25

$$M'_x(t) = \sum_{k=0}^{\infty} \frac{t^k}{k!} \left. \frac{d^k M'_x(t)}{dt^k} \right|_{t=0} \tag{A.29}$$

which is just the Taylor expansion for any function of one variable about $t = 0$.

A.3 Many cheerful facts

In fact the mathematics of the moment generating function itself is very interesting. We give further examples here without detailed proof.

(i) **Moment generating function about any origin a**

$$M'_{x-a}(t) = \langle e^{t(x-a)} \rangle = \sum_{k=0}^{\infty} \frac{t^k}{k!} \langle (x-a)^k \rangle, \tag{A.30}$$

$$M'_{x-a}(t) = e^{-at} M'_x(t). \tag{A.31}$$

(ii) **Moment generating function about the mean**

$$M_x(t) \equiv M'_{x-\mu}(t) = e^{-\mu t} M'_x(t), \tag{A.32}$$

$$M'_x(t) = e^{\mu t} M_x(t). \tag{A.33}$$

The central moment generating function also generates the central moments:

$$\mu_k = \langle (x - \langle x \rangle)^k \rangle = \left. \frac{d^k M_x(t)}{dt^k} \right|_{t=0}. \tag{A.34}$$

(iii) **Moment generating function of sums and averages of n independent variables**

$$\Sigma_x \equiv \sum_{i=1}^{n} x_i,$$

$$M'_{\Sigma_x}(t) = \left[M'_x(t) \right]^n, \tag{A.35}$$

$$M'_{\Sigma_x/n}(t) = \left[M'_x(t/n) \right]^n. \tag{A.36}$$

(iv) **The cumulant generating function**

$$g_x(t) \equiv \ln \langle e^{tx} \rangle = \ln M'_x(t) = \sum_{n=1}^{\infty} \kappa_n \frac{t^n}{n!} = -\sum_{n=1}^{\infty} \frac{1}{n} \left(-\sum_{m=1}^{\infty} \mu'_m \frac{t^m}{m!} \right)^n \tag{A.37}$$

where κ_n are the cumulants. Note that this is the ln of the non-central moment generating function. In general the cumulants κ_k represent the kth moment with all k-fold combinations of the lower order moments subtracted:

$$\kappa_k = \mu'_k - \sum_{m=1}^{k-1} \frac{(k-1)!}{(k-m)!(m-1)!} \kappa_k \, \mu'_{k-m}. \tag{A.38}$$

For instance, in terms of central moments

$$\begin{aligned}
\kappa_1 &= \mu_1' = \mu \\
\kappa_2 &= \mu_2' - \mu_1^2 = \mu_2 = \sigma^2 \\
\kappa_3 &= \mu_3 \\
\kappa_4 &= \mu_4 - 3\mu_2^2 \\
\kappa_5 &= \mu_5 - 10\mu_3\mu_2 \\
\kappa_6 &= \mu_6 - 15\mu_4\mu_2 - 10\mu_3^2 + 30\mu_2^3.
\end{aligned} \tag{A.39}$$

For more information about cumulants, see for example reference [789].

(v) **The factorial cumulant generating function** The factorial cumulants, also known as Mueller moments [790], are used in the study of multiparticle correlations, which is soft physics, possibly beyond the scope of this book; but for completeness, the factorial cumulant generating function [791] is:

$$\ln\left\langle (1+t)^x \right\rangle \equiv \sum_{m=1}^{\infty} \kappa_{[m]} \frac{t^m}{m!}. \tag{A.40}$$

For the all important negative binomial distribution (see below) the factorial cumulants are simple and elegant:

$$\kappa_{[m]} = \mu^m \frac{(1-m)!}{k^{m-1}} \tag{A.41}$$

where μ is the mean, and k is the NBD parameter (see below).

A.4 Statistics

Consider a sample $x_1, x_2, \ldots, x_n$, which is the result of repetitive independent trials from the same population or probability density function. The formal definition of a statistic is a quantity computed entirely from the sample. The principal statistics in common, sometimes everyday, use can be enumerated in a relatively short list.

(i) **Average** The average value of the sample is denoted $\overline{x}$:

$$\overline{x} \equiv \frac{1}{n} \sum_{i=1}^{n} x_i. \tag{A.42}$$

We can also calculate the mean and variance of the average of a repeated independent sample from a known distribution if we know the mean and variance of the distribution, μ and σ^2.

(a) Expectation value of the average:

$$\overline{x} = \frac{1}{n}\sum_{i=1}^{n} x_i$$

$$\langle\overline{x}\rangle = \frac{1}{n}\sum_{i=1}^{n} x_i = \frac{1}{n}\sum_{i=1}^{n} \langle x_i\rangle = \frac{n\,\langle x\rangle}{n}$$

$$\langle\overline{x}\rangle = \langle x\rangle \tag{A.43}$$

since the means of the independent x_i sampled from the same distribution are all the same, $\langle x_i\rangle = \langle x\rangle$.

(b) Variance of the average:

$$\overline{x}^2 = \frac{1}{n^2}\sum_{i=1}^{n} x_i \sum_{j=1}^{n} x_j$$

$$\overline{x}^2 = \frac{1}{n^2}\left[\sum_{i=1}^{n} x_i^2 + \sum_{(i,j,i\neq j)=1}^{n} x_i x_j\right]$$

$$\left\langle\overline{x}^2\right\rangle = \frac{n\left\langle x^2\right\rangle + (n^2 - n)\,\langle x\rangle^2}{n^2} \tag{A.44}$$

since $\left\langle x_i x_j\right\rangle = \langle x\rangle\,\langle x\rangle = \langle x\rangle^2$ if $i \neq j$.

Then

$$\sigma_{\overline{x}}^2 = \left\langle\overline{x}^2\right\rangle - \langle\overline{x}\rangle^2 = \frac{1}{n^2}(n\left\langle x^2\right\rangle + (n^2 - n)\,\langle x\rangle^2 - n^2\,\langle x\rangle^2)$$

$$\sigma_{\overline{x}}^2 = \frac{n\left\langle x^2\right\rangle - n\,\langle x\rangle^2}{n^2}$$

$$\sigma_{\overline{x}}^2 = \frac{\sigma_x^2}{n}. \tag{A.45}$$

(ii) **The kth moment of the sample** The kth moment of a sample is the average $\overline{x^k}$

$$\overline{x^k} \equiv \frac{1}{n}\sum_{i=1}^{n} x_i^k. \tag{A.46}$$

(iii) **The kth central moment** The central moments are the moments about the average:

$$\overline{(x-\overline{x})^k} \equiv \frac{1}{n}\sum_{i=1}^{n}(x_i - \overline{x})^k$$

$$\overline{(x-\overline{x})^k} = \frac{1}{n}\sum_{i=1}^{n}\sum_{m=2}^{k}\frac{k!}{(k-m)!m!}(-\overline{x})^{k-m}(x_i^m - \overline{x}^m). \tag{A.47}$$

The low moments are the most useful and have special names. For instance the following.

(iv) **The sample variance, s^2:**

$$s^2 \equiv \frac{1}{n}\sum_{i=1}^{n}(x_i - \overline{x})^2. \tag{A.48}$$

It is easy to see that

$$s^2 = \overline{x^2} - (\overline{x})^2 = \frac{1}{n}\sum_{i=1}^{n}(x_i - \overline{x})^2 = \frac{1}{n}\sum_{i=1}^{n}x_i^2 - \left(\frac{1}{n}\sum_{i=1}^{n}x_i\right)^2. \tag{A.49}$$

(a) Expectation value of the sample variance

$$\begin{aligned}\left\langle s^2\right\rangle &= \left(\frac{1}{n}\sum_{i=1}^{n}\left\langle x_i^2\right\rangle\right) - \left\langle\overline{x}^2\right\rangle \\ &= \frac{n\left\langle x^2\right\rangle}{n} - \left[\frac{\left\langle x^2\right\rangle + (n-1)\left\langle x\right\rangle^2}{n}\right] \\ &= \frac{n-1}{n}\left\langle x^2\right\rangle - \frac{n-1}{n}\left\langle x\right\rangle^2 \\ &= \frac{n-1}{n}\sigma_x^2,\end{aligned} \tag{A.50}$$

where we have used Eq. A.44 for $\left\langle\overline{x}^2\right\rangle$. This shows that the sample variance, s^2, is a biased estimator of the variance; and that the correct estimator for the variance is

$$\frac{n}{n-1}s^2 = \frac{1}{n-1}\sum_{i=1}^{n}(x_i - \overline{x})^2 \tag{A.51}$$

as we all learned in high school. Also see Appendix D.2.

A.5 Some useful probability distributions

A.5.1 Binomial distribution

The binomial distribution is the result of repeated independent trials, each with the same two possible outcomes: success, with probability p, and failure, with probability $q = 1 - p$. The probabilities must remain the same for all the independent trials. The probability for m successes on n trials is:

$$P(m)|_n = \frac{n!}{m!(n-m)!}\,p^m(1-p)^{n-m} \tag{A.52}$$

and the mean, standard deviation and $F_2 - 1$ of the distribution [789] are

$$\mu \equiv \langle m \rangle = np \qquad \sigma = \sqrt{np(1-p)} \qquad \frac{\sigma^2}{\mu^2} = \frac{1}{\mu} - \frac{1}{n} \qquad F_2 - 1 = -\frac{1}{n}. \tag{A.53}$$

A distinguishing feature of the binomial distribution is that $F_2 - 1$ is negative.

A.5.2 Multinomial distribution

A generalization of the binomial distribution is the case of n repeated independent trials, where there are k possible mutually independent and exclusive outcomes, with probabilities $p_1, p_2, \ldots, p_k$. The probability for the distribution of number of outcomes $m_1, m_2, \ldots, m_k$ on n trials is

$$P(m_1, m_2, \ldots, m_k)|_n = \frac{n!}{m_1!m_2!\cdots m_k!}\, p_1^{m_1} p_2^{m_2} \cdots p_k^{m_k} \tag{A.54}$$

where it is important to understand that

$$\sum_{i=1}^{k} p_i = 1 \qquad \text{and} \qquad \sum_{i=1}^{k} m_i = n \quad .$$

i.e. each trial must give a result p_1 or p_2 or $\cdots$ p_k.

For each outcome, there is a mean and standard deviation which is essentially binomial because for any single outcome i, the probability all the other outcomes is $\sum_{j(\neq i)=1}^{k} p_j = 1 - p_i \equiv q_i$, so that:

$$\mu_i \equiv \langle m_i \rangle = np_i \qquad \sigma_i = \sqrt{np_i(1-p_i)}. \tag{A.55}$$

A.5.3 Poisson distribution

The Poisson distribution is the limit of the binomial distribution for a large number of independent trials, n, with small probability of success, p, such that $\mu = np$, the expectation value of the number of successes m, remains fixed:

$$P(m)|_\mu = \frac{\mu^m e^{-\mu}}{m!} \tag{A.56}$$

$$\langle m \rangle = \mu \qquad \sigma = \sqrt{\mu} \qquad \frac{\sigma^2}{\mu^2} = \frac{1}{\mu}. \tag{A.57}$$

For the Poisson distribution, $F_2 - 1 = 0$, and, indeed, for all orders of normalized factorial moments: $F_q - 1 = 0$. The Poisson distribution is intimately linked to the exponential law of radioactive decay of nucleii [792, 793], the time distribution of nuclear disintegration counts, giving rise to the common usage of the term [792] "statistical fluctuations" to describe the Poisson statistics of such counts. The only assumptions required are the independence and constant probability of all trials – in other words, for a sample of radioactive material [792], the decay probability is the same for all atoms, is proportional to the time interval (for small intervals), is independent of time and is independent of the decay of other atoms.

A.5.4 Negative binomial distribution

The negative binomial distribution of an integer m is defined as

$$P(m) = \frac{(m+k-1)!}{m!(k-1)!} \frac{(\frac{\mu}{k})^m}{(1+\frac{\mu}{k})^{m+k}} \tag{A.58}$$

where $P(m)$ is normalized for $0 \leq m \leq \infty$, $\mu \equiv \langle m \rangle$, and some higher moments are:

$$\sigma = \sqrt{\mu(1+\frac{\mu}{k})} \qquad \frac{\sigma^2}{\mu^2} = \frac{1}{\mu} + \frac{1}{k} \qquad F_2 = 1 + \frac{1}{k}. \tag{A.59}$$

The normalized factorial moments (F_q) and normalized factorial cumulants (K_q) [789,790] of the NBD are particularly simple:

$$F_q = F_{(q-1)} \left(1 + \frac{q-1}{k}\right) \qquad\qquad K_q = \frac{(q-1)!}{k^{q-1}}. \tag{A.60}$$

The binomial distribution gives the probability of m successes on n repeated independent trials, each with the same probability p of success and $1-p$ of failure, while the negative binomial distribution gives the probability that the kth success occurs on the nth trial, where $m = n - k$ represents the number of trials more than the desired number of successes. Alternatively, the NBD is the distribution of the number of trials more than the number of successes, $m = n - k$, for a fixed number of successes, k:

$$P(n)|_k = P(m)|_k = \frac{(m+k-1)!}{m!(k-1)!} \, p^k \, (1-p)^m \tag{A.61}$$

and $P(m)$ is normalized for $0 \leq m \leq \infty$. This goes to the standard form (Eq. A.58) with the substitution

$$\langle m \rangle = \mu \quad p = \frac{1}{1+\frac{\mu}{k}} \quad 1 - p = \frac{\frac{\mu}{k}}{1+\frac{\mu}{k}}. \tag{A.62}$$

The NBD, with an additional parameter k compared to a Poisson distribution, becomes Poisson in the limit $k \to \infty$ and binomial for k equal to a negative integer (hence the name). The extra parameter has made the NBD useful to mathematical statisticians as a test for whether a distribution is Poisson – more precisely as a "test for independence in rare events" [794]. The test for a Poisson distribution consists of determining whether the NBD parameter $1/k$ is consistent with zero to within its error $s_{1/k}$, which is given [794] as:

$$s_{1/k} = \frac{s_k}{k^2} = \frac{1}{\mu}\sqrt{\frac{2}{N}} \tag{A.63}$$

where N is the total number of events. For statisticians, the NBD represents the first departure from a Poisson law. Physicists are more likely to describe the NBD as Bose–Einstein ($k = 1$) or generalized Bose–Einstein $k \neq 1$ distributions [795].

A.5.5 Gamma distribution

In distinction to the previous distributions which are defined for integers, the gamma distribution represents the probability density for a continuous variable x and has a parameter p (which is not to be confused with the symbol for probability used above):

$$f(x) = \frac{b}{\Gamma(p)}(bx)^{p-1}e^{-bx} \tag{A.64}$$

where

$$p > 0, \quad b > 0, \quad 0 \leq x \leq \infty,$$

$\Gamma(p) = (p-1)!$ if p is an integer, and $f(x)$ is normalized. The first few moments of the distribution are

$$\mu \equiv \langle x \rangle = \frac{p}{b} \qquad \sigma = \frac{\sqrt{p}}{b} \qquad \frac{\sigma^2}{\mu^2} = \frac{1}{p} \qquad F_2 - 1 = \frac{(1-b)}{p}. \tag{A.65}$$

The gamma distribution has an important property under convolution. Define the n-fold convolution of a distribution with itself as:

$$f_n(x) = \int_0^x dy\, f(y)\, f_{n-1}(x-y); \tag{A.66}$$

then for a gamma distribution (Eq. A.64), the n-fold convolution is simply given by the function

$$f_n(x) = \frac{b}{\Gamma(np)}(bx)^{np-1}e^{-bx} \tag{A.67}$$

i.e. $p \to np$ and b remains unchanged. Notice that the mean μ_n and standard deviation σ_n of the n-fold convolution obey the familiar rule

$$\mu_n = n\mu \qquad \sigma_n = \sigma\sqrt{n}. \tag{A.68}$$

A.6 Sums of independent and correlated random variables

In mathematical statistics [796], the probability distribution of a random variable S_n, which is itself the sum of n independent random variables with a common distribution $f(x)$:

$$S_n = x_1 + x_2 + \cdots + x_n \tag{A.69}$$

is just $f_n(x)$, the n-fold convolution of the distribution $f(x)$. This explains why convolutions, and the gamma distribution with its simple behavior, are so useful for E_T and multiplicity, which are variables of the form of Eq. A.69. There is a particularly interesting and direct application of the gamma distribution to the time interval between every nth count of radioactive decay, where the probability is exponential for the time interval between counts [792, 793, 797]. Since an exponential is just a gamma distribution (Eq. A.64) with $p = 1$, the distribution for the time x between n counts is just given by Eq. A.67, with $p = 1$, and $b = \lambda$, the normalized probability of decay per unit time.

Another complementary case is that of a random variable Z_n, which is the sum of n random variables with distribution $f(x)$ – which are themselves 100% correlated – for example:

$$Z_n = x + x + \cdots + x = nx. \tag{A.70}$$

This is just a scale transformation. The behavior of the mean and the standard deviation for a scale transformation is $\mu \to n\mu$, $\sigma \to n\sigma$, which is quite different than the more familiar behavior of the standard deviation under convolution (Eq. A.68). The result of the scale transformation $x \to nx$ for a gamma distribution (Eq. A.64) is simply $b \to b/n$, with p remaining unchanged. The most interesting example of a scale transformation for a gamma distribution is scaling by the mean value, $\mu = \langle x \rangle$, or $x \to x/\mu$, $p \to p$, $b \to \mu b = p$, with the result:

$$\psi(z) = \langle x \rangle f(x) = \frac{p}{\Gamma(p)}(pz)^{p-1}e^{-pz} \quad \text{where} \quad z = \frac{x}{\langle x \rangle} \tag{A.71}$$

and $\psi(z)$ is normalized for $0 \leq z \leq \infty$. Thus the gamma distribution has the property of "scaling in the mean," which means that the shape of the distribution Eq. A.71 is determined only by the parameter p, independently of the mean value μ. This property does not hold in general and is not satisfied for Poisson or negative binomial distributions. In particle physics, "scaling in the mean" is usually called KNO scaling [111].

A.6.1 Further properties of the negative binomial distribution

The negative binomial distribution bears a strong relationship to the gamma distribution, and becomes a gamma distribution in the limit $\mu \gg k > 1$. In fact, many times, gamma distributions are substituted for NBD to prove various theorems [798]. The convolution property of the gamma distribution also holds for the NBD. The probability distribution of the sum of n independent variables, each distributed as an NBD with mean μ and parameter k, is the n-fold convolution of the distribution, which is an NBD with mean $n\mu$ and parameter nk, so that the ratio μ/k remains constant for the convolutions exactly like the gamma distribution. Furthermore, the familiar rule for the mean and standard deviation (Eq. A.68) is satisfied. It is convenient, in analogy to the gamma distribution, to introduce the parameter

$$b \equiv \frac{k}{\mu} \quad \text{so that} \quad \langle m \rangle \equiv \mu = \frac{k}{b}, \tag{A.72}$$

and then to write the NBD, particularly for large k, as:

$$P(m) = \frac{1}{(1+\frac{1}{b})^k} \frac{k(k+1)\cdots(k+m-1)}{1\cdot 2\cdots m\ (b+1)^m}. \tag{A.73}$$

The only important difference between NBD and gamma distributions is in the limit m or $x \to 0$: for $p > 1$ the limit is always zero for a gamma distribution, whereas for the NBD it is always finite.

A.6.2 Relationship of the binomial, Poisson, gamma and negative binomial distributions

The history of the use of the Poisson distribution by statisticians includes the study of the number of accidental deaths by horse kicks in the Prussian army [794]. However, the Poisson distribution did not work for the case of factory accidents because different workers had different chances of having an accident. If the mean probability of an accident per worker for different workers is distributed according to a gamma distribution (also known as Pearson type III), then the NBD rather than the Poisson is the resultant distribution for all workers [794, 799–801].

In addition to the variation of a Poisson mean value leading to a NBD, any tendency of events to occur in groups instead of independently spreads the variance and makes the distribution more like a NBD. For instance, in repeated binomial trials, there could be some correlation such that some of the outcomes represent more than one success. If the relative probabilities p_1 for 1 success, p_2 for 2 successes, p_n for n successes on a trial form the series, $p_n = p_1 a^{n-1}/n$, then the overall distribution is again negative binomial [794]. Thus it has been stated that [794]

> the agreement of the data about deaths from kicks of a horse in the Prussian army may be taken to mean both (1) that nobody can be killed twice by the kick of a horse, (2) that the fact that one man has been so killed does not indicate an extra liability for others in the same unit to be. The agreement in the radioactivity law would mean that (1) the chances of disintegration of different atoms of the same radioactive substance are approximately equal, (2) the disintegration of one atom does not lead immediately to the disintegration of another.

A.6.3 Compound distributions

A compound distribution results from the sum S_n (Eq. A.69) of n independent random variables with a common distribution $f(x)$, when n is itself a random variable, independent of x. The two interesting examples for the present discussion concern the case where n is either Poisson or negative binomial and $f(x)$ is binomial ($x = 1$ for a success, with probability p; and $x = 0$ for a failure, with probability $q = 1 - p$). Thus, the probability $P(S_n = m)$ for a fixed n is given by Eq. A.52, and n varies randomly according to a distribution. This compound distribution is easier to visualize if one defines $A \equiv S_n$ as the number of successes on n trials, and B as the number of failures on n trials, so that the random variable n is the sum: $n = A + B$. The random variable n is composed of two distinct sub-populations: A, on the interval p, and B, on the interval $1 - p$.

If the random variable n (the number of trials) is Poisson distributed with mean μ, then the distribution of A, the number of successes, is also Poisson with mean $\langle A \rangle = p\mu$, where p is the binomial probability for success on a single trial. More importantly, the distributions of the two sub-populations A and B on the sub-intervals p and $1 - p$ are statistically independent. Again the Poisson distribution forms the "intuition" that the binomial division of a population gives two statistically independent (Poisson) sub-populations which can be summed to obtain the original population.

For the case where n is negative binomial with mean μ and parameter k, the compound distribution A is also NBD [801,802] with mean $p\mu$ and the same parameter k [803, 804]. However, for the NBD, the distributions on the two sub-intervals are *not* statistically independent – the distribution on one sub-interval depends explicitly on the result on the other sub-interval. In other words, if $z = A + B$, where the probability for z is NBD, and at fixed z the probability for A and B is binomial, then "the average of A is a linear function of B and vice-versa" [805]. A corollary of this result is that if A and B are independent random variables with a common NBD, and $z = A + B$, then the probability of A and B for fixed z is *not* binomial [806].

This characteristic property of a compound negative binomial distribution has important physical consequences – forward-backward (long-range) correlations [807–810]. Conversely, the search for a functional form for multiplicity distributions that supported the observed [807,808] forward-backward correlations (where the mean backward multiplicity is linearly proportional to the forward multiplicity) led to the negative binomial distribution [805, 811].

Appendix B

Methods of Monte Carlo calculations

B.1 Introduction

A Monte Carlo calculation is generally nothing other than a method to perform integrals numerically. The problem is to perform an integral over an r dimensional volume R:

$$I \equiv \int_R f(\boldsymbol{x}) d^r \boldsymbol{x} \tag{B.1}$$

where $\boldsymbol{x}$ indicates an r-dimensional vector, which represents the coordinates of a point $\boldsymbol{x}^\alpha$ in the volume R.

This is accomplished by slicing the volume R into M pieces $\Delta R_\alpha, \alpha = 1, \ldots M$, where

$$\Delta R_\alpha \equiv \Delta x_1^\alpha \cdot \Delta x_2^\alpha \cdots \Delta x_r^\alpha \tag{B.2}$$

at the point $\boldsymbol{x}^\alpha$.

Then,

$$I = \lim_{M \to \infty} \sum_{\alpha=1}^{M} f(\boldsymbol{x}^\alpha) \times \Delta R_\alpha. \tag{B.3}$$

B.2 The Monte Carlo method

In a Monte Carlo calculation, this sum or integral Eq. B.3 is performed by generating points $\boldsymbol{x}_i$ according to a probability distribution $W(\boldsymbol{x})$ which is normalized on R,

$$\int_R W(\boldsymbol{x}) d^r \boldsymbol{x} \equiv 1. \tag{B.4}$$

Then, for a total of n throws, the expected number of points lying in ΔR_α is

$$\langle n_\alpha \rangle = n\, W(\boldsymbol{x}^\alpha)\, \Delta R_\alpha \tag{B.5}$$

or

$$\Delta R_\alpha = \frac{\langle n_\alpha \rangle}{n\, W(\boldsymbol{x}^\alpha)}. \tag{B.6}$$

The integral can then be evaluated as:

$$I = \lim_{M\to\infty} \sum_{\alpha=1}^{M} \frac{f(\boldsymbol{x}^\alpha)}{W(\boldsymbol{x}^\alpha)} \frac{\langle n_\alpha \rangle}{n}. \tag{B.7}$$

In a real calculation, one never gets exactly the expected number of generated points on the region ΔR_α. The actual number of points n_α is in general not equal to $\langle n_\alpha \rangle$. Thus we approximate the integral as:

$$J = \lim_{M\to\infty} \sum_{\alpha=1}^{M} \frac{f(\boldsymbol{x}^\alpha)}{W(\boldsymbol{x}^\alpha)} \frac{n_\alpha}{n}, \tag{B.8}$$

and

$$\langle J \rangle = I. \tag{B.9}$$

B.3 The standard formula for the Monte Carlo integral

In practical terms, the limit $M \to \infty$ means that ΔR_α becomes so small that there is never more than one point thrown onto it, i.e. $n_\alpha = 0$ or 1 and $\langle n_\alpha \rangle \ll 1$. It is therefore worthwhile to rewrite the expression for the Monte Carlo integral J in the more intelligible form:

$$J = \frac{1}{n} \sum_{i=1}^{n} \frac{f(\boldsymbol{x}_i)}{W(\boldsymbol{x}_i)} \tag{B.10}$$

where $\boldsymbol{x}_i$ is defined as the value of $\boldsymbol{x}_i$ on the ith throw.

In order to elaborate further on the properties of the Monte Carlo integral J, it is convenient to define the relative weight function $g(\boldsymbol{x})$:

$$g(\boldsymbol{x}) \equiv \frac{f(\boldsymbol{x})}{W(\boldsymbol{x})}. \tag{B.11}$$

Then

$$J = \frac{1}{n} \sum_{i=1}^{n} g(\boldsymbol{x}_i) \equiv \overline{g}_R \tag{B.12}$$

where the subscript R indicates that the average value is taken over the whole region R.

B.4 The error in the Monte Carlo integral

In general, $g(\boldsymbol{x})$ will not have a constant value over the entire region R, but will vary from point to point. Define $\mathcal{F}(g)dg$ as the probability distribution of g, given that $\boldsymbol{x}$ is in R. Thus,

$$\int_{\boldsymbol{x}\,\text{in}\,R} \mathcal{F}(g)dg \equiv 1. \tag{B.13}$$

Then, using the distribution function $\mathcal{F}(g)$ and the rules of expectation values and variances, the following results are straightforward to obtain:

$$I = \langle J\rangle = \langle \overline{g}_R\rangle = \langle \overline{g}\rangle_R = \langle g\rangle_R \tag{B.14}$$

where

$$\langle g\rangle_R \equiv G_R = \int_R g\mathcal{F}(g)dg. \tag{B.15}$$

Similarly, for the variance:

$$\sigma_J^2 = \sigma_{\overline{g}_R}^2 = \frac{\sigma_{gR}^2}{n} = \frac{1}{n}\left(\langle g^2\rangle_R - \langle g\rangle_R^2\right) \tag{B.16}$$

or

$$\sigma_J^2 = \frac{1}{n}\left[\int_R g^2\mathcal{F}(g)dg \; - \; G_R^2\right]. \tag{B.17}$$

B.5 Binned Monte Carlo calculations

This is the most common type of Monte Carlo calculation because, in most cases, we are not only interested in the integral J over the entire region R, but also in the integrals over various subregions of R.

Let the region R consist of a finite number k of mutually exclusive and exhaustive subregions R_l, $l = 1, 2, \ldots, k$. It is instructive to think of the probability distribution of g (Eq. B.11) over the subregions R_l as related to the probability distribution $\mathcal{F}(g)$ over the whole region R (Eq. B.13) as follows.

Define:

$$\mathcal{F}(g)dg \equiv \sum_{l=1}^{k} p_l\mathcal{F}_l(g)dg \tag{B.18}$$

where p_l is the probability that the generated points $\boldsymbol{x}$ fall on the region R_l:

$$p_l = \frac{\int_{R_l} W(\boldsymbol{x})d^r\boldsymbol{x}}{\int_R W(\boldsymbol{x})d^r\boldsymbol{x}} \tag{B.19}$$

and $\mathcal{F}_l(g)dg$ is the conditional probability of g, given that $\boldsymbol{x}$ is on R_l.

According to this definition $\mathcal{F}_l(g)$ is independent of p_l, i.e. given that $\boldsymbol{x}$ is on R_l, the value of g is then randomly distributed according to the distribution $\mathcal{F}_l(g)dg$, where

$$\int_{R_l} \mathcal{F}_l(g)dg \equiv 1 \tag{B.20}$$

and

$$\sum_{l=1}^{k} p_l = 1. \tag{B.21}$$

This is just analogous to the wave function of a quantum mechanical system with k eigenstates:

$$\int_R \mathcal{F}(g)dg = \sum_{l=1}^{k} p_l \int_{R_l} \mathcal{F}_l(g)dg = \sum_{l=1}^{k} p_l = 1. \tag{B.22}$$

For the integral in individual bins, we can slice (Eq. B.12) into k independent regions (bins):

$$J = \frac{1}{n}\sum_{i=1}^{n} g(\boldsymbol{x}_i) \text{ all } \boldsymbol{x}_i \text{ on } R$$

$$J = \frac{1}{n}\sum_{l=1}^{k} \sum_{\boldsymbol{x}_j \text{ on } R_l} g(\boldsymbol{x}_j) \tag{B.23}$$

and define

$$J_l \equiv \frac{1}{n} \sum_{\boldsymbol{x}_j \text{ on } R_l} g(\boldsymbol{x}_j). \tag{B.24}$$

Then

$$J = \sum_{l=1}^{k} J_l \tag{B.25}$$

and

$$\langle J_l \rangle = \int_{R_l} f(\boldsymbol{x})d^r\boldsymbol{x}. \tag{B.26}$$

It is important to note that the integrals J_l (Eq. B.24) on the subregions R_l are no longer averages, but are more complicated. In particular, the value and therefore the variance of J_l depends both on the number of points $\boldsymbol{x}_j$ that happen to fall on the interval R_l, and a variation of $g(\boldsymbol{x}_j)$ on this interval.

Let $n_l \equiv$ the number of points $\boldsymbol{x}_j$ which have fallen on R_l. Then,

$$\langle n_l \rangle = n \; p_l$$

and the observed values of n_l, $l = 1, 2, \ldots, k$ are distributed according to the multinomial distribution. It is then possible to rewrite the expression for J_l:

$$J_l = \frac{1}{n} \sum_{x_j \text{ on } R_l} g(\boldsymbol{x}_j) = \frac{n_l}{n} \times \frac{1}{n_l} \sum_{\substack{j=1 \\ x_j \text{ on } R_l}}^{n_l} g(\boldsymbol{x}_j).$$

If we define $\overline{g}_l$ as the average value of g on the region R_l:

$$\overline{g}_l \equiv \frac{1}{n_l} \sum_{\substack{j=1 \\ x_j \text{ on } R_l}}^{n_l} g(\boldsymbol{x}_j). \tag{B.27}$$

Then we can write a simple equation for the integral on R_l:

$$J_l = \frac{n_l}{n} \overline{g}_l. \tag{B.28}$$

We can then write the expectation value and the variance of the Monte Carlo integral on the region R_l

$$I_l = \langle J_l \rangle = \left\langle \frac{n_l}{n} \overline{g}_l \right\rangle = \left\langle \frac{n_l}{n} \right\rangle \langle \overline{g}_l \rangle = \left\langle \frac{n_l}{n} \right\rangle G_l \tag{B.29}$$

where the subscript l indicates that the average value is taken over region R_l and

$$G_l = \langle \overline{g}_l \rangle = \langle \overline{g} \rangle_l = \langle g \rangle_l = \int_{R_l} g \mathcal{F}_l(g) dg.$$

The relation between Eq. B.29 for the subregion R_l and Eq. B.14 for the whole region R is evident. We can also use Eq. B.25 to give the explicit relationship between the overall integral and the integrals on the individual subregions:

$$I = \langle J \rangle = G_R = \sum_{l=1}^{k} \langle J_l \rangle = \sum_{l=1}^{k} p_l \, G_l. \tag{B.30}$$

We also give the error on the binned Monte Carlo integral:

$$\sigma_{J_l}^2 = \frac{1}{n} \left[p_l \, \sigma_{g_l}^2 + p_l (1 - p_l) G_l^2 \right] \tag{B.31}$$

where

$$\sigma_{g_l}^2 = n_l \sigma_{\overline{g}_l}^2 = \int_{R_l} g^2 \mathcal{F}_l(g) dg \; - G_l^2. \tag{B.32}$$

Similarly, we can relate the overall variance of the Monte Carlo integral (Eq. B.17) to the bin variances by substituting Eq. B.18

$$\sigma_J^2 = \frac{1}{n}\left[\int_R g^2 \mathcal{F}(g)dg \; - \; G_R^2\right]$$
$$= \frac{1}{n}\left[\sum_{l=1}^{k} p_l \int_{R_l} g^2 \mathcal{F}_l(g)dg \; - \; G_R^2\right]. \tag{B.33}$$

By use of the normalization condition, $\sum_{l=1}^{k} p_l = 1$, this expression reduces to

$$\sigma_J^2 = \frac{1}{n}\sum_{l=1}^{k} p_l\, \sigma_{g_l}^2 + \frac{1}{n}\sum_{l=1}^{k} p_l (G_l - G_R)^2. \tag{B.34}$$

This expression clearly shows that the variance of the Monte Carlo integral has two components, the weighted average of (i) the variances of the weights on the individual bins, plus (ii) the mean square deviations of the average weights on the individual bins from the overall average weight. Another important point is that if the randomness of n_l were removed, i.e. we threw a fixed number of points $n_l = np_l$, i.e. not randomly distributed over all regions, then the variance would be

$$\sigma_J^2 = \frac{1}{n}\sum_{l=1}^{k} p_l \sigma_{g_l}^2, \tag{B.35}$$

and the deviation of the average weights on each region from the overall average would not enter. It only enters due to the fluctuations in n_l. Also note that even with variable n_l, if the weighting function used is equally good over all the subregions, then this term will vanish also. This shows how the art of weighting function affects Monte Carlo calculations.

B.5.1 How Eqs. B.29 and B.31 were computed

The key to the calculation is that since $\mathcal{F}_l$ and p_l are independent, the distribution of $g(\boldsymbol{x}_j)$ for $\boldsymbol{x}_j$ on R_l is independent of how many points fall on R_l. Thus $\overline{g}_l$ and n_l are independent variates so that Eq. B.29 factorizes as shown and

$$\left\langle \frac{n_l}{n} \right\rangle = p_l.$$

However, it is important to realize that although $\overline{g}_l$ and n_l are statistically independent, the value of $\overline{g}_l$ depends on n_l (Eq. B.27) so that one has to be careful in multiplying out all terms before taking expectation values. Also the expectation values involve two independent varying distributions, one in n_l and one in $\mathcal{F}(g)$, so it is important to write out explicitly the integral over g and sum over n_l for the expectation value. Recall Eq. B.28:

$$J_l = \frac{n_l}{n}\,\overline{g}_l.$$

For fixed n_l, we take the expectation value for the variation of g:

$$\langle \overline{g}_l \rangle = \langle g \rangle_l = G_l$$

$$\sigma^2_{\overline{g}_l} = \frac{\sigma^2_{gl}}{n_l} \equiv \langle \overline{g}^2_l \rangle - \langle \overline{g}_l \rangle^2$$

or

$$\langle \overline{g}^2_l \rangle = \frac{\sigma^2_{gl}}{n_l} + G^2_l. \tag{B.36}$$

Thus, in general, when taking expectation values, we must first take the expectation value of $\overline{g}^k_l$ which may depend on n_l (as in Eq. B.36) and then take the expectation value over n_l.

For example:

$$\langle J^k_l \rangle = \sum_{n_l=0}^{n} P(n_l)|_n \int_{R_l} \left(\frac{n_l}{n}\right)^k (\overline{g}_l)^k\, \mathcal{F}_l(g)dg. \tag{B.37}$$

The integral over g cannot be factored out of the sum since $\langle \overline{g}^k_l \rangle$ depends on n_l. We write Eq. B.37 as follows:

$$\langle J^k_l \rangle = \left\langle \left(\frac{n_l}{n}\right)^k \langle \overline{g}^k_l \rangle \right\rangle \tag{B.38}$$

where $\langle \overline{g}^k_l \rangle$ is understood to be over the g variation and the other $\langle\rangle$ over the n_l variation.

Now we specifically evaluate the case for $k = 2$ using Eq. B.36:

$$\begin{aligned}\langle J^2_l \rangle &= \left\langle \left(\frac{n_l}{n}\right)^2 \left(\frac{\sigma^2_{gl}}{n_l} + G^2_l\right) \right\rangle \\ &= \frac{\sigma^2_{gl}}{n^2}\langle n_l \rangle + \frac{G^2_l}{n^2}\langle n^2_l \rangle.\end{aligned} \tag{B.39}$$

Then, using $\langle J_l \rangle$ from Eq. B.29:

$$\sigma^2_{J_l} = \langle J^2_l \rangle - \langle J_l \rangle^2 = \frac{\sigma^2_{gl}}{n^2}\langle n_l \rangle + \frac{G^2_l}{n^2}\left(\langle n^2_l \rangle - \langle n_l \rangle^2\right)$$

$$\sigma^2_{J_l} = \frac{\sigma^2_{gl}}{n^2}\langle n_l \rangle + \frac{G^2_l}{n^2}\sigma^2_{n_l}. \tag{B.40}$$

Since n_l is multinomial, $\langle n_l \rangle = np_l$, $\sigma^2_{nl} = np_l(1 - p_l)$; and substituting into Eq. B.40, we obtain Eq. B.31.

The importance of this method is that it allows the integrals in the individual bins and their errors to be evaluated while taking account of the multinomial constraint that $\sum_{l=1}^{k} n_l = n$, the sum of the events on the individual bins equals the total number of events thrown.

B.6 The overall variance is sometimes a bin variance

In many examples of Monte Carlo integrals, the integrand $f(\boldsymbol{x})$ in Eq. B.1 tends to have many regions in which it is zero. In such cases the overall integral may be considered as a bin, the region R_m on which m events have fallen, where $g(\boldsymbol{x}) \neq 0$, with the other $n - m$ events seemingly being irrelevant because their value of $g(\boldsymbol{x})$ (Eq. B.11) is zero. This is particularly true in acceptance calculations, where trials are discarded if they are not accepted.

If we define R_m as the region (of acceptance) on which $g(\boldsymbol{x}) \neq 0$, with probability p_m, and the rest of R as a region on which $g(\boldsymbol{x}) = 0$, then $\langle m \rangle = np_m$; and from Eq. B.30,

$$I = \langle J \rangle = G_R = p_m G_m, \tag{B.41}$$

which means that the overall Monte Carlo integral is smaller than the integral on R_m where $g(\boldsymbol{x}) \neq 0$ due to the zero contribution from the region $R - R_m$. [This is another way of saying that the acceptance is < 1 if there are regions with zero acceptance.] Similarly the error on the integral σ_J can be computed as if it were a binned calculation with two bins: the bin m, with p_m and G_m, and the other bin with $n - m$, $1 - p_m$ and $G_l = 0$. This is done using Eq. B.34, with result:

$$\sigma_J^2 = \frac{1}{n}\Big[p_m \sigma^2_{gm} + p_m(G_m - G_R)^2 + (1 - p_m)G_R^2\Big]. \tag{B.42}$$

This shows that the variance of a Monte Carlo integral actually has three components: (i) the variance on R_m for fixed number of throws m due to the variance of the weights on R_m; (ii) the mean square deviation of the average weight on the region R_m (where $g(\boldsymbol{x}) \neq 0$) from the average value over the entire region R. As noted above, this term is the contribution to the variance in the bin R_m because m is thrown randomly. The third term is the contribution to the overall variance from the region where $g(\boldsymbol{x}) = 0$. Thus it is important to note that this region is not irrelevant, it contributes to the variance of the Monte Carlo integral.

B.7 Example: five ways to find the area of a triangle

Consider a region in x, y space. We wish to find the area of the triangle:

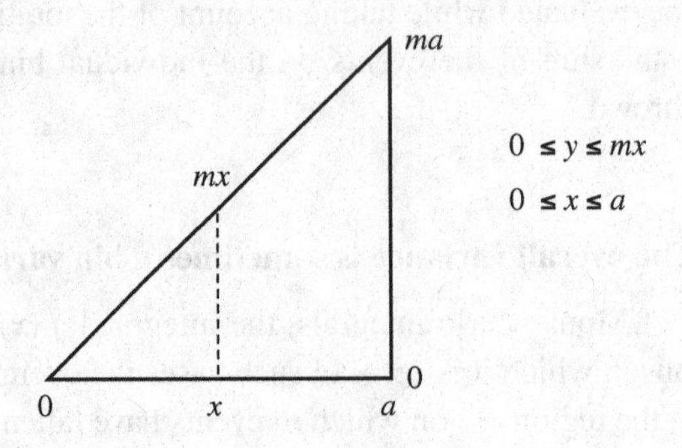

$$A_{R_{x,y}} = \int_{R_{xy}} dxdy = \frac{ma^2}{2} \tag{B.43}$$

by the Monte Carlo method.

B.7.1 Throw points uniformly onto the rectangle which contains the triangle

Let $R'_{x,y}$ be the rectangle $0 \le y \le ma$, $0 \le x \le a$. The area of the rectangle is $A_{R'_{x,y}} = ma^2$, while the area of the triangle (Eq. B.43) is $A_{R_{x,y}} = ma^2/2$. To find the area, throw points uniformly on $R'_{x,y}$ and see how many land inside the triangle, $R_{x,y}$:

$$A = \int_{R'_{x,y}} E(x, y)\, dxdy, \tag{B.44}$$

where $E(x, y) = 1$ on $R_{x,y}$ and zero otherwise.

Generate x and y uniformly over their ranges, i.e. with constant $W(x)$ and $W(y)$:

$$W(x) = C_x \quad W(y) = C_y \quad W(x, y) = W(x)W(y) = C_x C_y.$$

First generate x:

$$\int_0^a C_x dx = 1 = aC_x \qquad C_x = \frac{1}{a} \qquad W(x)dx = \frac{dx}{a}.$$

Find the random number between 0 and 1 distributed according to $W(x)$ by the standard method:

$$r_1 = I(x) = \int_0^x W(x)dx = \frac{x}{a} \qquad x = r_1 a. \tag{B.45}$$

Then generate y:

$$\int_0^{ma} C_y dy = 1 = maC_y \qquad C_y = \frac{1}{ma} \qquad W(y)dy = \frac{dy}{ma}.$$

Find the random number between 0 and 1 distributed according to $W(y)$ by the standard method:

$$r_2 = I(y) = \int_0^y W(y)dy = \frac{y}{ma} \qquad y = r_2 ma. \tag{B.46}$$

Throw n points with x_i, y_i on $R'_{x,y}$ by generating n pairs of random numbers r_1, r_2 according to Eqs. B.45, B.46. The density of generated points on an element of area $dxdy$ of the rectangle is uniform:

$$W(x, y)dxdy = \frac{dxdy}{ma^2}.$$

The Monte Carlo integral is

$$A = J = \frac{1}{n}\sum_{i=1}^{n} E(x_i, y_i) \times ma^2 = ma^2 \times \frac{1}{n}\sum_{i=1}^{n} E(x_i, y_i). \tag{B.47}$$

On the average only 1/2 the n thrown points will land on the triangle $R_{x,y}$. This is a binomial distribution, with the probability of success $p = 1/2$.

$$\left\langle \sum_{i=1}^{n} E(x_i, y_i) \right\rangle = np = \frac{n}{2}$$

and the expectation value of the Monte Carlo integral for the area A is

$$\langle A \rangle = ma^2 \times \frac{1}{n} \times \frac{n}{2} = \frac{ma^2}{2},$$

which is the correct answer.

Perhaps more interesting is the statistical error. Denote

$$E = \sum_{i=1}^{n} E(x_i, y_i) \qquad A = \frac{ma^2}{n} \times E.$$

For a binomial distribution:

$$\sigma_E^2 = \langle E^2 \rangle - \langle E \rangle^2 = np(1-p) = \frac{n}{4}$$

$$\sigma_A^2 = \left(\frac{ma^2}{n}\right)^2 \sigma_E^2 = \left(\frac{ma^2}{n}\right)^2 \frac{n}{4} = \frac{1}{n}\left(\frac{ma^2}{2}\right)^2.$$

Thus, the fractional statistical error for the area of the triangle by this method is:

$$\frac{\sigma_A}{A} = \frac{1}{\sqrt{n}}. \tag{B.48}$$

This result is a bit counter-intuitive because one would think that the fractional error would be the $1/\sqrt{n_s}$ where n_s is the number of successes, e.g. $n/2$ in the present example. It turns out that this reduction in error is a general property of the binomial distribution, and is a function of the value of p, the probability of success.

B.7.2 Throw points uniformly in x, and then for each x, throw points uniformly in y so that they all land in the triangle

$$A = \int_{R_{x,y}} E(x, y)\, dxdy \tag{B.49}$$

where $E(x, y) = 1$ since all the points land on $R_{x,y}$.

Generate x and y with constant $W(x)$ and $W(y)$ over their respective ranges, but now the range in y depends on x.

$$W(x) = C_x \quad W(y) = C_y \quad W(x, y) = W(x)W(y) = C_xC_y.$$

First generate x:

$$\int_0^a C_x dx = 1 = aC_x \qquad C_x = \frac{1}{a} \qquad W(x)dx = \frac{dx}{a}.$$

Find the random number between 0 and 1 distributed according to $W(x)$ by the standard method:

$$r_1 = I(x) = \int_0^x W(x)dx = \frac{x}{a} \qquad x = r_1 a. \tag{B.50}$$

Then generate y from 0 to mx:

$$\int_0^{mx} C_y dy = 1 = mxC_y \qquad C_y = \frac{1}{mx} \qquad W(y)dy = \frac{dy}{mx}.$$

Find the random number between 0 and 1 distributed according to $W(y)$ by the standard method:

$$r_2 = I(y)|_x = \int_0^y W(y)|_x dy = \frac{y}{mx} \qquad y = r_2 mx. \tag{B.51}$$

Throw n points with x_i, y_i on $R_{x,y}$ by generating n pairs of random numbers r_1, r_2 according to Eqs. B.50, B.51. The density of generated points on an element of area $dxdy$ of the rectangle is not uniform:

$$W(x, y)dxdy = \frac{dxdy}{max}.$$

The Monte Carlo integral is

$$A = J = \frac{1}{n}\sum_{i=1}^{n} max_i = ma \times \frac{1}{n}\sum_{i=1}^{n} x_i = ma\overline{x}$$
$$\langle A\rangle = \langle J\rangle = \langle ma\overline{x}\rangle = ma\,\langle x\rangle\,. \tag{B.52}$$

Since x_i is distributed uniformly between 0 and a, $\langle x\rangle = a/2$. More formally

$$\langle x\rangle = \int_0^a xW(x)dx = \int_0^a \frac{x}{a}dx = \frac{a^2}{2a} = \frac{a}{2}$$

$$\langle x^2\rangle = \int_0^a x^2W(x)dx = \int_0^a \frac{x^2}{a}dx = \frac{a^3}{3a} = \frac{a^2}{3}$$

$$\sigma_x^2 = \langle x^2\rangle - \langle x\rangle^2 = \frac{a^2}{3} - \frac{a^2}{4} = \frac{a^2}{12}.$$

Then from Eq. B.52:

$$\langle A\rangle = ma\,\langle x\rangle = \frac{ma^2}{2}$$

which is correct; and

$$\sigma_A^2 = m^2a^2\sigma_{\overline{x}}^2 = m^2a^2\frac{\sigma_x^2}{n} = m^2a^2\frac{a^2}{12n} = \frac{\langle A\rangle^2}{3n}.$$

Thus, the fractional statistical error for the area of the triangle by this method is:

$$\frac{\sigma_A}{A} = \frac{1}{\sqrt{3n}}, \tag{B.53}$$

which is less than the first two methods. For this method all the points land on A but the weight distribution $W(x, y) = 1/(max)$ is not equal to $E(x, y) = 1$ in Eq. B.49. In terms of Eq. B.16 this means that the distribution in $g(\boldsymbol{x})$ has a non-zero variance.

B.7.3 Throw points uniformly distributed in area dx, dy onto the triangle

Although we threw points onto the triangle uniformly distributed in both x and y, above, the weight distribution $W(x, y) = 1/(max)$ was not equal to the integrand $E(x, y) = 1$ for the area in Eq. B.49. We now generate x and y on the triangle so that the density $W(x, y) = C$ is uniform on $R_{x,y}$.

Recall that the integral we are evaluating is (Eq. B.43):

$$A_{R_{x,y}} = \int_{R_{xy}} dxdy = \frac{ma^2}{2}.$$

This is done by the Monte Carlo integral

$$J = \frac{1}{n} \sum_{i=1(\boldsymbol{x} \text{ in } R_{xy})}^{n} \frac{f(\boldsymbol{x}_i)}{W(\boldsymbol{x}_i)} = \frac{1}{n} \sum_{i=1(x,y \text{ in } R_{xy})}^{n} \frac{1}{W(x, y)}. \tag{B.54}$$

To generate points x, y on the triangle R_{xy} with uniform density, $W(x, y) = C_{xy}$, we must first normalize the weight on R_{xy}:

$$\int_0^a dx \int_0^{mx} dy W(x, y) = 1$$

$$\int_0^a dx \int_0^{mx} C_{xy}\, dy = C_{xy} \int_0^a dx\; y\Big]_0^{mx} = C_{xy} \int_0^a mx\, dx \tag{B.55}$$

$$C_{xy} \frac{ma^2}{2} = 1 \qquad C_{xy} = \frac{2}{ma^2}. \tag{B.56}$$

We generate first y then x by the same method:

$$I_1(y) = \int_0^y C_{xy}\, dy = C_{xy}\; y\Big]_0^y.$$

To use the random number $0 \leq r \leq 1$ to generate y, we have to use a normalized $I_1(y)$ for which x must be specified:

$$r_2 = \frac{I_1(y)}{I_1(y_{max} = mx)} = \frac{C_{xy}\, y}{C_{xy}\, mx} = \frac{y}{mx}$$

$$y = r_2 mx.$$

Then using Eq. B.55 which is normalized on $0 \leq x \leq a$

$$r_1 = I_2(x) = C_{xy} \int_0^x mx\, dx = C_{xy} \frac{mx^2}{2} = \frac{x^2}{a^2}$$

$$x^2 = r_1 a^2 \qquad x = a\sqrt{r_1}.$$

Since all of the generated x, y pairs are in R_{xy} with $W(x, y) = C_{xy} = 2/(ma^2)$, the integral (Eq. B.54) becomes:

$$A = J = \frac{1}{n} \sum_{i=1}^{n} \frac{1}{W(x, y)} = \frac{1}{n} \sum_{i=1}^{n} \frac{ma^2}{2} = \frac{1}{n} \frac{nma^2}{2} = \frac{ma^2}{2}. \tag{B.57}$$

Most importantly, since A is just the sum of n identical constant terms, independent of the points x_i, y_i

$$\sigma_A^2 = 0. \tag{B.58}$$

Thus the error in this Monte Carlo calculation is zero. This is due to the fact that $g_{x_i} = (ma^2/2) = A$ is the same constant for all terms in the sum, i.e. we already know the area A. This is general. If for any $f(x, y)$ that we are trying to integrate via Monte Carlo integration, we throw x, y according to $W(x, y) = f(x, y)$, we need to know the $\int f(x, y)dxdy$ in order to generate these points x, y. Hence there is zero error in the Monte Carlo integral, since all the terms in the sum in Eq. B.10 are 1.

B.7.4 Reduce to a lower dimensional problem

This is similar to the previous method, except we reduce the double integral to a single integral.

$$A_{R_{x,y}} = \int_{R_{xy}} dxdy = \int_0^a dx \int_0^{mx} dy = \int_0^a mx\, dx. \tag{B.59}$$

The simplest way to evaluate Eq. B.59 is to throw the single variable x according to $W(x) = f(x) = C\,mx$. First we have to normalize $W(x)$ in order to generate x_i:

$$\int_0^a W(x)\, dx = 1 = \int_0^a C\,mx\, dx \qquad C\frac{ma^2}{2} = 1 \qquad C = \frac{2}{ma^2}.$$

As above, we generate x according to $W(x)$ from a random number $0 \le r \le 1$ by solving the equation for the integral:

$$r = I(x) = \int_0^x W(x) = C\int_0^x mxdx = C\frac{mx^2}{2} = \frac{x^2}{a^2}$$

$$x^2 = ra^2 \qquad x = a\sqrt{r},$$

$$J = \frac{1}{n}\sum_{i=1}^{n}\frac{f(x_i)}{W(x_i)} = \frac{1}{n}\sum_{i=1}^{n}\frac{mx}{Cmx} = \frac{1}{n}\sum_{i=1}^{n}\frac{ma^2}{2} = \frac{ma^2}{2}. \tag{B.60}$$

The result is identical to Eq. B.57 above including $\sigma_A^2 = 0$ because we threw x with $W(x) = f(x)$.

B.7.5 Throw points uniformly onto a known area B in order to find the area of a region A contained in B

This is a straight exercise in binomial statistics. Throw n points uniformly distributed on known area B. The probability of a point landing on A is

$$p = \frac{A}{B}. \tag{B.61}$$

The expected number of points m which land on area A is:

$$\langle m \rangle = np$$

with variance

$$\sigma_m^2 = np(1-p)$$

$$\frac{\sigma_m^2}{\langle m \rangle^2} = \frac{1-p}{np}. \tag{B.62}$$

The area A is given by the integral Eq. B.44, with $E(x, y) = 1$ on A and zero otherwise. The Monte Carlo integral is, following Eq. B.47

$$A = J = \frac{B}{n} \sum_{i=1}^{n} E(x_i, y_i) = B\frac{m}{n}$$

$$\langle A \rangle = \langle J \rangle = B\frac{\langle m \rangle}{n} = B \times \frac{A}{B} = A \tag{B.63}$$

since $p = A/B$ (Eq. B.61) is the probability that a point thrown uniformly at random onto B lands on A.

Then from Eq. B.62, we get the general result:

$$\frac{\sigma_A}{\langle A \rangle} = \frac{1}{\sqrt{n}}\sqrt{\frac{1-p}{p}} = \frac{1}{\sqrt{n}}\sqrt{\frac{1-A/B}{A/B}}. \tag{B.64}$$

In the first case studied, we had chosen $p = A/B = 1/2$, with result $\sigma_A/A = 1/\sqrt{n}$ (Eq. B.48). If $p \ll 1$, then Eq. B.64 reduces to the intuitive result $\sigma_A/A = 1/\sqrt{n_s}$, where $n_s = np$. Another interesting thing to note is that as $p \to 1$, the error can be arbitrarily reduced, even to zero for $p = 1$. We saw this with two of the examples above. This shows the importance of matching the region sampled to the region of interest, as well as reducing the variance of the integrand relative to the weight of thrown points: $g(\mathbf{x}) = f(\mathbf{x})/W(\mathbf{x})$.

Appendix C
T_{AB} and the Glauber Monte Carlo calculation

C.1 T_{AB}

The profile function (or thickness function) of a nucleus A is defined as [812]:

$$T_A(\vec{s}) = \int dz \rho_A(z, \vec{s}), \tag{C.1}$$

which is the number of nucleons per unit area along a direction z at a point from the center of the nucleus represented by a two-dimensional vector $\vec{s}$, where z is perpendicular to $\vec{s}$ and $\rho_A(\vec{r})$ is the nuclear density distribution of nucleus A. For an interaction of nucleus A with nucleus B at impact parameter $\vec{b}$, the nuclear overlap integral $T_{AB}(\vec{b})$ is defined:

$$T_{AB}(\vec{b}) = \int d^2 s T_A(\vec{s}) T_B(\vec{b} - \vec{s}), \tag{C.2}$$

where $d^2 s = 2\pi s ds$ is the two-dimensional area element. The meaning of Eq. C.2 can be understood by considering a nucleon in A at $(z_A, \vec{s})$ and calculating the chance of a hard collision (equivalent to the expected number of hard collisions) with the nucleons from B at the same two-dimensional position ($\vec{b} - \vec{s}$ in coordinates from the center of B):

$$N_{coll}(\vec{s}, \vec{b}) = \sigma \times \int dz_B\, \rho_B(z_B, \vec{b} - \vec{s}) = \sigma \times T_B(\vec{b} - \vec{s}), \tag{C.3}$$

where σ is the hard cross section. Integrating over all the nucleons in A yields:

$$\begin{aligned} N_{coll}(\vec{b}) &= \sigma \times \int d^2 s T_A(\vec{s}) T_B(\vec{b} - \vec{s}) \\ &= \sigma \times T_{AB}(\vec{b}). \end{aligned} \tag{C.4}$$

Thus the nuclear overlap integral $T_{AB}(\vec{b})$ represents the expected number of hard collisions divided by the cross section per collision, assuming that all the nucleons in the overlap region contribute equally. This is true for hard scattering. For example, $N_{coll}(\vec{b})$ could represent the expected number of c–$\bar{c}$ pairs produced with cross section $\sigma_{c\bar{c}}$ at impact parameter $\vec{b}$. The cross section for hard scattering for an $A+B$ interaction at impact parameter b is found by multiplying $N_{coll}(\vec{b})$ by the area element $d^2b = 2\pi b db$:

$$d\sigma_{AB}(b) = N_{coll}(\vec{b})d^2b = d\sigma_{pp}\, T_{AB}(b)d^2b. \tag{C.5}$$

The integral of Eq. C.5 should give the inclusive hard cross section, $\sigma_{AB} = AB\sigma_{pp}$, so, it follows that

$$\int T_{AB}(b)\, d^2b = AB. \tag{C.6}$$

The total inelastic cross section for the interaction of nucleus A with nucleus B is defined as an integral over d^2b for regions of nuclear overlap where the expected number of inelastic nucleon–nucleon collisions (with inelastic cross section σ_{NN}) is not zero. Hence the equation for the total (geometrical) inelastic interaction cross section is

$$\sigma_{int}^{AB} = \int (1 - e^{-\sigma_{NN}\, T_{AB}(b)})\, d^2b, \tag{C.7}$$

where the integrand vanishes for any impact parameter b where $T_{AB}(b) = 0$.

We will need averages of T_{AB} for minimum bias and semi-inclusive cross sections as a function of centrality, typically defined by a certain fraction f of the geometrical interaction cross section. For the total interaction cross section (minimum bias):

$$\langle T_{AB}\rangle \equiv \frac{\int T_{AB}\, d^2b}{\int (1 - e^{-\sigma_{NN}\, T_{AB}(b)})\, d^2b} = \frac{AB}{\sigma_{int}^{AB}}. \tag{C.8}$$

Similarly, for any fraction of the interaction cross section f, defined by an integral over a certain impact parameter range, say b_1 to b_2, the average $\langle T_{AB}\rangle_f$ is defined:

$$\langle T_{AB}\rangle_f \equiv \frac{\int_{b_1}^{b_2} T_{AB}\, d^2b}{\int_{b_1}^{b_2} (1 - e^{-\sigma_{NN}\, T_{AB}(b)})\, d^2b} = \frac{\int_{b_1}^{b_2} T_{AB}\, d^2b}{f \times \sigma_{int}^{AB}}. \tag{C.9}$$

The key thing to realize from the definition of $\langle T_{AB}\rangle_f$ in Eqs. C.8, C.9 is that it corresponds to the Glauber estimate of the mean number of binary nucleon–nucleon inelastic collisions calculated with the cross section, σ_{NN}, divided by σ_{NN}:

$$\langle T_{AB}\rangle_f = \langle N_{coll}^{\sigma_{NN}}\rangle_f / \sigma_{NN}. \tag{C.10}$$

This is why T_{AB} scaling is also called binary or N_{coll} scaling, **although it is important to note that the correct scaling is $\langle N_{coll}^{\sigma_{NN}}\rangle / \sigma_{NN}$.**

C.2 Glauber Monte Carlo calculation

The Glauber model [630] for a nucleus–nucleus scattering is an "impulse approximation" [629] model in a similar spirit to the parton model. The nucleons in the two nuclei are assumed to be frozen in position (in their respective nuclei) during the collision; the individual nucleons travel in straight lines (in the z direction of Eq. C.1) during the collision; the interaction range is assumed to be small compared to the average spacing of the nucleons so that the nucleus–nucleus collision is composed of independent nucleon–nucleon collisions. This model is straightforward to implement in a Monte Carlo calculation [813–816] of for example Eqs. C.2, C.6–C.9. Typically the nuclear density for reasonably spherical nuclei such as Cu and Au is taken as a Woods–Saxon density function

$$\frac{d^3\mathcal{P}}{d^3r} = \frac{d^3\mathcal{P}}{r^2 dr \cos\theta d\theta d\phi} = \rho_{WS}(r) = \rho_0 \frac{1}{1+\exp(\frac{r-c}{a_0})} \tag{C.11}$$

where $c = \{1.18A^{1/3} - 0.48\}$ fm and the diffusivity $a_0 = 0.545$ fm (or use tabulated parameters [817] for the nuclei of interest). Also a hard-core or minimum distance between the centers of two nucleons in a nucleus with a value $r_c = 0.4$ fm is used. An inelastic N–N cross section σ_{NN} appropriate to the c.m. energy is used. The Monte Carlo integration proceeds as follows.

(1) Generate beam and target nuclei composed of A and B nucleons, respectively, with positions at random, uniform in $r^2\rho_{WS}(r)$, $\cos\theta$ and ϕ. If the distance between the generated and any existing nucleon is less than r_c, skip and generate another nucleon.
(2) For nucleus A generate a starting position in the x, y plane uniform in area over a square somewhat larger than $2R$, i.e. $2(c + 4a_0)$ on a side. This is the impact parameter $\vec{b}$.
(3) Follow each of the A nucleons along a straight line in z. Tag and record struck nucleons in B (and A) for which the nucleon from A passes within a projected distance squared $(\Delta x)^2 + (\Delta y)^2 \leq \sigma_{NN}/\pi$.
(4) For the $A + B$ collision calculate any geometrical property, e.g. eccentricity, principal axes, of the struck nucleons, number of binary collisions, etc.
(5) Repeat (2). The nuclei A and B can be regenerated occasionally by repeating (1).

The cross section σ^{AB}_{int} (Eq. C.7) is the fraction of the total number of throws in (2) resulting in at least one struck nucleon in each nucleus, times the area sampled in (2), etc. Calculations of the geometrical distributions of struck nucleons can be done as a function of impact parameter.

Appendix D

Fits including systematic errors

D.1 The maximum likelihood method and least squares fits

The likelihood function $\mathcal{L}$ is defined as the a priori probability of a given outcome. Since Gaussian probability distributions are common (as a consequence of the central limit theorem) and since there is also an important theorem regarding likelihood ratios for composite hypotheses, it is convenient to use the logarithm of the likelihood, $W = -2 \ln \mathcal{L}$.

There are generally two types of fits used for data from experiments. (i) The standard maximum likelihood method is to fit a function to n independent trials of the same distribution, for example the muon lifetime measured from the decay times t_i of individual decays. (ii) The method of least squares is also an extremum method which performs Gaussian least squares fits for binned data from histograms to predictions of the expectation values at each data point, which may be functions of parameters to be determined. This will be discussed following a review of the standard maximum likelihood method and the likelihood ratio test.

D.2 Maximum likelihood fit to a Gaussian, an instructive example

A nice example of the standard maximum likelihood method is a fit to a Gaussian with mean μ and variance σ^2 from the observation of n independent trials with values x_i from this distribution, in other words a sample of n trials from this population. The distribution to be determined is:

$$f(x, \mu, \sigma) = \frac{1}{\sqrt{2\pi\sigma^2}} \exp \frac{-(x-\mu)^2}{2\sigma^2}. \tag{D.1}$$

For n independent trials x_i from this distribution, the likelihood function is:

$$\mathcal{L} = \prod_i f(x_i, \mu, \sigma) = \frac{1}{\sigma^n} \frac{1}{\sqrt{2\pi}^n} \exp - \left[\sum_{i=1}^{n} \frac{(x_i - \mu)^2}{2\sigma^2} \right] \tag{D.2}$$

and

$$W(\boldsymbol{p}) = W(\mu, \sigma^2) \equiv -2\ln\mathcal{L} = \sum_{i=1}^{n} \frac{(x_i - \mu)^2}{\sigma^2} + n\ln\sigma^2 + n\ln 2\pi, \tag{D.3}$$

where $\boldsymbol{p}$ represents a vector of the two parameters μ and σ^2.

The maximum likelihood or the minimum of W for variation of μ and σ^2 is found the usual way by taking derivatives:

$$\begin{aligned}
\frac{\partial W}{\partial \mu} &= -2\sum_{i=1}^{n} \frac{(x_i - \mu)}{\sigma^2} \\
\frac{\partial W}{\partial \sigma^2} &= -\sum_{i=1}^{n} \frac{(x_i - \mu)^2}{\sigma^4} + \frac{n}{\sigma^2} \\
\frac{\partial^2 W}{\partial \mu \partial \sigma^2} &= 2\sum_{i=1}^{n} \frac{(x_i - \mu)}{\sigma^4} \\
\frac{\partial^2 W}{\partial^2 \mu} &= \frac{2n}{\sigma^2} \\
\frac{\partial^2 W}{\partial^2 \sigma^2} &= 2\sum_{i=1}^{n} \frac{(x_i - \mu)^2}{\sigma^6} - \frac{n}{\sigma^4}.
\end{aligned} \tag{D.4}$$

The values $\hat{\mu}$ and $\hat{\sigma}^2$ for the maximum likelihood are given by the solution of:

$$\left.\frac{\partial W}{\partial \mu}\right|_{\hat{\mu},\hat{\sigma}^2} = -2\sum_{i=1}^{n} \frac{(x_i - \hat{\mu})}{\hat{\sigma}^2} = 0 \qquad \hat{\mu} = \frac{1}{n}\sum_{i=1}^{n} x_i = \bar{x} \tag{D.5}$$

$$\left.\frac{\partial W}{\partial \sigma^2}\right|_{\hat{\mu},\hat{\sigma}^2} = -\sum_{i=1}^{n} \frac{(x_i - \hat{\mu})^2}{\hat{\sigma}^4} + \frac{n}{\hat{\sigma}^2} = 0 \qquad \hat{\sigma}^2 = \frac{1}{n}\sum_{i=1}^{n} (x_i - \hat{\mu})^2 = \overline{x^2} - \overline{x}^2 = s_x^2, \tag{D.6}$$

where the average, $\bar{x}$, and the sample variance, s^2, are standard statistics as we discussed in Appendix A.

The maximum likelihood estimators $\hat{\mu}$ and $\hat{\sigma}^2$ are a perfect illustration of the definition of a statistic: a quantity computed entirely from the sample. As such, they obviously vary from sample to sample of n independent trials from the Gaussian and we can compute the expectation values of the statistics $\langle\hat{\mu}\rangle$ and $\langle\hat{\sigma}^2\rangle$, assuming that the x_i are independent trials from Eq. D.1, so that $\langle x\rangle = \mu$, $\sigma_x^2 = \langle x^2\rangle - \langle x\rangle^2 = \sigma^2$. Then,

$$\langle\hat{\mu}\rangle = \langle\bar{x}\rangle = \mu \tag{D.7}$$

and

$$\langle\hat{\sigma}^2\rangle = \langle s^2\rangle = \langle\overline{x^2}\rangle - \langle\bar{x}^2\rangle = \sigma_x^2 + \mu^2 - (\sigma_{\bar{x}}^2 + \mu^2) = \sigma_x^2\left(1 - \frac{1}{n}\right) = \frac{n-1}{n}\sigma^2. \tag{D.8}$$

Thus the maximum likelihood estimator $\hat{\mu}$ is unbiased – its expectation value gives the correct value μ on the average; but the maximum likelihood estimator $\hat{\sigma}$ is biased, giving rise to the need for the famous replacement of n with $n - 1$ in the denominator of Eq. D.6 to get the correct estimate of the variance that we all learned in high school.

Equation D.7 is also a good illustration of the terminology of mathematical statistics: μ is the mean value or expectation value of the variable x which follows the distribution (or is drawn from the population) Eq. D.1 – i.e. the mean value, μ, is a property of the probability distribution. The "average," $\bar{x}$, which is computed entirely from the sample, is a statistic. The average-statistic has the nice property that its expectation or mean value is the mean, μ, of the distribution, i.e. $\langle\bar{x}\rangle = \mu$. This does not imply that the terms "mean" and "average" can be used interchangeably, since, for instance, $\langle x\rangle = \mu$, also. "Mean" is a probability issue, the property of the distribution or population, while "average" is a statistic, a property of the sample.

D.3 Errors of the fitted parameters, Taylor expansion about the minimum W

This method works for both maximum likelihood and least squares fits as well as any method that does a best fit for parameters by finding the minimum of a quantity $W(\boldsymbol{p})$.

For maximum likelihood estimators, the variances can be computed by the same methods that we just used to compute their mean values. However, a more general (and usually easier) method is to take the Taylor expansion of the function $W(\boldsymbol{p}) = W(\mu, \sigma^2)$ about its minimum:

$$\begin{aligned} W(\mu, \sigma^2) = {} & W(\hat{\mu}, \hat{\sigma}^2) + \left.\frac{\partial W}{\partial \mu}\right|_{\hat{\mu},\hat{\sigma}^2} (\mu - \hat{\mu}) + \left.\frac{\partial W}{\partial \sigma^2}\right|_{\hat{\mu},\hat{\sigma}^2} (\sigma^2 - \hat{\sigma}^2) \\ & + \frac{1}{2}\sum_{j=1}^{m}\sum_{k=1}^{m} \left.\frac{\partial^2 W(\boldsymbol{p})}{\partial p_j \partial p_k}\right|_{\hat{\boldsymbol{p}}} (p_j - \hat{p}_j)(p_k - \hat{p}_k) + \ldots \end{aligned} \tag{D.9}$$

where we have reverted to the general notation $\boldsymbol{p}$ for the $m = 2$ parameters, $p_1 = \mu$, $p_2 = \sigma^2$. Since $\partial W/\partial p_i = 0$ for the minimum, $\hat{\boldsymbol{p}}$, Eq. D.9 shows that for small variations of p_i from $\hat{p}_i$, the likelihood ratio $\lambda(\boldsymbol{p})/\lambda(\hat{\boldsymbol{p}})$ for the derived parameters, represented by

$$-2\ln\frac{\lambda(\boldsymbol{p})}{\lambda(\hat{\boldsymbol{p}})} = W(\boldsymbol{p}) - W(\hat{\boldsymbol{p}}) = \frac{1}{2}\sum_{j=1}^{m}\sum_{k=1}^{m}\left.\frac{\partial^2 W(\boldsymbol{p})}{\partial p_j \partial p_k}\right|_{\hat{\boldsymbol{p}}}(p_j - \hat{p}_j)(p_k - \hat{p}_k), \quad (D.10)$$

is a Gaussian with weight matrix (cf. Eq. D.25):

$$V_{jk}^{-1} = \frac{1}{2}\left.\frac{\partial^2 W(\boldsymbol{p})}{\partial p_j \partial p_k}\right|_{\hat{\boldsymbol{p}}}. \quad (D.11)$$

This is a general formula applicable to the errors on parameters for all maximum likelihood and least squares fits.

For the problem at hand, V_{jk}^{-1} is diagonal, since from Eqs. D.4 and D.5:

$$\left.\frac{\partial^2 W}{\partial \mu \partial \sigma^2}\right|_{\hat{\mu},\hat{\sigma}^2} = -\frac{1}{\hat{\sigma}^2}\left.\frac{\partial W}{\partial \mu}\right|_{\hat{\mu},\hat{\sigma}^2} = 0, \quad (D.12)$$

so that even we can invert the matrix V^{-1}:

$$V^{-1} = \begin{pmatrix} \frac{n}{\hat{\sigma}^2} & 0 \\ 0 & \frac{n}{2\hat{\sigma}^4} \end{pmatrix} \implies V = \begin{pmatrix} \frac{\hat{\sigma}^2}{n} & 0 \\ 0 & \frac{2\hat{\sigma}^4}{n} \end{pmatrix}. \quad (D.13)$$

The square roots of the variance matrix elements in Eq. D.13 give the errors of the maximum likelihood estimators:

$$\delta\hat{\mu} = \frac{\hat{\sigma}}{\sqrt{n}} \qquad \delta\hat{\sigma}^2 = \hat{\sigma}^2\sqrt{\frac{2}{n}} \Longrightarrow \delta\hat{\sigma} = \frac{\delta\hat{\sigma}^2}{2\hat{\sigma}} = \frac{\hat{\sigma}}{\sqrt{2n}}. \quad (D.14)$$

These errors are again too small (since it is easy to show that $\sigma_{\bar{x}}^2 = \sigma_x^2/n$) and should be multiplied by a factor of $\sqrt{n/n-1}$ to agree with the correct statistical estimators [84].

D.4 Fits to compare measurements to theoretical predictions

The maximum likelihood method described above, which uses the repeated trials x_i from a single population represented by the distribution $f(x, \mu, \sigma)$ (Eq. D.1) to derive the parent population or distribution, is quite different and distinct from the problem that we now address. This is a least squares fit to the theory or predictions of the expectation values $\mu_i(\boldsymbol{p})$ of measured quantities y_i, at point x_i, which may represent an interval or bin, including the systematic errors, where $\boldsymbol{p}$ represents a vector of m parameters to be tested or determined from the fit. In the case of only point-to-point uncorrelated uncertainties (statistical and/or systematic), if one assumes they are Gaussian distributed and characterized by the standard deviation σ_i, which may be different for each data point, the maximum likelihood and least squares methods happen to coincide. The methods also coincide so long as any point-to-point correlated systematic uncertainties leave the values of σ_i

unchanged. For such cases, calculating the best-parameter fit is straightforward via a log-likelihood (or, in the Gaussian limit, χ^2-least squares fit) method [84].

The likelihood function $\mathcal{L}$ is defined as the a priori probability of a given outcome. Let $y_1, y_2, \ldots, y_n$ be independent trials from different populations at n points x_i with normalized probability density function $f(y_i, \boldsymbol{p})$ where $\boldsymbol{p}$ represents a vector of m parameters. For instance y_i could represent a measurement of a cross section at the position x_i, where the probability density of the measurement is Gaussian distributed with standard deviation σ_i about the expectation value $\mu_i(\boldsymbol{p}) = \langle y_i(x_i) \rangle$:

$$f(y_i, \boldsymbol{p}) = \frac{1}{\sqrt{2\pi\sigma_i^2}} \exp \frac{-(y_i - \mu_i(\boldsymbol{p}))^2}{2\sigma_i^2}. \tag{D.15}$$

If the trials are independent, and at each position x_i, $f(y_i, \boldsymbol{p})$ is a Gaussian as Eq. D.15 with individual standard deviation, σ_i, and predicted expectation value, $\mu_i(\boldsymbol{p}) = \langle y_i(x_i) \rangle$, then the likelihood function is:

$$\mathcal{L} = \prod_i f(y_i, \boldsymbol{p}) = \frac{1}{\sigma_1 \sigma_2 \ldots \sigma_n} \frac{1}{\sqrt{2\pi}^n} \exp - \left[\sum_{i=1}^{n} \frac{(y_i - \mu_i(\boldsymbol{p}))^2}{2\sigma_i^2} \right]. \tag{D.16}$$

If the trials are correlated, then the full variance matrix must be used:

$$V_{ij} = \langle (y_i - \mu_i)(y_j - \mu_j) \rangle. \tag{D.17}$$

The likelihood function takes the more general form:

$$\mathcal{L} = \frac{1}{\sqrt{|V|}} \frac{1}{\sqrt{2\pi}^n} \exp - \left[\sum_{i=1}^{n} \sum_{j=1}^{n} \frac{(y_i - \mu_i) V_{ij}^{-1} (y_j - \mu_j)}{2} \right], \tag{D.18}$$

where $|V|$ is the determinant of the variance matrix V.

Note that Eq. D.18 reduces to Eq. D.16 if the correlations vanish so that the covariances are zero and V_{ij} is diagonal:

$$V_{ij} = \langle (y_i - \mu_i)(y_j - \mu_j) \rangle = \delta_{ij} \langle (y_i - \mu_i)^2 \rangle = \delta_{ij} \sigma_i^2. \tag{D.19}$$

D.4.1 Example: two variables

The case of two variables, y_1 and y_2, provides a simple example. If y_1 and y_2 are statistically independent, then $V_{11} = \sigma_1^2$, $V_{22} = \sigma_2^2$, $V_{12} = V_{21} = 0$, and

$$\mathcal{L} = \mathcal{P}(y_1, y_2) = \frac{1}{\sigma_1 \sigma_2} \frac{1}{2\pi} \exp - \left[\frac{(y_1 - \mu_1)^2}{2\sigma_1^2} + \frac{(y_2 - \mu_2)^2}{2\sigma_2^2} \right]. \tag{D.20}$$

However, if y_1 and y_2 are not statistically independent but are correlated, with covariance

$$V_{12} \equiv \langle (y_1 - \mu_1)(y_2 - \mu_2) \rangle \equiv \rho_{12}\sigma_1\sigma_2, \tag{D.21}$$

where ρ_{12} is commonly called the correlation coefficient, it is straightforward to show that

$$\begin{aligned} V_{11}^{-1} &= \frac{1}{\sigma_1^2(1-\rho_{12}^2)} \\ V_{22}^{-1} &= \frac{1}{\sigma_2^2(1-\rho_{12}^2)} \\ V_{12}^{-1} = V_{21}^{-1} &= \frac{-\rho_{12}}{\sigma_1\sigma_2(1-\rho_{12}^2)}; \end{aligned} \tag{D.22}$$

and the joint probability is

$$\mathcal{L} = \mathcal{P}(y_1, y_2) = \frac{1}{\sigma_1\sigma_2\sqrt{1-\rho_{12}^2}} \frac{1}{2\pi}$$
$$\exp -\frac{1}{2(1-\rho_{12}^2)}\left[\frac{(y_1-\mu_1)^2}{\sigma_1^2} + \frac{(y_2-\mu_2)^2}{\sigma_2^2} - 2\rho_{12}\frac{(y_1-\mu_1)}{\sigma_1}\frac{(y_2-\mu_2)}{\sigma_2}\right]. \tag{D.23}$$

It is important to note that the errors σ_1 and σ_2 represent statistical (i.e. random) variations of the measurements y_1 and y_2 about their respective means μ_1 and μ_2. These statistical errors are usually correlated because the quantities measured are correlated, typically linear combinations of some other statistically independent quantities. For the variance matrix (Eq. D.22) which is a real symmetric matrix, these linear combinations can be found by the unitary transformation which diagonalizes the matrix. For instance, correlated statistical errors are quite common for the best fit parameters from maximum likelihood and least squares fits (Eq. D.11).

D.4.2 Logarithm of the likelihood

Since Gaussian probability distributions are common (as a consequence of the central limit theorem) and since there is also an important theorem regarding likelihood ratios for composite hypotheses, it is convenient to use the logarithm of the likelihood:

$$-2\ln\mathcal{L} = 2\ln\left(\sigma_1\sigma_2\sqrt{1-\rho_{12}^2}\right) + 2\ln 2\pi$$
$$+\frac{1}{(1-\rho_{12}^2)}\left[\frac{(y_1-\mu_1)^2}{\sigma_1^2} + \frac{(y_2-\mu_2)^2}{\sigma_2^2} - 2\rho_{12}\frac{(y_1-\mu_1)}{\sigma_1}\frac{(y_2-\mu_2)}{\sigma_2}\right]. \tag{D.24}$$

For reference, we give $-2\ln\mathcal{L}$ for the general case (Eq. D.18):

$$-2\ln\mathcal{L} = \ln|V| + n\log 2\pi + \sum_{i=1}^{n}\sum_{j=1}^{n}(y_i-\mu_i)V_{ij}^{-1}(y_j-\mu_j). \qquad \text{(D.25)}$$

D.5 Correlated systematic uncertainties

Suppose there is a quantity a which is uncertain, e.g. momentum scale, which may cause a systematic variation Δy_i on the set of data points y_i around their nominal value. A simple way to express this correlation is to define:

$$\Delta y(sys)_i = a_i\Delta z_a + r_i, \qquad \text{(D.26)}$$

where z_a is a random variable which is the same for all i and represents the correlation, a_i is a constant of proportionality which may be different for each i, and r_i are random variables which are statistically independent for all i and statistically independent of Δz_a, i.e $\langle r_i r_j\rangle = \delta_{ij}\sigma_{r_i}^2$, $\langle r_i\Delta z_a\rangle = 0$. Also $\langle\Delta z_a^2\rangle = \sigma_a^2$. Then

$$\sigma(sys)_i^2 = \langle\Delta y(sys)_i^2\rangle = a_i^2\sigma_a^2 + \sigma_{r_i}^2 \qquad \text{(D.27)}$$

$$\langle\Delta y(sys)_j\Delta y(sys)_k\rangle = a_j a_k\sigma_a^2 \qquad j\neq k, \qquad \text{(D.28)}$$

$$\rho(sys)_{jk} \equiv \frac{a_j a_k\sigma_a^2}{\sigma(sys)_j\sigma(sys)_k} = \frac{a_j a_k\sigma_a^2}{\sqrt{a_j^2\sigma_a^2+\sigma_{r_j}^2}\sqrt{a_k^2\sigma_a^2+\sigma_{r_k}^2}} < 1. \qquad \text{(D.29)}$$

Note that the probability distribution $f(\Delta z_a)$ need not be Gaussian. It just has to have finite first and second moments so that the r.m.s. can be calculated. The existence of the random term r_i (Eq. D.26) and $\sigma_{r_i}^2$ (Eq. D.27) makes the correlated systematic errors not fully correlated.

An important point for the errors that are only partially correlated, i.e. $\rho_{ij} < 1$, is empirically to split off the random part of the systematic error so that the correlated systematic error becomes 100% correlated by definition, i.e.:

$$\Delta' y(sys)_i \equiv a_i\Delta z_a \qquad \text{(D.30)}$$

$$\sigma(sys)_i^2 = \langle\Delta' y(sys)_i^2\rangle = a_i^2\sigma_a^2 \equiv \sigma_{a_i}^2 \qquad \text{(D.31)}$$

$$\langle\Delta' y(sys)_j\Delta' y(sys)_k\rangle = a_j a_k\sigma_a^2 = \sigma_{a_j}\sigma_{a_k} \qquad j\neq k \qquad \text{(D.32)}$$

$$\rho(sys)_{jk} \equiv \frac{\langle\Delta' y(sys)_j\Delta' y(sys)_k\rangle}{\sigma(sys)_j\sigma(sys)_k} = \frac{a_j a_k\sigma_a^2}{\sqrt{a_j^2\sigma_a^2}\sqrt{a_k^2\sigma_a^2}} = 1. \qquad \text{(D.33)}$$

Note that $\sigma_{a_i} \equiv a_i\sigma_a$ represent the correlated r.m.s. systematic errors for each point y_i, and the sign of a_i, which may be positive or negative, is retained by σ_{a_i}.

Then, as there is no need to distinguish between random-systematic and statistical errors, define the total random error on each point as the quadrature sum of the random-systematic and statistical errors including any purely random-systematic error (i.e uncorrelated from point-to-point), which we call type A:

$$\sigma_i^2 \equiv \sigma_{r_i}^2 + \sigma_{A_i}^2 + \sigma(stat)_i^2. \tag{D.34}$$

These errors make up the diagonal elements of the variance matrix since they represent random variations which are independent for each data point.

The key issue (and a very important one) in separating the partially correlated systematic errors into a purely random component and a purely 100% correlated component, is that the random systematic errors act like statistical errors and so are added into the overall statistical error, σ_i, for each data point (Eq. D.34). The 100% correlated systematic errors, more properly called systematic uncertainties, act quite differently.

D.6 The meaning of the correlated systematic uncertainty

By correlated systematic uncertainty, we mean that all the data points together with their random errors σ_i may be systematically displaced from their nominal values by correlated amounts (repeating Eq. D.30):

$$\Delta' y(sys)_i \equiv a_i \Delta z_a \tag{D.35}$$

which correspond to some fixed value of Δz_a, which is the same for all data points, with a Gaussian probability, $f(\Delta z_a)$, for Δz_a, with r.m.s. σ_a, so that $\sigma_{a_i} \equiv a_i \sigma_a$ (Eq. D.31). We define Δz_a as a fraction ϵ_a of its r.m.s., or

$$\Delta' y(sys)_i \equiv a_i \Delta z_a = a_i \, \epsilon_a \, \sigma_a = \epsilon_a \, \sigma_{a_i}. \tag{D.36}$$

This makes it clear that all points move by the same fraction of their correlated systematic uncertainty σ_{a_i}, where the sign of a_i is retained by σ_{a_i} which may be positive or negative. This represents a systematic displacement of all the data points and their random errors together. This is quite different from the random errors σ_i for each data point (Eq. D.34) which imply a statistical uncertainty for each data point, independently, subject to the laws of statistics.

For the following discussion, we first assume that the systematic displacements do not affect the random errors σ_i of each point. This makes the maximum likelihood and likelihood ratio calculations with systematic uncertainties only marginally different from the standard calculations without systematic errors.

D.7 Three classes of systematic uncertaintiess

We usually discuss three kinds of systematic uncertainties:

(A) random systematic errors, uncorrelated (i.e. vary independently) from point-to-point;

(B) correlated systematic uncertainties, for which the correlation is 100% because the random part has been separated out and added to the statistical error according to Eqs. D.26, D.30 and D.34.

All points move by the same fraction of their type B error

$$\Delta' y^b(sys)_i \equiv b_i \Delta z_b = b_i \epsilon_b \sigma_b = \epsilon_b \sigma_{b_i} \tag{D.37}$$

where σ_{b_i} is known for all points and may be of either sign, as it is possible that one point could move up while its neighbor moves down;

(C) overall systematic errors (typically normalization) by which all the points move by the same fraction:

$$\Delta' y^c(sys)_i / y_i \equiv \Delta z_c = \epsilon_c \sigma_c, \tag{D.38}$$

by definition σ_c is the same for all points.

Then the likelihood function for any outcome, including the variation ϵ_b and ϵ_c would be:

$$\mathcal{L} = \prod_i f(y_i, \boldsymbol{p}) = \frac{1}{\sigma_1 \sigma_2 \ldots \sigma_n \sigma_b \sigma_c} \frac{1}{\sqrt{2\pi}^{(n+2)}}$$

$$\exp - \left[\sum_{i=1}^{n} \frac{(y_i + \epsilon_b \sigma_{b_i} + \epsilon_c y_i \sigma_c - \mu_i(\boldsymbol{p}))^2}{2\sigma_i^2} + \frac{\epsilon_b^2}{2} + \frac{\epsilon_c^2}{2} \right] \tag{D.39}$$

where the last two terms represent $\Delta^2 z_b / (2\sigma_b^2) = \epsilon_b^2 \sigma_b^2 / (2\sigma_b^2)$ and $\Delta^2 z_c / (2\sigma_c^2) = \epsilon_c^2 \sigma_c^2 / (2\sigma_c^2)$ since the probability of the systematic displacements $f(\Delta z_{b,c})$ was taken as Gaussian, for simplicity. Other probability distributions for the correlated systematic error could be used.

We then use the likelihood ratio test to establish the validity or the confidence interval of the theoretical predictions $\mu_i(\boldsymbol{p})$. One can use the modified log-likelihood function:

$$-2 \ln \mathcal{L} = \left[\sum_{i=1}^{n} \frac{(y_i + \epsilon_b \sigma_{b_i} + \epsilon_c y_i \sigma_c - \mu_i(\boldsymbol{p}))^2}{\sigma_i^2} + \epsilon_b^2 + \epsilon_c^2 \right] \equiv \chi^2(\epsilon_b, \epsilon_c, \vec{p}) \tag{D.40}$$

because we will eventually take the ratio of the likelihood of a given set of parameters $\boldsymbol{p}$ to the maximum likelihood when all the parameters ϵ_b, ϵ_c and $\boldsymbol{p}$ are varied (the minimum value of Eq. D.40) so that the terms preceding the exponential in

Eq. D.39 cancel because they are not varied. It is important to note that Eq. D.40 follows the χ^2-distribution with $n+2$ degrees of freedom because it is the sum of $n+2$ independent Gaussian distributed random variables (i.e. in statistical terminology $\chi^2(\epsilon_b, \epsilon_c, \vec{p})$ is $\chi^2_{(n+2)}$). This establishes Eq. D.40 as the χ^2-distributed quantity that we use for least squares fit to the theoretical predictions including the systematic errors.[1] The specific procedure is now described.

D.8 The likelihood ratio test

Let $\hat{\epsilon}_b, \hat{\epsilon}_c, \hat{\boldsymbol{p}}$ represent the values of the parameters which give the maximum likelihood and let $\boldsymbol{p}_0$ represent any other set of parameters whose significance you are trying to evaluate of which k are constrained to be specific values and the rest are allowed to take on any value by re-minimizing $-2\ln\mathcal{L}$. There is a theorem [819] that for large values of n, the "likelihood ratio," more precisely the quantity $-2\ln\left[\mathcal{L}(\boldsymbol{p}_0)/\mathcal{L}(\hat{\boldsymbol{p}})\right]$, is χ^2-distributed with k degrees of freedom. [Note that degrees of freedom are sometimes called constraints, and now you know why, if you did not know previously.]

D.9 Fits to data including systematic uncertainties – I

First "fit the theory" to the data by minimizing Eq. D.40 by varying all the parameters to find $\hat{\epsilon}_b, \hat{\epsilon}_c, \hat{\boldsymbol{p}}$. If the χ^2_{min} for this fit for the $n+2-(m+2)=n-m$ degrees of freedom, where m is the number of parameters in $\boldsymbol{p}$, is acceptable, then the theory is not rejected at this level. You can then find a confidence interval for testing any other set of k parameters constrained to specific values, $\boldsymbol{p}_0$, by again finding the minimum of Eq. D.40 for the k fixed values of $\boldsymbol{p}_0$, by letting all the other parameters including ϵ_b and ϵ_c vary. The "likelihood ratio" $-2\ln\left[\mathcal{L}(\boldsymbol{p}_0)/\mathcal{L}(\hat{\boldsymbol{p}})\right] = -2\left[\ln\mathcal{L}(\boldsymbol{p}_0) - \ln\mathcal{L}(\hat{\boldsymbol{p}})\right]$, colloquially $\chi^2(\boldsymbol{p}_0) - \chi^2_{min}$ is χ^2-distributed with k degrees of freedom, from which the confidence interval on the parameters can be evaluated.

D.9.1 A standard fitting program, such as MINUIT, can be used

It is important to note that if one finds the minimum of Eq. D.40 by varying all the parameters to find $\hat{\epsilon}_b, \hat{\epsilon}_c, \hat{\boldsymbol{p}}$, using the CERN program MINUIT [820], then you automatically get the correct errors of the parameters $\boldsymbol{p}$ and also of the parameters ϵ_b and ϵ_c, since MINUIT (MINOS) accounts correctly for the correlations of the best fit parameters $\hat{\epsilon}_b, \hat{\epsilon}_c, \hat{\boldsymbol{p}}$. This program can also be used to find the confidence intervals on the individual parameters.

[1] Note that Eq. D.40 agrees with Eq. 8 in section 6.5 of reference [818].

D.9.2 Full extent systematic errors

For full extent systematic errors (i.e. not Gaussian weighted about the best guess) just follow the same procedure as above with two changes:

(i) set the limits between which ϵ_b and/or ϵ_c are allowed to vary as -1 to $+1$;
(ii) drop the ϵ_b^2 and/or ϵ_c^2 terms in Eq. D.40 since the full extent systematic errors are presumed to be able to take on any value in the estimated range without penalty.

D.9.3 More complicated systematic error correlations

In general, correlated systematic uncertainties may be more complicated than discussed above. For instance, groups of points may have different correlations. We can easily adapt this to the method by rewriting Eq. D.40 as:

$$-2\ln\mathcal{L} = \left[\sum_{i=1}^{n} \frac{(y_i + \sum_j \epsilon_{bj}\sigma_{bj_i} + \sum_k \epsilon_{ck} y_i \sigma_c - \mu_i(\boldsymbol{p}))^2}{\sigma_i^2} + \sum_j \epsilon_{bj}^2 + \sum_k \epsilon_{ck}^2\right] \tag{D.41}$$

where ϵ_{bj} represent j different independent type b correlated systematic uncertainties, and ϵ_{ck} represent k different correlated type c systematic uncertainties, so that each of the j and k variations are independent of each other. A simple example of this would be if the first six points have correlated systematic errors, the next six points have a different correlation which is independent of the first six, etc. This allows groups of points to have different correlated errors among themselves but to be random with respect to the systematic variation of other groups of points. This same method could be used for all the separate systematic errors, i.e. enumerate each of them (scale, acceptance, peak extraction, ...) and their different correlations, but this might only make sense for a large number of data points ($n \gg j + k$). Once it gets too difficult to express the correlated systematic errors by something like Eq. D.41, then perhaps improving the systematic errors would be more important than trying to use the data for fitting. Alternatively one could take all the systematic errors as point-to-point random, absorb them into the σ_i and forget about the ϵ, and obtain the most conservative limits on the fit parameters, but probably without much sensitivity.

D.10 Fits to data including systematic uncertainties – II

Problems arise in applying Eq. D.40 to a data set for which the systematic corrections are dominantly multiplicative, since such corrections preserve the fractional statistical error of the data points, σ_i/y_i, when the data points and their errors are systematically displaced. This is generally the case for the statistical errors of most measurements, since they are typically the result of corrections of the form:

$$y = y_0 \times C_1 \times \cdots \times C_n$$

where y_0 is the raw yield with statistical + random systematic error σ_0 and the C_j are systematic corrections which have error σ_{C_j}, and the errors would be propagated as $\sigma_y^2/y^2 = \sigma_0^2/y_0^2 + \sigma_{C_1}^2/C_1^2 + \cdots + \sigma_{C_n}^2/C_n^2$. Then if there is a change in a correction, if we multiply y by a different factor, which is the result of different C_i (with the same fractional uncertainty), it is clear that σ_y/y would be unchanged so that σ_y would be multiplied by the same factor. For instance on a semi-log plot this means that the data points and their respective statistical errors remain the same size as they are displaced according to the systematic uncertainties, i.e. σ_{y_i}/y_i remains constant.

In this case the maximum likelihood and least squares methods no longer coincide and we use a least squares fit of Eq. D.42 instead of Eq. D.40 to estimate the best fit parameters

$$\tilde{\chi}^2 = \left[\sum_{i=1}^{n} \frac{(y_i + \epsilon_b \sigma_{b_i} + \epsilon_c y_i \sigma_c - \mu_i(\vec{p}))^2}{\tilde{\sigma}_i^2} + \epsilon_b^2 + \epsilon_c^2\right], \tag{D.42}$$

where $\tilde{\sigma}_i$ is the statistical error σ_i scaled by the multiplicative shift in y_i such that the fractional error is unchanged under shifts

$$\tilde{\sigma}_i = \sigma_i \cdot \left(\frac{y_i + \epsilon_b \sigma_{b_i} + \epsilon_c y_i \sigma_c}{y_i}\right), \tag{D.43}$$

and to evaluate the goodness of the best fit because Eq. D.42 follows the χ^2 distribution for $n+2$ degrees of freedom since for any given systematic shift it is the sum of $n+2$ independent Gaussian distributed random variables with standard deviations $\tilde{\sigma}_i$. The $\chi^2(\hat{\epsilon}_b, \hat{\epsilon}_c, \hat{\boldsymbol{p}})$ for this fit for the $n+2-(m+2) = n-m$ degrees of freedom, where m is the number of parameters in $\boldsymbol{p}$, is used to compute the "p-value"; and if this is acceptable, then the theory is not rejected at this level.

The procedure then follows the method of Sections and D.8 and D.9 with Eqs. D.42 and D.43 in place of Eq. D.40.

Appendix E

The shape of the x_E distribution triggered by a jet fragment, for example, π^0

As noted above (Section 9.1.1), the steeply falling power-law spectrum at a given $\sqrt{s}$ has many important and helpful consequences for single particle inclusive and two particle correlation measurements of hard scattering. The most famous properties in this regard are the "Bjorken parent–child relationship" [377] and the "leading particle effect," which also goes by the unfortunate name "trigger bias" [377,379,821]. We review this, adding some more details, and then proceed to two particle correlations.

E.1 Why single particle inclusive measurements accurately measure hard scattering

The reason why single particle inclusive measurements accurately measure hard scattering is the leading particle effect, also known as "trigger bias." Due to the steeply falling power-law transverse momentum ($\hat{p}_{T_t}$) spectrum of the scattered parton, the inclusive single particle (e.g. π) p_{T_t} spectrum from jet fragmentation is dominated by fragments with large z_t, where $z_t = p_{T_t}/\hat{p}_{T_t}$ is the fragmentation variable. The joint probability for a fragment pion, with $p_{T_t} = z_t \hat{p}_{T_t}$, originating from a parton with $\hat{p}_{T_t} = p_{T\,jet}$ is:

$$\begin{aligned}\frac{d^2\sigma_\pi(\hat{p}_{T_t}, z_t)}{\hat{p}_{T_t} d\hat{p}_{T_t} dz_t} &= \frac{d\sigma_q}{\hat{p}_{T_t} d\hat{p}_{T_t}} \times D_q^\pi(z_t) \\ &= f_q(\hat{p}_{T_t}) \times D_q^\pi(z_t),\end{aligned} \tag{E.1}$$

where $f_q(\hat{p}_{T_t})$ represents the final state scattered parton invariant spectrum $d\sigma_q/\hat{p}_{T_t} d\hat{p}_{T_t}$, and $D_q^\pi(z_t)$ represents the fragmentation function. The first term in Eq. E.1 is the probability of finding a parton with transverse momentum $\hat{p}_{T_t}$ and the second term corresponds to the conditional probability that the parton fragments into a particle of momentum $p_T = z_t \hat{p}_{T_t}$. A simple change of variables,

$\hat{p}_{T_t} = p_{T_t}/z_t$, $d\hat{p}_{T_t}/dp_{T_t}|_{z_t} = 1/z_t$, then gives the joint probability of a pion with transverse momentum p_{T_t} which is a fragment with momentum fraction z_t from a parton with $\hat{p}_{T_t} = p_{T_t}/z_t$:

$$\frac{d^2\sigma_\pi(p_{T_t}, z_t)}{p_{T_t} dp_{T_t} dz_t} = f_q\left(\frac{p_{T_t}}{z_t}\right) \times D_q^\pi(z_t) \times \frac{1}{z_t^2}. \tag{E.2}$$

The p_{T_t} and z_t dependences do not factorize. However, the p_{T_t} spectrum may be found by integrating over all values of $\hat{p}_{T_t}$ from $p_{T_t} \leq \hat{p}_{T_t} \leq \sqrt{s}/2$, which corresponds to values of z_t from $x_T = 2p_T/\sqrt{s}$ to 1:

$$\frac{1}{p_{T_t}}\frac{d\sigma_\pi}{dp_{T_t}} = \int_{x_T}^{1} f_q\left(\frac{p_{T_t}}{z_t}\right) D_q^\pi(z_t)\, \frac{dz_t}{{z_t}^2}. \tag{E.3}$$

Also, for any fixed value of p_{T_t} one can evaluate the $\langle z_t(p_{T_t})\rangle$, integrated over the parton spectrum:

$$\langle z_t(p_{T_t})\rangle = \frac{\int_{x_T}^{1} z_t\, D_q^\pi(z_t)\, f_q(p_{T_t}/z_t)\frac{dz_t}{z_t^2}}{\int_{x_T}^{1} D_q^\pi(z_t)\, f_q(p_{T_t}/z_t)\frac{dz_t}{z_t^2}}. \tag{E.4}$$

Since the observed π^0 spectrum is a power-law for $p_{T_t} \geq 3$ GeV/c, one can deduce from Eq. E.3 that the partonic $\hat{p}_{T_t}$ spectrum is also a power-law with the same power – this is the "Bjorken parent–child relationship" [377]. If we take:

$$\frac{d\sigma_q}{\hat{p}_{T_t} d\hat{p}_{T_t}} = f_q(\hat{p}_{T_t}) = A\hat{p}_{T_t}^{-n}, \tag{E.5}$$

then

$$\begin{aligned} \frac{1}{p_{T_t}}\frac{d\sigma_\pi}{dp_{T_t}} &= \int_{x_T}^{1} A\, D_q^\pi(z_t) \left(\frac{p_{T_t}}{z_t}\right)^{-n} \frac{dz_t}{z_t^2} \\ &= \frac{1}{p_{T_t}^n}\int_{x_T}^{1} A\, D_q^\pi(z_t)\, {z_t}^{n-2} dz_t, \end{aligned} \tag{E.6}$$

where the last integral depends only weakly on p_{T_t} due to the small value of x_T. Eq. E.6 also indicates that the effective fragmentation function for a detected inclusive single particle (with p_{T_t}) is weighted upward in z_t by a factor z_t^{n-2}, where n is the simple power-fall-off of the jet invariant cross section (i.e. not the $n(x_T, \sqrt{s})$ of Eq. 6.12). This is the so-called "trigger bias" although it does not actually involve a hardware trigger. Any particle selected from an inclusive p_{T_t} spectrum will most likely carry a large fraction of its parent parton transverse momentum; and it was commonly accepted that this would define the hard scattering kinematics ($\hat{s}$ in Eq. 9.5 or $\hat{p}_{T_t}$ in Eq. E.1) so that the jet from the other outgoing parton in the hard scattered parton pair would be unbiased [386, 391], so that its properties such as the fragmentation function and the fragmentation transverse momentum could be measured.

E.2 Fragmentation formalism – single inclusive

For an exponential fragmentation function,

$$D(z) = Be^{-bz}, \tag{E.7}$$

calculation of the "trigger bias" and the "parent–child" factor is straightforward [821]. The mean multiplicity of fragments in the jet is

$$\langle m \rangle = \int_0^1 D(z)dz = \frac{B}{b}(1 - e^{-b}) \tag{E.8}$$

and these fragments carry the total momentum of the jet:

$$\int_0^1 zD(z)dz = \frac{B}{b^2}(1 - e^{-b}(1 + b)) \equiv 1, \tag{E.9}$$

where the $\langle z \rangle$ per fragment is

$$\langle z \rangle = \frac{\int_0^1 zD(z)dz}{\int_0^1 D(z)dz} = \frac{1}{\langle m \rangle}. \tag{E.10}$$

The results are:

$$B = \frac{b^2}{1 - e^{-b}(1 + b)} \approx b^2 \tag{E.11}$$

$$\langle m \rangle = \frac{b(1 - e^{-b})}{1 - e^{-b}(1 + b)} \approx b \tag{E.12}$$

$$\langle z \rangle = \frac{1 - e^{-b}(1 + b)}{b(1 - e^{-b})} \approx \frac{1}{b}. \tag{E.13}$$

The mean multiplicity of particles in the jet is $\langle m \rangle \approx b$, which is 8–10 at RHIC (see below).

Here it is important to note that the normalization of the momentum of the fragments to carry the total momentum of the jet Eq. E.9 assumes ALL the particles in the jet and furthermore that all species of particles, e.g. charged hadrons (h), pizeroes (π^0), have the same shape fragmentation, i.e. the same b. Thus if we define the mean multiplicity of species h as a fraction ζ_h of the overall mean multiplicity:

$$\langle m \rangle_h \equiv \zeta_h \langle m \rangle . \tag{E.14}$$

Then the assumption of equal b for all species gives the results:

$$\begin{aligned} B^h &= \zeta_h B \\ \langle m \rangle_h &= \zeta_h b \\ \langle z \rangle_h &= \langle z \rangle \end{aligned} \tag{E.15}$$

and $\sum \zeta_i = 1$ for all species.

Substitution of Eq. E.7 into Eq. E.6 for the p_{T_t} spectrum of the π gives

$$\frac{1}{p_{T_t}} \frac{d\sigma_\pi}{dp_{T_t}} = \frac{AB^\pi}{p_{T_t}^n} \int_{x_{T_t}}^1 dz_t z_t^{n-2} \exp -bz_t, \tag{E.16}$$

which can be written as

$$\frac{1}{p_{T_t}} \frac{d\sigma_\pi}{dp_{T_t}} = \frac{AB^\pi}{p_{T_t}^n} \frac{1}{b^{n-1}} \left[\Gamma(n-1, bx_{T_t}) - \Gamma(n-1, b)\right], \tag{E.17}$$

where

$$\Gamma(a, x) \equiv \int_x^\infty t^{a-1} e^{-t} dt \tag{E.18}$$

is the complementary or upper incomplete gamma function, and $\Gamma(a, 0) = \Gamma(a)$ is the gamma function, where $\Gamma(a) = (a-1)!$ for a an integer.

A reasonable approximation for small x_T values is obtained by taking the lower limit of Eq. E.16 to zero and the upper limit to infinity, with the result that:

$$\frac{1}{p_{T_t}} \frac{d\sigma_\pi}{dp_{T_t}} \approx \frac{\Gamma(n-1)}{b^{n-1}} \frac{AB^\pi}{p_{T_t}^n}. \tag{E.19}$$

The parent–child ratio, the ratio of the number of π at a given p_{T_t} to the number of partons at the same p_{T_t}, is just given by the ratio of Eq. E.19 to Eq. E.5 at $\hat{p}_{T_t} = p_{T_t}$:

$$\left.\frac{\pi^0}{q}\right|_{\pi^0} (p_{T_t}) = \frac{B^\pi \, \Gamma(n-1)}{b^{n-1}} \approx \frac{\langle m \rangle_\pi \, \Gamma(n-1)}{b^{n-2}}. \tag{E.20}$$

Similarly, the same substitutions in Eq. E.4 for $\langle z_t(p_{T_t}) \rangle$ give:

$$\langle z_t(p_{T_t}) \rangle = \frac{\int_{x_{T_t}}^1 dz_t z_t^{n-1} \exp -bz_t}{\int_{x_{T_t}}^1 dz_t z_t^{n-2} \exp -bz_t} = \frac{1}{b} \frac{[\Gamma(n, bx_{T_t}) - \Gamma(n, b)]}{[\Gamma(n-1, bx_{T_t}) - \Gamma(n-1, b)]} \approx \frac{n-1}{b}. \tag{E.21}$$

This shows the "trigger bias" quantitatively. The $\langle z_t(p_{T_t}) \rangle$ of an inclusive single particle (e.g. π^0) with transverse momentum p_{T_t}, which is a fragment with momentum fraction z_t from a parent parton with $\hat{p}_{T_t} = p_{T_t}/z_t$ (Eq E.21), is $n-1$ times larger than the unconditional $\langle z \rangle$ of fragmentation (Eq. E.13) [822]. The prevailing opinion from the early 1970s until early 2006 was that although the inclusive single particle (e.g. pizero) spectrum from jet fragmentation is dominated by trigger fragments with large $\langle z_t \rangle \sim$ 0.7–0.8 the away jets should be unbiased and would measure the fragmentation function, once the correction is made for $\langle z_t \rangle$ and the fact that the jets do not exactly balance p_T due to the k_T smearing effect.

E.3 Fragmentation formalism – two particle correlations from a jet pair

Let x_h ($\hat{x}_h$) be the ratio of the associated particle (parton) transverse momentum to the trigger particle (parton) transverse momentum:

$$x_h \equiv \frac{p_{T_a}}{p_{T_t}} \qquad \hat{x}_h = \hat{x}_h(k_T, x_h) \equiv \frac{\hat{p}_{T_a}}{\hat{p}_{T_t}}. \tag{E.22}$$

Note that the hadronic variable x_h is measured on every event and that the partonic variable $\hat{x}_h$ is a function of both k_T and x_h, as is the "trigger bias" $\langle z_t \rangle$ [149].

Then, recall the joint probability for a fragment pion, with $p_{T_t} = z_t \hat{p}_{T_t}$, originating from a parton with $\hat{p}_{T_t}$ (Eq. E.1):

$$\begin{aligned}\frac{d^2\sigma_\pi(\hat{p}_{T_t}, z_t)}{\hat{p}_{T_t} d\hat{p}_{T_t} dz_t} &= \frac{d\sigma_q}{\hat{p}_{T_t} d\hat{p}_{T_t}} \times D_q^\pi(z_t) \\ &= f_q(\hat{p}_{T_t}) \times D_q^\pi(z_t).\end{aligned} \tag{E.23}$$

Here we make explicit that $f_q(\hat{p}_{T_t})$ represents the k_T smeared final state scattered parton invariant spectrum $d\sigma_q/\hat{p}_{T_t} d\hat{p}_{T_t}$ and $D_q^\pi(z_t)$ represents the fragmentation function. Due to the k_T smearing, the transverse momentum $\hat{p}_{T_a}$ of the away parton in the hard scattered parton pair is less than the transverse momentum of the trigger parton $\hat{p}_{T_t}$ [149]. The probability that the parton with $\hat{p}_{T_a}$ fragments to a hadron with $p_{T_a} = z_a \hat{p}_{T_a}$ in interval dz_a is given by $D_q^h(z_a)$. Thus, the joint probability for a fragment pion with $p_{T_t} = z_t \hat{p}_{T_t}$, originating from a parton with $\hat{p}_{T_t}$, and a fragment hadron with $p_{T_a} = z_a \hat{p}_{T_a}$, originating from the other parton in the hard scattered pair with $\hat{p}_{T_a}$ is:

$$\frac{d^3\sigma_\pi(\hat{p}_{T_t}, z_t, z_a)}{\hat{p}_{T_t} d\hat{p}_{T_t} dz_t dz_a} = \frac{d\sigma_q}{\hat{p}_{T_t} d\hat{p}_{T_t}} \times D_q^\pi(z_t) \times D_q^h(z_a), \tag{E.24}$$

where

$$z_a = \frac{p_{T_a}}{\hat{p}_{T_a}} = \frac{p_{T_a}}{\hat{x}_h \hat{p}_{T_t}} = \frac{z_t p_{T_a}}{\hat{x}_h p_{T_t}}$$

and $\hat{x}_h = \hat{p}_{T_a}/\hat{p}_{T_t}$ (Eq. E.22). Changing variables from $\hat{p}_{T_t}, z_t$ to p_{T_t}, z_t as above and similarly from z_a to p_{T_a} yields:

$$\frac{d^3\sigma_\pi}{dp_{T_t} dz_t dp_{T_a}} = \frac{1}{\hat{x}_h\, p_{T_t}} \frac{d\sigma_q}{d(p_{T_t}/z_t)} D_q^\pi(z_t) D_q^h\left(\frac{z_t p_{T_a}}{\hat{x}_h p_{T_t}}\right) \tag{E.25}$$

where for integrating over z_t or finding $\langle z_t \rangle$ for fixed p_{T_t}, p_{T_a}, the minimum value of z_t is $z_t^{min} = 2p_{T_t}/\sqrt{s} = x_{T_t}$ and the maximum value is:

$$z_t^{max} = \hat{x}_h \frac{p_{T_t}}{p_{T_a}} = \frac{\hat{x}_h}{x_h},$$

where $\hat{x}_h(p_{T_t}, p_{T_a})$ is also a function of k_T (Eq. E.22). Integrating over dz_t in Eq. E.25 gives the x_E distribution in the collinear limit, where $p_{T_a} = x_E\, p_{T_t}$, and it was thought [149] that a simply parameterized fragmentation function could be extracted from a joint fit to the measured x_E and inclusive p_{T_t} distributions (Eq. E.6). However, there were serious difficulties with convergence which took a while to sort out. Eventually, the x_E distributions were calculated from Eq. E.25 using LEP measurements [823, 824] for quark and gluon fragmentation functions, with shocking results (see Figure E.1) – the x_E distributions calculated with quark $D_q^h \approx \exp(-8.2 \cdot z)$ or gluon $D_g^h \approx \exp(-11.4 \cdot z)$ fragmentation functions do not differ significantly! Clearly, the x_E distributions are rather insensitive to the fragmentation function of the away jet in contradiction to the conventional wisdom dating from the early 1970s.

The evidence of this explicit counter example led to an attempt to perform the integral of Eq. E.25 analytically which confirmed straightforwardly that the shape of the x_E distribution is not sensitive to the shape of the fragmentation function. However, it was found that x_E distribution is sensitive to $\hat{x}_h$, the ratio of the

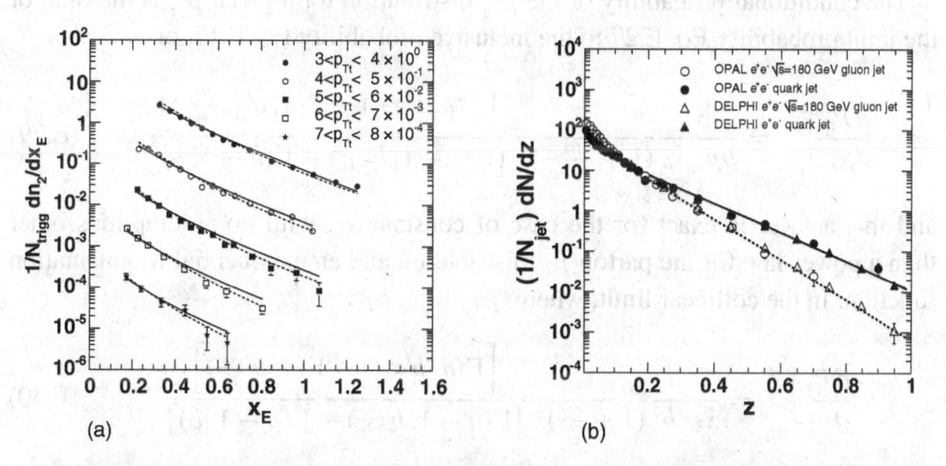

Figure E.1 (a) x_E distributions from PHENIX [149] in p–p collisions at $\sqrt{s}$ = 200 GeV for several values of p_{T_t}. The solid and dashed lines represent calculations of the distribution from the integral of Eq. E.25 for quark (solid lines) and gluon (dashed lines) fragmentation functions based on exponential fits [149] for $z \geq 0.2$ to the LEP measurements [823, 824] for charged hadrons, $h = h^+ + h^-$ shown in (b).

transverse momentum of the away-side jet ($\hat{p}_{T_a}$) to that of the trigger-side jet ($\hat{p}_{T_t}$). This can be put to use in $A+A$ collisions to measure the relative energy loss of the two jets from a hard scattering which escape from the medium.

E.4 Analytical formula for the x_E distribution

With a substitution of a power-law parton $\hat{p}_{T_t}$ spectrum (Eq. E.5) and an exponential fragmentation function (Eq. E.7), as in Section E.2, the integral of Eq. E.25 over z_t becomes:

$$\frac{d^2\sigma_\pi}{dp_{T_t}dp_{T_a}} = \frac{B^\pi B^h}{\hat{x}_h}\frac{A}{p_{T_t}^n}\int_{x_{T_t}}^{\hat{x}_h\frac{p_{T_t}}{p_{T_a}}} dz_t z_t^{n-1}\exp\left[-bz_t\left(1+\frac{p_{T_a}}{\hat{x}_h p_{T_t}}\right)\right]. \tag{E.26}$$

This is again an incomplete gamma function, if $\hat{x}_h$ is taken to be constant as a function of z_t for fixed p_{T_t}, p_{T_a}:

$$\frac{d^2\sigma_\pi}{dp_{T_t}dp_{T_a}} = \frac{B^\pi B^h}{\hat{x}_h}\frac{A}{p_{T_t}^n}\frac{1}{b'^n}\left[\Gamma(n, b'x_{T_t}) - \Gamma\left(n, b'\hat{x}_h\frac{p_{T_t}}{p_{T_a}}\right)\right], \tag{E.27}$$

where b' is given by

$$b' = b\left(1+\frac{p_{T_a}}{\hat{x}_h p_{T_t}}\right). \tag{E.28}$$

The conditional probability of the p_{T_a} distribution for a given p_{T_t} is the ratio of the joint probability Eq. E.27 to the inclusive probability Eq. E.17, or

$$\left.\frac{dP_\pi}{dp_{T_a}}\right|_{p_{T_t}} = \frac{B^h}{bp_{T_t}\hat{x}_h}\frac{1}{(1+\frac{p_{T_a}}{\hat{x}_h p_{T_t}})^n}\frac{\left[\Gamma(n, b'x_{T_t}) - \Gamma(n, b'\hat{x}_h\frac{p_{T_t}}{p_{T_a}})\right]}{\left[\Gamma(n-1, bx_{T_t}) - \Gamma(n-1, b)\right]}, \tag{E.29}$$

and this answer is exact for the case of constant $\hat{x}_h$, with no assumptions other than a power-law for the parton $\hat{p}_{T_t}$ distribution and an exponential fragmentation function. In the collinear limit, where, $p_{T_a} = x_E p_{T_t}$:

$$\left.\frac{dP_\pi}{dx_E}\right|_{p_{T_t}} = \frac{1}{\hat{x}_h}\frac{B^h}{b}\frac{1}{(1+\frac{x_E}{\hat{x}_h})^n}\frac{\left[\Gamma(n, b'x_{T_t}) - \Gamma(n, b'\frac{\hat{x}_h}{x_E})\right]}{\left[\Gamma(n-1, bx_{T_t}) - \Gamma(n-1, b)\right]}. \tag{E.30}$$

With the same approximation for the incomplete gamma functions used previously (Eq. E.19), namely taking the upper limit of the integral (Eq. E.26) to infinity and the lower limit to zero, the ratio of incomplete gamma functions in Eq. E.30 becomes equal to $n-1$ and the x_E distribution takes on a very simple and very interesting form:

$$\left.\frac{dP_\pi}{dx_E}\right|_{p_{T_t}} \approx \langle m\rangle_h (n-1)\frac{1}{\hat{x}_h}\frac{1}{(1+\frac{x_E}{\hat{x}_h})^n}, \tag{E.31}$$

where the only dependence on the fragmentation function is in the mean multiplicity of the species of detected particles in the jet $\langle m\rangle_h$. The dominant term in Eq. E.31 is the Hagedorn function $1/(1+x_E/\hat{x}_h)^n$ so that Eq. E.31 exhibits x_E scaling in the variable $x_E/\hat{x}_h$. The shape of the x_E distribution is given by the power n of the partonic and inclusive single particle transverse momentum spectra and does not depend on the exponential slope of the fragmentation function. However, the integral of the x_E distribution (from zero to infinity) is equal to $\langle m\rangle_h$, the mean multiplicity of the unbiased away jet for species h.

The reason that the x_E distribution is not very sensitive to the fragmentation function is that the integral over z_t for fixed p_{T_t} and p_{T_a} (Eqs. E.25, E.26) is actually an integral over the jet transverse momentum $\hat{p}_{T_t}$. However, since both the trigger and away jets are always roughly equal and opposite in transverse momentum, integrating over $\hat{p}_{T_t}$ simultaneously integrates over $\hat{p}_{T_a}$, and thus also integrates over the away jet fragmentation function. This can be seen directly by the presence of z_t in both the same-side and away-side fragmentation functions in Eq. E.25, so that the integral over z_t integrates over both fragmentation functions simultaneously.

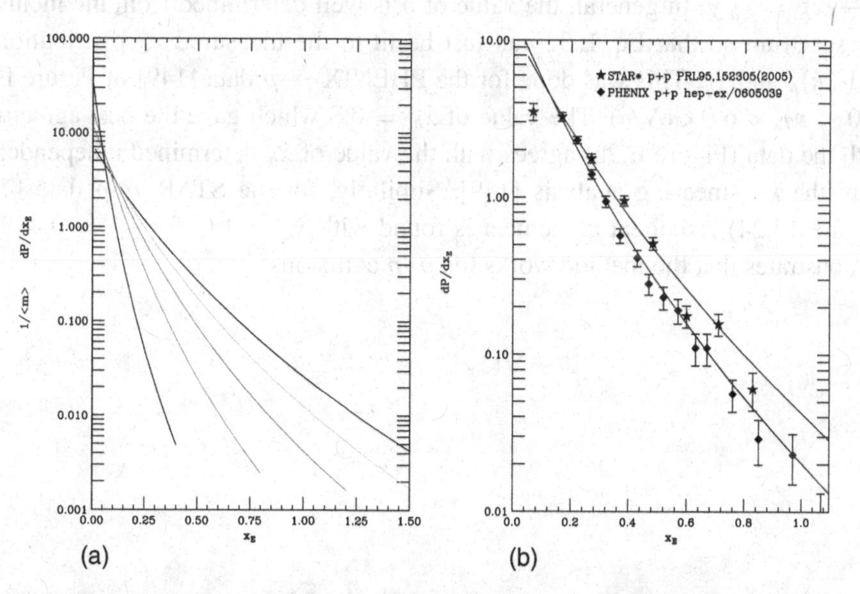

Figure E.2 (a) Equation E.31 for $n = 8.1$ divided by $\langle m\rangle_h$. The integral should be equal to 1. Curves are for $\hat{x}_h$ =1.0, 0.8, 0.6, 0.4, 0.2, with intercept $7.1/\hat{x}_h$ at $x_E = 0$. (b) PHENIX x_E distribution for $5 < p_{T_t} < 6$ GeV/c (Figure E.1a) with Eq. E.31 for $\hat{x}_h = 0.8$ (lower curve); STAR [766] x_E distribution (Figure 14.24) with Eq. E.31 for $\hat{x}_h = 1.0$ (upper curve). Curves and STAR data have been normalized to agree with the PHENIX data.

E.5 A very interesting formula

Equation E.31 (repeated below in a slightly different format) is very interesting:

$$\left.\frac{dP_\pi}{dy}\right|_{p_{T_t}} \approx \langle m\rangle_h (n-1)\frac{1}{(1+y)^n} \qquad \text{where} \qquad y = \frac{x_E}{\hat{x}_h}. \tag{E.32}$$

It relates the ratio of the transverse momenta of the away and trigger particles, $p_{T_a}/p_{T_t} = x_h \approx x_E$, which is measured, to the ratio of the transverse momenta of the away to the trigger jet, $\hat{p}_{T_a}/\hat{p}_{T_t}$, which can thus be deduced. Although derived for p–p collisions, Eq. E.31 (E.32) should work just as well in $A + A$ collisions since the only assumptions are independent fragmentation of the trigger and away jets with the same exponential fragmentation function and a power-law parton $\hat{p}_{T_t}$ distribution. The only other (and weakest) assumption is that $\hat{x}_h$ is constant for fixed p_{T_t} as a function of x_E. Thus in $A + A$ collisions, Eq. E.31 for the x_E distribution provides a method of measuring the ratio $\hat{x}_h = \hat{p}_{T_a}/\hat{p}_{T_t}$ and hence the relative energy loss of the away-side to the same-side jet assuming that both jets fragment outside the medium with the same fragmentation function as in p–p collisions.

A plot of Eq. E.31 is shown in Figure E.2a for $n = 8.1$ for various values of $\hat{x}_h$. Clearly, the smaller the value of $\hat{x}_h$, the steeper is the x_E distribution. However, all the curves in Figure E.2a are related by a simple scale transformation of Eq. E.32: $y \to x_E = \hat{x}_h\, y$. In general, the value of n is well determined from the inclusive p_{T_t} spectrum so that Eq. E.32 can just be fit to the measured x_E distribution to find $\langle m\rangle_h$ and $\hat{x}_h$. This was done for the PHENIX p–p data [149] of Figure E.1a ($5.0 < p_{T_t} < 6.0$ GeV/c). The value of $\hat{x}_h = 0.8$ which gave the best agreement with the data (Figure E.2b) agrees with the value of $\hat{x}_h$ determined independently from the k_T smearing analysis [149]. Similarly, for the STAR p–p data [766] (Figure 14.24), excellent agreement is found with $\hat{x}_h = 1.0$, for $x_E > 0.2$. This demonstrates that the method works for p–p collisions.

Appendix F

k_T phenomenology and Gaussian smearing

F.1 Definition of k_T

k_T denotes the magnitude of the effective transverse momentum vector $\boldsymbol{k}_T$ of each of the two colliding partons. The net p_T of the constituent c.m. system of the two colliding partons is

$$\boldsymbol{p}_T = \boldsymbol{k}_{T_1} + \boldsymbol{k}_{T_2}. \tag{F.1}$$

Or, more accurately

$$p_T^2 = k_{T_1}^2 + k_{T_2}^2 = 2k_T^2. \tag{F.2}$$

It is assumed that the two k_T act incoherently, or equivalently that one $\boldsymbol{k}_T$ acts in the scattering plane to make the energies of the two final state partons unequal, and one $\boldsymbol{k}_T$ acts perpendicular to the scattering plane to make the outgoing partons acoplanar with the direction of the colliding protons.

This simple concept leads to confusion since k_T is the magnitude of a vector, which is always positive like a radius, whereas the *components* of the vectors $\boldsymbol{p}_T$, $\boldsymbol{k}_T$, which act parallel or perpendicular to the scattering plane are more correctly written:

$$\langle p_{Tx}^2 \rangle = \langle k_{T1x}^2 \rangle + \langle k_{T2x}^2 \rangle \tag{F.3}$$

$$\langle p_{Tx}^2 \rangle = 2 \times \langle k_{Tx}^2 \rangle \tag{F.4}$$

$$\langle p_{Tx} \rangle = \sqrt{2} \times \langle k_{Tx} \rangle = 0 \tag{F.5}$$

where $k_T = k_{T_1} = k_{T_2}$.

The components k_{Tx} and k_{Ty} of the vector $\boldsymbol{k}_T$ are taken as Gaussian distributed, and following the notation of Apanasevich *et al.* [402]:

$$k_{Tx} = \sigma_{1parton,1d} \tag{F.6}$$

$$k_{Ty} = \sigma_{1parton,1d} \tag{F.7}$$

$$k_T = \sqrt{k_{Tx}^2 + k_{Ty}^2} = \sigma_{1parton,2d}. \tag{F.8}$$

F.2 Some Gaussian integrals

Recall how the integral of a one-dimensional Gaussian is obtained: take x to represent k_{Tx}, take y to represent k_{Ty}, and $r = \sqrt{x^2 + y^2}$ to represent k_T; and in two dimensions, $dx\,dy = r\,dr\,d\phi$

$$\text{Prob}(x)\,dx = \frac{1}{\sqrt{2\pi\sigma^2}} \exp -\frac{x^2}{2\sigma^2}\,dx \tag{F.9}$$

$$\text{Prob}(y)\,dy = \frac{1}{\sqrt{2\pi\sigma^2}} \exp -\frac{y^2}{2\sigma^2}\,dy \tag{F.10}$$

$$\text{Prob}(x, y)\,dx\,dy = \frac{1}{2\pi\sigma^2} \exp -\frac{x^2 + y^2}{2\sigma^2}\,dx\,dy \tag{F.11}$$

$$= \frac{1}{2\sigma^2} \exp -\frac{r^2}{2\sigma^2}\,dr^2. \tag{F.12}$$

This is just Eq. 4 of reference [402] and shows that the distribution of k_T^2 is exponential with mean value $\langle k_T^2 \rangle = 2\sigma^2 \equiv 2\sigma_{1d}^2 = \sigma_{2d}^2$.

A few more one and two-dimensional Gaussian identities can be derived but are just stated here (to understand why $\sqrt{\langle k_T^2 \rangle} = \langle k_T \rangle \times 2/\sqrt{\pi}$):

$$\langle r^2 \rangle = 2\sigma_{1d}^2 = \sigma_{2d}^2 \tag{F.13}$$

$$\langle x^2 \rangle = \langle y^2 \rangle = \sigma_{1d}^2 = \frac{\langle r^2 \rangle}{2} \tag{F.14}$$

$$\langle x \rangle = \langle y \rangle = 0 \tag{F.15}$$

$$\langle |x| \rangle = \langle |y| \rangle = \sqrt{\frac{2}{\pi}}\,\sigma_{1d} = \sqrt{\frac{\langle r^2 \rangle}{\pi}} \tag{F.16}$$

$$\langle r \rangle^2 = \frac{\pi}{4}\sigma_{2d}^2 = \frac{\pi}{4}\langle r^2 \rangle. \tag{F.17}$$

Finally a point made very clearly by reference [402]: k_T (really p_{Tx}) is the net momentum in the plane of the γ + jet pair. Thus the γ is smeared in momentum *by half this value*, i.e.

$$\sigma_{\gamma,1d} = k_T/2 = \sigma_{1parton,2d}/2. \tag{F.18}$$

F.3 Smearing of an exponential by a Gaussian

The k_T smearing of a p_T spectrum is very similar, if not identical, to the smearing effect of a Gaussian momentum resolution (Section 3.3.3).

Suppose that x_o is a quantity to be measured, e.g. p_T, which is distributed with a steeply falling distribution, exponential for example (since the integral can be done analytically):

$$d\mathcal{P}(x_o) = f(x_o)\,dx_o = e^{-bx_o}dx_o. \tag{F.19}$$

Further suppose that the true quantity x_o is measured with a Gaussian resolution function so that the result of the measurement is the quantity x, where

$$\mathcal{R}(x, x_o)\,dx = \text{Prob}(x)|_{x_o}\,dx = \frac{1}{\sqrt{2\pi\sigma^2}}\exp -\frac{(x-x_o)^2}{2\sigma^2}\,dx. \tag{F.20}$$

The result for the measured spectrum is simply

$$f(x)\,dx = \int_{x_o=x-\infty}^{x_o=x+\infty} dx_o\; f(x_o)\;\text{Prob}(x)|_{x_o}\,dx \tag{F.21}$$

$$f(x) = \frac{1}{\sqrt{2\pi\sigma^2}}\int dx_o\,\exp(-bx_o)\;\exp -\frac{(x-x_o)^2}{2\sigma^2}. \tag{F.22}$$

Complete the square:

$$f(x) = e^{-bx}\,e^{b^2\sigma^2/2}\times\frac{1}{\sqrt{2\pi\sigma^2}}\int_{x_o=x-\infty}^{x_o=x+\infty} dx_o\,\exp -\frac{(x-b\sigma^2-x_o)^2}{2\sigma^2}. \tag{F.23}$$

The result, since the Gaussian is normalized over ($-\infty$, $+\infty$), is simply

$$f(x) = e^{b^2\sigma^2/2}\times e^{-bx} = e^{-b(x-b\sigma^2/2)}. \tag{F.24}$$

This deceptively simple formula (compare Eq. A5 in reference [402], taking account of Eq. F.18) has important implications. The measured spectrum is shifted higher than the true spectrum by $\Delta x = b\sigma^2/2$; or equivalently, the measured spectrum at a true quantity x_o is higher than the true spectrum by a factor $\exp(b^2\sigma^2/2)$. Also, the steeper is the spectrum (larger b), the larger is the effect of the resolution smearing. This is a consequence of the fact that as the spectrum becomes steeper, it is relatively less probable to find larger values of the quantity of interest from the distribution itself, compared to the fluctuations due to resolution.

References

[1] E. V. Shuryak, Phys. Rep. **61**, 71 (1980).
[2] M. Breidenbach *et al.*, Phys. Rev. Lett. **23**, 935 (1969).
[3] F. W. Büsser *et al.*, Phys. Lett. **B46**, 471 (1973), see also reference [55].
[4] H. Fritzsch, M. Gell-Mann, and H. Leutwyler, Phys. Lett. **B47**, 365 (1973), introduces the term "color."
[5] C.-N. Yang and R. L. Mills, Phys. Rev. **96**, 191 (1954).
[6] For example, see references [7] and [8].
[7] P. A. M. Dirac, Proc. R. Soc. London **A112**, 661 (1926).
[8] W. Pauli, Nobel Lecture 1945, http://nobelprize.org/nobel_prizes/physics/ laureates/1945/pauli-lecture.html.
[9] E. Rutherford, Philos. Mag. **21**, 669 (1911).
[10] N. Bohr, Philos. Mag. **26**, 1 (1913).
[11] Sommerfeld extended Bohr's theory of circular orbits to elliptical orbits, which was of capital importance because it introduced the concept of space-quantization [12].
[12] A. Sommerfeld, Ann. Phys. **356**, 1 (1916).
[13] J. Chadwick, Nature **129**, 312 (1932).
[14] See, for example footnote 6, page 85 in reference [15].
[15] H. A. Bethe and R. F. Bacher, Rev. Mod. Phys. **8**, 82 (1936).
[16] An informative review of the development of the idea of the neutrino by Pauli in this period is given in reference [17].
[17] L. M. Brown, Physics Today **31N9**, 23 (1978).
[18] E. Fermi, Z. Phys. **88**, 161 (1934), see reference [19] for an English translation.
[19] F. L. Wilson, Am. J. Phys. **36**, 1150 (1968).
[20] R. Hofstadter and R. W. McAllister, Phys. Rev. **98**, 217 (1955).
[21] H. Fritzsch, M. Gell-Mann, and H. Leutwyler, Phys. Lett. **B47**, 365 (1973).
[22] H. D. Politzer, Phys. Rev. Lett. **30**, 1346 (1973).
[23] D. J. Gross and F. Wilczek, Phys. Rev. Lett. **30**, 1343 (1973).
[24] H. Yukawa, Nobel Lecture 1949, http://nobelprize.org/nobel_prizes/physics/ laureates/1949/yukawa.html.
[25] C. Peyrou, M. Gell-Mann, *et al.*, J. Phys. Coll. **43**, C8 (1982).
[26] R. N. Cahn and G. Goldhaber, *The Experimental Foundations of Particle Physics* (Cambridge University Press, Cambridge, 1989).
[27] M. Gell-Mann, Phys. Rev. **125**, 1067 (1962).
[28] Y. Ne'eman, Nucl. Phys. **26**, 222 (1961).
[29] M. Gell-Mann, Phys. Lett. **8**, 214 (1964).

[30] R. P. Feynman, M. Gell-Mann, and G. Zweig, Phys. Rev. Lett. **13**, 678 (1964).
[31] V. E. Barnes *et al.*, Phys. Rev. Lett. **12**, 204 (1964).
[32] O. W. Greenberg, Phys. Rev. Lett. **13**, 598 (1964).
[33] S. L. Glashow, J. Iliopoulos, and L. Maiani, Phys. Rev. **D2**, 1285 (1970).
[34] Peter Limon, quoting Ugo Camerini.
[35] T. D. Lee and C.-N. Yang, Phys. Rev. Lett. **4**, 307 (1960).
[36] T. D. Lee and C.-N. Yang, Phys. Rev. **119**, 1410 (1960).
[37] M. Schwartz, Phys. Rev. Lett. **4**, 306 (1960).
[38] B. Pontecorvo, Sov. Phys. JETP **9**, 1148 (1959).
[39] G. Danby, J.-M. Gaillard, K. Goulianos, L. M. Lederman, N. Mistry, M. Schwartz, and J. Steinberger, Phys. Rev. Lett. **9**, 36 (1962).
[40] J. Bernstein and G. Feinberg, Phys. Rev. **125**, 1741 (1962).
[41] J. Bernstein, Phys. Rev. **129**, 2323 (1963).
[42] R. Good, W. Melhop, O. Piccioni, and R. Swanson, in *Proc. XII Int. Conf. High Energy Physics, Dubna 1964* (Atomizdat, Moscow, 1966), Vol. 2, p. 32, see also footnote by A. Zichichi p. 35.
[43] G. Cocconi, L. J. Koester, and D. H. Perkins, Technical Report No. UCRL-10022 (1961), Lawrence Radiation Laboratory (unpublished), p. 167, as cited by reference [44].
[44] J. Orear, Phys. Rev. Lett. **12**, 112 (1964).
[45] F. Chilton, A. Saperstein, and E. Shrauner, Phys. Rev. **148**, 1380 (1966).
[46] R. C. Lamb, R. A. Lundy, T. B. Novey, D. D. Yovanovich, M. L. Good, R. Hartung, M. W. Peters, and A. Subramanian, Phys. Rev. Lett. **15**, 800 (1965).
[47] R. Burns, G. Danby, E. Hyman, L. M. Lederman, W. Lee, J. Rettberg, and J. Sunderland, Phys. Rev. Lett. **15**, 830 (1965).
[48] P. J. Wanderer, R. J. Stefanski, R. K. Adair, C. M. Ankenbrandt, H. Kasha, R. C. Larsen, L. B. Leipuner, and L. W. Smith, Phys. Rev. Lett. **23**, 729 (1969).
[49] Y. Yamaguchi, Nuovo Cimento **A43**, 193 (1966).
[50] J. H. Christenson, G. S. Hicks, L. M. Lederman, P. J. Limon, B. G. Pope, and E. Zavattini, Phys. Rev. Lett. **25**, 1523 (1970).
[51] E. Bloom *et al.*, Phys. Rev. Lett. **23**, 930 (1969).
[52] M. Breidenbach *et al.*, Phys. Rev. Lett. **23**, 935 (1969), see also reference [53].
[53] W. K. H. Panofsky, in *Proc. XIV Int. Conf. High Energy Physics (Vienna)* (CERN Scientific Information Service, Geneva, SZ, 1968), pp. 23–39.
[54] S. D. Drell and T.-M. Yan, Phys. Rev. Lett. **25**, 316 (1970).
[55] R. L. Cool *et al.*, in *Proc. XVI Int. Conf. High Energy Physics (Chicago-Batavia,1972)*, edited by J. D. Jackson and A. Roberts (NAL, Batavia, IL, 1973), Vol. 3, p. 317.
[56] M. Banner *et al.*, Phys. Lett. **B44**, 537 (1973).
[57] B. Alper *et al.*, Phys. Lett. **B44**, 521 (1973).
[58] J. A. Appel *et al.*, Phys. Rev. Lett. **33**, 719 (1974).
[59] J. W. Cronin *et al.*, Phys. Rev. Lett. **31**, 1426 (1973).
[60] J. J. Aubert *et al.*, Phys. Rev. Lett. **33**, 1404 (1974).
[61] J.-E. Augustin *et al.*, Phys. Rev. Lett. **33**, 1406 (1974).
[62] G. S. Abrams *et al.*, Phys. Rev. Lett. **33**, 1453 (1974).
[63] T. Appelquist and H. D. Politzer, Phys. Rev. Lett. **34**, 43 (1975).
[64] A. De Rujula and S. L. Glashow, Phys. Rev. Lett. **34**, 46 (1975).
[65] K. Eichten *et al.*, Phys. Rev. Lett. **34**, 369 (1975).
[66] K. Eichten *et al.*, Phys. Rev. **D17**, 3090 (1978).

[67] J. Kuti and V. F. Weisskopf, Phys. Rev. **D4**, 3418 (1971).
[68] F. E. Close, Rep. Prog. Phys. **42**, 1285 (1979), note, however that Close seemed to be unaware of the fact that the high p_T particle production at the CERN-ISR was consistent with QCD once the concept of "intrinsic transverse momentum" (k_T) had been introduced. See references [69–71], which were published after the submission date of Close's article.
[69] J. F. Owens, E. Reya, and M. Glück, Phys. Rev. **D18**, 1501 (1978).
[70] J. F. Owens and J. D. Kimel, Phys. Rev. **D18**, 3313 (1978).
[71] R. P. Feynman, R. D. Field, and G. C. Fox, Phys. Rev. **D18**, 3320 (1978).
[72] J. D. Bjorken, Phys. Rev. **179**, 1547 (1969).
[73] R. P. Feynman, Phys. Rev. Lett. **23**, 1415 (1969).
[74] J. Benecke, T. T. Chou, C. N. Yang, and E. Yen, Phys. Rev. **188**, 2159 (1969).
[75] An illuminating discussion on "hard" (Bjorken) versus "soft" (Feynman) partons is given in reference [76].
[76] J. D. Bjorken and R. P. Feynman, in *Proc. 1971 Int. Symposium on Electron and Photon Interactions at High Energies*, edited by N. B. Mistry (Cornell University, Ithaca, NY, 1971), pp. 295–297.
[77] A. Chodos *et al.*, Phys. Rev. **D9**, 3471 (1974).
[78] J. C. Collins and M. J. Perry, Phys. Rev. Lett. **34**, 1353 (1975).
[79] E. V. Shuryak, Phys. Rep. **61**, 71 (1980).
[80] W. J. Willis, in *Proc. XVI Int. Conf. High Energy Physics (Chicago-Batavia, 1972)*, edited by J. D. Jackson and A. Roberts (NAL, Batavia, IL, 1973), Vol. 4, pp. 321–331, note particularly the suggestion attributed to G. Cocconi on page 323 that "new states of matter might be produced when hadrons could be concentrated at high densities for a sufficiently long time."
[81] G. F. Chapline, M. H. Johnson, E. Teller, and M. S. Weiss, Phys. Rev. **D8**, 4302 (1973).
[82] T. D. Lee and G. C. Wick, Phys. Rev. **D9**, 2291 (1974).
[83] *Report of the Workshop on BeV/Nucleon Collisions of Heavy Ions–How and Why* (BNL-50445, Upton, NY, 1975), Bear Mountain, NY, 29 November–1 December 1974.
[84] K. Nakamura *et al.*, J. Phys. G **37**, 075021 (2010), Particle Data Group, Review of Particle Physics.
[85] G. Policastro, D. T. Son, and A. O. Starinets, Phys. Rev. Lett. **87**, 081601 (2001).
[86] H. Nastase, (2005), Preprint, hep-th/0501068.
[87] It should be noted that the requirement of specificity to A+A collisions immediately rules out the QGP in p-p collisions, which is disputable. See references [88, 89].
[88] R. M. Weiner, Int. J. Mod. Phys. **E15**, 37 (2006), Preprint arXiv:hep-ph/0507115.
[89] T. Alexopoulos *et al.*, Phys. Lett. **B528**, 43 (2002).
[90] A. S. Kronfeld and C. Quigg, Am. J. Phys. **78**, 1081 (2010).
[91] W. Heisenberg, Z. Phys. **77**, 1 (1932).
[92] E. Wigner, Phys. Rev. **51**, 106 (1937).
[93] Wikimedia Commons, Creative Commons Attribution 3.0 Unported License. Author: MissMJ 27 June 2006 (after reference [94]), version June 2009, http://en.wikipedia.org/wiki/File:Standard_Model_of_Elementary_Particles.svg.
[94] The science of matter, space and time: What is the world made of? Fermilab, http://www.fnal.gov/pub/inquiring/matter/madeof/index.html.
[95] S. L. Glashow, Nucl. Phys. **22**, 579 (1961).
[96] A. Salam, Nobel Lecture 1979, http://nobelprize.org/nobel_prizes/physics/laureates/1979/salam-lecture.html.

[97] S. Weinberg, Phys Rev. Lett. **19**, 1264 (1967).
[98] N. Cabibbo, Phys. Rev. Lett. **10**, 531 (1963).
[99] D. Diakonov, Prog. Part. Nucl. Phys. **36**, 1 (1996).
[100] S. Weinberg, Eur. Phys. J. **C34**, 5 (2004).
[101] O. W. Greenberg, Annu. Rev. Nucl. Part. Sci. **28**, 327 (1978).
[102] C. Amsler, T. DeGrand, and B. Krusche, Quark Model, in reference [84], pp. 184–192.
[103] L. Van Hove, Nucl. Phys. **A461**, 3c (1987).
[104] R. E. Marshak, *Meson Physics* (McGraw-Hill, New York, 1952), chapter 8, also see reference [105].
[105] U. Camerini, W. O. Lock, and D. Perkins, in *Progress in Cosmic Ray Physics*, edited by J. G. Wilson (North Holland, Amsterdam, 1952), Vol. I, pp. 1–34.
[106] W. B. Fowler *et al.*, Phys. Rev. **95**, 1026 (1954).
[107] 20 inch bubble chamber photograph BNL-1–132–62 used with permission.
[108] R. P. Feynman, in *High Energy Collisions Third International Conference at Stony Brook, NY*, edited by J. A. Cole *et al.* (Gordon and Breach, New York, 1969).
[109] A. H. Mueller, Phys. Rev. **D2**, 2963 (1970).
[110] C. E. DeTar, Phys. Rev. **D3**, 128 (1971).
[111] Z. Koba, H. B. Nielsen, and P. Olesen, Nucl. Phys. **B40**, 317 (1972).
[112] W. Thomé *et al.*, Nucl. Phys. **B129**, 365 (1977).
[113] F. Abe *et al.*, Phys. Rev. Lett. **61**, 1819 (1988).
[114] G. Giacomelli and M. Jacob, Phys. Rep. **55**, 1 (1979).
[115] P. Capiluppi *et al.*, Nucl. Phys. **B79**, 189 (1974).
[116] P. Capiluppi *et al.*, Nucl. Phys. **B70**, 1 (1974).
[117] H. Bøggild *et al.*, Nucl. Phys. **B57**, 77 (1973).
[118] V. Blobel *et al.*, Nucl. Phys. **B69**, 454 (1974).
[119] E. Shuryak and O. V. Zhirov, Phys. Lett. **B171**, 99 (1986).
[120] For the flavor of the period, see Maurice Jacob's Eulogy for the ISR in CERN report 84-13 (CERN, SIS, Geneva, 1984).
[121] K. Alpgård *et al.*, Phys. Lett. **B123**, 361 (1983).
[122] K. Guettler *et al.*, Nucl. Phys. **B116**, 77 (1976).
[123] G. J. Alner *et al.*, Z. Phys. **C33**, 1 (1986).
[124] L. Van Hove, in *Niels Bohr Centennial Symposium – The lesson of quantum theory*, edited by E. Dal, J. De Boer, and O. Ulfbeck (North Holland, Amsterdam, 1986), pp. 167–179, Preprint CERN-TH-4353-85.
[125] See, for example, reference [126] and references therein.
[126] M. J. Tannenbaum, Int. J. Mod. Phys. **A4**, 3377 (1989).
[127] C. De Marzo *et al.*, Phys. Lett. **B112**, 173 (1982).
[128] C. De Marzo *et al.*, Nucl. Phys. **B211**, 375 (1982).
[129] G. Arnison *et al.*, Transverse energy distributions in the central calorimeters, presented to 21st Int. Conf. High Energy Physics, Paris, France, Jul 26-31, 1982, CERN-EP-82-122 (unpublished), also see reference [130].
[130] G. Arnison *et al.*, Phys. Lett. **B107**, 320 (1981).
[131] A more comprehensive treatment of this subject is found, for example, in references [84, 132, 133].
[132] R. Fernow, *Introduction to Experimental Particle Physics* (Cambridge University Press, Cambridge, 1989).
[133] R. M. Sternheimer, in *Methods of Experimental Physics*, edited by L. C. L. Yuan and C. S. Wu (Academic Press, New York, 1963), Vol. 5A, p. 1.
[134] G. Diambrini-Palazzi, Rev. Mod. Phys. **40**, 611 (1968).

[135] P. L. Anthony *et al.*, Phys. Rev. **D56**, 1373 (1997).
[136] K. Kleinknecht, *Detectors for Particle Radiation* (Cambridge University Press, Cambridge, 1986).
[137] T. Ypsilantis and J. Seguinot, Nucl. Instrum. Methods **A343**, 30 (1994).
[138] J. D. Jackson, *Classical Electrodynamics*, second edition (John Wiley & Sons, Inc., New York, 1975), Chapters 13, 14.
[139] A. Andronic and J. P. Wessels, Nucl. Instrum. Methods **A666**, 130 (2012).
[140] Y.-S. Tsai, Rev. Mod. Phys. **46**, 815 (1974).
[141] C. W. Fabjan and T. Ludlam, Annu. Rev. Nucl. Part. Sci. **32**, 335 (1982).
[142] R. L. Gluckstern, Nucl. Instrum. Methods **24**, 381 (1963).
[143] G. D. Lafferty and T. R. Wyatt, Nucl. Instrum. Methods **A355**, 541 (1995).
[144] This is the same convention used in Perkins' textbook [145].
[145] D. H. Perkins, *Introduction to High Energy Physics*, third edition (Addison-Wesley, Reading, MA, 1987).
[146] H. Poincaré, Rend. Circ. Matem. Palermo **21**, 129 (1906).
[147] S. Mandelstam, Phys. Rev. **112**, 1344 (1958).
[148] R. M. Sternheimer, Phys. Rev. **99**, 277 (1955).
[149] S. S. Adler *et al.*, Phys. Rev. **D74**, 072002 (2006).
[150] M. J. Tannenbaum *et al.*, in *Proc. 20th Winter Workshop on Nuclear Dynamics, March 15–20, 2004, Trelawny Beach, Jamaica WI*, edited by W. Bauer *et al.* (EP Systema, Budapest, Hungary, 2004), pp. 245–252, also appeared as arXiv:nucl-ex/0406024v3.
[151] K. Reygers *et al.*, Nucl. Phys. **A715**, 683c (2003).
[152] S. S. Adler *et al.*, Phys. Rev. **C75**, 024909 (2007).
[153] S. S. Adler *et al.*, Phys. Rev. Lett. **98**, 012002 (2007).
[154] S. S. Adler *et al.*, Phys. Rev. **D71**, 071102(R) (2005).
[155] H. Becquerel, Compt. Rend. **122**, 501 (1896).
[156] E. Rutherford, Philos. Mag. **47**, 109 (1899).
[157] E. Rutherford, Philos. Mag. **11**, 166 (1906).
[158] E. Rutherford, Philos. Mag. **12**, 134 (1906).
[159] H. Geiger, Proc. R. Soc. **81**, 174 (1908).
[160] H. Geiger and E. Marsden, Proc. R. Soc. **82**, 495 (1909).
[161] J. Campbell, *Rutherford Scientist Supreme* (AAS Publications, Christchurch, NZ, 1999).
[162] J. J. Thomson, Philos. Mag. **7**, 237 (1904).
[163] M. Goldhaber, A Physicist's Journey: Reminiscences from the Cavendish Laboratory, Pegram Lecture, BNL, April 27, 1992.
[164] J. A. Crowther, Proc. R. Soc. **84**, 226 (1910).
[165] E. Rutherford, J. Chadwick, and C. D. Ellis, *Radiations from Radioactive Substances* (Cambridge University Press, Cambridge, 1951).
[166] H. Geiger and E. Marsden, Philos. Mag. **25**, 604 (1913).
[167] R. March, *Physics for Poets* (McGraw-Hill, New York, 2003).
[168] E. M. Lyman, A. O. Hanson, and M. B. Scott, Phys. Rev. **84**, 626 (1951).
[169] R. Hofstadter, The electron-scattering method and its application to the structure of nuclei and nucleons., Nobel Lecture 1961, http://nobelprize.org/nobel_prizes/physics/laureates/1961/hofstadter-lecture.html.
[170] R. Sherr, Phys. Rev. **68**, 240 (1945).
[171] L. R. B. Elton, Phys. Rev. **79**, 412 (1950).
[172] M. E. Rose, Phys. Rev. **73**, 279 (1948).
[173] R. Hofstadter, Rev. Mod. Phys. **28**, 214 (1956).

[174] M. N. Rosenbluth, Phys. Rev. **79**, 615 (1950).
[175] N. F. Mott, Proc. R. Soc. **124**, 425 (1929).
[176] N. F. Mott, Proc. R. Soc. **135**, 429 (1932).
[177] A. F. Sill *et al.*, Phys. Rev. **D48**, 29 (1993).
[178] R. Hofstadter and R. W. McAllister, Phys. Rev. **98**, 217 (1955).
[179] P. N. Kirk *et al.*, Phys. Rev. **D8**, 63 (1973).
[180] R. Hofstadter, F. Bumiller, and M. R. Yerian, Rev. Mod. Phys. **30**, 482 (1958).
[181] L. H. Chan *et al.*, Phys. Rev. **141**, 1298 (1966).
[182] R. W. Ellsworth *et al.*, Phys. Rev. **165**, 1449 (1968).
[183] X. Zhan *et al.*, High precision measurement of the proton elastic form factor ratio $\mu_p G_E/G_M$ at low Q^2, ArXiv:1102.0381v1 [nucl-ex] 1 Feb 2011.
[184] J. Litt *et al.*, Phys. Lett. **B31**, 40 (1970).
[185] I. A. Qattan *et al.*, Phys. Rev. Lett. **94**, 142301 (2005).
[186] O. Gayou *et al.*, Phys. Rev. Lett. **88**, 092301 (2002).
[187] L. M. Lederman and M. J. Tannenbaum, in *Advances in Particle Physics*, edited by R. L. Cool and R. E. Marshak (Wiley Interscience, New York, 1968), Vol. 1, pp. 1–70.
[188] C. E. Carlson and M. Vanderhaeghen, Annu. Rev. Nucl. Part. Sci. **57**, 171 (2007).
[189] D. W. Dupen, H. A. Hogg, G. A. Loew, and R. B. Neal, in *The Stanford Two-Mile Accelerator*, edited by R. B. Neal (W. A. Benjamin, Inc., New York, 1968).
[190] For example, see L. Hand, references [191,192].
[191] L. N. Hand, Phys. Rev. **29**, 1834 (1963).
[192] L. Hand, in *Proc. 1967 Int. Symposium on Electron and Photon Interactions at High Energies*, edited by S. M. Berman (Stanford Linear Accelerator Center, Stanford, CA, 1967), pp. 128–155.
[193] N. Dombey, Rev. Mod. Phys. **41**, 236 (1969).
[194] S. D. Drell and J. D. Walecka, Ann. Phys. (NY) **28**, 18 (1964).
[195] For example, see reference [196], equation 2.4.
[196] J. D. Bjorken and E. A. Paschos, Phys. Rev. **185**, 1975 (1969).
[197] J. D. Bjorken, Phys. Rev. **163**, 1767 (1967).
[198] M. Gourdin, Nuovo Cimento **37**, 208 (1965).
[199] G. Miller *et al.*, Phys. Rev. **D5**, 528 (1972).
[200] J. I. Friedman, Rev. Mod. Phys. **63**, 615 (1991).
[201] K. Gottfried, Phys. Rev. Lett. **18**, 1174 (1967).
[202] For example, see reference [203].
[203] F. Gilman, in *Proc. 4th Int. Symposium on Electron and Photon Interactions at High Energies (Liverpool, September 14th–20th 1969)*, edited by D. W. Braben (Daresbury Nuclear Physics Laboratory, Daresbury, UK, 1969), pp. 177–192.
[204] For example see discussion in reference [200].
[205] J. D. Bjorken and M. Nauenberg, Annu. Rev. Nucl. Part. Sci. **18**, 229 (1968).
[206] S. L. Adler, Phys. Rev. **143**, 1144 (1966).
[207] N. Cabibbo, Phys. Rev. Lett. **10**, 531 (1963).
[208] J. D. Bjorken, Phys. Rev. Lett. **16**, 408 (1966).
[209] J. D. Bjorken, Phys. Rev. **163**, 1767 (1967).
[210] J. D. Bjorken, in *Proc. 1967 Int. Symposium on Electron and Photon Interactions at High Energies*, edited by S. M. Berman (Stanford Linear Accelerator Center, Stanford, CA, 1967), pp. 109–127.
[211] R. P. Feynman, *Photon-Hadron Interactions* (Addison-Wesley, Redwood City, CA, 1989).
[212] C. G. Callan and D. J. Gross, Phys. Rev. Lett. **22**, 156 (1969).

[213] For example, see reference [214] and citations therein.
[214] J. T. Dakin *et al.*, Phys. Rev. **D10**, 1401 (1974).
[215] H. W. Kendall, in *Proc. 1971 Int. Symposium on Electron and Photon Interactions at High Energies*, edited by N. B. Mistry (Cornell University, Ithaca, NY, 1971), pp. 247–261.
[216] W. L. Lakin *et al.*, Phys. Rev. Lett. **26**, 34 (1971).
[217] M. J. Tannenbaum, in *Proc. 6th Int. Symposium on Electron and Photon Interactions at High Energies (Bonn, August 27–31, 1973)*, edited by H. Rollnik and W. Pfeil (North-Holland, Amsterdam–London, 1974), pp. 310–311, also see reference [218].
[218] M. May *et al.*, Phys. Rev. Lett. **35**, 407 (1975).
[219] R. P. Feynman, Science **183**, 601 (1974).
[220] J. D. Bjorken, in *Proc. 1971 Int. Symposium on Electron and Photon Interactions at High Energies*, edited by N. B. Mistry (Cornell University, Ithaca, NY, 1971), pp. 281–297.
[221] D. H. Perkins, in *Proc. XVI Int. Conf. High Energy Physics (Chicago-Batavia,1972)*, edited by J. D. Jackson and A. Roberts (NAL, Batavia, IL, 1973), Vol. 4, pp. 189–247.
[222] H. Yukawa, Proc. Phys.-Math. Soc. Jpn. **17**, 48 (1935).
[223] R. Burns *et al.*, Phys. Rev. Lett. **15**, 42 (1965).
[224] M. M. Block *et al.*, Phys. Lett. **12**, 281 (1964).
[225] G. Bernardini *et al.*, Phys. Lett. **13**, 86 (1964).
[226] A. Zichichi, in *Proc. 12th Int. Conf. High Energy Physics August 5–15* (Atomizdat, Moscow, 1964), Vol. 2, p. 35.
[227] G. Arnison *et al.*, Phys. Lett. **B122**, 103 (1983).
[228] M. Banner *et al.*, Phys. Lett. **B122**, 476 (1983).
[229] J. A. Appel *et al.*, Z. Phys. **C30**, 1 (1986).
[230] B. Aubert and C. Rubbia, Phys. Rep. **239**, 215 (1994).
[231] J. Bernstein, Phys. Rev. **129**, 2323 (1963).
[232] J. Nearing, Phys. Rev. **132**, 2323 (1963), also see the erratum [233].
[233] J. Nearing, Phys. Rev. **135**, AB2 (1964).
[234] O. Piccioni, in *Proc. 12th Int. Conf. High Energy Physics August 5–15* (Atomizdat, Moscow, 1964), Vol. 2, pp. 32–36. It is in a comment to this talk that Nino Zichichi [226] described the method by which the W was actually discovered.
[235] A. Zichichi, S. M. Berman, N. Cabibbo, and R. Gatto, Nuovo Cimento **24**, 170 (1962).
[236] M. Conversi, T. Massam, T. Muller, and A. Zichichi, Nuovo Cimento **A 40**, 690 (1965).
[237] G. Altarelli, R. A. Brant, and G. Preparata, Phys. Rev. Lett. **26**, 42 (1971).
[238] L. M. Lederman *et al.*, NAL Proposal No. 70, 1970, *Study of Lepton Pairs from Proton-Nuclear Interactions: Search for Intermediate Bosons and Lee-Wick Structure.*
[239] R. L. Cool *et al.*, Preliminary Proposal ISR/69-43, *ISR Study of Dileptons*, June 1969; B. J. Blumenfeld *et al.*, CCR Proposal, ISRC/69-43/Add, *Search for Massive Dileptons*, September 1970.
[240] K. Johnsen, Nucl. Instrum. Methods **108**, 205 (1973).
[241] J. R. Sanford, Annu. Rev. Nucl. Part. Sci. **26**, 151 (1976).
[242] R. R. Wilson, Rev. Mod. Phys. **51**, 259 (1979).
[243] W. Jentschke, CERN/ISRC/69-1, Letter from Chairman of the ISRC to CERN users, 15 January 1969.

[244] R. R. Wilson, Notice to NAL Users, March 26, 1970.
[245] M. Jacob and K. Johnsen, Yellow Report CERN-84-13, CERN, Geneva, Switzerland (unpublished).
[246] L. Camilleri, Phys. Rep. **144**, 51 (1987).
[247] L. Evans, Yellow Report CERN-88-01, CERN, Geneva, Switzerland (unpublished).
[248] L. Van Hove and M. Jacob, Phys. Rep. **62**, 1 (1980).
[249] J. D. Jackson, J. Phys. Coll. **34**, C1 (1973).
[250] G. Giacomelli, A. F. Greene, and J. R. Sanford, Phys. Rep. **19**, 169 (1975).
[251] R. R. Wilson, Letter to Members of the NAL Users Organization, February 10, 1971.
[252] L. M. Lederman, in *1968 Summer Study*, edited by A. Roberts (NAL, Batavia, IL, 1968), Vol. 2, pp. 55–58, also appeared as NAL TM-0132.
[253] M. J. Tannenbaum, in *1968 Summer Study*, edited by A. Roberts (NAL, Batavia, IL, 1968), Vol. 2, pp. 49–53, also appeared as NAL TM-0067.
[254] E. A. Choban, Sov. J. Nucl. Phys. **7**, 245 (1968).
[255] F. A. Berends and G. B. West, Phys. Rev. **D1**, 122 (1970).
[256] J. W. Cronin and P. A. Piroué, NAL Proposal No. 100, December 1, 1970, *A proposal to study particle production at high transverse momenta.*
[257] D. Antreasyan, J. W. Cronin, *et al.*, Phys. Rev. **D19**, 764 (1979).
[258] A. Russo, in *History of CERN*, edited by J. Krige (North Holland, Elsevier Science, Amsterdam, 1996), Vol. 3, pp. 97–170.
[259] J. Billan, R. Perin, and V. Sergo, in *4th International Conference on Magnet Technology, Upton, NY, USA, 19–22 Sept. 1972*, edited by Y. Winterbottom (Brookhaven National Laboratory, Upton, NY, USA, 1972), pp. 433–443.
[260] R. Bouclier, G. Charpak, *et al.*, Nucl. Instrum. Methods **115**, 235 (1974).
[261] G. Ciapetti, V. Lüth, P. Steffen, and J. Steinberger, *"Acceptance" tables for three alternative proposals for an ISR general purpose detector*, CERN-NP-Internal-Report-68-34, November 1968.
[262] E. Keil, *ISR parameter list (revision 5)*, Internal Report CERN-ISR-TH-68-32, February 1976.
[263] See, for example, reference [283].
[264] S. M. Berman, J. D. Bjorken, and J. B. Kogut, Phys. Rev. **D4**, 3388 (1971).
[265] F. W. Büsser *et al.*, Phys. Lett. **B46**, 471 (1973).
[266] R. Blankenbecler, S. J. Brodsky, and J. F. Gunion, Phys. Lett. **B42**, 461 (1972).
[267] R. Blankenbecler, S. J. Brodsky, and J. F. Gunion, Phys. Rev. **D12**, 3469 (1975).
[268] R. F. Cahalan, K. A. Geer, J. B. Kogut, and L. Susskind, Phys. Rev. **D11**, 1199 (1975).
[269] A. L. S. Angelis *et al.*, Phys. Lett. **B79**, 505 (1978).
[270] F. W. Büsser *et al.*, Nucl. Phys. **B106**, 1 (1976).
[271] C. Kourkoumelis *et al.*, Phys. Lett. **B84**, 271 (1979).
[272] R. D. Field, Phys. Rev. Lett. **40**, 997 (1977).
[273] D. Antreasyan *et al.*, Phys. Rev. Lett. **38**, 112 (1977).
[274] C. Bromberg *et al.*, Phys. Rev. Lett. **43**, 561 (1979).
[275] H. J. Frisch *et al.*, Phys. Rev. Lett. **44**, 511 (1980).
[276] D. Jones and J. F. Gunion, Phys. Rev. **D19**, 867 (1979).
[277] R. D. Field, Phys. Rev. **D27**, 546 (1983).
[278] M. K. Chase and W. J. Stirling, Nucl. Phys. **B133**, 157 (1978).
[279] S. J. Brodsky, H. J. Pirner, and J. Raufeisen, Phys. Lett. **637**, 58 (2006).
[280] J. W. Cronin *et al.*, Phys. Rev. **D11**, 3105 (1975).

[281] M. Banner *et al.*, Measurement of High Transverse Momentum Charged Particles, Letter of intent to CERN-ISRC, May 16, 1972, CERN/ISRC/72-13.
[282] M. Banner *et al.*, Addendum 2 to Proposal: Measurement of High Transverse Momentum Charged Particles, April 1973, CERN/ISRC/72-13/Add. 2.
[283] F. W. Büsser *et al.*, Nucl. Phys. **B113**, 189 (1976).
[284] V. V. Abramov *et al.*, Phys. Lett. **B64**, 365 (1976).
[285] H. A. Bethe and J. Ashkin, in *Experimental Nuclear Physics*, edited by E. Segrè (Wiley, New York, 1953), Vol. 1, p. 166.
[286] H. Bethe and W. Heitler, Proc. R. Soc. London **A146**, 83 (1934).
[287] R. H. Dalitz, Proc. Phys. Soc. **A64**, 667 (1951).
[288] N. M. Kroll and W. Wada, Phys. Rev. **98**, 1355 (1955).
[289] D. W. Joseph, Nuovo Cimento **16**, 997 (1960).
[290] *Proc. XVII Int. Conf. High Energy Physics (London, July 1974)*, edited by J. R. Smith (Rutherford Laboratory, Chilton, Didcot, UK, 1975), pp. V-41–V-56.
[291] J. P. Boymond *et al.*, Phys. Rev. Lett. **33**, 112 (1974).
[292] F. W. Büsser *et al.*, in *Proc. XVII Int. Conf. High Energy Physics (London, July 1974)*, edited by J. R. Smith (Rutherford Laboratory, Chilton, Didcot, UK, 1975), pp. V-41–V-43, presented by S. Segler.
[293] F. W. Büsser *et al.*, Phys. Lett. **B53**, 212 (1974).
[294] J. A. Appel *et al.*, Phys. Rev. Lett. **33**, 722 (1974).
[295] V. Abramov *et al.*, in *Proc. XVII Int. Conf. High Energy Physics (London, July 1974)*, edited by J. R. Smith (Rutherford Laboratory, Chilton, Didcot, UK, 1975), pp. V-53–V-54, presented by S. Nurushev.
[296] See a full list of citations in reference [283].
[297] B. Alper *et al.*, Nucl. Phys. **B87**, 19 (1975).
[298] G. R. Farrar and S. C. Frautschi, Phys. Rev. Lett. **36**, 1017 (1976).
[299] C. O. Escobar, Nucl. Phys. **B98**, 173 (1975).
[300] E. L. Feinberg, Nuovo Cimento **A34**, 391 (1976).
[301] H. Fritzsch and P. Minkowski, Phys. Lett. **B69**, 316 (1977).
[302] I. Hinchliffe and C. H. Llewellyn-Smith, Nucl. Phys. **B114**, 45 (1976).
[303] G. Goldhaber *et al.*, Phys. Rev. Lett. **37**, 255 (1976).
[304] F. W. Büsser *et al.*, in *Proc. 1975 Int. Symposium on Lepton and Photon Interactions at High Energies*, edited by W. T. Kirk (Stanford Linear Accelerator Center, Stanford, CA, 1975), paper 202.
[305] M. Bourquin and J.-M. Gaillard, Nucl. Phys. **B114**, 334 (1976).
[306] L. Baum *et al.*, Phys. Lett. **B60**, 485 (1976).
[307] R. Vogt, Z. Phys. **C71**, 475 (1996).
[308] K. Adcox *et al.*, Phys. Rev. Lett. **88**, 192303 (2002).
[309] S. P. K. Tavernier, Rep. Prog. Phys. **50**, 1439 (1987).
[310] J. A. Appel, Annu. Rev. Nucl. Part. Sci. **42**, 367 (1992).
[311] M. Basile *et al.*, Nuovo Cimento **A65**, 421 (1981).
[312] B. Alper *et al.*, Nucl. Phys. **B100**, 237 (1975).
[313] D. Drijard *et al.*, Phys. Lett. **B81**, 250 (1979).
[314] J. Appel, Hadroproduction of Charm Particles, 1992, Fermilab-Pub-92/49.
[315] S. C. C. Ting, in *Nobel Lectures, Physics 1971–1980*, edited by S. Lundqvist (World Scientific Publishing Co., Singapore, 1992).
[316] J. W. Cronin, in *Proc. 1977 Int. Symposium on Lepton and Photon Interactions at High Energies*, edited by F. Gutbrod (DESY, Hamburg, 1977), pp. 579–598.
[317] M. R. Jane *et al.*, Phys. Lett. **B73**, 503 (1978).
[318] T. Åkesson *et al.*, Phys. Lett. **B192**, 463 (1987).

[319] U. Goerlach, in *Proc. Intl. Workshop on Quark Gluon Plasma Signatures (Strasbourg, Oct. 1–4, 1990)*, edited by V. Bernard, A. Capella, W. Geist, P. Gorodetsky, R. Selz, and C. Voltolini (Editions Frontieres, Gif-sur-Yvette, France, 1991), pp. 305–315.
[320] U. Goerlach, Nucl. Phys. **A544**, 109c (1992).
[321] J. H. Christenson, G. S. Hicks, L. M. Lederman, P. J. Limon, B. G. Pope, and E. Zavattini, Bull. Am. Phys. Soc. **15**, 579 (1970).
[322] A. S. Ito *et al.*, Phys. Rev. **D23**, 604 (1981).
[323] P. J. Limon, AGS Proposal #549, September 14, 1970, *A high resolution study of the production of electron pairs in hadronic interactions covering the mass range 0.7 GeV/*$c^2 \leq M_{e^+e^-} \leq 4.5$ *GeV/*c^2.
[324] J. G. Asbury *et al.*, Phys. Rev. Lett. **18**, 65 (1967).
[325] J. G. Asbury *et al.*, Phys. Rev. Lett **19**, 869 (1967).
[326] J. J. Aubert *et al.*, Nucl. Phys. **B89**, 1 (1975).
[327] J. J. Aubert *et al.*, Phys. Rev. Lett. **33**, 1624 (1974).
[328] F. W. Büsser *et al.*, Phys. Lett. **B56**, 482 (1975).
[329] A. G. Clark *et al.*, Nucl. Phys. **B142**, 29 (1978).
[330] D. C. Hom, L. M. Lederman, *et al.*, Phys. Rev. Lett. **36**, 1236 (1976).
[331] H. D. Snyder, D. C. Hom, L. M. Lederman, *et al.*, Phys. Rev. Lett. **36**, 1415 (1976).
[332] J. F. Gunion, Phys. Rev. **D14**, 1400 (1976).
[333] D. C. Hom, L. M. Lederman, *et al.*, Phys. Rev. Lett. **37**, 1374 (1976).
[334] M. L. Good, L. M. Lederman, and T. Yamanouchi, NAL Proposal No. 494, May 11,1976, *Di-Hadron Mass Search; II Dimuon Phase*.
[335] S. W. Herb *et al.*, Phys. Rev. Lett. **39**, 252 (1977).
[336] W. R. Innes *et al.*, Phys. Rev. Lett. **39**, 1240 (1977).
[337] K. Ueno *et al.*, Phys. Rev. Lett. **42**, 486 (1979).
[338] R. N. Cahn and S. D. Ellis, Phys. Rev. **D16**, 1484 (1977).
[339] J. Ellis, M. K. Gaillard, D. V. Nanopoulos, and S. Rudaz, Nucl. Phys. **B131**, 285 (1977).
[340] E. Eichten and K. Gottfried, Phys. Lett. **B66**, 286 (1977).
[341] C. Berger *et al.*, Phys. Lett. **B76**, 243 (1978).
[342] J. K. Bienlein *et al.*, Phys. Lett. **B78**, 360 (1978).
[343] L. M. Lederman private communication to M. J. Tannenbaum.
[344] J. K. Yoh *et al.*, Phys. Rev. Lett. **41**, 684 (1978).
[345] M. Grossmann-Handschin *et al.*, Phys. Lett. **B179**, 170 (1986).
[346] J. B. Adams, *The CERN Accelerators*, CERN Annual Report 1976, pp. 11–18.
[347] B. Aebischer *et al.*, Letter of intent, CERN/SPSC 77-20, March 1, 1977, *Inclusive production of massive muon pairs with intense pion beams.*
[348] B. Aebischer *et al.*, Proposal to SPSC, CERN/SPSC 77-110, November 17, 1977, *High resolution study of the inclusive production of massive muon pairs by intense pion beams.*
[349] L. Anderson *et al.*, Nucl. Instrum. Methods **A223**, 26 (1984).
[350] L. Camilleri *et al.*, Nucl. Instrum. Methods **156**, 275 (1978).
[351] A. Yamamoto, Nucl. Instrum. Methods **A453**, 445 (2000).
[352] A. L. S. Angelis *et al.*, Phys. Lett. **B87**, 398 (1979).
[353] J. H. Cobb *et al.*, Phys. Lett. **B72**, 273 (1977).
[354] C. Kourkoumelis *et al.*, Phys. Lett. **B91**, 481 (1980).
[355] J. H. Cobb *et al.*, Phys. Lett. **B68**, 101 (1977).
[356] C. Kourkoumelis *et al.*, Phys. Lett. **B91**, 475 (1980).

[357] D. Antreasyan *et al.*, Phys. Rev. Lett. **45**, 863 (1980).
[358] J. Badier *et al.*, Nucl. Instrum. Methods **175**, 319 (1980).
[359] W. Kienle, A. Michelini, *et al.*, Proposal to CERN-SPSC, CERN/SPSC/74-90/P24, October 2, 1974, *Proposal to measure production of high-p_T leptons and hadrons*.
[360] NA3 Collaboration, Memorandum to Prof. I. Butterworth, Chairman of the SPSC, CERN/SPSC/77-57, SPSC/M 83, June 20, 1977, *Physics of the NA3 experiment; request for beam and computer time for 1978*.
[361] J. Badier *et al.*, Phys. Lett. **B86**, 98 (1979).
[362] D. M. Alde *et al.*, Phys. Rev. Lett. **66**, 133 (1991).
[363] D. M. Alde *et al.*, Phys. Rev. Lett. **66**, 2285 (1991).
[364] J. C. Peng, P. L. McGaughey, and J. M. Moss, Talk presented at the RIKEN-BNL Workshop on Hard Parton Physics in Nucleus–Nucleus Collisions, March, 1999, *Dilepton Production at Fermilab and RHIC*, Preprint arXiv:hep-ph/9905447v1.
[365] M. J. Leitch, Talk given at RHIC & AGS Annual User's Meeting, Workshop 4 - RHIC Future Strategy, June 21, 2011, *Ongoing PHENIX upgrades + science strategy (HI and spin) for coming 5 years*, http://www.bnl.gov/aum/content/past/2011/content/workshops/4.asp.
[366] J. Moss, G. Garvey, J.-C. Peng, *et al.*, Fermilab Proposal P-772, March 11, 1986, *Study of the Nuclear Antiquark Sea via $p+N \rightarrow$ Dimuons*.
[367] L. M. Lederman, J. P. Rutherfoord, *et al.*, Fermilab Proposal P-605 plus Addendum, *Study of Leptons and Hadrons near the Kinematic Limits*.
[368] D. E. Jaffe *et al.*, Phys. Rev. **D40**, 2777 (1989).
[369] C. Alff-Steinberger, W. Heuer, K. Kleinknecht, C. Rubbia, A. Scribano, J. Steinberger, M. J. Tannenbaum, and K. Tittel, Phys. Lett. **20**, 207 (1966).
[370] J. G. Branson *et al.*, Phys. Rev. Lett. **38**, 1334 (1977).
[371] D. M. Kaplan *et al.*, Phys. Rev. Lett. **40**, 435 (1978).
[372] A. L. S. Angelis *et al.*, Nucl. Phys. **B348**, 1 (1991).
[373] A. Kulesza, G. Sterman, and W. Vogelsang, Phys. Rev. **D66**, 014011 (2002).
[374] L. Camilleri *et al.*, CERN/ISRC/78-21, July 10, 1978, *Proposal for a study of large transverse momentum phenomena using the sueconducting low-β*.
[375] CCOR Collaboration, CERN/ISRC/79-7, February 28, 1979, *Proposal for a five-fold increase in the acceptance of R-108 for electron pairs*.
[376] A. L. S. Angelis *et al.*, Phys. Lett. **B126**, 132 (1983), also see figure 8 in reference [412].
[377] J. D. Bjorken, Phys. Rev. **8**, 4098 (1973).
[378] However, this delicacy is often glibly ignored.
[379] S. D. Ellis, M. Jacob, and P. V. Landshoff, Nucl. Phys. **B108**, 93 (1976).
[380] M. Jacob and P. V. Landshoff, Nucl. Phys. **B113**, 395 (1976), see also reference [379].
[381] H. Boggild, in *Proc. XIV Roncontre de Moriond–Session I–"Quarks, Gluons and Jets" (Les Arcs, Savoie, France–March 11–17, 1979)*, edited by J. Tran Thanh Van (Editions Frontières, Dreux, France, 1979), p. 321.
[382] M. J. Tannenbaum, in *Proc. XIV Roncontre de Moriond–Session I–"Quarks, Gluons and Jets" (Les Arcs, Savoie, France–March 11–17, 1979)*, edited by J. Tran Thanh Van (Editions Frontières, Dreux, France, 1979), pp. 351–391.
[383] M. Della Negra *et al.*, Nucl. Phys. **B127**, 1 (1977).
[384] B. Alper *et al.*, Nucl. Phys. **B87**, 19 (1975).
[385] M. G. Albrow *et al.*, Nucl. Phys. **B145**, 305 (1978).
[386] P. Darriulat *et al.*, Nucl. Phys. **B107**, 429 (1976).
[387] A. L. S. Angelis *et al.*, Physica Scripta **19**, 116 (1979).

[388] A. L. S. Angelis *et al.*, Phys. Lett. **B97**, 163 (1980).
[389] M. Jacob, in *Proc. EPS Int. Conf. High Energy Physics (Geneva, June 27–July 4, 1979)*, edited by W. S. Newman (Scientific Information Service, CERN, Geneva, Switzerland, 1979), Vol. 2, pp. 473–522.
[390] A. L. S. Angelis *et al.*, Nucl. Phys. **B209**, 284 (1982).
[391] R. P. Feynman, R. D. Field, and G. C. Fox, Nucl. Phys. **B128**, 1 (1977).
[392] See p. 512 in reference [389].
[393] See p. 511 in reference [389].
[394] A. G. Clark *et al.*, Nucl. Phys. **B160**, 397 (1979).
[395] P. Darriulat, Annu. Rev. Nucl. Part. Sci. **30**, 159 (1980).
[396] P. Darriulat, Large Transverse Momentum Hadronic Processes, 1980, CERN-EP/80-16.
[397] E. M. Levin and M. G. Ryskin, Sov. Phys. JETP **42**, 783 (1975).
[398] C. Kourkoumelis *et al.*, Nucl. Phys. **B158**, 39 (1979).
[399] M. Althoff *et al.*, Z. Phys. **C22**, 307 (1984).
[400] D. de Florian and W. Vogelsang, Phys. Rev. **D71**, 114004 (2005).
[401] L. G. Almeida, G. Sterman, and W. Vogelsang, Phys. Rev. **D80**, 074016 (2009).
[402] L. Apanasevich *et al.*, Phys. Rev. **D59**, 074007 (1999).
[403] M. J. Tannenbaum, in *Particles and Fields-1979*, edited by B. Margolis and D. G. Stairs (American Institute of Physics, New York, 1980), Vol. 59, pp. 263–309.
[404] G. Arnison *et al.*, Phys. Lett. **B132**, 214 (1983).
[405] T. Åkesson *et al.*, Z. Phys. **C34**, 163 (1987).
[406] T. Sjöstrand and M. van Zijl, Phys. Rev. **D36**, 2019 (1987).
[407] CMS collaboration, S. Chatrchyan, *et al.*, arXiv:1107.0330, subm. JHEP.
[408] M. D. Corcoran *et al.*, Phys. Rev. Lett. **41**, 9 (1978).
[409] M. D. Corcoran *et al.*, Phys. Rev. **D21**, 641 (1980).
[410] H. Jöstlein *et al.*, Phys. Rev. **D20**, 53 (1979).
[411] R. Baier, J. Engels, and B. Petersson, Z. Phys. **C2**, 265 (1979).
[412] M. J. Tannenbaum, in *Proc. 21st Int. Conf. High Energy Physics (Paris)*, edited by P. Petiau and M. Porneuf (Journal de Physique Colloques, Paris, 1982), Vol. 43, pp. C3-134–C3-139.
[413] B. L. Combridge, J. Kripfganz, and J. Ranft, Phys. Lett. **B70**, 234 (1977).
[414] R. Cutler and D. W. Sivers, Phys. Rev. **D17**, 196 (1978).
[415] D. Treille, in *Europhysics Conference on High-Energy Physics (Bari, Italy, July 18–24, 1985)*, edited by L. Nitti and G. Preparata (Laterza, Bari, 1985), pp. 793–835.
[416] J.-P. Repellin *et al.*, in *Proc. 21st Int. Conf. High Energy Physics (Paris)*, edited by P. Petiau and M. Porneuf (Journal de Physique Colloques, Paris, 1982), Vol. 43, pp. C3-571–C3-578.
[417] M. Banner *et al.*, Phys. Lett. **B118**, 203 (1982).
[418] G. Wolf, in *Proc. 21st Int. Conf. High Energy Physics (Paris)*, edited by P. Petiau and M. Porneuf (Journal de Physique Colloques, Paris, 1982), Vol. 43, pp. C3-525–C3-568.
[419] J. F. Owens, Rev. Mod. Phys. **59**, 465 (1987).
[420] L. DiLella, Annu. Rev. Nucl. Part. Sci. **35**, 107 (1985).
[421] *Proceedings of the 1982 DPF Summer Study on Elementary Particle Physics and Future Facilities (June 28–July 16, 1982, Snowmass, CO)*, edited by R. Donaldson, R. Gustafson, and F. Paige (American Institute of Physics, New York, 1983).
[422] A. Chilingarov *et al.*, Nucl. Phys. **B151**, 29 (1979).
[423] J. H. Cobb *et al.*, Phys. Lett. **B78**, 519 (1978).
[424] T. Ferbel and W. R. Molzon, Rev. Mod. Phys. **56**, 181 (1984).

[425] *Proc. 1979 Int. Symposium on Lepton and Photon Interactions at High Energies*, edited by T. B. W. Kirk and H. D. I. Abarbanel (Fermi National Accelerator Laboratory, Batavia, IL, 1979), pp. 589–621.
[426] P. Darriulat *et al.*, Nucl. Phys. **B110**, 365 (1976).
[427] K. Eggert *et al.*, Nucl. Phys. **B98**, 49 (1975).
[428] E. Amaldi *et al.*, Nucl. Phys. **B150**, 326 (1979).
[429] R. D. Field, Physica Scripta **19**, 131 (1979).
[430] D. d'Enterria, J. Phys. **G31**, S491 (2005).
[431] E. Amaldi *et al.*, Phys. Lett. **B77**, 240 (1978).
[432] M. Diakonou *et al.*, Phys. Lett. **B87**, 292 (1979).
[433] M. Diakonou *et al.*, Phys. Lett. **B91**, 296 (1980).
[434] E. Anassontzis *et al.*, Z. Phys. **C13**, 277 (1982).
[435] A. L. S. Angelis *et al.*, Phys. Lett. **B94**, 106 (1980).
[436] A. P. Contogouris, S. Papadopoulos, and J. Ralston, Phys. Lett. **B104**, 70 (1981).
[437] A. P. Contogouris, S. Papadopoulos, and M. Hongoh, Phys. Rev. **D19**, 2607 (1979).
[438] A. L. S. Angelis *et al.*, Nucl. Phys. **B327**, 541 (1989).
[439] T. Åkesson *et al.*, Z. Phys. **C34**, 293 (1987).
[440] H. Abramowicz *et al.*, Z. Phys. **C12**, 289 (1982).
[441] P. Aurenche *et al.*, Phys. Rev. **D73**, 094007 (2006).
[442] G. Sterman and W. Vogelsang, Phys. Rev. **D71**, 014013 (2005).
[443] L. Apanasevich *et al.*, Phys. Rev. **D70**, 092009 (2004).
[444] P. Slattery, Phys. Rev. **D7**, 2073 (1973).
[445] For example, see references [446,447].
[446] A. H. Mueller, in *Proc. XVI Int. Conf. High Energy Physics (Chicago-Batavia, 1972)*, edited by J. D. Jackson and A. Roberts (NAL, Batavia, IL, 1973), Vol. 1, pp. 347–388, M. Jacob, ibid., Vol. 3, pp. 373–458.
[447] L. Van Hove, Phys. Rep. **1**, 347 (1971).
[448] A. H. Mueller, Phys. Rev. **D4**, 150 (1971), see also reference [449].
[449] L. Caneschi, Nucl. Phys. **B35**, 406 (1971).
[450] L. Foa, Phys. Rep. **22**, 1 (1975), see also, reference [451].
[451] R. Panvini, in *Proc. XVI Int. Conf. High Energy Physics (Chicago-Batavia, 1972)*, edited by J. D. Jackson and A. Roberts (NAL, Batavia, IL, 1973), Vol. 1, pp. 330–346.
[452] Z. Koba, H. B. Nielsen, and P. Olesen, Nucl. Phys. **B40**, 317 (1972).
[453] G. J. Alner *et al.*, Phys. Lett. **B138**, 304 (1984).
[454] G. J. Alner *et al.*, Phys. Lett. **B160**, 193 (1985).
[455] G. J. Alner *et al.*, Phys. Lett. **B160**, 199 (1985).
[456] G. J. Alner *et al.*, Phys. Lett. **B167**, 476 (1986).
[457] K. Alpgård *et al.*, Phys. Lett. **B123**, 361 (1983).
[458] G. Ekspong, Nucl. Phys. **A461**, 145c (1987).
[459] W. J. Willis, in *ISABELLE Physics Prospects (BNL-17522)*, edited by R. B. Palmer (Brookhaven National Laboratory, Upton, NY, 1972), pp. 207–234, (same report appears as CRISP-72-15, BNL-16841).
[460] J. D. Bjorken, Phys. Rev. **D8**, 4098 (1973).
[461] This "parent-child effect" is a consequence of the steeply falling spectrum and is the cause of the so-called "trigger bias" (see reference [462]).
[462] This consequence of the steeply falling spectrum has been given the name "trigger bias," since the condition of requiring a large transverse momentum single particle distorts the observed jet fragmentation from that of the unconditional case [380].

[463] C. Bromberg *et al.*, Phys. Rev. Lett. **38**, 1447 (1977).
[464] C. Bromberg *et al.*, Nucl. Phys. **B134**, 189 (1978).
[465] M. D. Corcoran *et al.*, Phys. Rev. Lett. **44**, 514 (1980).
[466] M. A. Dris, Nucl. Instrum. Methods **158**, 89 (1979).
[467] W. Ochs and L. Stodolsky, Phys. Lett. **B69**, 225 (1977).
[468] H. Fesefeldt, W. Ochs, and L. Stodolsky, Phys. Lett. **B74**, 389 (1978).
[469] W. Ochs, Physica Scripta **19**, 127 (1979).
[470] P. V. Landshoff and J. C. Polkinghorne, Phys. Rev. **D18**, 3344 (1978).
[471] It was realized that the energy flow for a pair of high p_T jets from constituent scattering might be related to that of the ordinary forward-backward "jet" in soft $p-p$ collisions and the "radiation profile" of the soft collisions from 12 and 24 GeV/c proton interactions was measured [467–469]. It is interesting to note that the definition of "radiation profile" was a bit obscure but turned out to be equal to the "rapidity density of particles weighted by their transverse mass" or $dm_T/dy = \sum_k m_T^k(y)\, dn^k/dy$, where $m_T = \sqrt{m^2 + p_T^2}$.
[472] M. Deutschmann *et al.*, Nucl. Phys. **B155**, 307 (1979).
[473] A multiparticle system M was defined as the group of all charged particles that have the same sign of the p_T component along a principal axis in the transverse plane. This axis was obtained by a rotation in the transverse plane about the beam direction such that the sum of the squares of all momentum components along that axis was a maximum. The four-vector of the multiparticle system was then the c.m. vector sum of its constituents. Since the data were obtained with a 4π detector, the conditions of other experiments could be simulated for comparison purposes.
[474] J. D. Bjorken and S. J. Brodsky, Phys. Rev. **D1**, 1416 (1970), see also reference [475].
[475] G. Hanson *et al.*, Phys. Rev. **D26**, 991 (1982).
[476] V. Cook *et al.*, Nucl. Phys. **B186**, 219 (1981).
[477] R. W. Williams *et al.*, in *Proc. XX Int. Conf. High Energy Physics (Madison, WI, 1980)*, edited by L. Durand and L. G. Pondrom (AIP, New York, 1981), pp. 98–100.
[478] K. Pretzl, C. Favuzzi, *et al.*, in *Proc. XX Int. Conf. High Energy Physics (Madison, WI, 1980)*, edited by L. Durand and L. G. Pondrom (AIP, New York, 1981), pp. 92–97.
[479] B. Brown *et al.*, Phys. Rev. Lett. **49**, 711 (1982).
[480] R. Singer, T. Fields, and W. Selove, Phys. Rev. **D25**, 2451 (1982).
[481] R. Odorico, Phys. Lett. **B118**, 151 (1982).
[482] R. Odorico, Nucl. Phys. **B228**, 381 (1983).
[483] R. D. Field, G. C. Fox, and R. L. Kelly, Phys. Lett. **B119**, 439 (1982), see also reference [484].
[484] G. C. Fox and R. Kelly, in *Proc. Proton-Antiproton Collider Physics-1981 (Madison, WI)*, edited by V. Barger, D. Cline, and F. Halzen (AIP, New York, 1982), pp. 435–468.
[485] H. Bøggild *et al.*, CERN-EP-82-104, submitted to 21st Int. Conf. High Energy Physics, Paris, France, July 26–31, 1982 (unpublished).
[486] T. Åkesson and H. U. Bengtsson, Phys. Lett. **B120**, 233 (1983).
[487] J. A. Appel *et al.*, Phys. Lett. **B165**, 441 (1985).
[488] Clear experimental evidence of QCD jets in a small aperture calorimeter ($\Delta\phi^* = 69°$, $\Delta\theta^* = \pm 46°$) was first demonstrated at the CERN-ISR. See reference [489].
[489] T. Åkesson *et al.*, Phys. Lett. **B118**, 185 (1982).

[490] A nice discussion on the problem of trying to draw physics conclusions from all-encompassing Monte Carlo calculations in the absence of clear, unambiguous effects is given in reference [491].
[491] M. D. Corcoran, Phys. Rev. **D32**, 592 (1985).
[492] G. Marchesini and B. R. Webber, Nucl. Phys. **B238**, 1 (1984).
[493] B. R. Webber, Nucl. Phys. **B238**, 492 (1984), see also reference [494].
[494] G. Sterman and S. Weinberg, Phys. Rev. Lett. **39**, 1436 (1977).
[495] M. Della Negra *et al.*, Nucl. Phys. **B127**, 1 (1977).
[496] G. Altarelli, R. K. Ellis, and G. Martinelli, Phys. Lett. **B151**, 457 (1985).
[497] R. D. Field, Phys. Rev. Lett. **40**, 997 (1978).
[498] W. B. Fowler, R. P. Shutt, A. M. Thorndike, and W. L. Whittemore, Phys. Rev. **95**, 1026 (1954). In the very early days it was not at all clear that a nucleon–nucleon collision could lead to the production of more than one meson. See reference [104].
[499] T. Åkesson *et al.*, Z. Phys. **C38**, 383 (1988).
[500] R. Wigmans, Nucl. Instrum. Methods **A259**, 389 (1987).
[501] H. Gordon *et al.*, Phys. Rev. **D28**, 2736 (1983).
[502] B. C. Brown *et al.*, Phys. Rev. **D29**, 1895 (1984).
[503] G. Arnison *et al.*, Phys. Lett. **B118**, 167 (1982).
[504] G. N. Fowler, E. M. Friedlander, M. Plumer, and R. M. Weiner, Phys. Lett. **B145**, 407 (1984).
[505] A. Breakstone *et al.*, Phys. Lett. **B132**, 463 (1983).
[506] One of the proposed signatures for a phase transition to a quark–gluon plasma would be an initial increase of $\langle p_T \rangle$ with dn/dy, followed by a levelling off (as indicated in Figure 11.13), followed by another sharp increase in $\langle p_T \rangle$ as dn/dy increased further. It is important that dn/dy be used for this test and not dE_T/dy to avoid the induced correlation effect shown in Figure 11.16. See L.Van Hove, references [507,508].
[507] L. Van Hove, Nucl. Phys. **A461**, 3c (1987).
[508] L. Van Hove, Phys. Lett. **B118**, 138 (1982).
[509] For example, see references [114–116].
[510] In detail the situation is somewhat more complicated. The many subtleties and details of "soft" physics [509] have been glossed over in this discussion but they do not change the essential relationship between E_T and multiplicity.
[511] T. Åkesson *et al.*, Phys. Lett. **B128**, 354 (1983).
[512] For a contemporary view of the excitement of this period and some more details, see reference [403].
[513] T. Åkesson *et al.*, Phys. Lett. **B128**, 354 (1983).
[514] R. Barlow, Rep. Prog. Phys. **56**, 1067 (1993).
[515] A. A. Bhatti and D. Lincoln, Annu. Rev. Nucl. Part. Sci. **60**, 267 (2010).
[516] S. D. Ellis, Z. Kunszt, and D. E. Soper, Phys. Rev. Lett. **64**, 2121 (1990).
[517] D. Acosta *et al.*, Phys. Rev. **D68**, 012003 (2003).
[518] F. Abe *et al.*, Phys. Rev. **D45**, 1448 (1992).
[519] For example, see reference [520].
[520] M. J. Tannenbaum, PoS (**LHC07**), 004 (2007).
[521] Michael Begel, Physics Colloquium, *Recent results from ATLAS*, Brookhaven National Laboratory, April 26, 2011.
[522] R. D. Field, in *Proceedings of Snowmass 2001, Snowmass Village CO, June 30–July 21*, edited by N. Graf (eConf C010630, 2001), p. P501, also available as arXiv: hep-ph/0201192v1.
[523] For a book with such details, see, for example, reference [524].

[524] R. K. Ellis, W. J. Stirling, and B. R. Webber, *QCD and Collider Physics* (Cambridge University Press, Cambridge, 1996).
[525] S. D. Ellis and D. E. Soper, Phys. Rev. **D48**, 3160 (1993).
[526] M. Cacciari, G. P. Salam, and G. Soyez, JHEP **04**, 063 (2008).
[527] Z. Kunszt and D. E. Soper, Phys. Rev. **D46**, 192 (1992).
[528] V. M. Abazov *et al.*, Phys. Rev. Lett. **101**, 062001 (2008).
[529] S. Chatrchyan *et al.*, Phys. Rev. Lett. **107**, 132001 (2011).
[530] A. L. S. Angelis *et al.*, Phys. Lett. **B105**, 233 (1981).
[531] G. Arnison *et al.*, Phys. Lett. **B158**, 494 (1985).
[532] J. A. Appel *et al.*, Z. Phys. **C30**, 341 (1986).
[533] T. Åkesson *et al.*, Z. Phys. **C32**, 317 (1986).
[534] J. C. Collins, D. E. Soper, and G. Sterman, Adv. Ser. Direct. High Energy Phys. **5**, 1 (1988), available on-line as arXiv:hep-ph/0409313v1.
[535] S. J. Brodsky and S. D. Drell, Annu. Rev. Nucl. Part. Sci. **20**, 147 (1970).
[536] Y. L. Dokshitzer, JETP **46**, 641 (1977).
[537] G. Altarelli and G. Parisi, Nucl. Phys. **B126**, 298 (1977).
[538] P. Aurenche, R. Baier, M. Fontannaz, J. F. Owens, and M. Werlen, Phys. Rev. **D39**, 3275 (1989).
[539] I. Hinchliffe and A. Kwiatkowski, Annu. Rev. Nucl. Part. Sci. **46**, 609 (1996).
[540] J. Pumplin, D. R. Stump, J. Huston, H.-L. Lai, P. Nadolsky, and W.-K. Tung, JHEP **07**, 012 (2002).
[541] S. Alekhin *et al.*, The PDF4LHC Working Group Interim Report, arXiv:1101.0536v1.
[542] F. D. Aaron *et al.*, JHEP **01**, 109 (2010).
[543] J. J. Aubert *et al.*, Z. Phys. **C18**, 189 (1983).
[544] B. Adeva *et al.*, Phys. Lett. **B259**, 199 (1991).
[545] R. Sassot, M. Stratmann, and P. Zurita, Phys. Rev. **D81**, 054001 (2010).
[546] G. Hanson *et al.*, Phys. Rev. Lett. **35**, 1609 (1975).
[547] B. Andersson, G. Gustafson, G. Ingelman, and T. Sjöstrand, Phys. Rep. **97**, 31 (1983).
[548] J. Kogut and L. Susskind, Phys. Rev. **D9**, 3501 (1974).
[549] E. Eichten, K. Gottfried, T. Kinoshita, J. Kogut, K. D. Lane, and T. M. Yan, Phys. Rev. Lett. **34**, 369 (1975).
[550] E. Eichten, K. Gottfried, T. Kinoshita, K. D. Lane, and T. M. Yan, Phys. Rev. **D17**, 3090 (1978).
[551] T. Appelquist and H. D. Politzer, Phys. Rev. Lett. **34**, 43 (1975).
[552] For example, see reference [492].
[553] Y. L. Dokshitzer, V. S. Fadin, and V. A. Khoze, Phys. Lett. **B115**, 242 (1982).
[554] A. H. Mueller, Phys. Lett. **B104**, 161 (1981).
[555] B. I. Ermolaev and V. S. Fadin, JETP Lett. **33**, 269 (1981).
[556] A. H. Mueller, Nucl. Phys. **B213**, 85 (1983).
[557] Y. I. Azimov, Y. L. Dokshitzer, V. A. Khoze, and S. I. Troyan, Z. Phys. **C27**, 65 (1985).
[558] Y. L. Dokshitzer, A. H. Mueller, V. A. Khoze, and S. Troyan, *Basics of Perturbative QCD* (Editions Frontières, Gif-sur-Yvette, France, 1991).
[559] D. de Florian, R. Sassot, and M. Stratmann, Phys. Rev. **D75**, 114010 (2007).
[560] D. de Florian, R. Sassot, and M. Stratmann, Phys. Rev. **D76**, 074033 (2007).
[561] S. Albino, B. A. Kniehl, and G. Kramer, Nucl. Phys. **B803**, 42 (2008).
[562] G. Marchesini, B. R. Webber, G. Abbiendi, I. G. Knowles, M. H. Seymour, and L. Stanco, Comp. Phys. Commun. **67**, 465 (1992).

[563] T. Sjöstrand, S. Mrenna, and P. Skands, JHEP **05**, 026 (2006).
[564] T. Sjöstrand, in *Workshop on Z physics at LEP 1*, Vol. 3: Event generators and software of *CERN 89-08*, edited by G. Altarelli, R. Kleiss, and C. Verzegnassi (CERN, Geneva, 1989), pp. 143–179.
[565] R. D. Field and R. P. Feynman, Nucl. Phys. **B136**, 1 (1978).
[566] D. O. Caldwell *et al.*, Phys. Rev. Lett. **23**, 1256 (1969).
[567] J. J. Aubert *et al.*, Phys. Lett. **B123**, 275 (1983).
[568] P. Amaudruz *et al.*, Nucl. Phys. **B441**, 3 (1995).
[569] K. J. Eskola, H. Paukkunen, and C. A. Salgado, JHEP **04**, 065 (2009).
[570] S. S. Adler *et al.*, Phys. Rev. Lett. **98**, 172302 (2007).
[571] L. McLerran and R. Venugopalan, Phys. Rev. **D49**, 2233 (1994).
[572] J. Ashman *et al.*, Z. Phys. **C52**, 1 (1991).
[573] L. S. Osborne *et al.*, Phys. Rev. Lett. **40**, 1624 (1978).
[574] R. N. Cahn, Phys. Lett. **B78**, 269 (1978).
[575] A. Airapetian *et al.*, Nucl. Phys. **B780**, 1 (2007).
[576] P. B. Straub *et al.*, Phys. Rev. Lett. **68**, 452 (1992).
[577] H. B. White *et al.*, Phys. Rev. **D48**, 3996 (1993).
[578] C. N. Brown *et al.*, Phys. Rev. **C54**, 3195 (1996).
[579] A. Accardi, F. Arleo, W. K. Brooks, D. d'Enterria, and V. Muccifora, Rev. Nuovo Cimento **032**, 439 (2010), also available as arXiv:0907.3534v1.
[580] F. Arleo, Eur. Phys. J. **C61**, 603 (2009).
[581] R. Sassot, M. Stratmann, and P. Zurita, Phys. Rev. **D81**, 054001 (2010).
[582] G. Sterman *et al.*, Rev. Mod. Phys. **67**, 157 (1995).
[583] A. D. Rujula, H. Georgi, and H. D. Politzer, Ann. Phys. **103**, 315 (1977).
[584] S. Bethke, J. Phys. **G17**, 1455 (1991).
[585] S. Bethke, J. Phys. **G26**, R27 (2000).
[586] S. Bethke, Eur. Phys. J. **64**, 689 (2009).
[587] E. J. Eichten, K. D. Lane, and M. E. Peskin, Phys. Rev. Lett. **50**, 811 (1983).
[588] M. Abolins *et al.*, Testing the compositeness of quarks and leptons, pp 274–287, in reference [421].
[589] R. Longacre and M. J. Tannenbaum, QCD Tests and Large Momentum-Transfer Reactions at CBA, Informal Report, BNL 32888, March, 1983, Brookhven National Laboratory, Upton, NY 11973, avalible online at http://www.osti.gov/energycitations/product.biblio.jsp?osti_id=6211488.
[590] P. Bagnaia *et al.*, Phys. Lett. **B144**, 283 (1984).
[591] G. Arnison *et al.*, Phys. Lett. **B136**, 294 (1984).
[592] D. Drijard *et al.*, Phys. Lett. **B121**, 433 (1983).
[593] R. Ansari *et al.*, Z. Phys. **C41**, 395 (1988).
[594] G. Arnison *et al.*, Phys. Lett. **B177**, 244 (1986).
[595] G. Ballocchi *et al.*, Phys. Lett. **B436**, 222 (1998).
[596] B. L. Combridge and C. J. Maxwell, Nucl. Phys. **B239**, 429 (1984).
[597] G. Aad *et al.*, New J. Phys **13**, 50344 (2011).
[598] K. Adcox *et al.*, Phys. Rev. Lett. **88**, 022301 (2002).
[599] S. S. Adler *et al.*, Phys. Rev. **C69**, 034910 (2004).
[600] L. R. Cormell *et al.*, Phys. Lett. **B150**, 322 (1985).
[601] C. De Marzo *et al.*, Phys. Rev. **D36**, 8 (1987).
[602] M. Bonesini *et al.*, Z. Phys. **C38**, 371 (1988).
[603] T. Åkesson *et al.*, Phys. Lett. **B123**, 133 (1983).
[604] P. Bagnaia *et al.*, Phys. Lett. **B138**, 430 (1984).

[605] K. Okada and PHENIX Collab., Phys. Rev. **D**, submitted (2012), preprint available as arXiv:1205.5533v1 [hep-ex].
[606] A. Akopian, A. Bhatti, B. Flaugher, *et al.*, Scaling Violation in Inclusive Jet Production, Internal CDF Note: 4890, 1999, available on-line at http://www-cdf.fnal.gov /physics /new /qcd/qcd_plots/inclusive_jets /public /cdf4890 /scaling.html.
[607] F. Abe *et al.*, Phys. Rev. Lett. **77**, 438 (1996).
[608] G. C. Blazey and B. L. Flaugher, Annu. Rev. Nucl. Part. Sci. **49**, 633 (1999).
[609] J. Huston *et al.*, Phys. Rev. Lett. **77**, 444 (1996).
[610] J. Pumplin, J. Huston, H. L. Lai, P. M. Nadolski, W.-K. Tung, and C.-P. Yuan, Phys. Rev. **D80**, 014019 (2009).
[611] A. Adare *et al.*, Phys. Rev. **C83**, 064903 (2011).
[612] A. Adare *et al.*, Phys. Rev. **D79**, 012003 (2009).
[613] S. J. Brodsky, M. Burkhardt, and I. Schmidt, Nucl. Phys. **441**, 197 (1995).
[614] F. Arleo, S. J. Brodsky, D. S. Hwang, and A. M. Sickles, Phys. Rev. Lett. **105**, 062002 (2010).
[615] H. Satz, Rep. Prog. Phys. **63**, 1511 (2000).
[616] H. Satz, Nucl. Phys. **A418**, 447 (1984).
[617] T. Matsui and H. Satz, Phys. Lett. **B178**, 416 (1986).
[618] R. P. Feynman, Phys. Rev. **91**, 1291 (1953).
[619] C. W. Bernard, Phys. Rev. **D9**, 3312 (1974).
[620] A. Mocsy and P. Petreczky, Phys. Rev. Lett. **99**, 211602 (2007).
[621] M. Creutz, Phys. Rev. **D15**, 1128 (1977).
[622] O. Kaczmarek, F. Karsch, P. Petreczky, and F. Zantow, Phys. Lett. **B543**, 41 (2002).
[623] F. Karsch, Lect. Notes Phys. **583**, 209 (2002).
[624] F. Karsch, Eur. Phys. J. **C43**, 35 (2005).
[625] H. Satz, J. Phys. **G32**, R25 (2006).
[626] O. Kaczmarek, PoS (**CPOD07**), 043 (2007).
[627] S. S. Adler and PHENIX Collab., Phys. Rev. **C71**, 034908 (2005).
[628] D. P. Morrison, as emphasized in early PHENIX internal memos.
[629] G. F. Chew and G. C. Wick, Phys. Rev. **85**, 636 (1952).
[630] R. J. Glauber, Phys. Rev. **100**, 242 (1955).
[631] A. Białas, A. B. Błeszynski, and W. Czyż, Nucl. Phys. **B111**, 461 (1976).
[632] W. Busza *et al.*, Phys. Rev. Lett. **34**, 836 (1975).
[633] K. Adcox and PHENIX Collab., Phys. Rev. Lett. **86**, 3500 (2001).
[634] K. Aamodt and ALICE Collab., Phys. Rev. Lett. **106**, 032301 (2011).
[635] S. Eremin and S. Voloshin, Phys. Rev. **C67**, 064905 (2003).
[636] M. J. Tannenbaum, PoS (**CERP2010**), 019 (2010).
[637] B. De and S. Bhattacharyya, Phys. Rev. **C71**, 024903 (2005).
[638] R. Nouicer, Eur. Phys. J. **C49**, 281 (2007).
[639] A. Bialas, W. Czyz, and L. Lesniak, Phys. Rev. **D25**, 2328 (1982).
[640] A. Bialas, J. Phys. **G35**, 044053 (2008).
[641] R. A. Lacey, Nucl. Phys. **A774**, 199 (2006).
[642] K. Aamodt and ALICE Collab., Phys. Rev. Lett. **105**, 252302 (2010).
[643] J.-Y. Ollitrault, Phys. Rev. **D46**, 229 (1992).
[644] H. Heiselberg and A.-M. Levy, Phys. Rev. **C59**, 2716 (1999).
[645] M. Kaneta and PHENIX Collab., J. Phys. **G30**, S1217 (2004).
[646] A. Adare and PHENIX Collab., Phys. Rev. Lett. **98**, 162301 (2007).
[647] S. A. Voloshin, Nucl. Phys. **A715**, 379c (2003).
[648] D. Teaney, Phys. Rev. **C68**, 034913 (2003).

[649] P. K. Kovtun, D. T. Son, and A. O. Starinets, Phys. Rev. Lett. **94**, 111601 (2005).
[650] D. Rischke and G. Levin, eds., Nucl. Phys. **A750**, 1 (2005).
[651] B. H. Alver, C. Gombeaud, M. Luzum, and J. Y. Ollitrault, Phys. Rev. **C82**, 034913 (2010).
[652] B. Alver and G. Roland, Phys. Rev. **C81**, 054905 (2010).
[653] J. Takahashi *et al.*, Phys. Rev. Lett. **103**, 242301 (2009).
[654] A. P. Mishra *et al.*, Phys. Rev. **C77**, 064902 (2008).
[655] Y. Schutz and U. A. Wiedemann, eds., J. Phys. **G38**, 124001 (2011).
[656] S. Esumi and PHENIX Collab., J. Phys. **G38**, 124010 (2011).
[657] R. P. Crease, Phys. Perspect. **7**, 330 (2005).
[658] R. P. Crease, Phys. Perspect **7**, 404 (2005).
[659] L. M. Lederman, *Large Q^2 experiments at Isabelle*, CRISP 71-29, ISABELLE–Physics Prospects, BNL-17522 (1972), pp. 396–405.
[660] C. Rubbia, Phys. Rep. **239**, 241 (1994).
[661] Accelerator History-Main Ring transition to Energy Douber/Saver, Fermilab History and Archives Project, http://history.fnal.gov/transition.html.
[662] L. Hoddeson, Hist. Studies Phys. Bio. Sci. **18**, 25 (1987), also available on-line at http://lss.fnal.gov/archive/other/fprint-87-01.pdf.
[663] L. Hoddeson and A. Kolb, Phys. Perspect. **5**, 67 (2003).
[664] For a review, see reference [665].
[665] D. Haidt, Eur. Phys. J. **C34**, 25 (2004).
[666] L. Camilleri *et al.*, *Physics with very high-energy e^+e^- colliding beams*, CERN 76-18, 1976.
[667] H. Schopper, *LEP – The Lord of the Collider Rings at CERN 1980–2000* (Springer-Verlag, Berlin, 2009), available on-line at http://www.springerlink.com/content/978-3-540-89300-4/contents/.
[668] B. Richter, *The Future of Electron-Positron Colliders*, pp. 128–133 in reference [421], available on-line as http://lss.fnal.gov/conf/C8206282/pg128.pdf.
[669] H. Mieras and G. A. Voss, in *Proc. VIth Int. Conf. High Energy Accelerators*, edited by R. A. Mack (Cambridge Electron Accelarator Lab., Cambridge, MA, 1967), pp. 119–122.
[670] G. Arnison *et al.*, Phys. Lett. **B126**, 398 (1983).
[671] P. Bagnaia *et al.*, Phys. Lett. **B129**, 130 (1983).
[672] CERN - LEP: the Z factory, http://public.web.cern.ch/public/en/research/lep-en.html.
[673] M. Jacob, ed., *Proc. ECFA-CERN Workshop: Large Hadron Collider in the LEP Tunnel (21–27 March 1984, Lausanne)*, CERN-84-10-V-1; CERN-84-10-V-2, September 1984.
[674] CERN/DG-RB 83-45, 4 February, 1983, *Minutes of the 61st Meeting of the Research Board.*
[675] R. Klapisch, Nucl. Phys. **A418**, 347c (1984).
[676] CERN/DG-RB 82-36, 24 June 1982, *Decisions of the 58th Meeting of the Research Board.* It is important to note the correction to the report from G. Bellettini, Chair of the ISRC, on the results obtained with beams of α particles in DG-RB 82-36, as given in CERN/DG-RB 82-39, namely: "Page 3, line 16, after 'new and unexpected' add 'in particular the excess rate at large p_T compared to that predicted by incoherent production.' "
[677] L. Evans and P. Bryant, eds., Jinst **3**, S08001 (2008).
[678] E. J. Bleser *et al.*, Nucl. Instrum. Methods **A 235**, 435 (1985).

[679] R. R. Wilson, *Superferric Magnets for 20 TeV*, pp. 330–334 in reference [421], available on-line as http://lss.fnal.gov/conf/C8206282/pg330.pdf.
[680] M. Riordan, Phys. Perspect. **2**, 411 (2000).
[681] R. M. Jones, *U.S. Contribution to LHC: On Budget and Ahead of Schedule*, FYI: The AIP Bulletin of Science Policy News, July 2, 2008, available on-line at http://www.aip.org/fyi/2008/073.html.
[682] G. Baym, Nucl. Phys. **A698**, xxiii (2002).
[683] DOE/NSF Nuclear Science Advisory Committee, *A Long Range Plan for Nuclear Science*, December 1983, available on-line as http://science.energy.gov/~/media/np/nsac/pdf/docs/lrp_1983.pdf.
[684] R. P. Crease, Hist. Stud. Nat. Sci. **38**, 535 (2008).
[685] T. W. Ludlam and H. E. Wegner, eds., Nucl. Phys. **A418**, 3c (1984).
[686] N. Angert *et al.*, *Study of relativistic nucleus-nucleus reactions induced by ^{16}O beams of 9–13 GeV per Nucleon at the CERN PS*, Proposal to CERN PSCC, January 26, 1982, CERN/PSCC/82-1;PSCC-P-53.
[687] H. H. Gutbrod *et al.*, *Study of Relativistic Nucleus-Nucleus Collisions at the CERN SPS*, Proposal to SPSC, August 26, 1985, CERN/SPSC 85-33; SPSC/M 406.
[688] C. Chasman and P. Thieberger, Nucl. Instrum. Methods **B10/11**, 347 (1985).
[689] M. J. Tannenbaum, Rep. Prog. Phys. **69**, 2005 (2006).
[690] M. A. Faessler, *A Study of Inelastic p–α Collisions with the SFM*, Letter of intent to CERN-ISRC, January 8, 1979, CERN/ISRC/79-3.
[691] M. A. Faessler *et al.*, *p–α and α–α Collisions in the ISR*, Proposal to CERN-ISRC, April 10, 1979, CERN/ISRC/79-10; ISRC/P101.
[692] M. A. Faessler, in *Proc. 1st Int. Conference on Physics in Collision (Blacksburg, VA, May 28–31, 1981)*, edited by G. Bellini and W. P. Trower (Plenum, New York, 1982), pp. 281–296.
[693] A. L. S. Angelis *et al.*, Phys. Lett. **B185**, 213 (1987).
[694] A. L. S. Angelis *et al.*, Phys. Lett. **B116**, 379 (1982).
[695] M. A. Faessler, Phys. Repts. **115**, 1 (1984).
[696] M. Fatyga and B. Moskowitz, eds., *Fourth Workshop on Experiments and Detectors for a Relativistic Heavy Ion Collider (July 2–7, 1990)*, BNL-52262, available on-line at http://www.osti.gov/energycitations/basicsearch.jsp.
[697] K. Kadija *et al.*, *An Experiment on Particle and Jet Production at Midrapidity*, LBL-29651, September, 1990, letter of intent for experiment at RHIC, available on-line at http://www.osti.gov/bridge/product.biblio.jsp?osti_id=6366571.
[698] M. Schwartz, letters to spokespersons of RHIC LOIs, September 9, 1991.
[699] T. Ludlam, in *Proc. General Meeting on LHC Physics & Detectors: Towards the LHC Experimental Programme (5-8 March 1992, Evian-les-Bains, France)*, edited by G. Flügge (CERN, Geneva, 1992), pp. 539–554.
[700] M. Harrison, T. Ludlam, and S. Ozaki, eds., Nuc. Instrum. Methods **A499**, 235 (2003).
[701] *Polarized Collider Workshop (University Park, PA November 15–17, 1990)*, AIP Conference proceedings No. 223., edited by J. C. Collins, S. F. Heppelman, and R. W. Robinett (American Institute of Physics, New York, 1991).
[702] G. Bunce, N. Saito, J. Soffer, and W. Vogelsang, Annu. Rev. Nucl. Part. Sci. **50**, 525 (2000).
[703] G. Bunce *et al.*, *W, Z^0 Production at a pp Collider*, pp. 489–499 in reference [421], available on-line as http://lss.fnal.gov/conf/C8206282/pg489.pdf.
[704] L. Teng *et al.*, *Report of the RHIC Polarized Proton Review*, Letter to M. Schwartz, June 1993.

[705] A. Breskin and R. Voss, eds., Jinst **3**, S08002, S08003, S08004 (2008).
[706] O. Kodolova and M. Murray, Nucl. Phys. **A830**, 97c (2009).
[707] D. V. Aleksandrov *et al.*, Nucl. Inst. Meth. **A550**, 169 (2005).
[708] J. D. Bjorken, *Energy Loss of Energetic Partons in Quark–Gluon Plasma: Possible Extinction of High p_T Jets in Hadron–Hadron Collisions*, FERMILAB-Pub-82/59-THY, August, 1982, available on-line at http://lss.fnal.gov/archive/preprint/fermilab-pub-82-059-t.shtml.
[709] M. Gyulassy and M. Plümer, Phys. Lett. **B243**, 432 (1990).
[710] X.-N. Wang and M. Gyulassy, *Jets in Relativistic Heavy Ion Collisions*, pp. 79–102 in reference [696], available on-line as LBL-29390 at http://www.osti.gov/bridge/product.biblio.jsp?osti_id=6414092.
[711] M. Gyulassy and X.-N. Wang, Nucl. Phys. **B420**, 583 (1994).
[712] *Proc. Int. Wks. Quark Gluon Plasma Signatures (October 1–4, 1990, Strasbourg, France)*, edited by V. Bernard, A. Capella, W. Geist, P. Gorodetsky, R. Seltz, and C. Voltolini (Editions Frontieres, Gif-sur-Yvette, France, 1991), table of contents available on-line at http://www.slac.stanford.edu/spires/find/hep/wwwbrief?cnum=C90-10-02.
[713] M. Lev and B. Petersson, Z. Phys. **C21**, 155 (1983).
[714] B. I. Abelev and STAR Collab., Phys. Rev. Lett. **97**, 252001 (2006).
[715] For a review, see reference [716].
[716] R. Baier, D. Schiff, and B. G. Zakharov, Annu. Rev. Nucl. Part. Sci. **50**, 37 (2000).
[717] Y. L. Dokshitzer and D. E. Kharzeev, Phys. Lett. **B519**, 199 (2001).
[718] A. Adare and PHENIX Collab., Phys. Rev. **D76**, 051106(R) (2007).
[719] A. Adare and PHENIX Collab., Phys. Rev. Lett. **97**, 252002 (2006).
[720] A. Adare and PHENIX Collab., Phys. Rev. **C84**, 044905 (2011).
[721] E. Wang and X.-N. Wang, Phys. Rev. **C64**, 034901 (2001).
[722] S. S. Adler *et al.*, Phys. Rev. Lett. **91**, 072303 (2003).
[723] A. Adare and PHENIX Collab., Phys. Rev. Lett. **101**, 162301 (2008).
[724] F. Arleo and D. d'Enterria, Phys. Rev. **D78**, 094004 (2008).
[725] A. Adare and PHENIX Collab., Phys. Rev. **D79**, 012003 (2009).
[726] D. d'Enterria, J. Phys. **G31**, S491 (2005).
[727] S. S. Adler and PHENIX Collab., Phys. Rev. **C76**, 034904 (2007).
[728] V. S. Pantuev, JETP Lett. **85**, 104 (2007).
[729] S. Afanasiev and PHENIX Collab., Phys. Rev. **C80**, 054907 (2009).
[730] A. Adare and PHENIX Collab., Phys. Rev. Lett. **105**, 142301 (2010).
[731] S. S. Adler and PHENIX Collab., Phys. Rev. **C69**, 034910 (2004).
[732] S. S. Adler and PHENIX Collab., Phys. Rev. Lett. **91**, 172301 (2003).
[733] S. S. Adler and PHENIX Collab., Phys. Rev. **C69**, 034909 (2004).
[734] A. Adare and PHENIX Collab., Phys. Lett. **B649**, 359 (2007).
[735] V. Greco, C. M. Ko, and P. Levai, Phys. Rev. Lett. **90**, 202302 (2003).
[736] R. J. Fries, B. Müller, C. Nonaka, and S. A. Bass, Phys. Rev. Lett. **90**, 202303 (2003).
[737] S. S. Adler and PHENIX Collab., Phys. Rev. **C71**, 051902(R) (2005).
[738] N. M. Kroll and W. Wada, Phys. Rev. **98**, 1355 (1955).
[739] A. Adare and PHENIX Collab., Phys. Rev. Lett. **104**, 132301 (2010).
[740] A. S. Ito *et al.*, Phys. Rev. **D23**, 604 (1981).
[741] Thanks to Sasha Milov for the plot of $R_{AA}(p_T)$ for all PHENIX published and preliminary measurements. With the exception of the internal-conversion direct-γ where the fit to the p–p data is used to compute R_{AA}, all the other values of R_{AA} are computed from Eq. 14.1 using the measured Au+Au and p–p data points.

[742] A. Adare and PHENIX Collab., Phys. Rev. **81**, 034911 (2010).
[743] A. Adare and PHENIX Collab., Phys. Rev. Lett. **101**, 232301 (2008).
[744] A. Adare and PHENIX Collab., Phys. Rev. **C82**, 011902(R) (2010).
[745] B. Alessandro *et al.*, Eur. Phys. J. **C39**, 335 (2005).
[746] A. Adare and PHENIX Collab., Phys. Rev. **C**, submitted (2012), available on-line as http://arxiv.org/abs/1204.0777v1.
[747] T. Gunji and PHENIX Collab., J. Phys. **G 34**, S749 (2007), figure was in the talk, but did not appear in the proceedings.
[748] A. Adare and PHENIX Collab., Phys. Rev. Lett. **98**, 232301 (2007).
[749] A. Adare and PHENIX Collab., Phys. Rev. **C84**, 054912 (2011).
[750] X. Zhao and R. Rapp, Phys. Lett. **B664**, 253 (2008), also see the list of citations to previous work.
[751] B. Abelev and ALICE Collab., Phys. Rev. Lett. submitted (2012), available on-line at http://arxiv.org/abs/1202.1383v1. Also see the list of citations to previous work on coalescence.
[752] M. J. Tannenbaum, Heavy Ion Physics **4**, 139 (1996).
[753] M. Cacciari, P. Nason, and R. Vogt, Phys. Rev. Lett. **95**, 122001 (2005).
[754] A. Adare and PHENIX Collab., Phys. Rev. Lett. **98**, 172301 (2007).
[755] A. Adare and PHENIX Collab., Phys. Rev. Lett. **103**, 082002 (2009).
[756] For example, see reference [754] for a discussion with citations.
[757] A. Zichichi, Nucl. Phys. **A805**, 36c (2008).
[758] S. Chatrchyan and CMS Collab., Phys. Rev. **C84**, 024906 (2011).
[759] B. Wyslouch and CMS Collab., J. Phys. **G38**, 124005 (2011).
[760] A. Adare and PHENIX Collab., Phys. Rev. **D82**, 072001 (2010).
[761] W. Braunschweig and TASSO Collab., Z. Phys. **C47**, 187 (1990).
[762] D. Hardtke and STAR Collab., Nucl. Phys. **A715**, 272c (2003).
[763] J. Adams and STAR Collab., Phys. Rev. Lett. **91**, 072304 (2003).
[764] C. Adler and STAR Collab., Phys. Rev. Lett. **90**, 082302 (2003).
[765] F. Wang and STAR Collab., J. Phys. **G30**, S1299 (2004).
[766] J. Adams and STAR Collab., Phys. Rev. Lett. **95**, 152301 (2005).
[767] M. J. Tannenbaum, PoS (**CFRNC2006**), 001 (2006).
[768] For a recent critique and review of Mach cones in RHI Physics see reference [769] and references therein.
[769] I. Bouras *et al.*, Phys. Lett. **B**, in press (2012), corrected proof available on-line as http://dx.doi.org/10.1016/j.physletb.2012.03.040.
[770] A. Adare and PHENIX Collab., Phys. Rev. Lett. **104**, 252301 (2010).
[771] M. J. Tannenbaum, in *Proc. 6th Int. Workshop on High-p_T physics at the LHC (04–07 April 2011, Utrecht, The Netherlands), CERN-Proceedings-2012-001*, edited by K. Eskola *et al.* (CERN, Geneva, 2012), pp. 37–42, also available on-line as http://arxiv.org/abs/1109.0760.
[772] H. Zhang, J. F. Owens, E. Wang, and X.-N. Wang, Phys. Rev. Lett. **98**, 212301 (2007).
[773] B. I. Abelev and STAR Collab., Phys. Rev. **C82**, 034909 (2010).
[774] M. Connors and PHENIX Collab., Nucl. Phys. **A855**, 335 (2011).
[775] CERN press release 19 March 2010, available on-line at http://press.web.cern.ch/press/PressReleases/Releases2010/PR05.10E.html.
[776] G. Aad and ATLAS Collab., Phys. Rev. Lett. **105**, 252303 (2010).
[777] S. Chatrchyan and CMS Collab., Phys. Lett. **B**, submitted (2012), available on-line as http://arxiv.org/abs/1202.5022.
[778] K. Aamodt and ALICE Collab., Phys. Lett. **B696**, 30 (2011).

[779] M. L. Purschke and PHENIX Collab., J. Phys. **G38**, 124016 (2011).
[780] S. Chatrchyan and CMS Collab., Eur. Phys. J. **C72**, 1945 (2012).
[781] C. Klein-Bösing and ALICE Collab., Nuovo Cimento **C034**, 41 (2011).
[782] K. Aamodt and ALICE Collab., Phys. Rev. Lett. **108**, 092301 (2012).
[783] S. Chatrchyan and CMS Collab., Phys. Lett. **B710**, 256 (2012).
[784] H. K. Wöhri and CMS Collab., Nucl. Phys. **A855**, 23 (2011).
[785] B. Abelev and ALICE Collaboration, *Suppression of high transverse momentum D mesons in central Pb–Pb collisions at* $\sqrt{s_{NN}}$ = 2.76 *TeV*, preprint arXiv:1203.2160v1 [nucl-ex] 9 Mar 2012.
[786] S. Chatrchyan and CMS Collab., JHEP submitted (2012), available on-line as arXiv:1201.5069v1 [nucl-ex].
[787] S. Chatrchyan and CMS Collab., Phys. Rev. Lett. **107**, 052302 (2011).
[788] E. Fermi, in *Nuclear Physics*, edited by J. Orear, A. H. Rosenfeld, and R. A. Schluter (University of Chicago Press, Chicago, IL, 1950).
[789] M. G. Kendall and A. Stuart, *The Advanced Theory of Statistics* (Hafner, New York, 1969).
[790] A. H. Mueller, Phys. Rev. **D4**, 150 (1971).
[791] F. J. Anscombe, Biometrika **37**, 358 (1950).
[792] R. D. Evans, *The Atomic Nucleus* (McGraw Hill, New York, 1955).
[793] A. C. Melissinos, *Experiments in Modern Physics* (Academic Press, New York, 1966).
[794] H. Jeffreys, *Theory of Probability* (Clarendon Press, Oxford, 1961).
[795] P. Carruthers and C. C. Shih, Phys. Lett. **127**, 242 (1983).
[796] W. Feller, *An Introduction to Probability Theory and Its Applications* (Wiley, New York, 1966).
[797] J. Rainwater and C. S. Wu, Nucleonics **1**, 62 (1947).
[798] A. Bialas and R. Peschanski, Phys. Lett. **207**, 59 (1988).
[799] K. Werner and M. Kutschera, Phys. Lett. **B220**, 243 (1989), as noted by Ekspong [805].
[800] For an arbitrary distribution of Poisson means, the result is known as a Poisson transform, see reference [801].
[801] P. Carruthers and C. C. Shih, Int J. Mod. Phys. **A2**, 1447 (1987).
[802] M. I. Adamovich *et al.*, Phys. Lett. **B242**, 512 (1990).
[803] Here again, any clustering will cause k to vary, see reference [804].
[804] W. Q. Chao, T. C. Meng, and J. C. Pan, Phys. Lett. **B 176**, 211 (1986).
[805] G. Ekspong, in *Festschrift for Leon Van Hove and Proceedings Multiparticle Dynamics*, edited by A. Giovannini and W. Kittel (World Scientific, Singapore, 1990), and references therein.
[806] The distribution can be found in M. J. Tannenbaum, E-802-MEM-58, internal memorandum, Brookhaven National Laboratory (1993).
[807] S. Uhlig, I. Derado, R. Meinke, and H. Preissner, Nucl. Phys. **B132**, 15 (1978).
[808] G. J. Alner *et al.*, Phys. Lett. **B160**, 193 (1985), see also UA5 collaboration [825] and references therein.
[809] P. Carruthers and C. C. Shih, Phys. Lett. **165**, 209 (1985).
[810] W. A. Zajc, Phys. Lett. **B175**, 219 (1986).
[811] Previously, the search for multiplicity distributions that obeyed KNO scaling [111] had led to gamma distributions, see P. Slattery [826].
[812] K. J. Eskola, R. Vogt, and X.-N. Wang, Int. J. Mod. Phys. **A10**, 3087 (1995).
[813] W. A. Zajc, private communication, also see reference [814].
[814] H. Brody, S. Frankel, W. Frati, and I. Otterlund, Phys. Rev. **28**, 2334 (1983).

[815] L. P. Remsberg, M. J. Tannenbaum, and E802 Collab., Z. Phys. **C38**, 35 (1988).
[816] M. L. Miller, K. Reygers, S. J. Sanders, and P. Steinberg, Annu. Rev. Nucl. Part. Sci. **57**, 205 (2007).
[817] H. De Vries, C. W. De Jager, and C. De Vries, Atomic Data and Nuclear Data Tables **36**, 495 (1987).
[818] J. Heinrich and L. Lyons, Annu. Rev. Nucl. Part. Sci. **57**, 145–169 (2007).
[819] S. S. Wilks, Ann. Math. Statist. **9**, 60 (1938).
[820] F. James and M. Roos, Comput. Phys. Commun. **10**, 342 (1975).
[821] M. Jacob and P. V. Landshoff, Phys. Rep. **48**, 285 (1978).
[822] Note that n in the equations in Sections E.1 and E.2 is the partonic power whereas in practice the power is measured from the π^0 spectrum, n_π. To the extent that x_T is small, the approximation of Eq. E.17 to Eq. E.19 is sufficiently accurate that $n_\pi \approx n$. Furthermore, the neat results described in these sections plus Section 9.1.1 depend on a partonic power law spectrum for all $\hat{p}_{T_t}$ up to the endpoint $\sqrt{s}/2$, which may not be strictly true. Nevertheless, the simplifications based on the power-law work very well.
[823] G. Alexander *et al.*, Z. Phys. **C69**, 543 (1996).
[824] P. Abreu *et al.*, Eur. Phys. J. **C13**, 573 (2000).
[825] G. J. Alner *et al.*, Phys. Rep. **154**, 247 (1987).
[826] P. Slattery, Phys. Rev. **D7**, 2073 (1973).

Index

Printed in the United States
By Bookmasters